Modelling and Control of Mechatronic and Robotic Systems, Volume II

Modelling and Control of Mechatronic and Robotic Systems, Volume II

Editors

Alessandro Gasparetto
Stefano Seriani
Lorenzo Scalera

MDPI • Basel • Beijing • Wuhan • Barcelona • Belgrade • Manchester • Tokyo • Cluj • Tianjin

Editors
Alessandro Gasparetto
University of Udine
Italy

Stefano Seriani
University of Trieste
Italy

Lorenzo Scalera
University of Udine
Italy

Editorial Office
MDPI
St. Alban-Anlage 66
4052 Basel, Switzerland

This is a reprint of articles from the Special Issue published online in the open access journal *Applied Sciences* (ISSN 2076-3417) (available at: https://www.mdpi.com/journal/applsci/special_issues/Mechatronic_Robotic_Systems_2).

For citation purposes, cite each article independently as indicated on the article page online and as indicated below:

LastName, A.A.; LastName, B.B.; LastName, C.C. Article Title. *Journal Name* **Year**, *Volume Number*, Page Range.

ISBN 978-3-0365-4843-2 (Hbk)
ISBN 978-3-0365-4844-9 (PDF)

Cover image courtesy of Jeffery Ho

Contents

About the Editors vii

Alessandro Gasparetto, Stefano Seriani and Lorenzo Scalera
Modelling and Control of Mechatronic and Robotic Systems, Volume II
Reprinted from: *Appl. Sci.* **2022**, *12*, 5922, doi:10.3390/app12125922 1

Carlos Prados Sesmero, Luca Rosario Buonocore and Mario Di Castro
Omnidirectional Robotic Platform for Surveillance of Particle Accelerator Environments with Limited Space Areas
Reprinted from: *Appl. Sci.* **2021**, *11*, 6631, doi:10.3390/app11146631 5

Lucian Ștefăniță Grigore, Ionica Oncioiu, Iustin Priescu and Daniela Joița
Development and Evaluation of the Traction Characteristics of a Crawler EOD Robot
Reprinted from: *Appl. Sci.* **2021**, *11*, 3757, doi:10.3390/app11093757 29

Xing Chi and Qiang Zhan
Design and Modelling of an Amphibious Spherical Robot Attached with Assistant Fins
Reprinted from: *Appl. Sci.* **2021**, *11*, 3739, doi:10.3390/app11093739 61

Miguel Hernando, Mercedes Alonso, Carlos Prados and Ernesto Gambao
Behavior-Based Control Architecture for Legged-and-Climber Robots
Reprinted from: *Appl. Sci.* **2021**, *11*, 9547, doi:10.3390/app11209547 81

Kai Hu, Lang Tian, Chenghang Weng, Liguo Weng, Qiang Zang, Min Xia and Guodong Qin
Data-Driven Control Algorithm for Snake Manipulator
Reprinted from: *Appl. Sci.* **2021**, *11*, 8146, doi:10.3390/app11178146 97

Xiangming Liu, Hongxu Ma, Lin Lang and Honglei An
Online Foot Location Planning for Gait Transitioning Using Model Predictive Control
Reprinted from: *Appl. Sci.* **2021**, *11*, 7866, doi:10.3390/app11177866 115

Jaehoon Sim, Seungyeon Kim, Suhan Park, Sanghyun Kim, Mingon Kim and Jaeheung Park
Design of JET Humanoid Robot with Compliant Modular Actuators for Industrial and Service Applications
Reprinted from: *Appl. Sci.* **2021**, *11*, 6152, doi:10.3390/app11136152 141

Chun-Tse Lee and Jen-Yuan (James) Chang
A Workspace-Analysis-Based Genetic Algorithm for Solving Inverse Kinematics of a Multi-Fingered Anthropomorphic Hand
Reprinted from: *Appl. Sci.* **2021**, *11*, 2668, doi:10.3390/app11062668 157

Valentina Mattioni, Edoardo Ida' and Marco Carricato
Design of a Planar Cable-Driven Parallel Robot for Non-Contact Tasks
Reprinted from: *Appl. Sci.* **2021**, *11*, 9491, doi:10.3390/app11209491 175

Ming-Yen Wei
Design and Implementation of Inverse Kinematics and Motion Monitoring System for 6DoF Platform
Reprinted from: *Appl. Sci.* **2021**, *11*, 9330, doi:10.3390/app11199330 193

Tu Thi Cam Duong, Charles C. Nguyen and Thien Duc Tran
Synchronization Sliding Mode Control of Closed-Kinematic Chain Robot Manipulators with Time-Delay Estimation
Reprinted from: *Appl. Sci.* **2022**, *12*, 5527, doi:10.3390/app12115527 213

Ekaterina Kopets, Artur Karimov, Lorenzo Scalera and Denis Butusov
Estimating Natural Frequencies of Cartesian 3D Printer Based on Kinematic Scheme
Reprinted from: *Appl. Sci.* **2022**, *12*, 4514, doi:10.3390/app12094514 . . . . **231**

Hossein Azarmdel, Seyed Saeid Mohtasebi, Ali Jafary, Hossein Behfar and Alfredo Rosado Muñoz
Design and Simulation of a Vision-Based Automatic Trout Fish-Processing Robot
Reprinted from: *Appl. Sci.* **2021**, *11*, 5602, doi:10.3390/app11125602 . . . . **259**

Yuta Sueki and Yoshiyuki Noda
Experimental Verification of Real-Time Flow-Rate Estimations in a Tilting-Ladle-Type Automatic Pouring Machine
Reprinted from: *Appl. Sci.* **2021**, *11*, 6701, doi:10.3390/app11156701 . . . . **289**

Ionel-Alexandru Gal, Alexandra-Cătălina Ciocîrlan and Mihai Mărgăritescu
State Machine-Based Hybrid Position/Force Control Architecture for a Waste Management Mobile Robot with 5DOF Manipulator
Reprinted from: *Appl. Sci.* **2021**, *11*, 4222, doi:10.3390/app11094222 . . . . **315**

Cristian Napole, Oscar Barambones, Mohamed Derbeli and Isidro Calvo
Advanced Trajectory Control for Piezoelectric Actuators Based on Robust Control Combined with Artificial Neural Networks
Reprinted from: *Appl. Sci.* **2021**, *11*, 7390, doi:10.3390/app11167390 . . . . **339**

Sang Yong Park, Buchun Song and Yoon Su Baek
A Theoretical Method for Designing Thin Wobble Motor Using an Electromagnetic Force and an Electropermanent Magnet for Application in Portable Electric Equipment
Reprinted from: *Appl. Sci.* **2021**, *11*, 881, doi:10.3390/app11020881 . . . . **355**

About the Editors

Alessandro Gasparetto

Alessandro Gasparetto (MSc in Electronic Engineering, University of Padova, Italy, 1992; MSc in Mathematics, University of Padova, Italy, 2003; PhD in Mechanics of Machines, University of Brescia, Italy, 1996) is a Full Professor of Mechanics of Machines at the Polytechnic Department of Engineering and Architecture, University of Udine (Udine, Italy), where he is the head of the research group in Mechatronics and Robotics. He has been included in the ranking of the top 2% most quoted and authoritative scientists in the world, published by researchers at Stanford University. He is the Deputy Rector for Research at the University of Udine and Chair of IFToMM Italy, the Italian branch of IFToMM (the International Federation for the Promotion of Mechanism and Machine Science). His research interests are in the fields of modelling and control of mechatronic systems, robotics, mechanical design, industrial automation, and mechanical vibrations. He is the author of more than 200 international publications.

Stefano Seriani

Stefano Seriani received his MSc in Mechanical Engineering from the University of Trieste, Italy. He received his PhD at the University of Trieste in 2016. In 2016, he was a research fellow at the Institute of Robotics and Mechatronics of the German space agency (DLR). From 2017 to 2022, he was a research fellow at the University of Trieste; since 2022, he has been a tenure track assistant professor at the Department of Engineering and Architecture of the University of Trieste. His research interests include space robotics, mobile robotics, applied mechanics, robotic space exploration and settlement.

Lorenzo Scalera

Lorenzo Scalera achieved his BSc in Industrial Engineering and MSc in Mechanical Engineering (both cum laude) at the University of Trieste (Trieste, Italy) in 2012 and 2015, respectively, and his PhD in Industrial and Information Engineering at University of Udine (Udine, Italy) in 2019. In 2018, he was a visiting PhD student at the Stevens Institute of Technology in Hoboken (NJ, USA). In 2019, he was a Post Doc Research Fellow at the Free University of Bozen-Bolzano (Bolzano, Italy). In 2019, he was a visiting Research Fellow for one month at the Chiang Mai University (Chiang Mai, Thailand). Since 2021, he has been a tenure track Assistant Professor of Mechanics Applied to Machines at the Polytechnic Department of Engineering and Architecture of the University of Udine. His research interests include modelling and control of mechatronic and robotic systems, trajectory planning, collaborative robotics, and mechanical vibrations.

applied sciences

MDPI

Editorial

Modelling and Control of Mechatronic and Robotic Systems, Volume II

Alessandro Gasparetto [1], Stefano Seriani [2] and Lorenzo Scalera [1,*]

1 Polytechnic Department of Engineering and Architecture, University of Udine, 33100 Udine, Italy; alessandro.gasparetto@uniud.it

2 Department of Engineering and Architecture, University of Trieste, 34127 Trieste, Italy; sseriani@units.it

* Correspondence: lorenzo.scalera@uniud.it

1. Introduction

In modern times, mechatronic and robotic systems are developing at a faster pace than in the past [1], and research on novel solutions and applications of such devices are studied in both industrial and academic environments [2]. The modelling and control of mechatronic and robotic systems is a fundamental field of investigation, especially within the context of Industry 4.0, in which novel scenarios for manufacturing processes and production lines are strictly related to the implementation of robots, automatic machines, and cyber-physical systems.

In this scenario, it is crucial to develop kinematic and dynamic models that can predict the behavior of a mechatronic or robotic system over time, and can support the planning of the path and the trajectory that the system needs to track during its operation [3]. Furthermore, the modelling, design, and control of mechatronic devices and robots play an increasingly central role in enhancing their performance with respect to different objectives, such as energy efficiency [4] or vibration suppression, when the flexibility of the mechanical structure cannot be neglected [5].

The second volume of this Special Issue of *Applied Sciences* aims to disseminate the latest research achievements, ideas, and applications of the modelling and control of mechatronic and robotic systems, with particular emphasis on novel trends and challenges. We invited contributions to this Special Issue on topics including (but not limited to): modelling and control, path and trajectory planning, optimization problems, collaborative robotics, mechatronics, flexible multi-body systems, mobile robotics, and manufacturing applications.

Citation: Gasparetto, A.; Seriani, S.; Scalera, L. Modelling and Control of Mechatronic and Robotic Systems, Volume II. *Appl. Sci.* **2022**, *12*, 5922. https://doi.org/10.3390/app12125922

Received: 3 June 2022
Accepted: 9 June 2022
Published: 10 June 2022

2. Modelling and Control of Mechatronic and Robotic Systems, Volume II

The papers collected in this Special Issue refer to a broad range of disciplines, such as robotic manipulation, mobile robots, service and social robots, cable-driven robotic systems, biomimetic robots, manufacturing, trajectory planning, and control. In most of the papers, numerical and simulation results are corroborated by experimental tests on real prototypes.

Several papers discuss mobile and autonomous systems. In [6], the kinematic and dynamic modelling, and the design of an omni-directional robotic platform for tunnel inspection is presented. That robot was built with the aim of automating the surveillance of a particle acceleration environment characterised by remaining radiations and spatial limitations. The authors of [7] illustrate the development and the experimental evaluation of a crawling terrestrial robot capable of rapidly adapting to the mission it must perform. The proposed prototype can provide a basis for future crawler robots used to detect, disarm, and dispose of explosive threats in extreme environments. Furthermore, the design and modelling of an amphibious spherical robot with fins is proposed in [8]. That robotic system is capable of both terrestrial and aquatic locomotion by exploiting the rolling motion of a spherical shell. Moreover, the spinning motion of the spherical shell is used to steer the robot efficiently and with agility.

The paper [9] presents a behaviour-control architecture for legged and climbing robots by considering their ability to perform self-reconfiguration after unforeseen leg failures. The proposed control approach is suitable for robots with different numbers of legs that move in any direction and inclination planes. Another control algorithm for biomimetic robots is illustrated in [10]. The authors provide a more detailed description of a data-driven control approach for snake manipulators composed of several modules coupled with universal joints. That control algorithm solves the problem of model uncertainty when the number of connecting rods of the serpentine manipulator increases and the environment becomes complex. Moreover, the model of a quadruped robot is considered in [11] to demonstrate the performance of an online foot-location-planning approach for legged robots. In particular, the illustrated strategy leverages model predictive control to solve the problem of large posture changes during gait transitions.

The work of [12] presents the design of the JET humanoid robot, characterized by low stiffness of the actuator modules, high motion capability, and wide range of motion of each joint. Experimental results, including stair climbing, egress from a car, and object manipulation, verify the robot performance and design concepts. In [13], the inverse kinematic problem of a multi-fingered anthropomorphic hand is solved using a genetic algorithm based on workspace analysis. Results show the effectiveness of the proposed approach and its potential application to many industrial robots.

Other works within this Special Issue focus on robotic systems with parallel kinematics. The authors of [14] present the design of a cable-driven parallel robot for non-contact tasks on planar surfaces, such as laser engraving on a paper sheet, inspection, and thermal treatment. A novel cable guidance system is illustrated, which allows for a simple kinematic model to control the manipulator. The work of [15] describes the design and implementation of a platform with six degrees of freedom used as a motion simulator. The inverse kinematic model of the robotic platform and its position-control system are implemented and verified with experimental results. Moreover, the authors of [16] present the control of closed-kinematic chain manipulators based on the concept of sliding mode control. The proposed controller is tested numerically on a planar manipulator with two degrees of freedom.

The paper [17] presents a method for estimation of the natural frequencies of a robotic Cartesian 3D printer based on the kinematics of the system. The approach can help the development of preliminary mechanical design of 3D printers and promises to be useful for emerging 3D printing technologies, which allows for new and sustainable manufacturing paradigms, especially within the framework of Industry 4.0 [18].

The papers [19,20] deal with automation applications in industry. In [19], the design and simulation of a fish-processing machine is shown. The system is introduced to process trout fish in four steps thanks to a vision-based approach. The effectiveness of the proposed design solution is verified through the fabrication of a physical prototype. The article in [20] presents a flow-rate estimation method for an automatic pouring machine for the casting industry. The approach is applied to a laboratory pouring machine to verify its performance in the case of uncertainties in the system model parameters.

The paper in [21] describes a hybrid position and force control architecture based on a finite state machine, which is applied to a robotic manipulator with five degrees of freedom. That control approach is tested on a waste management robotic systems adopted to the selective recycling of different types of materials.

In [22], a trajectory control for piezoelectric actuators based on artificial neural network is presented. The proposed scheme allows for compensating the unmodelled dynamics, uncertainties and perturbations.

Finally, the paper in [23] describes a theoretical method for designing thin motors using electromagnetic forces and electropermanent magnets for applications in portable electrical equipment. A prototype of a motor is fabricated to verify the results obtained with the theoretical approach.

3. Final Remarks

The second volume of this Special Issue collects interesting research papers focused on the modelling and control of mechatronic and robotic systems. This collection of works covers a wide range of applications with both numerical and experimental results. This Special Issue "*Modelling and Control of Mechatronic and Robotic Systems, Volume II*" demonstrates the level of interest in these topics and hints at future developments and challenges.

Author Contributions: Conceptualisation, A.G., S.S. and L.S.; writing—original draft preparation, L.S.; writing—review and editing, A.G., S.S. and L.S.; supervision, project administration, A.G. All authors have read and agreed to the published version of the manuscript.

Funding: This research received no external funding.

Conflicts of Interest: The authors declare no conflict of interest.

References

1. Gasparetto, A.; Scalera, L. From the Unimate to the Delta robot: The early decades of Industrial Robotics. In *Explorations in the History and Heritage of Machines and Mechanisms*; Springer: Berlin/Heidelberg, Germany, 2019; pp. 284–295.
2. Gasparetto, A.; Seriani, S.; Scalera, L. Modelling and Control of Mechatronic and Robotic Systems. *Appl. Sci.* **2021**, *11*, 3242. [CrossRef]
3. Trigatti, G.; Boscariol, P.; Scalera, L.; Pillan, D.; Gasparetto, A. A look-ahead trajectory planning algorithm for spray painting robots with non-spherical wrists. In *IFToMM Symposium on Mechanism Design for Robotics*; Springer: Berlin/Heidelberg, Germany, 2018; pp. 235–242.
4. Carabin, G.; Scalera, L. On the trajectory planning for energy efficiency in industrial robotic systems. *Robotics* **2020**, *9*, 89. [CrossRef]
5. Boscariol, P.; Gallina, P.; Gasparetto, A.; Giovagnoni, M.; Scalera, L.; Vidoni, R. Evolution of a dynamic model for flexible multibody systems. In *Advances in Italian Mechanism Science*; Springer: Berlin/Heidelberg, Germany, 2017; pp. 533–541.
6. Prados Sesmero, C.; Buonocore, L.R.; Di Castro, M. Omnidirectional Robotic Platform for Surveillance of Particle Accelerator Environments with Limited Space Areas. *Appl. Sci.* **2021**, *11*, 6631. [CrossRef]
7. Grigore, L.S.; Oncioiu, I.; Priescu, I.; Joita, D. Development and Evaluation of the Traction Characteristics of a Crawler EOD Robot. *Appl. Sci.* **2021**, *11*, 3757. [CrossRef]
8. Chi, X.; Zhan, Q. Design and Modelling of an Amphibious Spherical Robot Attached with Assistant Fins. *Appl. Sci.* **2021**, *11*, 3739. [CrossRef]
9. Hernando, M.; Alonso, M.; Prados, C.; Gambao, E. Behavior-Based Control Architecture for Legged-and-Climber Robots. *Appl. Sci.* **2021**, *11*, 9547. [CrossRef]
10. Hu, K.; Tian, L.; Weng, C.; Weng, L.; Zang, Q.; Xia, M.; Qin, G. Data-Driven Control Algorithm for Snake Manipulator. *Appl. Sci.* **2021**, *11*, 8146. [CrossRef]
11. Liu, X.; Ma, H.; Lang, L.; An, H. Online Foot Location Planning for Gait Transitioning Using Model Predictive Control. *Appl. Sci.* **2021**, *11*, 7866. [CrossRef]
12. Sim, J.; Kim, S.; Park, S.; Kim, S.; Kim, M.; Park, J. Design of JET Humanoid Robot with Compliant Modular Actuators for Industrial and Service Applications. *Appl. Sci.* **2021**, *11*, 6152. [CrossRef]
13. Lee, C.T.; Chang, J.Y.J. A Workspace-Analysis-Based Genetic Algorithm for Solving Inverse Kinematics of a Multi-Fingered Anthropomorphic Hand. *Appl. Sci.* **2021**, *11*, 2668. [CrossRef]
14. Mattioni, V.; Ida, E.; Carricato, M. Design of a Planar Cable-Driven Parallel Robot for Non-Contact Tasks. *Appl. Sci.* **2021**, *11*, 9491. [CrossRef]
15. Wei, M.Y. Design and Implementation of Inverse Kinematics and Motion Monitoring System for 6DoF Platform. *Appl. Sci.* **2021**, *11*, 9330. [CrossRef]
16. Duong, T.T.C.; Nguyen, C.C.; Tran, T.D. Synchronization Sliding Mode Control of Closed-Kinematic Chain Robot Manipulators with Time-Delay Estimation. *Appl. Sci.* **2022**, *12*, 5527. [CrossRef]
17. Kopets, E.; Karimov, A.; Scalera, L.; Butusov, D. Estimating Natural Frequencies of Cartesian 3D Printer Based on Kinematic Scheme. *Appl. Sci.* **2022**, *12*, 4514. [CrossRef]
18. Carabin, G.; Scalera, L.; Wongratanaphisan, T.; Vidoni, R. An energy-efficient approach for 3D printing with a Linear Delta Robot equipped with optimal springs. *Robot. Comput.-Integr. Manuf.* **2021**, *67*, 102045. [CrossRef]
19. Azarmdel, H.; Mohtasebi, S.S.; Jafary, A.; Behfar, H.; Rosado Muñoz, A. Design and Simulation of a Vision-Based Automatic Trout Fish-Processing Robot. *Appl. Sci.* **2021**, *11*, 5602. [CrossRef]
20. Sueki, Y.; Noda, Y. Experimental Verification of Real-Time Flow-Rate Estimations in a Tilting-Ladle-Type Automatic Pouring Machine. *Appl. Sci.* **2021**, *11*, 6701. [CrossRef]

21. Gal, I.A.; Ciocîrlan, A.C.; Mărgăritescu, M. State Machine-Based Hybrid Position/Force Control Architecture for a Waste Management Mobile Robot with 5DOF Manipulator. *Appl. Sci.* **2021**, *11*, 4222. [CrossRef]
22. Napole, C.; Barambones, O.; Derbeli, M.; Calvo, I. Advanced Trajectory Control for Piezoelectric Actuators Based on Robust Control Combined with Artificial Neural Networks. *Appl. Sci.* **2021**, *11*, 7390. [CrossRef]
23. Park, S.Y.; Song, B.; Baek, Y.S. A Theoretical Method for Designing Thin Wobble Motor Using an Electromagnetic Force and an Electropermanent Magnet for Application in Portable Electric Equipment. *Appl. Sci.* **2021**, *11*, 881. [CrossRef]

applied sciences

MDPI

Article

Omnidirectional Robotic Platform for Surveillance of Particle Accelerator Environments with Limited Space Areas

Carlos Prados Sesmero, Luca Rosario Buonocore * and Mario Di Castro

European Organization for Nuclear Research, 1217 Meyrin, Switzerland; carlos.prados.sesmero@cern.ch (C.P.S.); mario.di.castro@cern.ch (M.D.C.)
* Correspondence: luca.rosario.buonocore@cern.ch; Tel.: +39-334-331-2833

Abstract: Intelligent robotic systems are becoming essential for inspections and measurements in harsh environments. This article presents the design of an omnidirectional robotic platform for tunnel inspection with spatial limitations. This robot was born from the need to automate the surveillance process of the Super Proton Synchrotron (SPS) accelerator of the European Organization for Nuclear Research (CERN), where there is remaining radiation. The accelerator is located within a tunnel that is divided by small doors of 400 × 200 mm dimensions, through which the robot has to cross. The designed robot brings a robotic arm, and the needed devices to carry out the inspection. Thanks to this design, the robot application may vary by replacing certain devices and tools. In addition, this paper presents the kinematic and dynamic control models for the robotic platform.

Keywords: dynamic model; harsh environment; kinematic model; mecanum wheel; omnidirectional robot; robotic platform; surveillance

Citation: Prados Sesmero, C.; Buonocore, L.R.; Di Castro, M. Omnidirectional Robotic Platform for Surveillance of Particle Accelerator Environments with Limited Space Areas. *Appl. Sci.* **2021**, *11*, 6631. https://doi.org/10.3390/app11146631

Academic Editors: Alessandro Gasparetto, Stefano Seriani and Lorenzo Scalera

Received: 29 May 2021
Accepted: 16 July 2021
Published: 19 July 2021

1. Introduction

One of the most significant problems in underground tunnels is the survey of the proper performance of the security sensors available all along the whole corridor. Inspection of large tunnels can be laborious when performed by operators. However, robots may help with performing 4D tasks (dirty, dangerous, difficult, and dull), reducing risks for the personnel. Inspection in underground tunnels use to be complicated because of (a) the long time to access the facilities, (b) the long time to escape the facilities in case of evacuation, (c) the strong safety protocols, and (d) the lack of GPS signal to locate where you are. Usually used tunnels have other ways to communicate with the outside, such as 4G and a WiFi network.

To enhance the process of inspection in long tunnels, a robotic platform is presented in this paper. This mobile robot has been focused in the surveillance of the SPS where a large amount of restriction are present due to its actual state. However, the platform could be used in any other underground tunnel, where the only requirement is found in the floor material, which should guarantee that the omnidirectional wheels work properly.

The SPS is the second largest machine in CERN's accelerator complex, with a circumfer ence of approximately 7 km. Like the rest of tunnels of these characteristics, the corridor of the tunnel becomes completely monotonous, making navigation and SLAM an authentic challenge. Environmental sensors, which measure the radiation, temperature and oxygen concentration among others, are essential to guarantee material and personal safety. Due to the need of periodic check-ups of the sensors, inspections have to be carried out every month. Thus, in this paper we propose the design of an omnidirectional wheeled robot. The robot should be stored in a charging station, located in one of the entrances of the tunnel, where the batteries are charged. When required, it has to go around the ring and record data about the environmental variables. During its journey, the robot brings the needed sensors to record these variables and locate itself in the environment.

One of the most challenging features of the SPS is the fact that it has 19 partition doors which separate sections of the tunnel. These doors have an aperture of 400 × 200 mm, allowing the robot to cross them. Furthermore, the charging station is located in an area of difficult access. It means that the robot has to do several maneuvers to enter and go out of the station. In conclusion, the wheeled robot has the following requirements:

- The survey should take around 80 min. To achieve that, the robot should drive at a speed of 5.5 km/h, or 1.5 m/s, approximately.
- The robot should be able to deal with small spaces, such as doors of 200 × 400 mm.
- The motors should guarantee the nominal velocity of the robot of 5.5 km/h.
- The robot's autonomy should be of 3 h at least, guaranteeing that the robot completes the surveillance even with the presence of unforeseen circumstances.
- The robot should be equipped with the required devices to perform the surveillance autonomously, or through tele-operation if the situation requires it. Autonomous inspection is understood as the fact that the robot travels through the tunnel without direct control by an operator. In such a way, that the robot is able to leave the charging station until the start of the journey, navigate from door to door, cross the door to go from one section to another, and go back to the charging station to charge the batteries. In this line, it is expected that the robot will need external sensors, such as cameras, a 3D LIDAR, and distance sensors to be placed in the middle of the tunnel.

The paper is organized as follows: In Section 2, an overview of wheeled robots is presented. In Section 3, we present the reasons which justify the selected locomotion arrangement. Later, we explain the selection of the locomotion actuators, the electronic and electrical design, the selection of the needed sensors to comply with the robot applications, and the mechanical design, where all the previous is included. Section 4 includes the kinematic and dynamic models of the robot for the locomotion arrangement selected, which will help us move the robot as desired. Section 6 includes the results of the design. Lastly, in Section 7, we present the conclusions of the developed work.

2. Related Work

Wheeled mobile robots are widely used in a large variety of applications, such as in industries, inspection, domestic tasks, rescue, planetary exploration, mining, hazardous waste clean-up, and medical assistance, among others. These robot may have different arrangements, such as differential, synchronous, tricycle, ackerman, or omnidirectional.

Differential drive mobile robots use two steering wheels with a free balancing wheel, called "castor". These robots are controlled by two motors independently, are non-holonomic, have good manoeuvring capacities and work well in indoors environments [1]. However, the speed of these robot is very limited and the odometry estimation is very sensitive. To solve the last problem, backstepping methods for posture tracking [2], adaptive controllers to compensate errors and improve stability [3], and non- linear controllers [4] are used. Furthermore, a wheel synchronization is a critical point for the orientation control of differential mobile robots. Sun et al. propose in [5] a control approach for improving the orientation control, focusing in the coordination of the wheels instead on the robot configuration.

Synchronous drive mobile robots use chains or belts to orientate the wheels at the same time to move the robot in the same direction. This kind of locomotion arrangement needs a lot of space to include all the needed devices. Even with the easiness of its control and the good accuracy of its proprioceptive sensors, its use is not very widespread in robotics mainly because of design problems. They may have as many wheels as desired, but usually they have three or four. Zaman et al. present in [6] a model for translation of a synchoronous drive mobile robot of three wheels, whereas Wada presents in [7] the concept of a variation of this kind of arrangement, making use of four "synchro-caster" wheels.

Tricycle mobile robots have never attracted attention in the scientific community due to its structure, bad stability, and complexity in the path planning in narrow environments.

Kamga et al. present in [8] a path tracking controller for this kind of robot, where problematic dynamic effects are neglected.

Ackerman vehicles, as the one presented in [9], are required when an aggressive steering is needed while maintaining a reasonable velocity when positioned in flat surfaces with obstacles. However, this locomotion arrangement increases the design components (and design complexity), as well as it reduces the manoeuvrability. This kind of locomotion arrangement is widely used in agriculture.

Lastly, omnidirectional robots are capable of moving in arbitrary directions without changing the direction of the wheels, thanks to the smart design of the mecanum wheels, ball wheels, or off-centered wheels among others. In order to increase stability and enhance manoeuvrability of omnidirectional robots with high heights and small bases, a reconfigurable footprint mechanism is developed in [10], which allows varying the ratio of wheel base to tread so that the vehicle could go through a narrow doorway, ensuring the stability. Similarly, the design of a four wheeled omnidirectional mobile robot with a variable wheel arrangement mechanism is presented in [11], where they take advantage of the redundant d.o.f. to drive the mechanism enabling the wheel arrangement to vary. Similarly, Byun et al. present in [12] an omnidirectional robotic platform with the capability of modifying the orientation of the wheels in order to save energy under some circumstances. These approaches would help a robot cross a door as described in Section 1; however, they only allow the platform to move at very slow velocities. Moreover, they do not have into account the viability of using the systems in machines with spatial constraints, since the systems require a mechanical connection between the four wheels, throughout a free joint.

Similarly to our requirements, in [13] an omnidirectional wheeled robot is developed with the objectives of manipulating objects and navigating in indoor environments. They mount two 3D LIDARs, one in front and one in the back, in order to have autonomous navigation in indoors environments. They focus the design on a robotic arm with high load capabilities, without focusing on spatial constraints. In addition, the disposal of the 3D LIDARs produces (a) the necessity of a very accurate calibration to join the generated point clouds, (b) the limitation of navigation only in indoors environments with a lot of features (unable to implement in tunnels), and (c) high costs due to the location of two sensors. Jia et al. present in [14] a kinematic model of a mecanum wheeled robot of four wheels. Later they design and develop the robot. The design approach does not take advantage of the free space between the wheels, raising the load within the body of the platform. Equally, Liu et al. present in [15] an omnidirectional robotic platform with for mecanum wheels and a manipulator. By mounting a manipulator in this kind of arrangement, the working space is expanded. Lastly, since this locomotion arrangement produces an unavoidable glide, ref. [16,17] study the movement performance of mecanum wheeled omnidirectional mobile robots, trying to solve the problem of determining the robot pose even though the robot slides.

About trading platforms, many large manufacturers sell omnidirectional robotic platforms, such as KUKA with the Kuka Mobile Platform 1500, Stanley with Flex Omni, or Nexus with Mecanum 4WD. However, all these platforms are much larger than ours required, in addition to not leaving free the possibility of including a robotic arm to carry out maintenance and handling tasks.

3. Omnidirectional Robotic Platform. Hardware Design

According to the requirements described in Section 1, we have decided to select an omnidirectional locomotion arrangement, with four mecanum wheels located in parallel. Mecanum wheels are composed by passive rollers that transform the force to other directions, giving the robot the capability to be omnidirectional. The selected wheels have a frame that covers the rubber rollers. They work properly on smooth and hard surfaces, such as the tunnel floor, excluding in the same way the necessity of damping devices due to the regularity of the surface.

With the chosen locomotion structure, where the wheels are located in a rectangle with straight forward orientation, we denote important advantages. First of all, (a) the design is more simple, since each wheel has only one associated motor set. (b) The robot is more compact, since the wheels are mounted in the covering box of the whole robot, giving important advantages when crossing the doors or when entering in the charger station. (c) The risk when a motor fails is lower, giving more robustness against foreseen failures in one of the wheels. This kind of structure makes the system redundant, this is, there are more modifiable d.o.f., four in this case, than motion d.o.f, three in this case (x, y, θ). That means that in case a motor fails, the system would be able to finish its task through the implementation of fault tolerance techniques and control architecture approaches. However, this locomotion arrangement presents some disadvantages that should be taken into account: (a) the rollers of the wheels usually slide, spoiling slightly the odometry estimation of the wheels, and (b) the energetic efficiency is low. These problems are not critical, since (a) a SLAM system may correct the wrong odometry, (b) the wheeled robots allow enough space for the batteries to cover the task energetic requirements, and (c) the required speed is low for safety reasons, since the robot is expected to be autonomous in an environment where crashes are not allowed.

Intending to comply with the requirements (surveillance time, robot speed, spatial constraints, autonomy time and tele-operation availability), we present along this section the selection of the locomotion actuators, the electronic and electrical design, the selection of the needed sensors to comply the robot applications, and the mechanical design, where all the previous is included.

3.1. Locomotion Calculation

We present the guidelines to determine the proper motor, gearhead, and encoder. It is important to highlight that the wheels have a known size. They have been selected during the mechanical design process, and they have a crucial role in the selection of the motor set since the most critical part is found in the acceleration torques, which have a strong dependency on the moments of inertia. Firstly, we calculate the nominal force, torque, and power that each motor set has to provide, this is, when the robot is driving at the desired nominal speed. Secondly, we find out the maximum force, torque, and power, which are highly related to the inertial moments, this is, when the robot is under acceleration. With all these calculations, we choose the required devices, which eventually are tested in simulation.

3.1.1. Nominal Torque and Power

Equation (1) relates the desired linear velocity of the robot, V_{nom}, with the angular velocity of the wheels in r.p.m., n_{nom}. Here, a factor η is applied to ensure good performance even with the inevitable presence of friction.

$$n_{nom} = \frac{60}{\pi} \cdot \frac{V_{nom}}{d \cdot \eta} \tag{1}$$

On the other hand, (2) calculates the nominal torque, M_{nom}, to move the entire robot at nominal velocity. This torque depends proportionally on the force necessary to apply, F_{nom}, which is determined in (3), and inversely on the bearing factor, δ.

$$M_{nom} = \frac{d}{2} \cdot \frac{F_{nom}}{\delta} \tag{2}$$

The nominal force is the needed force to keep the robot at the nominal velocity. Typically, three main friction forces appear during the displacement of a vehicle: (a) rolling resistance, (b) aerodynamic resistance, and (c) slope resistance. Both aerodynamic and slope resistances may be neglected in our case. Aerodynamic resistance does not contribute because of the following reasons: (a) the air in the tunnel is static, without the presence of air currents, (b) the robot moves at low velocity, and (c) the air is clean and dry. Slope resistance does not contribute because the tunnel is mainly flat. Thus, the rolling force is calculated in (3) through (a) the rolling resistance coefficient, or the coefficient of rolling friction (CRF), C_{rr}, and (b) the normal force, F_N, this is, the perpendicular force to the surface on which the wheel is rolling. The coefficient of rolling friction is given in (4) [18] by the sinking depth, z, which depends on the elastic features of the material of the rollers, and the wheel diameter, d. On the other hand, F_N depends on the robot weight, m_R.

$$F_{nom} = C_{rr} \cdot F_N = C_{rr} \cdot m_R \cdot g \tag{3}$$

$$C_{rr} = \frac{z}{d} \tag{4}$$

The calculated nominal torque has to be split in the number of traction wheels, N, in the robot, as expressed in (5), resulting in the needed nominal torque in each motor set, $M_{nom/m}$.

$$M_{nom/m} = \frac{M_{nom}}{N} \tag{5}$$

So far, the torque when the nominal velocity is reached has been calculated. However, in order to calculate the acceleration, average and maximum torques, a velocity profile has to be created, as shown in Figure 1, where the time to reach the nominal velocity, t_{acc}, and the time of nominal velocity, t_{nom}, are indicated. For simplicity, we have assumed constant acceleration, a, and equal acceleration and deceleration time. Thus, acceleration is calculated in (6).

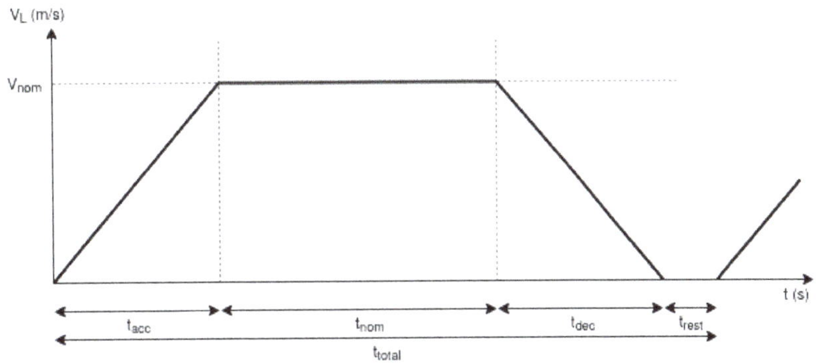

Figure 1. Velocity profile of the robotic platform.

$$a = \frac{V_{nom}}{t_{acc}} \tag{6}$$

The last variable to figure out is the power that the motor set has to provide while the robot is driving at nominal velocity. Thus, in (7) we calculate the necessary nominal power of the motor set, $P_{nom/m}$, where a confidence factor, κ, and a power factor, PF, have been set to slightly oversize the set. Anyway, the chosen devices have to provide more power than calculated during the nominal stage.

$$P_{nom/m} = PF \cdot \frac{m_R \cdot V_{nom} \cdot (a + g \cdot \sin\theta)}{2 \cdot \pi \cdot \kappa \cdot N} \tag{7}$$

3.1.2. Maximum Torque and Power

One of the most important parts in the selection of the locomotion actuators is found in the inertial forces, this is, the force that the locomotion system has to apply during the acceleration and deceleration. With a preliminary motor selection, it is possible to know its moment of inertia. Since the wheels and shaft ones are well known, the one of the full system, this is motor/shaft/wheel, may be calculated. Both wheel and shaft ones, J_w and

J_s, respectively, are calculated through (9), where the inertial mass, m_i, is calculated in (8). Here, we define (a) ρ as the component density, (b) r_a as its external radius, (c) r_i as its internal (if applicable), and (d) h the width.

Thus, the robot acceleration torque, M_{acc}, is calculated in (10), where the moment of inertia of the motor, J_m, is one of the motors that satisfies the previous specifications, and where d is the diameter of the wheels. Although the moment of inertia of the motor may change with the final decision, it is practically negligible concerning that of the set of wheel, shaft, motor, and robot weight (each one is indicated in (10).

$$m_i = \rho \cdot \pi \cdot ({r_a}^2 - {r_i}^2) \cdot h \tag{8}$$

$$J_w = \frac{1}{2} \cdot m_i \cdot ({r_a}^2 + {r_i}^2) \tag{9}$$

$$M_{acc} = (N \cdot (\underbrace{J_w}_{wheel} + \underbrace{J_m}_{motor} + \underbrace{J_s}_{shaft}) + \underbrace{\frac{m_R}{\eta} \cdot \frac{d^2}{4}}_{robot}) \cdot a \cdot PF \tag{10}$$

Then, taking into account the torque during the nominal moment (2) and during the acceleration (10), it is possible to calculate the maximum and the root mean square, RMS, torque in each motor, $M_{max/m}$ and $M_{rms/m}$, respectively, according to (11), (13) and (14), respectively. For a wheeled robot, the decelerate torque, M_{dec}, is determined by the difference between the nominal and acceleration torques, as indicated in (12). For a robotic platform that is continuously working, the resting torque, denoted as M_{rest} is null, since the robot is expected to perform the survey without stops.

$$M_{max/m} = \frac{M_{max}}{N} = \frac{M_{nom} + M_{acc}}{N} \tag{11}$$

$$M_{dec} = M_{nom} - M_{acc} \tag{12}$$

$$M_{rms} = \left\{ \frac{1}{t_{total}} \cdot (t_{acc} \cdot M_{max}^2 + t_{nom} \cdot M_{nom}^2 + t_{dec} \cdot M_{dec}^2 + t_{rest} \cdot M_{rest}^2) \right\}^{1/2} \tag{13}$$

$$M_{rms/m} = \frac{M_{rms}}{N} \tag{14}$$

After all, the maximum and RMS power may be calculated according to (15) and (16), respectively.

$$P_{max/m} = M_{max/m} \cdot n_{nom} \cdot \frac{\pi}{30} \tag{15}$$

$$P_{rms/m} = M_{rms/m} \cdot n_{rms} \cdot \frac{\pi}{30} \tag{16}$$

3.1.3. Motor and Gearhead Selection

Thanks to the equations described before, we should select a gearhead with a continuous torque of at least $M_{rms/m}$, and an intermittent torque at gear output of at least $M_{max/m}$. In our case, we have chosen the model GP 26A of Maxon, which has a reduction of K_g of 1:27 and a maximum efficiency η of 80%. It is required to select the immediately inferior reduction that fulfills the requirements.

After that, we should chose a motor that provide (a) an angular velocity of at least $n_{nom} \cdot K_g$ (r.p.m.), (b) a continuous torque of at least $M_{rms/m'}$, calculated in (17), (c) a maximum torque of at least $M_{max/m'}$, calculated in (18), and (d) a continuous operation power of at least $P_{rms/m}$.

$$M_{rms/m'} = \frac{M_{rms/m}}{K_g \cdot \eta} \tag{17}$$

$$M_{max/m'} = \frac{M_{max/m}}{K_g \cdot \eta} \tag{18}$$

Then, it is needed to check that the speed constant of the motor is at least C_v, calculated in (19). Here, V_0 denotes the no-load speed calculated in (20), whereas v denotes the voltage of the motor. Regarding the no-load speed equation, τ denotes the speed/torque gradient of the motor.

$$C_v = \frac{V_0}{v} \tag{19}$$

$$V_0 = V_{nom} + \tau \cdot M_{max/m'} \tag{20}$$

In the end, we have chosen the motor model RE 25, Graphite Brushes, 20 W of 24 V, which fulfills the previous specifications. Additionally, we select the encoder model Encoder MR Type ML, which allows 500 counts per turn, high enough to guarantee real-time behavior for a wheeled robot.

For mechanical reasons, it is needed a motion transmission between the motor system and the wheels. To solve this problem, a couple of pulleys, connected by a belt, are used. Taking advance of this restriction, a velocity reduction is applied to get the nominal robot velocity. The relationship of the pulleys in our case is 1:1.1.

3.1.4. Electronic and Electrical Design of the Motor Set

Since the motor, the gearhead, the encoder, and the controller (EPOS4 Compact 50/5) have been selected from Maxon Motor seller, the cables have been chosen as well. In order to communicate the controller with the computer, we have used CAN bus, as shown in the electronic and electrical scheme shown in Figure 2. The electronic and electrical scheme shows each cable with the corresponding identifier of the seller.

3.1.5. Motor Validation

In order to guarantee the proper performance of the selected motor and gearhead, we have simulated the motor set behavior under the expected circumstances during the driving. Thus, the simulation shows the performance of the motor set when the robot follows the velocity profile described before.

In this way, simulating the behavior of the motor, we have developed the block diagram illustrated in Figure 3, where we include a PID controller that directly control the motor model (defined by its features such as the terminal resistance, terminal inductance, torque constant, inertial moment and viscous constant).

M_{in} denotes the load reaction torque, this is the torque produced by the robot weight, friction, slope of the terrain, etc. This torque is related to the force specified in (3). Thus, the simulation process follows the following guidelines:

- Guided by the high level controller, the desired velocity, V_d, is inserted.
- The PID controller takes the difference between the desired velocity and the current robot velocity, V_R, to generate the control signal, v_c, which is inserted in the model of the motor.
- The angular velocity of the motor is transformed to the robot velocity (linear) through the reduction generated by the gearhead and the pulleys, K_n. On the way, a friction constant is applied, η, as well as a transformation between angular to linear.

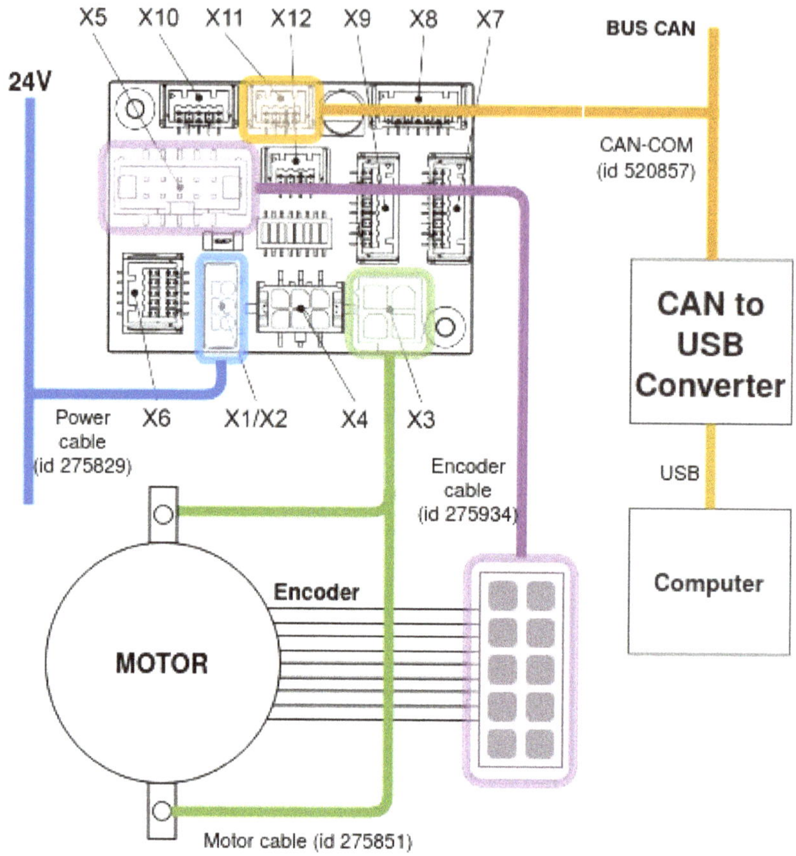

Figure 2. Electronic and electrical scheme to control the motors. Blue wires represent power cables, yellow ones represent USB cables, green ones indicate the motor specific cables, purple one is the encoder specific cable, and light brown is the BUS CAN, which collect the data from all the motor sets.

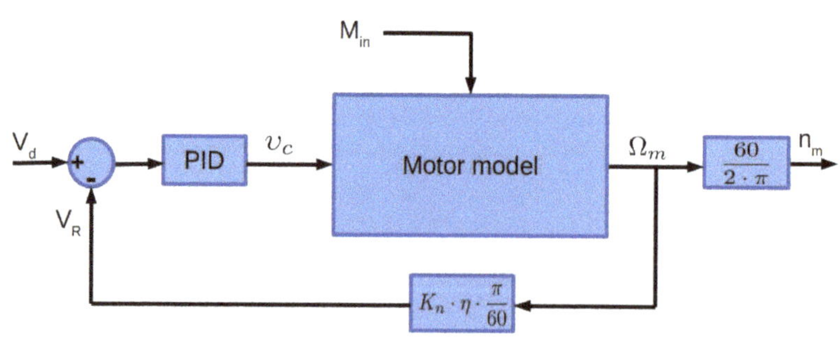

Figure 3. Block diagram for the motor validation.

The obtained results for the required velocity show us that the motor does not reach its limits, concluding that the selected devices are proper for the first and third requirements described in Section 1.

3.2. Device Selection

In the final version, the robot will be equipped with the following sensors:

- Two 2D LIDARs for object detection. They will be used to prevent the robot from crashing with obstacles, with the walls, and with the accelerator. The model UST-20LX, by Hokuyo, is used.
- Cameras for teleoperation and optionally for visual odometry. The model UI-3080CP Rev. 2, by IDS, is used.
- A point cloud generator sensor for the self-localization. It can be a 3D LIDAR or an RGBd camera (only for small environments). The model HDL-32E of Velodyne has been used for localization and environment reconstruction tests with good results. However, it is recommendable to use another sensor like Puck Lite or Ultra Puck from the same manufacturer, since the size of this sensor is too big for our spatial constraints, the vertical field of view is focused down (problematic for low and small robots like ours) and it has moving parts (recommendable to avoid them). In this paper, we include the 3D LIDAR Puck Lite, by Velodyne.
- An inertial measurement unit, IMU, for the localization system. The model VMU931 is used.
- A radiation sensor.
- An ultrasonic sensor, SONAR, to determine the distance between the radiation sensor and the radiation source. The model MB7040 I2CXL-MaxSonar-WR Ultra compact Housing, by MaxBotix is used.
- A WiFi antenna for communication.
- A robotic arm to bring the radiation sensor in the particular application of the radiation survey, and to bring a specific tool in other cases. The Jaco by Kinova is used.

Furthermore, the onboard computer is the NUC i5-8259U, 8 GB, DDR4-SDRAM, 256 GB SSD Mini, by Intel, which is the computer that supports all the logic behavior. The devices are connected to the NUC as shown in Figure 4. We have decided to connect the cameras through Ethernet, whereas the robotic arm through USB drivers. However, both the cameras and the arm may use Ethernet or USB. With these devices, the system fulfills the needs to comply with the last requirement of Section 1.

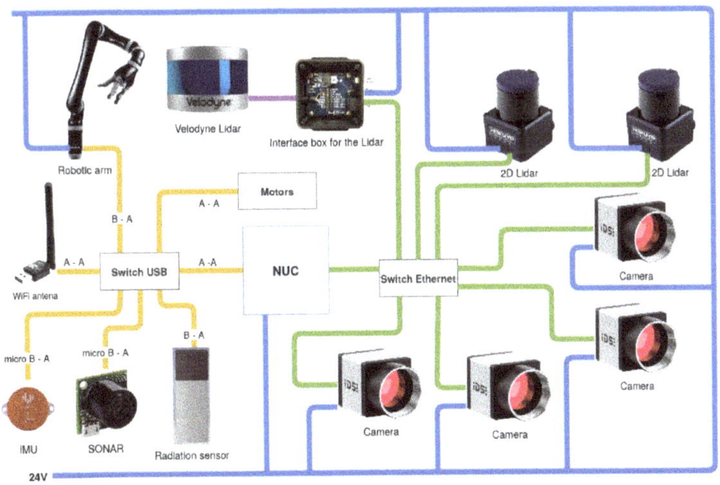

Figure 4. Electric diagram where all the devices are connected to the NUC for its processing. Blue wires represent power cables, yellow ones represent USB cables, where both termination types are indicated, green ones indicate Ethernet alternatives, and the purple one is a specific cable from Velodyne that communicates the LIDAR with the interface box throughout a given protocol.

3.3. Mechanical Design

We present a mechanical design based on the use of aluminum profiles for the main structure, giving consistency and strength to the robot body. Furthermore, the bottom plates are also of aluminum material, since this material is proper for radioactive environments. The design (Figure 5) has the following characteristics:

Figure 5. Design of the robotic platform for the survey in underground tunnels with space constraints. Adapted from: [19].

- Four lead–acid batteries, two located at each side. Each provides $7A \cdot$ h. In total they provide energy so that all the robot devices are working at their maximum power for 3 h and a quarter, enough to guarantee the forth requirement of Section 1.
- A magnetic connector for charging the batteries is localized on one side.
- Four possible localization for cameras behind the wheels, place where the field of view is good enough.
- Two 2D LIDAR for obstacle and wall detection (one in front and one in the back). Both of them are linked to the structure with a piece that has to be manufactured in aluminum material.

- A 3D LIDAR for the SLAM algorithm.
- A support for the radiation sensor in the end-effector of the robotic arm.
- A support in the radiation sensor to place the SONAR.

The main dimensions are 707 × 350 × 184 mm, a proper size to cross the doors, complying with the second requirement of Section 1. The estimation of the weight is around 45 kg, taking into account all the components. The reach of the sensor is around 1100 mm at its main point.

The robot adopts the position shown in Figure 6 when it is crossing the door, where it is appreciated that space while doing so is enough. In this configuration, the robotic arm is set horizontally and the 3D LIDAR is folded against the other face of the plate that supports the robotic arm.

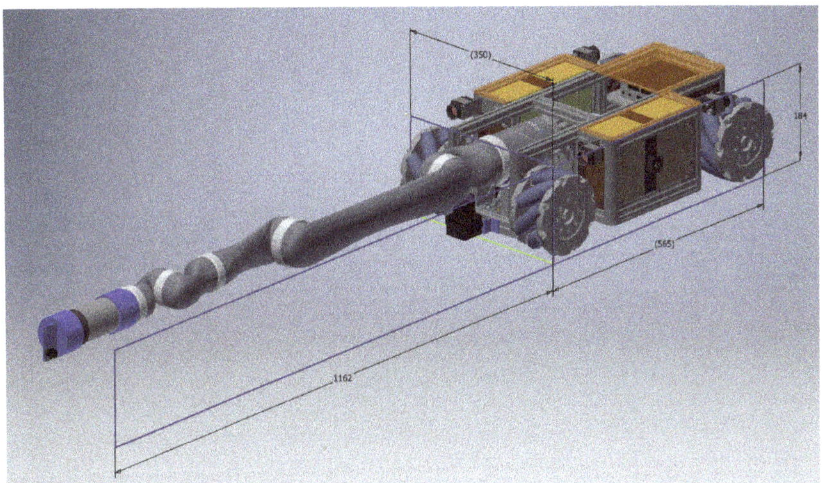

Figure 6. Robot configuration while crossing the doors.

The robot has been designed with an eye on the main application, this is, the surveillance of the SPS accelerator. However, with the cameras and with the robotic arm, the application range increases abruptly, having the possibility of using the robot in teleoperations (drilling, leak repair, component replacement, cable welding, etc.), visual surveillance of other environments, variable checking (temperature, oxygen concentration, etc.), and many more, with the single replacement of the end-effector tool.

4. Kinematic and Dynamic Models of the Robotic Platform

In this section, the kinematic and dynamic models of the platform are presented from a theoretical point of view, to include them into the CERNTAURO framework [20], which contains all the robotic software of CERN.

4.1. Kinematic Model

Firstly, to describe the robot posture in the ground plane, the reference systems shown in Figure 7 have been created, where we find reference systems of: (a) the world, $\sum_W$, (b) the robot, $\sum_R$, (c) the wheel i with the same orientation that the robot one, $\sum_{Ri}$, and (d) the wheel i with the orientation of the mecanum wheel rollers, $\sum_{ri}$.

In that picture, we describe: (a) the distance between $\sum_R$ and $\sum_{Ri}$, L_i, (b) the angle between the axis R_x and the contact point of the wheel i rollers with the floor, α_i, (c) the angle between the axis r_{ix} and the axis R_{iy}, β_i, and (d) the angle between the axis R_{iy} and the axis r_{iy}, γ_i. All of them are well known: (a) L_i, (b) α_i and (c) β_i are determined by the dimensions of the robot design, and (d) γ_i is determined by the wheel design (usually 45°). Moreover, we describe (e) ϕ_i as a sum of α_i, β_i and γ_i, and (f) δ_i as a sum of β_i and γ_i.

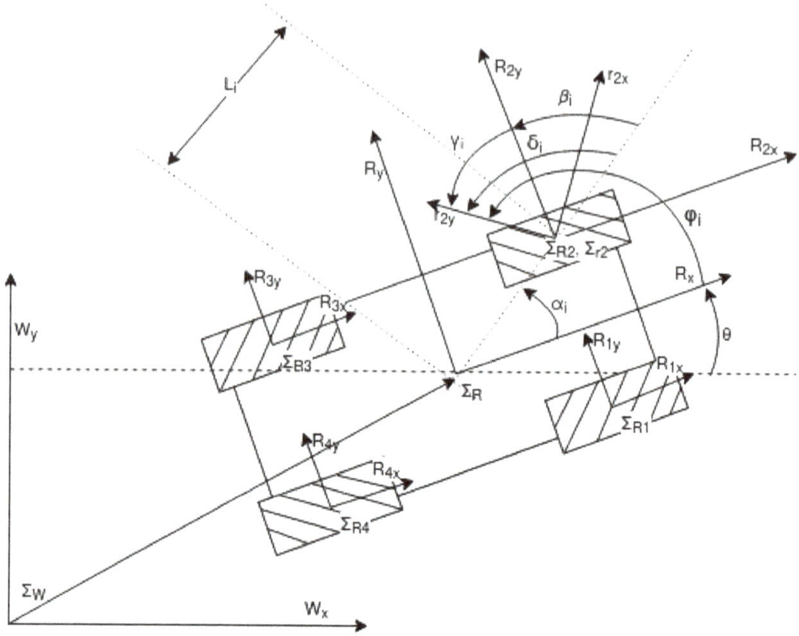

Figure 7. Overall of the robot reference systems ($\sum_W$, $\sum_R$, $\sum_{Ri}$ and $\sum_{ri}$).

Like the vast majority of wheeled robots, the position of the robot is described by 3 d.o.f., this is, position and orientation x, y and θ. Thus, the vector ξ which defines the posture of the robot is presented in (21). Furthermore, we define (a) $\dot{\xi}$ as the velocity vector of the robot in $\sum_W$, and (b) η as the velocity vector of the robot in $\sum_R$, calculated from $\dot{\xi}$ in (22), where $R_w^R(\theta)$ is the rotation matrix between $\sum_R$ and $\sum_W$.

$$\xi = \begin{bmatrix} x \\ y \\ \theta \end{bmatrix} \in \Re^3 \tag{21}$$

$$\eta = R_w^R(\theta) \cdot \dot{\xi} = \begin{bmatrix} c_\theta & s_\theta & 0 \\ -s_\theta & c_\theta & 0 \\ 0 & 0 & 1 \end{bmatrix} \cdot \begin{bmatrix} \dot{x} \\ \dot{y} \\ \dot{\theta} \end{bmatrix} \tag{22}$$

The required velocity in each wheel i, $\dot{O}_{Ri}$, to move the robot at a given velocity η is described in (23). In addition, in η the velocities are referred to $\sum_R$, so in (24) we displace them to $\sum_{ri}$ through the rotation matrix $R_R^{R_i}(\phi_i)$, which does so.

$$\begin{bmatrix} \dot{O}_{R_{i1}} \\ \dot{O}_{R_{i2}} \\ 0 \end{bmatrix}^R = \begin{bmatrix} \dot{x}^R - L_i \cdot s_{\alpha_i} \cdot \dot{\theta} \\ \dot{y}^R - L_i \cdot c_{\alpha_i} \cdot \dot{\theta} \\ 0 \end{bmatrix} \tag{23}$$

$$\begin{bmatrix} \dot{O}_{R_{i1}} \\ \dot{O}_{R_{i2}} \\ 0 \end{bmatrix}^{R_i} = R_R^{R_i}(\phi_i) \cdot \begin{bmatrix} \dot{O}_{R_{i1}} \\ \dot{O}_{R_{i2}} \\ 0 \end{bmatrix}^R = \begin{bmatrix} s_{\phi_i} & -c_{\phi_i} & 0 \\ c_{\phi_i} & s_{\phi_i} & 0 \\ 0 & 0 & 1 \end{bmatrix} \cdot \begin{bmatrix} \dot{O}_{R_{i1}} \\ \dot{O}_{R_{i2}} \\ 0 \end{bmatrix}^R \tag{24}$$

According to [21], we define: (a) the wheel speed, $\dot{\varphi}_{R_i}$, (b) the wheel radius, r_{R_i}, (c) the roller speed, $\dot{\varphi}_{r_i}$, and (d) the roller radius r_{r_i}. In our case, the robot values are shown in Table 1.

Table 1. Kinematic parameters of our robot. Figure 8 shows each variable in the robot frame.

Wheel i	L_i	α_i	β_i	γ_i	r_{R_i}
1	L	$360-\alpha$	$\frac{\pi}{2}-\alpha_1$	$\frac{\pi}{4}$	r
2	L	α	$\frac{\pi}{2}-\alpha_2$	$-\frac{\pi}{4}$	r
3	L	$180-\alpha$	$\frac{\pi}{2}-\alpha_3$	$\frac{\pi}{4}$	r
4	L	$180+\alpha$	$\frac{\pi}{2}-\alpha_4$	$-\frac{\pi}{4}$	r
	L			274.06 mm	
	α			29.5°	
	r			76.2 mm	

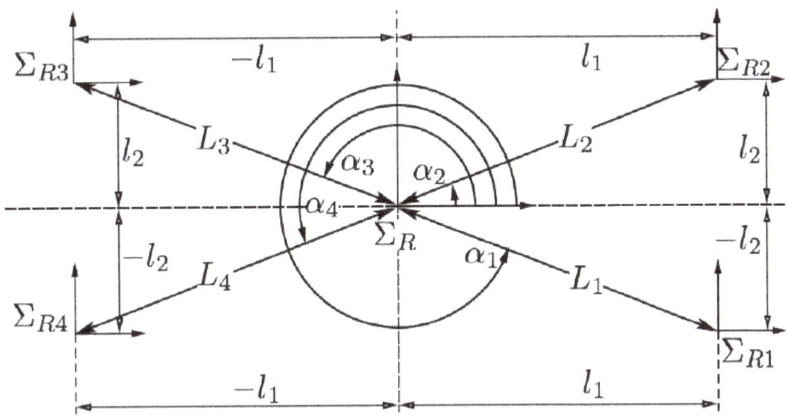

Figure 8. Wheel parameters relative to the robot.

For omnidirectional robots with mecanum wheels, the delivered speed by the wheel is $r_{R_i}\cdot\varphi_{R_i}$ in the R_{ix} direction, whereas the roller lineal velocity is $r_{r_i}\cdot\varphi_{r_i}$ in the r_{ix} direction. Thus, making use of (24), we transform the delivered speed by the wheel to $\sum_{r_i}$ in (25).

$$\begin{bmatrix} r_{R_i}\cdot\dot{\varphi}_{R_i}\cdot c_{\gamma_i} \\ r_{r_i}\cdot\dot{\varphi}_{r_i}+r_{R_i}\cdot\dot{\varphi}_{R_i}\cdot s_{\gamma_i} \\ 0 \end{bmatrix} = \begin{bmatrix} s_{\phi_i} & -c_{\phi_i} & L_i\cdot c_{\delta_i} \\ c_{\phi_i} & s_{\phi_i} & L_i\cdot c_{\delta_i} \\ 0 & 0 & 0 \end{bmatrix}\cdot\eta \tag{25}$$

For each wheel, a motion constraint, shown in (26), is obtained from the first row of the equation system of (25) [21].

$$\begin{bmatrix} -s_{\phi_i} & c_{\phi_i} & L_i\cdot c_{\delta_i} \end{bmatrix}\cdot R_w^R(\theta)\cdot\dot{\xi}+r_{R_i}\cdot\dot{\varphi}_{R_i}\cdot c_{\gamma_i}=0 \tag{26}$$

Thus, the motion constraints equation of the entire robot is shown in (27), where (a) J_1 is composed by the motion constraint rows of each wheel, as described in (26), and where (b) J_2 is defined by the wheel design. The results are the Equations (28) and (29), respectively.

$$J_1\cdot R_w^R(\theta)\cdot\dot{\xi}+J_2\cdot\dot{\varphi}_i=0 \tag{27}$$

$$J_1 = \begin{bmatrix} -s_{\phi_1} & c_{\phi_1} & L_1 \cdot c_{\delta_1} \\ -s_{\phi_2} & c_{\phi_2} & L_2 \cdot c_{\delta_2} \\ -s_{\phi_3} & c_{\phi_3} & L_3 \cdot c_{\delta_3} \\ -s_{\phi_4} & c_{\phi_4} & L_4 \cdot c_{\delta_4} \end{bmatrix} \tag{28}$$

$$J_2 = \begin{bmatrix} r_{R_1} . c_{\gamma_1} & 0 & 0 & 0 \\ 0 & r_{R_2} . c_{\gamma_2} & 0 & 0 \\ 0 & 0 & r_{R_3} . c_{\gamma_3} & 0 \\ 0 & 0 & 0 & r_{R_4} . c_{\gamma_4} \end{bmatrix} \tag{29}$$

The kinematic configuration model for that robot is given by (30) [21], where $S(q)$ is decomposed in the high and low part, relating to the kinematic posture model and the wheels velocities, respectively.

$$\dot{q} = \begin{bmatrix} \dot{\xi} \\ \dot{\varphi}_1 \\ \dot{\varphi}_2 \\ \dot{\varphi}_3 \\ \dot{\varphi}_4 \end{bmatrix} = \begin{bmatrix} \dot{x} \\ \dot{y} \\ \dot{\theta} \\ \dot{\varphi}_1 \\ \dot{\varphi}_2 \\ \dot{\varphi}_3 \\ \dot{\varphi}_4 \end{bmatrix} = S(q) \cdot \eta = \begin{bmatrix} R_w^R(\theta) \\ E \end{bmatrix} \cdot \begin{bmatrix} \dot{x}^R \\ \dot{y}^R \\ \dot{\theta}^R \end{bmatrix} \tag{30}$$

The high part of the previous equation is determined by the kinematic posture model for an omnidirectional robot, which is given by (31). The low part of the equation is given by the isolation of $\dot{\varphi}_i$ from (27), in such a way that for all the wheels we obtained (32), which is related to E from (30).

$$\dot{\xi} = R_R^w(\theta) \cdot \eta \tag{31}$$

$$\dot{\varphi}_i = -J_2^{-1} \cdot J_1 \cdot R_w^R(\theta) \cdot \dot{\xi} = -J_2^{-1} \cdot J_1 \cdot \eta \tag{32}$$

Then, substituting the values of Table 1, we obtain our kinematic model, which is presented in (33).

$$\dot{q} = \begin{bmatrix} \dot{x} \\ \dot{y} \\ \dot{\theta} \\ \dot{\varphi}_1 \\ \dot{\varphi}_2 \\ \dot{\varphi}_3 \\ \dot{\varphi}_4 \end{bmatrix} = \begin{bmatrix} c_\theta & -s_\theta & 0 \\ s_\theta & c_\theta & 0 \\ 0 & 0 & 1 \\ \frac{1}{r} & \frac{1}{r} & \frac{(l_1+l_2)}{r} \\ \frac{1}{r} & -\frac{1}{r} & -\frac{(l_1+l_2)}{r} \\ \frac{1}{r} & \frac{1}{r} & -\frac{(l_1+l_2)}{r} \\ \frac{1}{r} & -\frac{1}{r} & \frac{(l_1+l_2)}{r} \end{bmatrix} \cdot \begin{bmatrix} \dot{x}^R \\ \dot{y}^R \\ \dot{\theta}^R \end{bmatrix} \tag{33}$$

4.2. Dynamic Model

The dynamic model for a wheeled omnidirectional robot can be calculated through (34), where (a) the sum of kinetic energies of the robot, T, is calculated in (35), and where (b) the applied torque in the wheel i, φ, is figured out in (36) [21].

$$R_w^R(\theta)[T]_\xi + E^T[T]_\varphi = E^T\tau_\varphi \tag{34}$$

$$[T]_\psi = \frac{d}{dt}\left(\frac{\partial T}{\partial \dot{\psi}}\right) - \frac{\partial T}{\partial \psi} \tag{35}$$

$$\tau_\varphi = \begin{bmatrix} \tau_{\varphi 1} & \tau_{\varphi 2} & \tau_{\varphi 3} & \tau_{\varphi 4} \end{bmatrix}^T \tag{36}$$

We denote (a) m_R as the robot total mass, (b) I_{Rz} as the robot total moment of inertial, and (c) $I_{\varphi y}$ a the wheels moment of inertial. For an omnidirectional robot, T can be expressed as shown in (37), where M, I_φ and $\dot{\varphi}$ have the values shown in (38)–(40).

$$T = \dot{\xi}^T \left(R_w^R(\theta)\right)^T \cdot M \cdot R_w^R(\theta) \cdot \dot{\xi} + \varphi^T \cdot \dot{I}_\varphi \cdot \dot{\varphi} \tag{37}$$

$$M = \frac{1}{2} diag\{m_R, m_R, I_{Rz}\} \tag{38}$$

$$I_\varphi = \frac{1}{2} diag\{I_{\varphi y}, I_{\varphi y}, I_{\varphi y}, I_{\varphi y}\} \tag{39}$$

$$\dot{\varphi} = \begin{bmatrix} \dot{\varphi}_1 & \dot{\varphi}_2 & \dot{\varphi}_3 & \dot{\varphi}_4 \end{bmatrix}^T \tag{40}$$

Developing (37), (41) is obtained.

$$\begin{aligned} T = \frac{m_R}{2}\left(\dot{x}^2 + \dot{y}^2\right) + \frac{I_{Rz}}{2}\dot{\theta}^2 + \\ \frac{I_{\varphi y}}{2}\left(\dot{\varphi}_1^2 + \dot{\varphi}_2^2 + \dot{\varphi}_3^2 + \dot{\varphi}_4^2\right) \end{aligned} \tag{41}$$

$[T]_\xi$ and $[T]_\varphi$ terms are given using (35), whose results are reflected in (42) and (44).

$$[T]_\xi = M_R \cdot \ddot{\xi} \tag{42}$$

$$M_R = diag\{m_R, m_R, I_{Rz}\},\ \ddot{\xi} = \begin{bmatrix} \ddot{x} \\ \ddot{y} \\ \ddot{\theta} \end{bmatrix} \tag{43}$$

$$[T]_\varphi = M_\varphi \cdot \ddot{\varphi} \tag{44}$$

$$M_\varphi = diag\{I_{\varphi y}\},\ \ddot{\varphi} \begin{bmatrix} \ddot{\varphi}_1 \\ \ddot{\varphi}_2 \\ \ddot{\varphi}_3 \\ \ddot{\varphi}_4 \end{bmatrix} \tag{45}$$

Upgrading (34) with the previous calculations, (46) is obtained. In order to transform the accelerations from the world frame to the robot frame, (47) and (48) are presented. The second one uses and derives the high part of (33).

$$R_w^R(\theta) \cdot M_R \cdot \ddot{\xi} + E^T \cdot M_\varphi \cdot \ddot{\varphi} = E^T \tau_\varphi \tag{46}$$

$$\ddot{\xi} = R_R^w(\theta) \cdot \dot{\eta} + \dot{R}_R^w(\theta) \cdot \eta \tag{47}$$

$$\ddot{\varphi} = E \cdot \dot{\eta} \tag{48}$$

The result is shown in (49), with (50) and (51), where the dynamic model of the omnidirectional robot is presented as the final model related to the operational space in (49)–(51).

$$\bar{M} \cdot \dot{\eta} + \bar{C} \cdot \eta = E^T \cdot \tau_\varphi \tag{49}$$

$$\bar{M} = R_w^R(\theta) \cdot M_R \cdot R_R^w(\theta) + E^T \cdot M_\varphi \cdot E \tag{50}$$

$$\bar{C} = R_w^R(\theta) \cdot M_R \cdot \dot{R}_R^w(\theta) \tag{51}$$

Since the compact form described in [22] (specified in (52)) works in the robot space (unlike η that is described in the operational space), it is needed to apply the inverse kinematic transformation calculated in (33) and specified in (53).

$$M(q) \cdot \ddot{q} + C(q, \dot{q}) \cdot \dot{q} + g(q) = \tau \tag{52}$$

$$\dot{q} = \begin{bmatrix} \dot{\varphi}_1 \\ \dot{\varphi}_2 \\ \dot{\varphi}_3 \\ \dot{\varphi}_4 \end{bmatrix} = \begin{bmatrix} \frac{1}{r} & \frac{1}{r} & \frac{(l_1+l_2)}{r} \\ \frac{1}{r} & -\frac{1}{r} & -\frac{(l_1+l_2)}{r} \\ \frac{1}{r} & \frac{1}{r} & -\frac{(l_1+l_2)}{r} \\ \frac{1}{r} & -\frac{1}{r} & \frac{(l_1+l_2)}{r} \end{bmatrix} \cdot \begin{bmatrix} \dot{x}^R \\ \dot{y}^R \\ \dot{\theta}^R \end{bmatrix} = K \cdot \eta \tag{53}$$

Since K is not quadratic, the transformation should use the pseudo-inverse matrix (54), as shown in (55).

$$\eta = K^+ \cdot \dot{q} \tag{54}$$

$$K^+ = \frac{r}{4} \begin{bmatrix} 1 & 1 & 1 & 1 \\ 1 & -1 & 1 & -1 \\ \frac{1}{l_1+l_2} & \frac{-1}{l_1+l_2} & \frac{-1}{l_1+l_2} & \frac{1}{l_1+l_2} \end{bmatrix} \tag{55}$$

For the same reason (transform the data from the operational to the robot space), the frame transformation matrices $R_w^R(\theta)$ and $R_R^w(\theta)$, have to be transformed through K^+. In order to calculate the transformation, $\dot{\theta}$ is figured out in (56).

$$\dot{\theta} = \frac{r}{4 \cdot (l_1 + l_2)} \cdot (\dot{\varphi}_1 - \dot{\varphi}_2 - \dot{\varphi}_3 + \dot{\varphi}_4) \tag{56}$$

Lastly, to convert the dynamic model to the compact form [22], the final model in the robot space is shown in (57), where $\bar{M}$ and $\bar{C}$ are defined in (57) and (59).

$$\bar{M} \cdot \ddot{q} + \bar{C} \cdot \dot{q} = \tau_\varphi \tag{57}$$

$$\bar{M} = \left(E^T\right)^{-1} \cdot R_w^R(q) \cdot M_R \cdot R_R^w(q) \cdot K^+ + M_\varphi \cdot E \cdot K^+ \tag{58}$$

$$\bar{C} = \left(E^T\right)^{-1} \cdot R_w^R(q) \cdot M_R \cdot \dot{R}_R^w(q) \cdot K^+ \tag{59}$$

In the compact form, $\bar{M}$ is the inertial matrix, $\bar{C}$ is the centrifuge and Coriolis matrix, $g(q)$ is the gravity vector and τ is the torque applied by the motors. The content of these matrices is strongly important and relevant in the design of robotic arms, not so much for the platform. Its use is quite relevant in the design of joint controllers. Even so, it is possible to detail some important features about these matrices [22]:

- The inertial matrix is very important for (a) the dynamic model (related to the kinematic energy), and for (b) the robot controller design (related to stability studies for robotic arms mainly).
- The centrifuge and Coriolis matrix are also important to perform stability studies for control systems in robotic arms.
- The gravity vector is present in non-designed robots, this is, a robot whose design does not provide gravity torques of compensation. In addition, for robotic platforms, which move in the horizontal plane, this vector does not appear in the dynamic model.

5. Control Architecture

As shown in Figure 9, the proposed control architecture is divided in two levels, the *Decisional level*, and the *Executive level*. The *Decisional level* includes:

- The graphical user interface (*GUI*), through which the user sets the robot's goals and the task. In addition, the user can control the robot, observe its status and view the information from the sensors.
- The *Supervisor* is responsible for observing the robot status and take the needed decisions when an internal error occurs.
- The *Task manager* is a finite state machine that breaks down the main objective into small tasks.

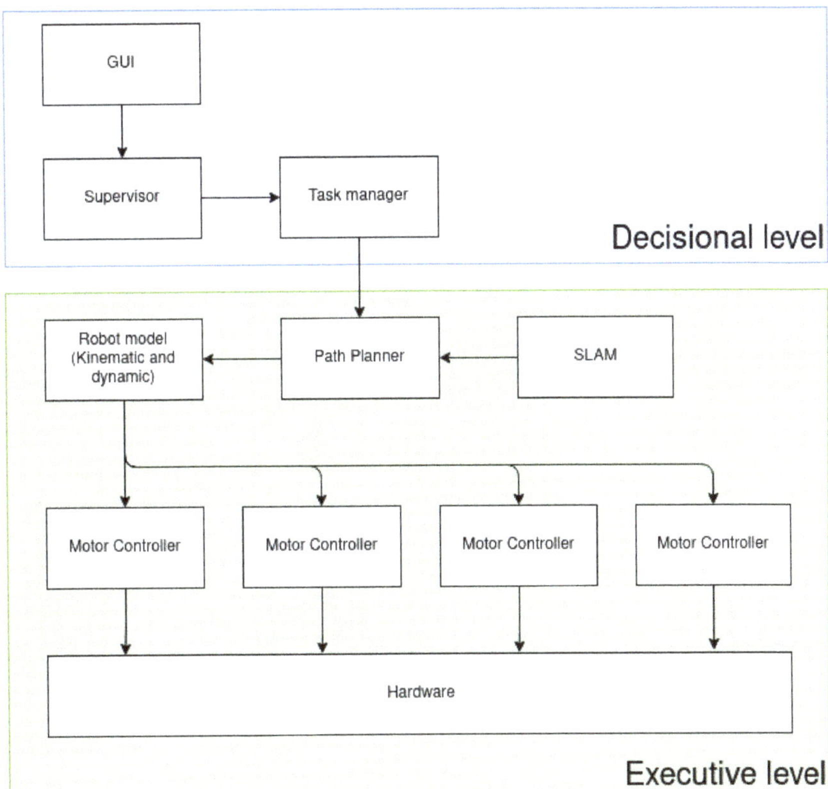

Figure 9. Control architecture of the robot.

On the other hand, the *Executive level* includes:

- *Hardware*: It is composed of motors, drivers, etc.
- Four *Motor controllers*, which consequently move the motors according to the desired position, velocity, and speed.
- The *Robot model (Kinematic and dynamic)* is composed by the models described previously. According to the desired trajectory, the motor state is calculated.
- The *Path planner* takes into account the goal task to generate the trajectory.
- The *SLAM* system makes the simultaneous location and mapping of the environment. The reconstruction will be used by the path planner to generate the trajectory.

6. Results

The result of the work presented in this paper is an omnidirectional robotic platform, whose first prototype still does not count all the required sensors for autonomous surveillance. As shown in Figure 10, a monitored surveillance was perform in the SPS accelerator by members of the robotic group of CERN. Throughout the tests, the operator remotely navigated the robotic platform from the control room. In this case, the operator had the information from the cameras; however, in future interventions a point cloud-based reconstruction system for three-dimensional environments will be available, thanks to the installed 3D LIDAR. This addition will help the operator and control to cross the gates along the tunnel. Due to the developed tests, it was tested that the robot design comply with the first, third and forth requirements of Section 1, which are related with surveillance time, robot speed and autonomy.

The robot was tested in its most critical point, this is, during crossing the section doors. As shown in Figure 11, the robot was able to cross it, complying in this way with the spatial requirements (the second requirement).

Figure 10. SPS robot driving during the first test.

As explained in Section 1, the robot charges its batteries in one of the entrances of the SPS tunnel, where a charging station is available. In that moment, the robot folds the robotic arm over its top plate, as shown in Figure 12.

Figure 11. SPS robot crossing the door during the first test.

Figure 12. Robot in the charging station with the robotic arm folded over its top end plate.

The kinematic and dynamic models were tested in a wide environment, obtaining the results shown in Figure 13. In that picture, the blue line represents the desired trajectory or ground truth, the red line represents the trajectory that the robot followed using the kinematic model, and the green line represents the trajectory using the dynamic model. The test was done within a wide environment, traveling a distance of approximately 90 m and changing the orientation in the meantime. It is calculated that the maximum error by the kinematic model during the given trajectory was 2.78 m and 5.1°, while the dynamic model reduced this error to 0.89 m and 0.45°.

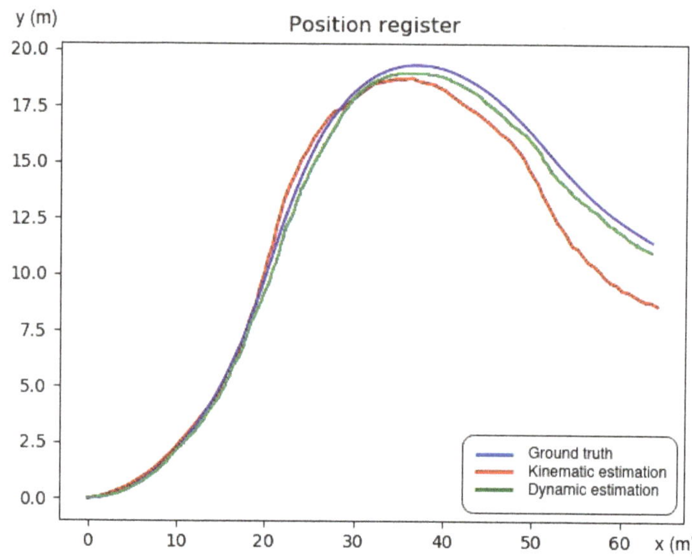

Figure 13. Behaviour of the kinematic and dynamic models. The robot was given a trajectory twice (blue) and the kinematic and dynamic model calculated the needed robot inputs to follow that trajectory.

7. Conclusions

In this paper, we have presented an original design of an omnidirectional robotic platform under very high spatial restrictions. The robot will perform the surveillance of the tunnel Super Proton Synchrotron, SPS, an accelerator of the European Organization for Nuclear Research, CERN. However, its applications are not only included in this environment, since it can be used in any installation where maneuverability is an essential requirement. Thanks to the compact design, this robot can deal with small spaces making use of the feature of setting horizontally the robotic arm that is mounted on it, together with the feature of folding the 3D LIDAR plate while crossing the doors and extract it while the surveillance. The robot is stably tested from a theoretical point of view, guaranteeing in this way the performance of the robot under foreseen circumstances. In addition, the final design of the robot is equipped with all the needed sensors and actuators, allowing not only the autonomous survey of the tunnel but also allowing it to be used in teleoperation tasks, where the end-effector tool may be changed.

The work presented represents a robot with very particular features. In the first place, the great distances that it must travel accompany the design of a large robotic platform, with the possibility of housing a large number of devices for measuring environmental variables and for intervention on the environment. However, the strict spatial restrictions that arise make it necessary to design a very small platform, even requiring physical robustness, high computing capacity and high load capacity for the portability of measurement and actuation devices. In addition, the design developed presents a novel ability to fold the robotic arm when crossing the doors, maintaining the stability of the small platform. In this way, the robot becomes the first interchangeable tool robot capable of performing tasks in large environments, such as tunnels, which have narrow passages 400 mm wide and 200 mm high.

The prototype of the platform has shown high capabilities to cross the small doors or narrow passages with the help of operators that control the motion, as well as good performance when the robot is driven to the charging station. The platform shows high maneuverability, allowing the robot to move within the narrow corridors and spaces that communicate the charging station with the tunnel.

Furthermore, to achieve proper control of the robot, the kinematic and dynamic models have been obtained. These models have been included in a user interface, which allows the operator to move the robot as desired, according to the information sent by the cameras and sensors.

Author Contributions: Conceptualization, C.P.S., L.R.B. and M.D.C.; methodology, C.P.S. and L.R.B.; software, C.P.S.; validation, C.P.S. and L.R.B.; formal analysis, C.P.S.; investigation, C.P.S.; resources, C.P.S., L.R.B. and M.D.C.; data curation, C.P.S. and L.R.B.; writing—original draft preparation, C.P.S.; writing—review and editing, C.P.S. and M.D.C.; visualization, C.P.S.; supervision, L.R.B.; project administration, L.R.B. and M.D.C.; funding acquisition, M.D.C. All authors have read and agreed to the published version of the manuscript.

Funding: This research was funded by European Organization for Nuclear Research (CERN).

Institutional Review Board Statement: Not applicable.

Informed Consent Statement: Not applicable.

Data Availability Statement: Software is released under a CERN proprietary software licence. Any permission to use it shall be granted in writing. Requests shall be addressed to CERN through mail-KT@cern.ch.

Acknowledgments: This work is supported by the European Organization for Nuclear Research, which has allowed us to make use of all the needed resources and facilities. In particular, the authors would like to thank Valentina Baldassarre for her precious suggestions, and revisions of the presented work. Also, the authors would like to thank the mechanics of the CERN robotics group for the work carried out in the construction of the prototype, and the corresponding tests in the real environment.

Conflicts of Interest: The authors declare no conflict of interest.

Abbreviations

The following abbreviations are used in this manuscript:

CERN	European Organization for Nuclear Research
GPS	Global Positioning System
SPS	Super Proton Synchrotron
SLAM	Simultaneous Localization And Mapping
LHC	Large Hadron Collider
TIM	Train Inspection Monorail
USB	Universal Serial Bus

References

1. Sandeep Kumar Malu, J.M. Kinematics, Localization and Control of Differential Drive Mobile Robot. *Glob. J. Res. Eng.* **2014**, *14*
2. Chwa, D. Tracking Control of Differential-Drive Wheeled Mobile Robots Using a Backstepping-Like Feedback Linearization. *IEEE Trans. Syst. Man Cybern. Part Syst. Hum.* **2010**, *40*, 1285–1295. [CrossRef]
3. Feng, L.; Koren, Y.; Borenstein, J. A model-reference adaptive motion controller for a differential-drive mobile robot. In Proceedings of the 1994 IEEE International Conference on Robotics and Automation, San Diego, CA, USA, 8–13 May 1994. [CrossRef]
4. He, S. Feedback control design of differential-drive wheeled mobile robots. In Proceedings of the ICAR'05, 12th International Conference on Advanced Robotics, Seattle, WA, USA, 18–20 July 2005. [CrossRef]
5. Sun, D.; Feng, G.; Lam, C.M.; Dong, H. Orientation control of a differential mobile robot through wheel synchronization. *IEEE/ASME Trans. Mechatron.* **2005**, *10*, 345–351. [CrossRef]

6. Zaman, M.; Liao, I.Y. A Generic Model, and its Validation, for the Translational Systematic Errors in Synchronous Drive Robots. In Proceedings of the 2010 Fourth Asia International Conference on Mathematical/Analytical Modelling and Computer Simulation, Kota Kinabalu, Malaysia, 26–28 May 2010; pp. 106–111. [CrossRef]
7. Wada, M. A synchro-caster drive system for holonomic and omnidirectional mobile robots. In Proceedings of the 2000 26th Annual Conference of the IEEE Industrial Electronics Society. IECON 2000. 2000 IEEE International Conference on Industrial Electronics, Control and Instrumentation. 21st Century Technologies, Nagoya, Japan, 22–28 October 2000; Volume 3, pp. 1937–1942. [CrossRef]
8. Kamga, A.; Rachid, A. A simple path tracking controller for car-like mobile robots. In Proceedings of the 1997 European Control Conference (ECC), Brussels, Belgium, 1–4 July 1997; pp. 138–143. [CrossRef]
9. Orozco-Velázquez, R.A.; Silva-Ortigoza, R.; Márquez-Sánchez, C.; García-Sánchez, J.R.; Taud, H.; Sosa-Cervantes, C.Y.; Antonio-Cruz, M.; Marcelino-Aranda, M.; Carrizosa-Corral, F. Ackerman Mobile Robot with Arm. In Proceedings of the 2016 International Conference on Mechatronics, Electronics and Automotive Engineering (ICMEAE), Cuernavaca, Mexico, 22–25 November 2016; pp. 55–60. [CrossRef]
10. Wada, M.; Asada, H. Design and control of a variable footprint mechanism for holonomic omnidirectional vehicles and its application to wheelchairs. *IEEE Trans. Robot. Autom.* **1999**, *15*, 978–989. [CrossRef]
11. Byun, K.S.; Kim, S.J.; Song, J.B. Design of a four-wheeled omnidirectional mobile robot with variable wheel arrangement mechanism. In Proceedings of the 2002 IEEE International Conference on Robotics and Automation (Cat. No.02CH37292), Washington, DC, USA, 11–15 May 2002; Volume 1, pp. 720–725. [CrossRef]
12. Byun, K.-S.; Song, J.-B. CVT control of an omnidirectional mobile robot with steerable omnidirectional wheels for energy efficient drive. In Proceedings of the 2003 IEEE International Conference on Robotics and Automation (Cat. No.03CH37422), Taipei, Taiwan, 14–19 September 2003; Volume 1, pp. 503–508. [CrossRef]
13. Kebritchi, A.; Hosseiniakram, P.; Havashinezhadian, S.; Rostami, M. Design and Development of an Omnidirectional Mobile Manipulator for Indoor Environment. In Proceedings of the 2018 6th RSI International Conference on Robotics and Mechatronics (IcRoM), Tehran, Iran, 23–25 October 2018; pp. 152–158. [CrossRef]
14. Jia, Q.; Wang, M.; Liu, S.; Ge, J.; Gu, C. Research and development of mecanum-wheeled omnidirectional mobile robot implemented by multiple control methods. In Proceedings of the 2016 23rd International Conference on Mechatronics and Machine Vision in Practice (M2VIP), Nanjing, China, 28–30 November 2016; pp. 1–4. [CrossRef]
15. Liu, Y.; Li, H.; Ding, L.; Liu, L.; Liu, T.; Wang, J.; Gao, H. An omnidirectional mobile operating robot based on mecanum wheel. In Proceedings of the 2017 2nd International Conference on Advanced Robotics and Mechatronics (ICARM), Tai'an, China, 27–31 August 2017; pp. 468–473. [CrossRef]
16. Lu, X.; Zhang, X.; Zhang, G.; Jia, S. Design of Adaptive Sliding Mode Controller for Four-Mecanum Wheel Mobile Robot. In Proceedings of the 2018 37th Chinese Control Conference (CCC), Wuhan, China, 25–27 July 2018; pp. 3983–3987. [CrossRef]
17. Xu, H.; Yu, D.; Wang, Q.; Qi, P.; Lu, G. Current Research Status of Omnidirectional Mobile Robots with Four Mecanum Wheels Tracking based on Sliding Mode Control. In Proceedings of the 2019 IEEE International Conference on Unmanned Systems and Artificial Intelligence (ICUSAI), Xi'an, China, 22–24 November 2019; pp. 13–18. [CrossRef]
18. Hibbeler, R.C.; Fan, S.C. *Engineering Mechanics: Dynamics*; Prentice Hall: Hoboken, NJ, USA, 2002.
19. Prados Sesmero, C. Mechatronic Design of an Omnidirectional Robotic Platform and Its Navigation Based on graphSLAM Implementation. Ph.D. Thesis, Polytechnic University, Madrid, Spain, 2020.
20. Di Castro, M.; Ferre, M.; Masi, A. CERNTAURO: A modular architecture for robotic inspection and telemanipulation in harsh and semi-structured environments. *IEEE Access* **2018**, *6*, 37506–37522. [CrossRef]
21. Sáenz, A.; Santibánez, V.; Bugarin, E. Modelado cinemático y dinámico de un robot móvil omnidireccional de 4 ruedas considerando dinámica de actuadores. *AMRob J. Robot. Theory Appl.* **2016**, *4*, 1–6.
22. Kelly, R.; Davila, V.S.; Perez, J.A.L. *Control of Robot Manipulators in Joint Space*; Springer Science & Business Media: Berlin/Heidelberg, Germany, 2006.

Short Biography of Authors

Carlos Prados Sesmero was born in Sanchidrián, Ávila, Spain, in 1996. He received the degree in electronic and automatic engineering from the Universidad de Valladolid, UVa, Valladolid, Spain, in 2018. He received the master degree in automatic and robotic from the Universidad Politécnica de Madrid (UPM), Madrid, Spain, in 2020. In 2018, he was a Robotic Engineer with the Research and Development Department, Astibot. From 2019 to 2021, he was a Robotic Engineer with the Engineering Department, European Organization for Nuclear Research, CERN, Geneva, Switzerland. From 2021, he is doing his Ph.D. in Modular Robotics in Universidad Politécnica de Madrid (UPM).

Luca Rosario Buonocore received the Ph.D. degree in Computer and Automation Engineering at the University of Naples Federico II in 2015. His main research interests are: mechatronic design of novel robotic solutions, like mobile robotic platforms and ultralight robotic arms for areal manipulation. Currently, he is a research Staff at CERN in R&D robotic division of engineering group.

Mario Di Castro received the M.Sc. degree in electronic engineering from the University of Naples "Federico II", Italy. From 2007 to 2011, he was with DESY in charge of advanced mechatronics solutions for synchrotron beamlines and industrial controls. Since 2011, he has been with CERN. Since 2018, he leads the Measurements, Robotics and Operation Section, Survey, Mechatronics and Measurements Group. The section is responsible for the design, installation, operation, and maintenance of control systems on different platforms (PLC, PXI, and VME) for all the equipment under the group's responsibility, mainly movable devices characterized by few um positioning accuracy (e.g., scrapers, collimators, shielding, and target) in hard radioactive environment. Important section activities are the robotic support in hazardous environments for the whole CERN accelerators. His research interests are mainly focused on automatic controls, mechatronics, motion control in harsh environment, and robotics.

applied sciences

MDPI

Article

Development and Evaluation of the Traction Characteristics of a Crawler EOD Robot

Lucian Ștefăniță Grigore [1], Ionica Oncioiu [2,*], Iustin Priescu [3] and Daniela Joița [3]

1 Center of Excellence in Robotics and Autonomous Systems—CERAS, Military Technical Academy, "FERDINAND I", 39–49 George Coșbuc Av., 050141 Bucharest, Romania; lucian.grigore@mta.ro
2 Accountancy and Business Administration, Faculty of Finance-Banking, Titu Maiorescu University, 040051 Bucharest, Romania
3 Department of Informatics, Faculty of Informatics, Titu Maiorescu University, 040051 Bucharest, Romania; iustin.priescu@prof.utm.ro (I.P.); daniela.joita@prof.utm.ro (D.J.)
* Correspondence: ionica.oncioiu@prof.utm.ro; Tel.: +40-372-710-962

Abstract: Today, terrestrial robots are used in a multitude of fields and for performing multiple missions. This paper introduces the novel development of a family of crawling terrestrial robots capable of changing very quickly depending on the missions they have to perform. The principle of novelty is the use of a load-bearing platform consisting of two independent propulsion systems. The operational platform, which handles the actual mission, is attached (plug and play) between the two crawler propulsion systems. The source of inspiration is the fact that there are a multitude of intervention robots in emergency situations, each independent of the other. In addition to these costs, there are also problems with the specialization of a very large number of staff. The present study focused on the realization of a simplified, modular model of the kinematics and dynamics of the crawler robot, so that it can be easily integrated, by adding or removing the calculation modules, into the software used. The designed model was integrated on a company controller, which allowed us to compare the results obtained by simulation with those obtained experimentally. We appreciate that the analyzed Explosive Ordnance Disposal (EOD) robot solution represents a premise for the development of a family of EOD robots that use the same carrier platform and to which a multitude of operational platforms should be attached, depending on the missions to be performed.

Keywords: robot; crawler; traction; kinematics; EOD Robot; terrorist attacks

Citation: Grigore, L.Ș.; Oncioiu, I.; Priescu, I.; Joița, D. Development and Evaluation of the Traction Characteristics of a Crawler EOD Robot. *Appl. Sci.* **2021**, *11*, 3757. https://doi.org/10.3390/app11093757

Academic Editor: Alessandro Gasparetto

Received: 31 March 2021
Accepted: 19 April 2021
Published: 21 April 2021

1. Introduction

The most serious threat to the civilized world, following nuclear threats, is terrorist threats [1]. In recent decades, due to the expansion of civilization across the globe, and the ease of access to information, technology and culture, extremist actions and reactions have proliferated [2]. They are symmetrical with extremism in the process of globalization and aim to stop and destabilize the offensive growth of the civilized world against increased fragmentation, promiscuity, poverty and violence [3].

Terrorism does not have a generally accepted definition. The difficulty of defining it stems both from its complexity and from a wide divergence of positions of individuals, organizations or states involved in the fight against terrorism [4]. The simplest definition is "purpose and method": Terrorism is an unconventional fighting tactic used to achieve strictly political goals that are based on acts of violence, sabotage or threat, executed against a state, organizations, social groups or against a group of civilians, with the precise purpose of producing a generalized psychological effect of fear and intimidation [5]. The ultimate goal is to put pressure on that entity to get it to act in accordance with the wishes of the terrorists if that goal cannot be achieved by conventional means [6].

The correct and complete definition and understanding of the phenomenon of terrorism are absolutely necessary for the elaboration of effective strategies and tactics in the fight against it [7]. As there is no valid general model, the definition and analysis of the terrorist act must be done on a case-by-case basis. The main defining elements of terrorism [8] are: *Means:* violent actions [9]; *Method:* inducing fear for citizens [10]; *Target:* civilians (non-combatants) [11]; *Purpose:* to bring about a major political change [12]; *Actors:* individuals or non-state groups [13].

On the other hand, terrorism uses only one type of action—surprise attacks of small magnitude but of maximum intensity on people, infrastructures and institutions. The procedures for such attacks are endless [14]. Among the most common terrorist acts, we mention: surprise attack with firearms; bomb attack; attack of infrastructures of any kind (sophisticated artisanal systems) [15]; attack on public places and public institutions [16]; radiological attack [17]; chemical attack [18]; biological attack [19].

There are, of course, many other forms of terrorist acts. Therefore, the main forms and procedures of counter-terrorism actions must respond promptly to these challenges, and must be anticipatory, preventive, effective and dissuasive [20].

In previous studies, no specific forms and procedures in the war on terrorism have been found, but actions and measures have been initiated by all structures, forces and institutions to protect themselves against terrorist attacks. That is why we followed the description, quite exhaustively, of the testing procedures, which allows for the validation of the developed analytical–numerical model.

In this stage of project development, the formulation of research hypotheses for our model refers to the validation of the analytical–numerical model regarding the ability of the robot to move in different environments (structured/unstructured), the validation being experimentally performed. The solution proposed in the article has already been implemented, and the obtained results encourage us to move towards the development of operational platforms with multi-role destinations. The software architecture is modular, and the proposed algorithm can be upgraded relatively easily, by simply inserting or removing modules. Additionally, the proposed mechanism can provide a basis for discussion/development for those interested in developing EOD robots. On this basis, the results showed that the robot can operate on an operational platform specific to the actions of taking over and relocating objects that are considered to be suspicious.

Regarding the novelty of the project, we consider that the possibility to assemble an intervention robot at the crime scene, according to the identified specific, is much more useful than more expensive and complex logistics. On top of that, the price of the prototype robot (designed and made) is 1/10 compared to a similar robot in the same class, respectively compared to TELEROB (https://www.telerob.com/de/ (accessed on 18 February 2021)); most robots in the same class are made to perform a single mission, it is a robot that at the maximum opening of the arm can support 15 kg, even when there is no power supply. This could be achieved by introducing electromagnetic brakes in each joint from R0 to R5; with the available resources we were able to make a multifunctional robot, only by changing the operational platform, given that the design and assembly were performed entirely in ATM, only the mechanical elements were made at the industrial partner.

The paper is structured as follows: Section 2 highlights the state of the art in the field of EOD; Section 3 shows how the geometric configuration of the crawler robot was established; Section 4 details the performance of simulations based on the defined analytical–numerical model; Section 5 details the evaluation tests of the propulsion system in order to validate the analytical–numerical model; Section 6 presents the conclusions.

2. State of the Art

Currently, mobile robotics is at the intersection of several fields of science. These systems include: *Knowledge representation* [21]: the field of artificial intelligence helps to obtain an internal model of the world associated with the task performed by the robot, and uses data structures and specific algorithms to represent the characteristics of objects in the environment; *Natural language* [22]: natural language changes quite frequently, so simple word recognition is no longer enough to understand the real meaning of a processed text; *Learning* [23]: ideally, a robot could learn the tasks it has to perform by simply repeating similar actions performed by a human operator; *Planning and problem solving* [24]: intelligence is inherently associated with the ability to plan actions needed to achieve a particular goal, as well as the ability to solve problems that arise when these action plans do not work; *Inference* [25]: by inference, it is possible to fill in the missing information to solve a certain problem; *Search* [26]: in terms of artificial intelligence, search means the efficient examination of a representation of knowledge, specific to a certain problem (search in a "space") in order to determine a solution; Vision [27]: by creating artificial vision systems, the actions of the mobile robot can be made more precise and complex. Sometimes, mobile robots appear in the form of vehicles capable of operating autonomously on the ground (UGVs—Unmanned Ground Vehicles), in the air (UAVs—Unmanned Aerial Vehicles) or in the water (UUVs—Unmanned Undersea Vehicles). In the broadest sense of the definition, a mobile robot is a real-time integrated system (embedded real-time system), that has sensors [28] for environmental perception, execution elements for performing actions on the environment and a control system for mapping perception in action [29]. Perception and action are the major aspects that define a mobile robot. Depending on the organization of these three functional aspects, it can be considered that the mobile robot has a greater or lesser degree of autonomy.

The best-known type of mobile robot is the Automated Guided Vehicle (AGV—Automated Guided Vehicle). AGVs operate in pre-programmed environments and are inflexible and "fragile" in operation; thus, any unforeseen route changes (objects on the runway) can lead to a compromise of the entire mission [30].

The alternative to AGV is to build mobile robots with autonomy in motion to travel in the environment to perform various tasks; adapt to changes in the environment; learn from experience and change their behavior accordingly (how to act); build internal representations of the world around them that can be used for decision-making processes (for example, for navigation).

These characteristics allow for the use of these robotic structures in specific applications (Figure 1): transport operations, exploration, surveillance, orientation, inspection, evolution in environments inaccessible or hostile to the human operator such as robots intended to combat terrorist actions, or robots operating in contaminated environments, underwater robots, planetary exploration vehicles, etc. [31].

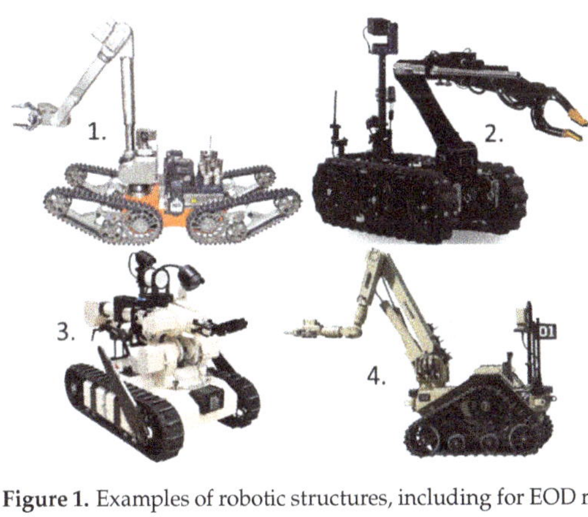

Figure 1. Examples of robotic structures, including for EOD missions: (**1**). https://www.nides.cz/en/telerob/ (accessed on 18 February 2021); (**2**). https://icortechnology.com/robots/caliber-t5/caliber-t5-swat-eod-robot-leveled/ (accessed on 18 February 2021); (**3**). https://www.rvconnex.com/eod-robot/ (accessed on 18 February 2021); (**4**). https://www.publictechnology.net/articles/news/introducing-harris-t7-%E2%80%93-mod%E2%80%99s-new-%C2%A31m-bomb-disposal-robot (accessed on 18 February 2021).

2.1. The Functions of a Crawler Mobile Robot

From a functional point of view, a mobile robot consists of three modules: the role of the sensor module is to estimate the speed and position of the mobile robot [32–34]; the orientation module is the reference generator for the orientation of the mobile robot in space [33]; the control module is the subsystem that generates the control quantities for the execution elements based on the data provided by the orientation module [34].

The following essential functions of a mobile robot can be defined according to the functional modules:

A. Navigation function [35]: workspace representation; avoidance of obstacles; location function; motion planning; motion control.
B. Task Planning Function [36]:
C. Interaction and communication function [37].

2.2. Justification for the Use of a Crawler Mobile Robot for EOD Actions

Anti-terrorism intervention requires the use of robots to limit risks for the following reasons: performing procedures during IED intervention missions, which include observation and inspection, X-ray analysis, disruption, etc.; no matter how careful the operator is in responding to the IED, there is always the possibility that the attacker may be nearby so that he can operate the explosion-proof device remotely; even if a robot cannot neutralize an IED, it can still be used to provide information on the choice of procedure and devices needed to evacuate/eliminate the dangerous object from the threatened area.

2.3. Features of Crawler Mobile Robots

Named mobile robots and controlled/autonomous vehicles, robot trailers or robotics, have a structure consisting of (Figure 2):

Figure 2. The general structure of an EOD mobile robot [https://www.telerob.com/en/products/teodor-evo (accessed on 18 February 2021)] mechanical and drive system [38]; location system [39]; environmental perception and security system [40]; data interpretation and task management system [41]. Operation in the environment is defined by agility and maneuverability: agility is the ability of the mobile robot to overcome obstacles [42]; maneuverability is defined by the minimum area required for a maneuver [43].

Mobile intervention robots have the following main features: *degree of mobility:* following a varied trajectory in terms of shape, length, and number of parking/stopping points [44]; *speed:* it is between 1–10 m/s, depending on the environment in which it moves inside/outside [45]; *autonomy:* is the dependence on own power sources or the length of the power cord [46]; *remote control* and *automation of operations* [47].

2.4. Requirements for Crawler Mobile Robots, Resulting from an Anonymized Market Study

The robot should be affordable, if the designers meet the following requirements, simultaneously or individually: is intended only for specific missions and threats, for example pyrotechnic intervention [48]; is built modularly [49]; subsystems to be redesigned so that the solution is as simple and reliable as possible [50]; can use a disruptor [51]; should be equipped with video cameras [52]; must have an adequate lifting capacity [53]; can use X-ray systems [54]; should be equipped with omnidirectional cameras [55]; the robot must be able to communicate with an operator [56]; can easily climb stairs it can be said that only 10% of the threats targeted easily accessible open areas; should be easy to maintain and repair; must operate in a wide range of terrains [57]; arm should have high mobility [58].

3. Geometric Configuration of the Crawler Propeller

In order to achieve the geometric configuration of the crawler engine, one must consider the influence of the geometric characteristics of the tracked engine on the following: mobility of the robot on terrain with various configurations and consistencies; stability of the robot during movement and during the execution of the specific mission. The effects of the terrain on the robot can be grouped into the categories presented in Figure 3.

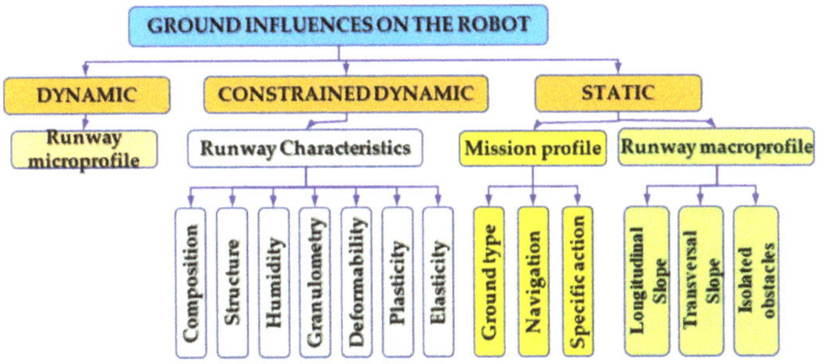

Figure 3. Representation of different types of environmental effects on the crawler engine [59]. This representation aims to allow us to establish algorithms for defining the kinematic and dynamic model depending on the type of terrain on which the crawler robot will move.

Analyzing the requirements formulated by the potential beneficiaries, as well as the current stage of technological development of mobile robots, the main functional parameters were quantified, and the main functional characteristics were estimated (Table 1).

The global analysis of the problem of locomotion on surfaces leads to the identification of four types of forces, among them including the propulsive force: Propulsion; Adhesion; Cohesion; Support.

Table 1. Type-dimensional and actuation requirements for the EOD robot under analysis.

Characteristics	U.M.	Amount
Fully equipped mobile robot table	kg	250
Maximum speed on horizontal terrain	m/s	1
Maximum slope when walking straight	deg	30
The height of the steps	mm	150
Average ground pressure	MPa	0.05
The height of the stepped obstacle crossed	mm	225
The width of the ditch crossed	mm	800
Depth of water on hard horizontal ground	mm	100
Race of turns	mm	continuously variable
Minimum turning radius	mm	0–rotation
Travel control	-	wireless
The mass of the manipulated object	kg	25
Range of the manipulator arm	mm	1000
Rotate the manipulator arm to the side	deg	30
The gripping length of the gripping mechanism	mm	150
Arm displacement position	-	folded/extended
Omnidirectional video camera for travel	buc.	2
Video camera for inspection	buc.	2
Riding the platform while working	-	no
Working autonomy	h	6
Construction type	-	modular

3.1. Conditions Regarding the Crossing of the Step Type Obstacle

The step type obstacle is often encountered in urban environments (curbs, stairs in buildings, etc.), which requires a detailed study of this process for an optimal geometric configuration of the crawler engine (Figure 4). The crossing of the step-type obstacle presupposes the obligatory completion of two successive stages: the approach to the obstacle; crossing the obstacle.

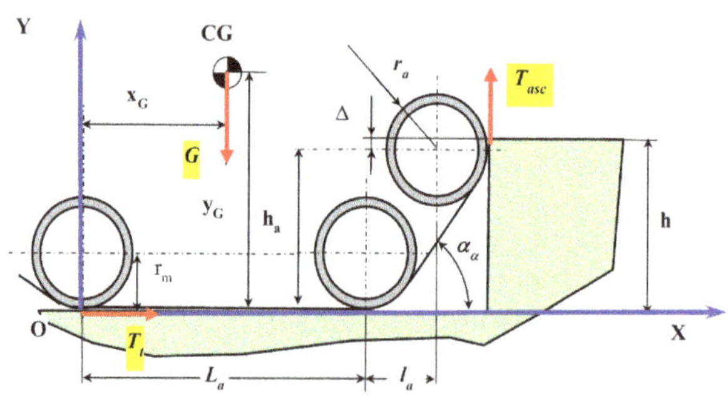

Figure 4. Geometric diagramming of the robot while approaching the step-type obstacle.

The ascending force arises T_{asc} [N]. Its value is limited by: φ_t [−] and φ_{obs} [−].

$$T_t = \varphi_t \cdot G\ [N], \quad (1)$$

where: $T_t[N]$—traction force, manifested at the wheel; $G[N]$—the weight of the robot.

The ascending force is given by the following relation:

$$T_{asc} = \varphi_{obs} \cdot Z = \varphi_{obs} \cdot T_t = \varphi_{obs} \cdot \varphi_t \cdot G\ [N], \quad (2)$$

where: $T_{asc}[N]$—the ascending force, manifested in front of the tensioning roller, at the contact of the track with the obstacle.

The limiting condition for them of climbing the obstacle is:

$$T_{asc}(L_a + l_a) = G \cdot x_G \;[N], \tag{3}$$

$$\varphi_{obs}\varphi_t \geq \frac{x_G}{L_a + l_a} \equiv \varphi_{obs}\varphi_t \geq \frac{x_G}{L}, \tag{4}$$

where: CG—weight center; $y_G[m]$—distance from the CG to the running surface, on the axle Oy; $r_m[m]$—drive wheel radius; $r_a[m]$—the radius of the tension roller; $h_a[m]$—distance from the axis of the tension roller to the tread, on the axis Oy; $\alpha_a[\mathrm{deg}]$—the angle of inclination of the ascending branch of the track to the axis Ox; $x_G[m]$—distance from the CG to the axle of the drive wheel, on the axle Ox; $L_a[m]$—the distance from the axle of the drive wheel to the last roller, the one at the beginning of the ascending branch of the track, on the axis Ox; $l_a[m]$—distance from the center of the last roller to the center of the tension roller, on the axis Ox.

The coefficient of friction between the tracked thruster and the obstacle (Figure 5), μ [−], is given by the following calculation relation [60–62]:

$$\mu \geq \frac{L - L_1}{L_1 + R/\sqrt{1 - \left(\frac{h-R}{L}\right)^2}} \;[-], \tag{5}$$

where: $h[m]$—the height of the obstacle step, on the axis Oy; $\Delta = h - h_a\;[m]$—the difference between the height of the step obstacle and the height of the axis of the tension roller, on the axis Oy; $L_1[m]$—the distance from the axis of the drive wheel to the CG, along the longitudinal axis of the robot; $L - L_1[m]$—the distance from the CG to the axis of the tension roller, along the longitudinal axis of the robot; $\beta[\mathrm{deg}]$—the angle of inclination of the longitudinal axis of the robot with respect to the axis Ox.

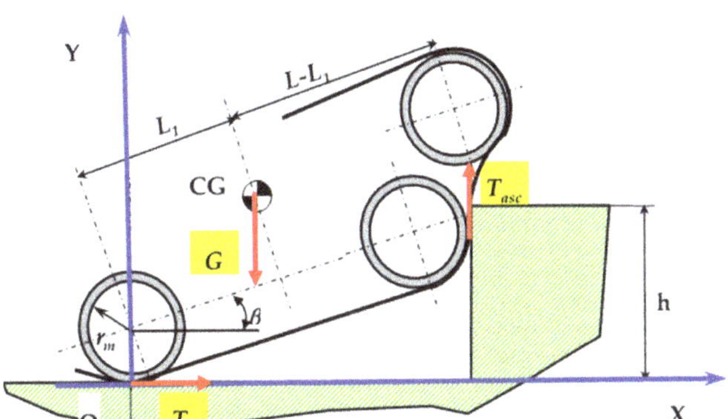

Figure 5. Representation of the beginning of the climbing of high obstacles by the robot.

Which is denoted by λ, the ratio:

$$\lambda = \frac{L_1}{L} \;[-], \tag{6}$$

Thus, (5) becomes:

$$\mu \geq \frac{1 - \lambda}{\lambda + R/L\sqrt{1 - \left(\frac{h-R}{L}\right)^2}} \;[-], \tag{7}$$

In order to assess the influence of the length of grip and the position of the center of gravity, the height of the obstacle shall be considered to be: $h = 1,5 \cdot r_m$. It is denoted as $\gamma = r_m/L$, so that inequality (7) becomes:

$$\mu(\lambda, \gamma) \geq \frac{1-\lambda}{\lambda + \frac{2\gamma}{\sqrt{4-\gamma^2}}} \, [-], \tag{8}$$

The above relation allows for the evaluation of the influence that the position of the center of mass has on the conditions of escalation of the obstacle (Figure 6).

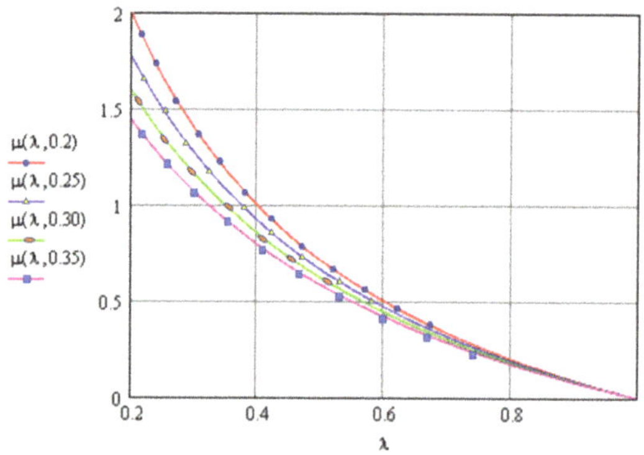

Figure 6. Variation in the minimum coefficient of friction required for escalating the step obstacle. Four different values were considered for the report $\gamma = r_m/L\,[-]$ to highlight how the CG position leads to a change in the coefficient of friction between the tracked thruster and the obstacle μ.

Given the notation presented in Figure 7, the following equations result:

$$s_G = r\sin\theta + x_G\cos\theta - y_G\sin\theta\ [m], \tag{9}$$

$$h = r - r\cos\theta + d\sin\theta\ [m], \tag{10}$$

$$d = \frac{(h - r + r\cos\theta)}{\sin\theta}\ [m], \tag{11}$$

$$s = r\sin\theta + d\cos\theta = r\sin\theta + \frac{(h - r + r\cos\theta)}{\sin\theta}\ [m], \tag{12}$$

where: r $[m]$—the running radius of the support wheel (bucket), including the thickness of the track; h [m]—the height of the step obstacle.

Analyzing the way in which the vehicle climbs onto the obstacle, the following distinct situations can be distinguished: $s_G < s$—the vehicle can climb over the obstacle; $s_G > s$—the vehicle cannot climb over the obstacle; $s_G = s$—critical condition, at the limit.

From the analysis of the Equations (9)–(12), from the graphical representation of these equations (Figure 8) and of the presented limit conditions, it results in (13):

$$r\sin\theta + x_G\cos\theta - y_G\sin\theta = r\sin\theta + \frac{(h - r + r\cos\theta)\cos\theta}{\sin\theta}\ [m]. \tag{13}$$

$$x_G\sin\theta\cos\theta - y_G\sin^2\theta = (h - r)\cos\theta + r\cos^2\theta\ [m]. \tag{14}$$

Solving Equations (13)–(14) was performed for a running radius of 0.0425 m (Figure 9).

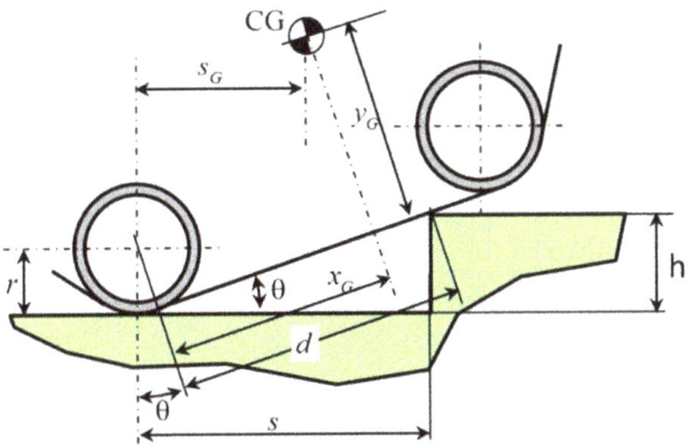

Figure 7. Schematic of the ascent of the step-type obstacle, considering that the lower branch of the track is already ascended on the obstacle. Explaining notations: $d[m]$—distance from the wheel axle to the point where the lower track branch is tangent to the obstacle edge; $s[m]$ —distance from the wheel axle to the point where the lower track branch is tangent to the obstacle edge; θ[deg] —the angle of inclination of the longitudinal axis (of the track) with respect to the running surface in the conditions in which the lower branch of the track has already *"climbed"* on the obstacle; $s_G[m]$ —the horizontal distance between the rear point of contact and the edge of the obstacle.

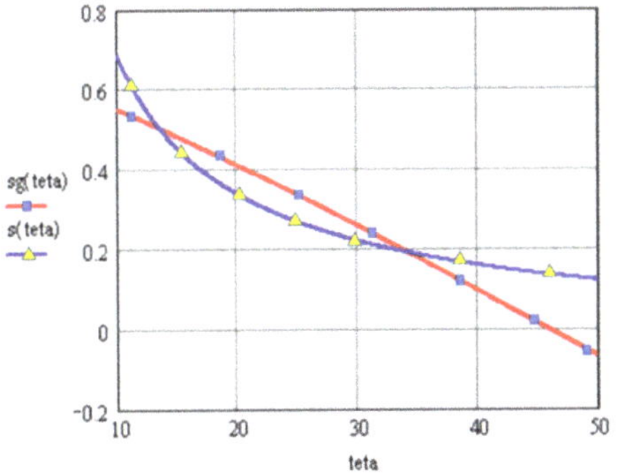

Figure 8. Graphical representation of Equations (9)–(12). The variations inf the two graphs mean that, as the obstacle escalates, the distance increases s_G horizontally between the rear point of contact and the edge of the obstacle decreases to zero, at which point it can be stated that the robot remains stable at the obstacle stage and the distance s from the center of the driving wheel to the point where the lower branch of the track is tangent to the edge of the obstacle is moving towards the point of minimum, and "0", followed with increasing distance from the edge of the obstacle, the proximity of the edge following the corresponding allure.

How to vary the maximum angle of arrangement of the support branch of the track to the ground θ is shown in Figures 9 and 10. It can be seen that the angle of inclination decreases as the center of mass moves forward.

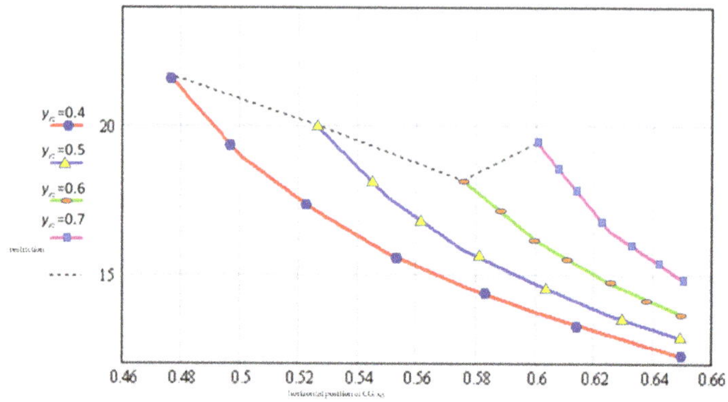

Figure 9. Angle variation with the horizontal position of the center of mass for h = 0.20 m, depending on y_G. The variation in the inclination angle of the longitudinal axis (of the caterpillar) is obviously θ compared to the running surface in the conditions in which the lower branch of the track has already "climbed" on the obstacle, in the sense that the increase in the CG height, compared to the axle Oy, leads to increased angle θ, enabling (exaggerating) the robot to overturn.

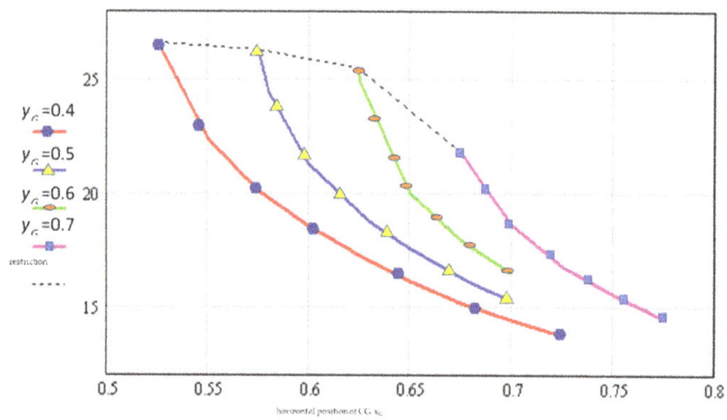

Figure 10. Angle variation with the horizontal position of the center of mass for h = 0.15 m. Additionally, in this case, in which the height of the step obstacle has changed (reducing it), the allure of variation of θ according to $y_G[m]$ is similar; only the dimensions differ.

The schematic of the tracked platform geometry for the trench crossing limit situation is presented in Figure 11. *In the initial phase*, the vehicle (drawn in blue) remains horizontal until the vertical passing through the center of mass coincides with the edge of the ditch. The vehicle starts to oscillate around the point of contact with the edge of the ditch (point marked with P). *The limit position*, at which it is still possible to cross the ditch (drawn in red) occurs when contact is made with the other edge of the ditch at a point on the circumference of the front wheel on its horizontal axis of symmetry. The above-mentioned approach scenario neglects inertial forces due to the low travel speeds of the mobile robot platform. For the limit situation, the following geometric equations can be written:

$$(L - L_1)\cos\varphi + h_r \sin\varphi + R\cos\varphi = L_s\ [m]. \tag{15}$$

$$(L - L_1)\sin\varphi - h_r \cos\varphi + R\sin\varphi = 0\ [m]. \tag{16}$$

for the general case, in which it was assumed that $h_r > r_m\ h_r > r_m$.

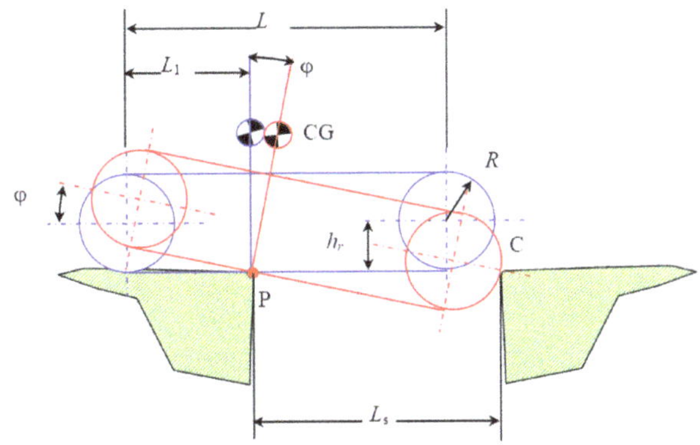

Figure 11. Geometric schematization of the limit situation when crossing the ditch. The explanations of the notation in the figure are: $L_S[m]$—the width of the ditch; φ[deg] —the angle of rotation of the CG relative to the vertical line passing through the CG and the point of contact with the first edge of the ditch; $h_r[m]$ —the radius of a bucket.

Eliminating the unknown φ, the width of the ditch that can be crossed was calculated according to the horizontal position of the center of mass and using the notation $\gamma = L/L_1$, the results shown in Figure 12 are obtained.

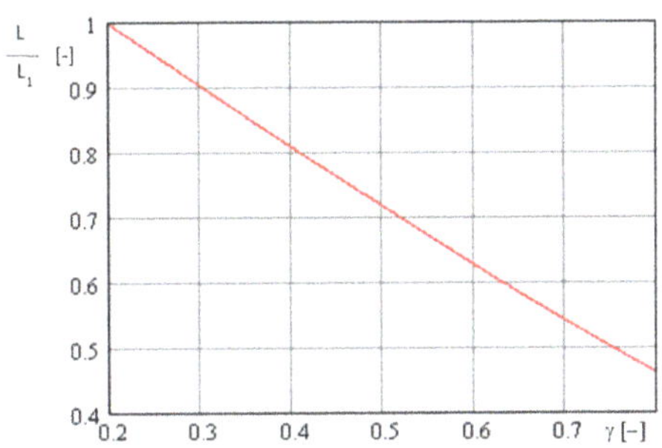

Figure 12. Variation of the maximum width of the ditch that can be crossed depending on the horizontal position of the center of mass.

There is an inversely proportional dependence between the maximum width of the ditch that can be crossed and the horizontal distance from the rear wheel axle and the center of mass. It turns out that the forward movement of the center of mass leads to an increase in the height of the step that can be climbed, but also to a decrease in the width of the ditch that can be crossed.

3.2. Conditioning Imposed by the Average Pressure on the Ground

The average pressure exerted by the tracks on the ground is given by the following relation:

$$p_{med} = \frac{G}{2 \cdot b \cdot L_a} \left[\frac{N}{m^2}\right]. \tag{17}$$

where: b [m]—width of the track; L_a [m]—adhesion length of the track to the ground.

Using the relation (17) the "*Contour Plot*" type graph was obtained (Figure 13).

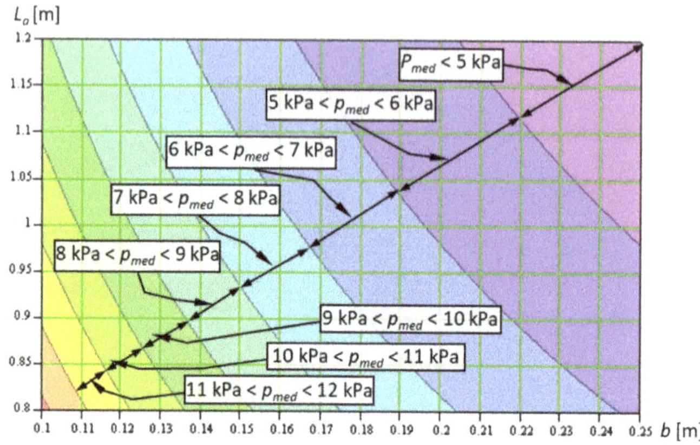

Figure 13. Graph of the variation of the average pressure on the ground depending on the width of the track and the length of adhesion. The simple explanation of the graph is as follows: as the width of the track b (m) and the length of the adhesion (track) increase L_a (m), the average pressure on the ground decreases.

In Figure 13, it can be naturally concluded that the area located in the upper right, for which $p_{med} < 6$ [kPa], is the most advantageous. An average–low ground pressure can be obtained by increasing the adhesion length and/or by increasing the track width.

However, reporting to military vehicles is not able to generate consistent values for the recommended average ground pressure due to the "*scale effect*" [62,63] (Figure 14).

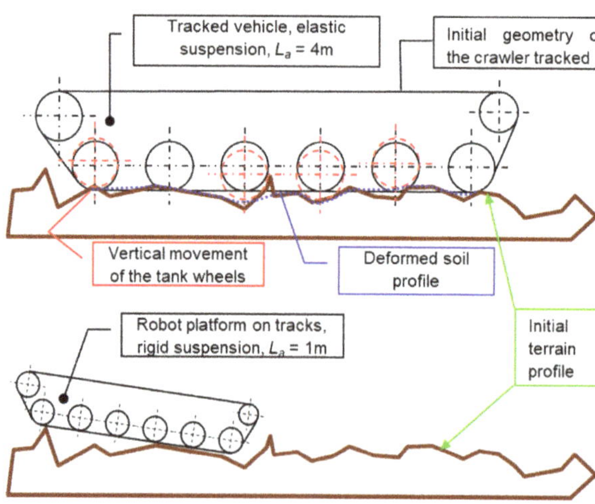

Figure 14. Behavior of a crawler vehicle with elastic suspension versus one with rigid suspension [64]; Figure 8. We cannot draw an exhaustive conclusion on which suspension system.

3.3. Geometric Conditions Imposed by the Execution of the Turn

From the point of view of cornering, there are two major conditioning factors: the position of the center of mass and the gauge (the distance between the planes of symmetry of the two tracks).

The position of the center of mass (CG) directly conditions the distribution of the pressure exerted by the support branch of the track on the ground. The value of the displacement of the turning poles is obtained by solving the following equation [64]:

$$\frac{12x}{L_a^2}x_0^2 + 2x_0 - 3x = 0, \tag{18}$$

and the moment of resistance in the turn is determined by the relationship:

$$M_{re} = \mu\frac{GL_a}{4}k\,[Nm], \tag{19}$$

where μ [−] = the steering resistance coefficient, and k [−] = a coefficient that takes into account the displacement of the turning poles.

$$k = \left(1 + \frac{4x_0^2}{L_a^2}\right)\cdot\left(1 + \frac{4x_0x}{L_a^2}\right) - \frac{16x_0x}{L_a^2}\,[-]. \tag{20}$$

It can also be seen that a relative displacement of a maximum of 0.27 to the front contributes to the decrease in the resistant movement in cornering Figure 15. The track gauge, denoted by B, determines the magnitude of the traction forces on the two tracks:

$$F_1 = f\frac{G}{2} - \mu\frac{G}{4}\frac{L_a}{B}k\,[N], \tag{21}$$

$$F_2 = f\frac{G}{2} + \mu\frac{G}{4}\frac{L_a}{B}k\,[N], \tag{22}$$

where $F_{1,2}$ $[N]$ = the traction forces at the inner and outer tracks of the turn, respectively; f [−] = the coefficient of resistance to progress in rectilinear gait; μ [−] = the steering resistance coefficient,

$$\mu = \frac{\mu_{\max}}{0.85 + 0.15\frac{R_2}{B}}\,[-], \tag{23}$$

in which $\mu_{\max}$ [−] = the steering resistance coefficient for the turning radius equal to the track gauge; R_2 $[m]$ = the turning radius measured at the outer track of the turn.

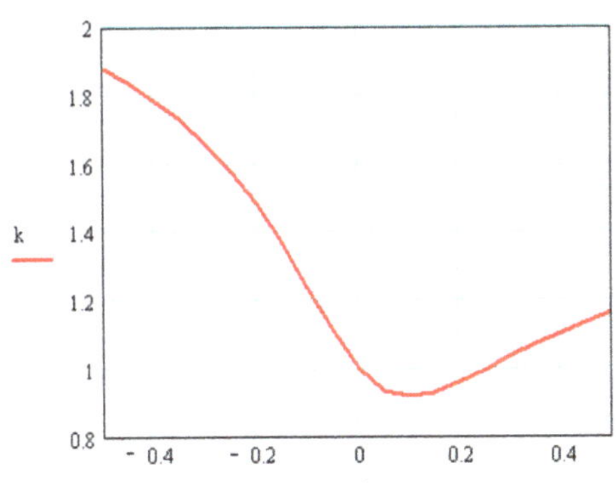

Figure 15. Variation of the coefficient k as a function of the relative displacement of the center of mass $x_r = x/L_a$, for L_a = 1.0 m. The significance of the graph lies in the fact that, with the change in the CG, in the longitudinal plane, the turning radius increases.

For constant values of the coefficient k, the weight of the platform and the maximum coefficient of cornering resistance, the value of the traction forces at the two tracks depends on the ratio L_a/B and the specific turning radius (Figure 16):

$$\rho = \frac{R_2}{B}\,[-]. \tag{24}$$

It follows that it is rational to adopt a track gauge value as close as possible to the value of the track length of the track.

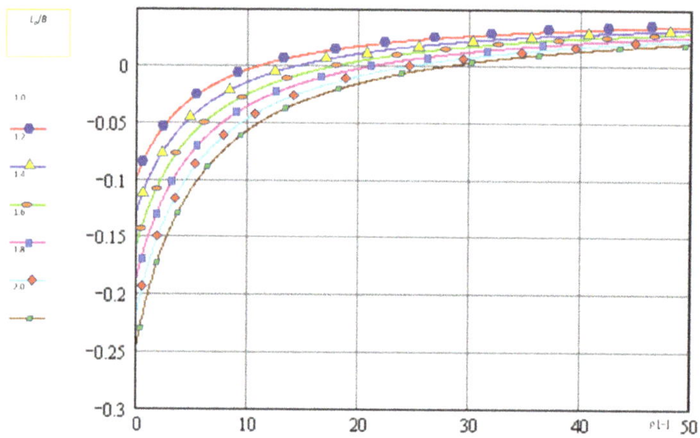

Figure 16. Variation of the specific force at the inner track of the turn. The representations of the five graphs show that, as the ratio increases $\frac{L_a}{B}$ [−], the traction forces required to make the turn decrease. The observation itself is useful for determining energy consumption, f = 0.1 [−]; G = 1962 [N]; μ_{max} = 0.5 [−].

3.4. Static Stability Study of the Crawler Engine

The geometric characteristics of the crawler propeller, as well as the position of the center of mass of the mobile robot, are critical for ensuring stability in static mode: movement on the longitudinal slope; crossing the slope; turning with a stable radius; cargo handling.

From the point of view of the stability of the tracked platform when moving on the longitudinal slope, the most difficult situation occurs when braking during the descent of the slope (Figure 17), which is denoted for the deceleration of the platform during braking: $a = \frac{d^2x}{dt^2}\left[\frac{m}{s^2}\right]$. At the limit, the misalignment x of the ground pressure center becomes equal to L-L1, so that the equation of equilibrium of the moments with respect to the point P of the forces acting on the platform results in the expression:

$$a = \frac{d^2x}{dt^2} = g\left(\frac{L - L_1}{h_g}\cos\alpha - \sin\alpha\right)\left[\frac{m}{s^2}\right]. \tag{25}$$

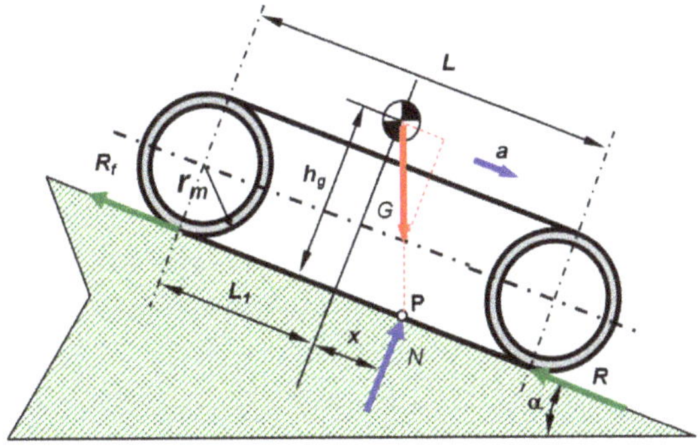

Figure 17. Schematic of the forces acting on the braking during the descent of the longitudinal slope. R_f— the force of resistance corresponding to the contact area of the track with the ground in front of the drive wheel; R — the force of resistance corresponding to the contact area of the track with the ground in front of the bucket; N —the normal force; α —the angle of the slope.

It can be concluded that the increase in the height of the center of gravity, as well as its movement towards the front of the platform leads to a substantial decrease in the braking deceleration at which the overturning occurs (Figure 18).

$$a = \frac{d^2x}{dt^2} = g\left(\frac{L - L_1}{h_g}\right) \left[\frac{m}{s^2}\right]. \tag{26}$$

Resuming the calculations that led to the plot of the graph presented above, we obtain the variation of the maximum deceleration when braking on the horizontal terrain, graphically presented in Figure 19.

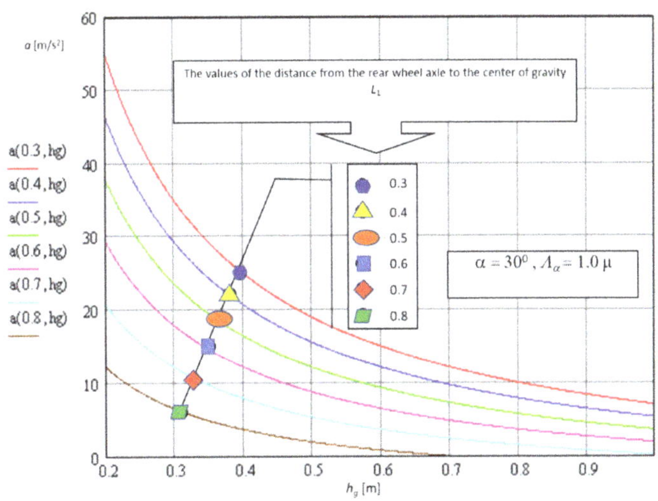

Figure 18. The variation of the maximum deceleration when braking on the slope of 30° depending on the position of the center of mass. It is observed that with enlargement y_G and the movement of CG towards the front of the robot in the longitudinal plane, when descending the slope, we will obtain a lower deceleration until the moment that coincides with the overturning of the robot.

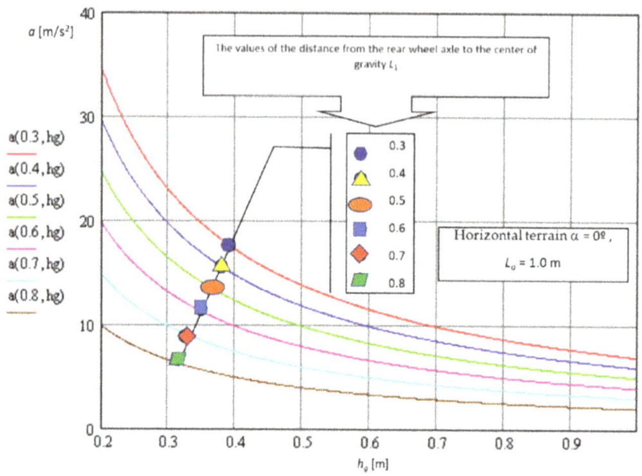

Figure 19. Variation in the maximum deceleration when braking in horizontal terrain (30° slope) depending on the position of the center of mass. These simulations were performed to predict how the robot would react if it was equipped with an arm used to transport suspicious objects.

The deceleration during braking is constant and the initial speed is v; it follows that the minimum time required for braking is given by the relation: $t_{\min} = v/a$ $[s]$.

Stability on the transverse slope is a particular case of the issue of lateral stability. The geometric scheme, as well as the forces acting on the platform during the crossing of the slope, are shown in Figure 20. The overturning stability condition is satisfied as long as the resultant of the ground reactions is inside the support polygon:

$$y \leq \frac{B+b}{2} = h_g \frac{G\sin\alpha + F_c}{G\cos\alpha}\ [m], \tag{27}$$

where F_c $[N]$—the centrifugal force, b $[m]$—track width; B $[m]$—track gauge; F_C $[N]$—centrifugal force.

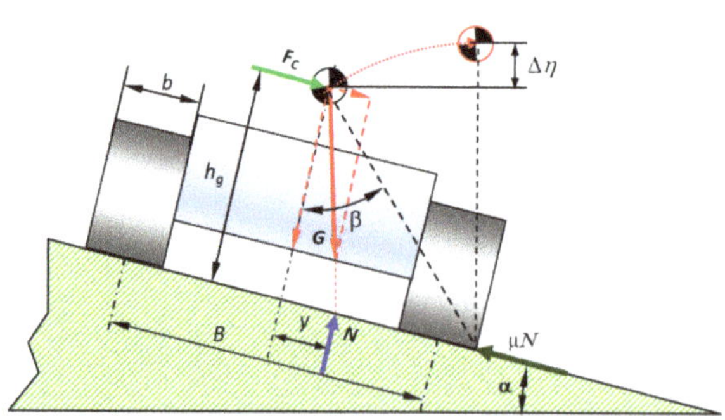

Figure 20. Schematic of the forces acting on the robot's platform when crossing the slope.

For a transverse slope angle of 30°, it results in $h_g = 0.866 \cdot (B+b)$, which does not imply severe conditions.

4. Simulation Singularity Crossing Obstacles

4.1. Simulation of the Slope Ascent

In order to evaluate the simulation potential of the simulation performed on the basis of a simple 2D model, a two-axle vehicle was schematized in the situation of longitudinal displacement on the slope Figure 21.

$$\sum F_X = 0 = \mu\gamma_r F_{Nf} - mg\sin\theta\ [N], \tag{28}$$

$$\sum F_Y = 0 = F_{Nr} + F_{Nf} - mg\cos\theta\ [N], \tag{29}$$

$$\sum M = 0 = F_{Nr}L_w - mg(L_{cg}\cos\theta + h_{mg}\sin\theta)\ [Nm]. \tag{30}$$

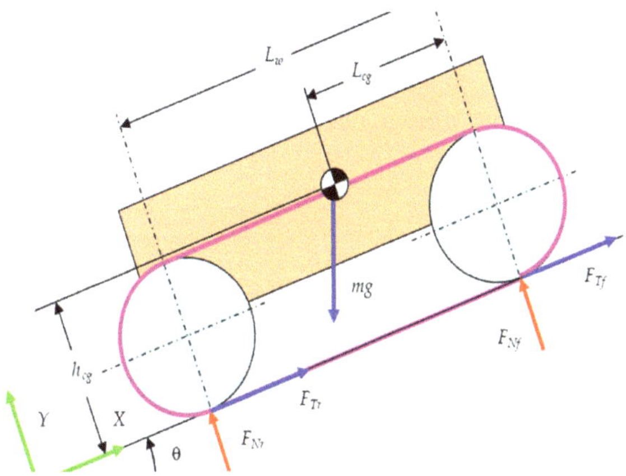

Figure 21. Schematic diagram of the vehicle moving on the longitudinal slope.

Equations (28)–(30) allow for the determination of the expressions for the normal forces acting on the drive wheels and of the minimum coefficient of adhesion that ensures the capitalization of the traction forces:

$$F_{Nr} = \frac{mg(L_{cg}\cos\theta + h_{cg}\sin\theta)}{L_w}\ [N], \tag{31}$$

$$F_{Nf} = \frac{mg[(L_w - L_{cg})\cos\theta - h_{cg}\sin\theta]}{L_w}\ [N], \tag{32}$$

$$\mu \geq \frac{L_w\sin\theta}{(\gamma_r - \gamma_f)(h_{cg}\sin\theta + L_{cg}\cos\theta) + \gamma_f L_w\cos\theta}\ [-], \tag{33}$$

$$\theta = \mathrm{arctg}\left(\frac{L_w - L_{cg}}{h_{cg}}\right)\ [\mathrm{deg}]. \tag{34}$$

where: m [kg]—weight of the vehicle; $F_{Tr,f}$ [N]—rear and front traction force, respectively; $F_{Nr,\,Nf}$ [N]—normal force back, respectively front; θ [deg]—angle of the longitudinal slope; L_{cg} [m]—distance of the center of mass from the axis of the front axle; h_{cg} [m]—distance of the center of mass from the running surface; L_w, [m]—the distance between the axes of the axles; γ [−]—coefficient of indication of the driving axis, with value 1 if the axis is driving and 0 if the axis is not driving.

The data that were considered for the simulation are contained in Table 2.

Table 2. Characteristics of the crawler robot used for simulation.

Characteristics	U.M.	Amount
Rear wheel diameter, D_{rw}	mm	350
Front wheel diameter, D_{fw}	mm	350
Wheelbase, L_w	mm	1200
Distance between front axle and center of gravity, L_{cg}	mm	600
Height of the center of gravity relative to the running track, h_{mg}	mm	650
The mass of the vehicle, m	kg	250
The weight of the vehicle, mg	N	2450
Coefficient of adhesion, μ	-	0.8

From the analysis of the obtained results, it results that the all-wheel drive solution is the most favorable.

4.2. Simulation of Crossing the Step-Type Obstacle

The simulation of crossing a step-type obstacle was performed in two variants (Figure 22): when climbing the obstacle; and suspended on the obstacle.

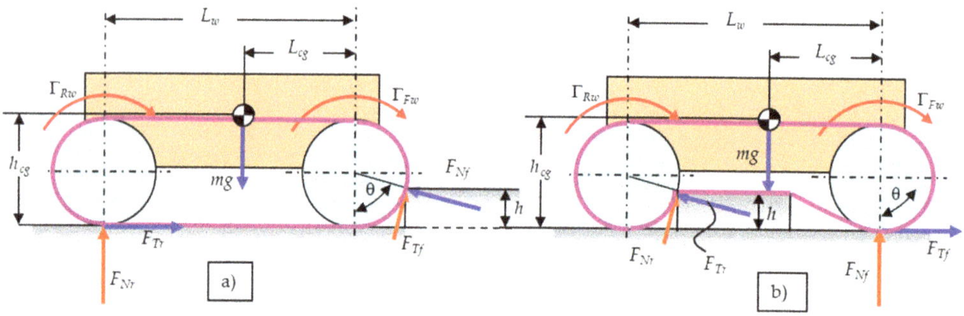

Figure 22. Schematic of crossing a step-type obstacle: (**a**) shows the passage over the obstacle with the first cheek, in front of the robot; (**b**) follows the approach of the obstacle with the driving wheel.

Using the schematic in Figure 22, the following geometric relationships are calculated:

$$\cos\theta = \frac{R-h}{R}\ [-],\ \sin\theta = \frac{\sqrt{h(2R-h)}}{R}\ [-] \tag{35}$$

The equilibrium equations for the forces and moments acting on the vehicle are as follows:

$$\sum F_X = 0 = \mu\gamma_r F_{Nr} + \mu\gamma_f F_{Nf}\cos\theta - F_{Nf}\sin\theta\ [N], \tag{36}$$

$$\sum F_Y = 0 = F_{Nr} - mg + F_{Nf}\cos\theta + \mu\gamma_f F_{Nf}\sin\theta\ [N], \tag{37}$$

$$\sum M = 0 = -F_{Nr}(L_w + R\sin\theta) + mg\left(L_{cg} + R\sin\theta\right)\ [Nm]. \tag{38}$$

The following notations are introduced:

$$A = L_{CG} + R\sin\theta\ [m],\ B = L_w - L_{cg}\ [m],\ S = \sin\theta\ [-],\ C = \cos\theta\ [-] \tag{39}$$

From Equations (35)–(42), the following second-degree equation is finally obtained:

$$\mu^2\gamma_r\gamma_f AS + \mu C\left(A\gamma_r + B\gamma_f\right) - SB = 0, \tag{40}$$

with the next positive solution:

$$\mu \geq \frac{-C\left(A\gamma_r + B\gamma_f\right) + \sqrt{C^2\left(A\gamma_r + B\gamma_f\right)^2 + 4\gamma_r\gamma_f ABS^2}}{2\gamma_r\gamma_f AS}. \tag{41}$$

Relationship (41) allows one to determine the coefficient of adhesion required to cross the obstacle. From Equations (36)–(38) result the expressions of the normal forces acting on the drive wheels:

$$F_{Nr} = \frac{mgA}{L_w + RS}\,[N],\; F_{Nf} = \frac{mgB}{(L_w + RS)\left(C + \mu\gamma_f S\right)}\,[N] \tag{42}$$

If the obstacle is crossed by a single axis, the following equations are used:

$$\sum F_X = 0 = \mu\gamma_r F_{Nr} + \mu\gamma_f F_{Nf}\cos\theta - F_{Nf}\sin\theta/2\;[N]. \tag{43}$$

$$\mu \geq \frac{-C\left(A\gamma_r + B\gamma_f\right) + \sqrt{C^2\left(A\gamma_r + B\gamma_f\right)^2 + 2\gamma_r\gamma_f ABS^2}}{2\gamma_r\gamma_f AS}\,[-]. \tag{44}$$

$$F_{Nr} = \frac{mgA}{L_w + RS}\,[N].,\; F_{Nf} = \frac{mgB}{(L_w + RS)\left(C + \mu\gamma_f S\right)}\,[N] \tag{45}$$

The calculations were performed tabularly in Microsoft Excel. The results obtained are presented in Tables 3 and 4.

Table 3. The results of the simulation for crossing the step-type obstacle with a single track.

h, mm	50	75	100	125	150
Normal force at the rear axle wheels, N	669.2	678	684	687	689.8
Minimum adhesion coefficient, μ_2, μ_{min2}	0.2	0.26	0.33	0.41	0.51
Normal force on the front axle wheels, N	650	693	741	788	829.3
Total traction force, N	2514	2514	2514	2514	2171
Limited tensile strength of grip, N	1070	1085	1094	1100	1104

Table 4. The results of the simulation for crossing the step-type obstacle with a single track.

h, mm	50	75	100	125	150
Normal force at the rear axle wheels, N	669.2	678	684	688	689.8
Minimum adhesion coefficient, μ_2, μ_{min2}	0.37	0.47	0.56	0.66	0.76
Total traction force, N	2514	2514	2514	2514	2171
Limited tensile strength of grip, N	1070	1085	1094	1100	1104

The data obtained allow for an assessment of the demands of the driving wheels that attack the obstacle; the coefficient of adhesion required for the development of the traction force; the requirements imposed on the energy aggregate (engine and transmission) to overcome the obstacle.

From the analysis of the obtained data, it can be seen that the most important variation is the coefficient of adhesion necessary to allow for the crossing of the obstacle. For the two situations studied, Figures 23 and 24 show the variation in the minimum adhesion coefficient that ensures the crossing of the obstacle, depending on its height.

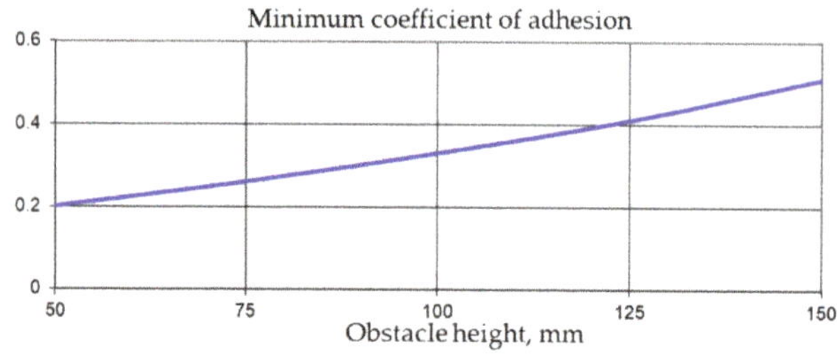

Figure 23. Variation in the minimum coefficient of adhesion when crossing the obstacle with a single track. The obstacle was positioned so that only one track could be passed over it.

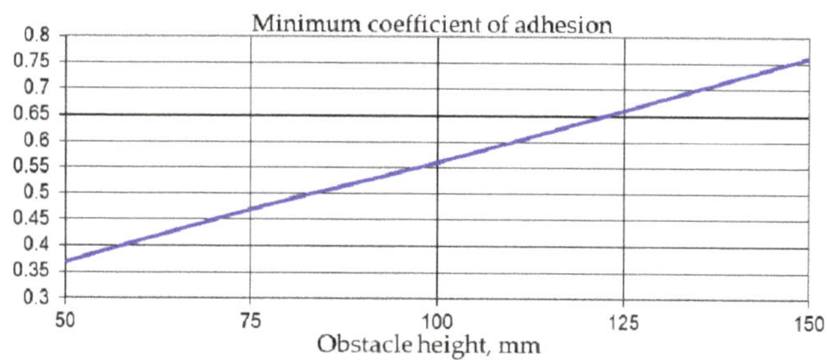

Figure 24. Minimum variation coefficient of adhesion crossing an obstacle with both tracks.

As the height of the obstacle increases, so does the need for traction. The variation in the traction force required to cross the obstacle is shown in Figures 25 and 26.

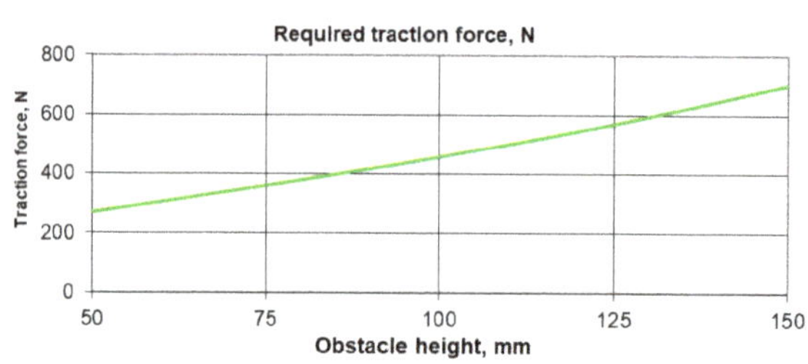

Figure 25. Variation in traction force when crossing the obstacle with a single track.

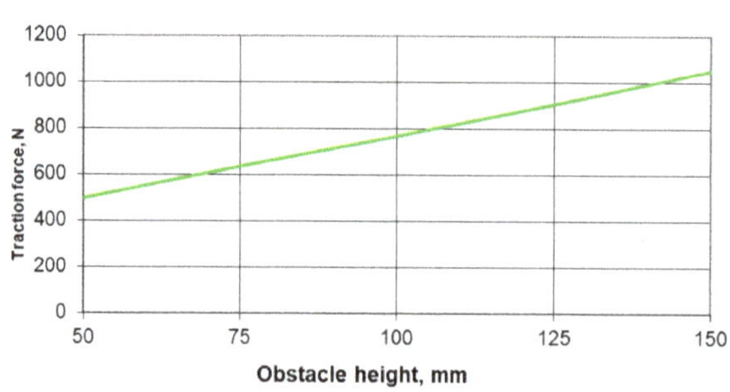

Figure 26. Variation in traction force when crossing the obstacle with both tracks.

From the analysis of the presented graphs, it can be concluded that crossing the obstacle with both tracks leads to the increase in the required traction force with a value of about 250–350 N, approximately constant depending on the height of the obstacle.

The simulations were performed for a disposition of the center of mass, arranged symmetrically with respect to the two axes, and the traction force was evenly distributed over the entire adhesion length of the track: the model is simple and easy to operate, providing important and relevant information; the correlation between the constructive parameters and the functional characteristics is easy to highlight and to be optimized at a coarse level; the simplicity of the model makes it quickly integrated into complex models for simulating mobility.

4.3. Simulation of Crossing the Step-Type Obstacle with Recurdyn

The virtual model of a crawler engine was made using the specialized program Recurdyn. After making the virtual model, we proceeded to check the arrangement of all elements in order to eliminate interference or incorrect positioning (Figure 27).

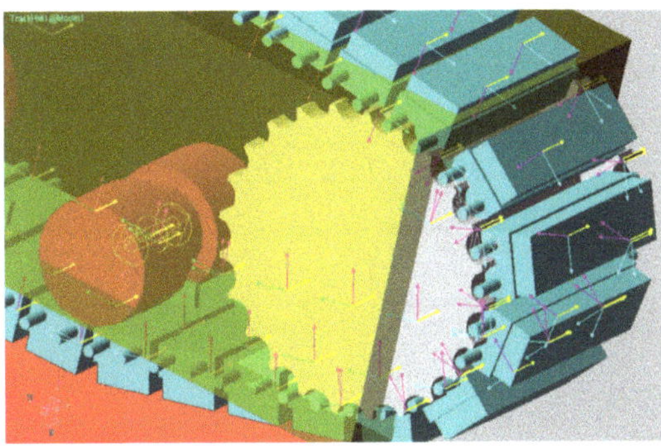

Figure 27. Detail representing the crawler gearing with the drive wheel. The simulation was performed using Recurdyn software.

These simulations refer to the decomposition of the traction forces on the three directions of the orthogonal axis system. Each track element (patina) displaces the tread in the three directions. The graphs in Figure 28 refer to those stated.

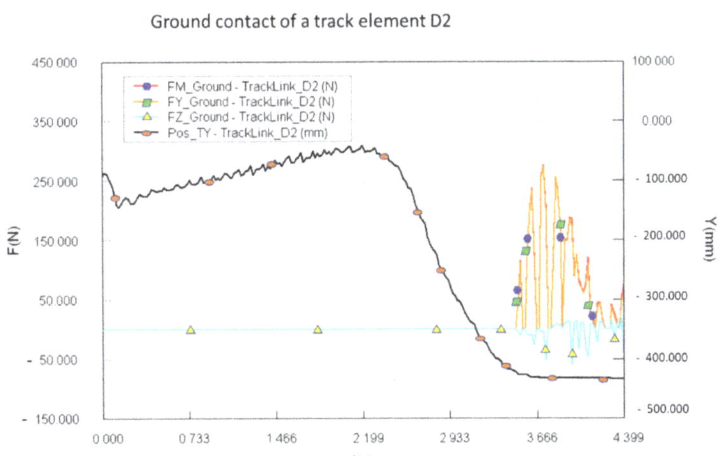

Figure 28. Simulation study of track contact with the ground.

Simulating the contact of a skid in the composition of the track with the ground allows for the extraction of information on vertical stresses on the ground (ground pressure), as well as the evaluation of reactions in the direction of travel, which allows for the study of the ground lift in the direction of traction.

The study of the contact between the skate and the buckets allows for the evaluation of the contact force with implications for the dimensioning of the rubber bandage, but also for the identification of possible shocks when the bucket passes over two skates due to the intersection between them (Figure 29).

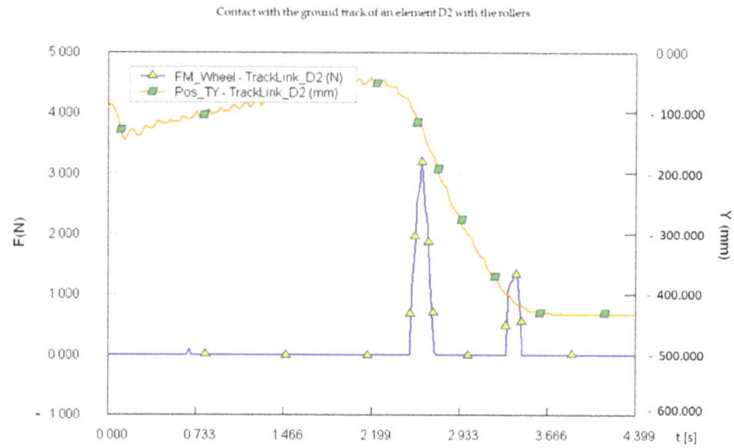

Figure 29. Study of the contact of a patina in the composition of the caterpillar with the buckets.

The simulation study of the torque variation at the driving wheel indicates its important variations, in the conditions of a stationary movement, with constant speed (Figure 30).

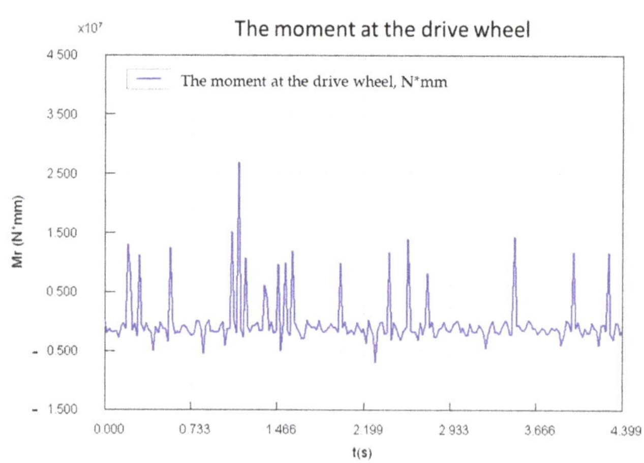

Figure 30. Simulation study of the moment at the drive wheel.

Studying the simulation of travel speed for a value of about 45% of the maximum travel speed indicates the existence of irregularities that can have disruptive adverse effects on the image stability recorded by the video cameras installed on the platform (Figure 31).

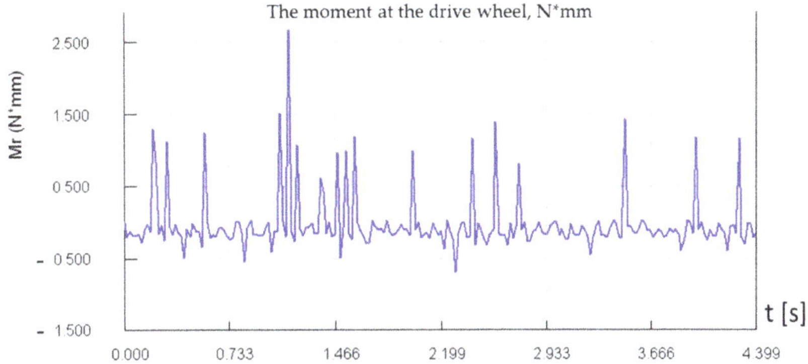

Figure 31. Study by simulation of travel speed.

An attenuation of these variations will be recorded by taking into account the moment of inertia of the electric motor and gearboxes.

5. Testing and Evaluation Propulsion Systems

The following main objectives of the experimental research were determined: the functional characteristics of the mobile platform; the functional characteristics of some main components of the mobile platform;

The testing of the mobility of the mobile robot had the following important objectives: the operation of the propulsion system when moving rectilinearly; when cornering; the control mode of the controller; the operation of the engine: in reverse and reverse; at rapid changes of the operating mode; the road holding of the mobile robot.

5.1. Testing and Evaluation on the Stand

The testing of electric motors had the following objectives: the operation of motors at supply voltages of 12 Vcc and 24 Vcc; the controller and the electrical connections between the motor and the controller; the way of simultaneously changing the speed of electric motors; how to differentiate the speed of electric motors.

Stand-up testing of electric motors followed, checking the way of making the electrical connections between the motor and the controller; the RS232 and ANALOG control modes of the motors; the operation of the motors in simultaneous and differentiated speed regime; the efficiency of the engine failure stop system.

The following operations were performed prior to the actual execution of the tests (Figure 32): the smooth rotation of the shafts of the electric motors was verified; integrity of the electrical conductors connecting the electric motors was checked; motors were mounted on the test stand. The schematic of the electrical connections is included in Figure 33. The electrical connections were made with conductors of large section so as to avoid voltage drops when high currents occur.

In the case of designing the DasyLab application, a decision was made to monitor and record the measured values. Butherworth type 3 digital filters were used to reduce the noise associated with the acquired signals. A first set of data was purchased during the stand test of electric motors. Next, a vacuum test was performed at different speeds of the two engines. The results are shown in Figure 32 For these tests, the digital control of the controller was used through the RS232 interface of the computer.

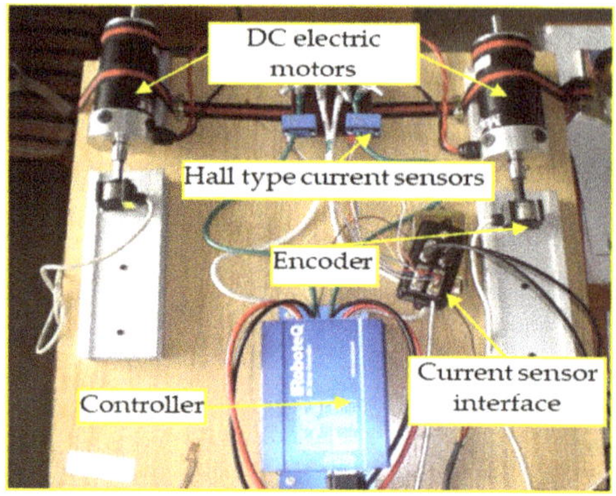

Figure 32. How to mount the motors and the controller on the stand.

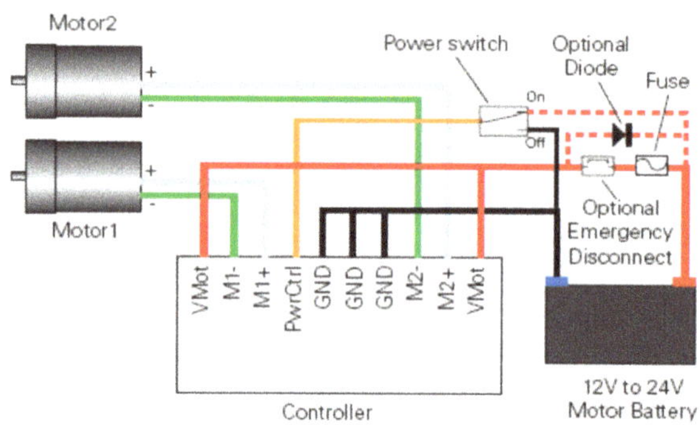

Figure 33. Diagram of electrical connections for connecting motors.

From the analysis of the presented graph, the following can be noticed (Figure 34): the value of the current varies relatively little with the speed due to the fact that the mechanical load of the electric motor is zero, so that the absorbed current has relatively small values; control through the RS232 interface provided by the RoboteQ program leads to the execution of the turn by differentiating the speeds as follows: the outer track of the turn: $v + \Delta v$; and the inner track of the turn: $v - \Delta v$; where Δv it is proportional to the turn command (Figure 35).

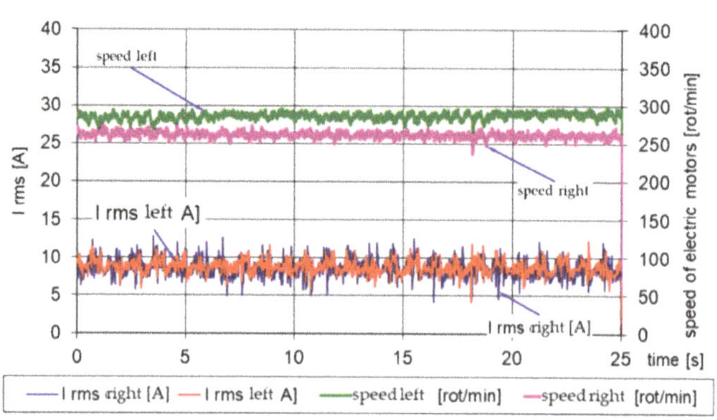

Figure 34. Vacuum testing at constant speed of electric motors.

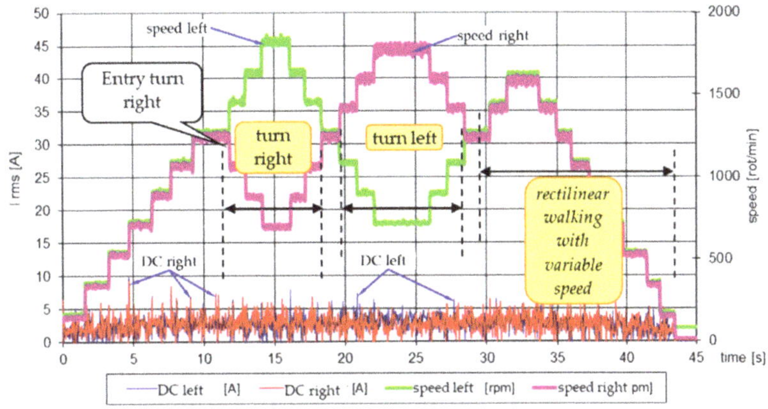

Figure 35. Vacuum testing at variable speed of electric motors.

5.2. Engine Testing by Suspension Test

Suspended testing of electric motors followed the same things as in Section 5.1. The results of the test for simulating rectilinear gait are shown in Figure 36.

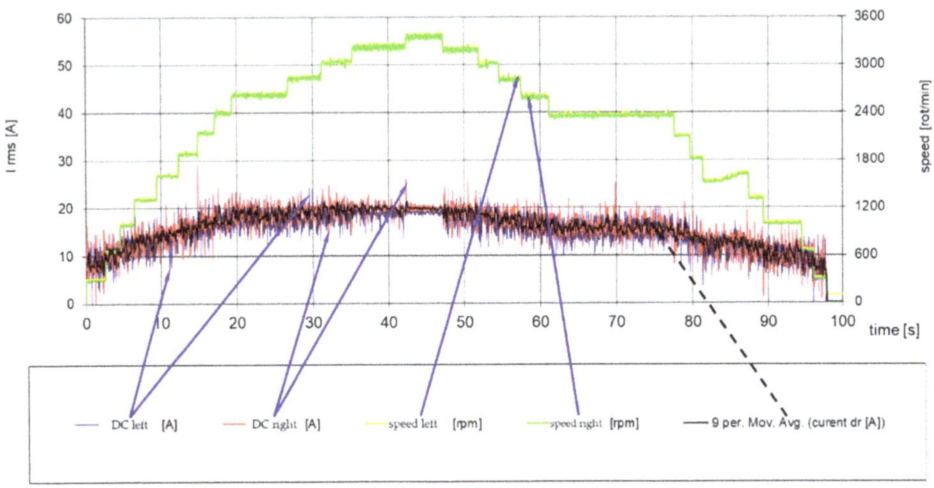

Figure 36. Example of recording for the simulation of rectilinear motion in the suspended sample.

The analysis of the graphically represented data shows the dependence between speed and absorbed current, which can be explained by the increased level of friction torque in the gearboxes.

Figure 37 shows examples of results obtained when simulating the turn in the suspended test.

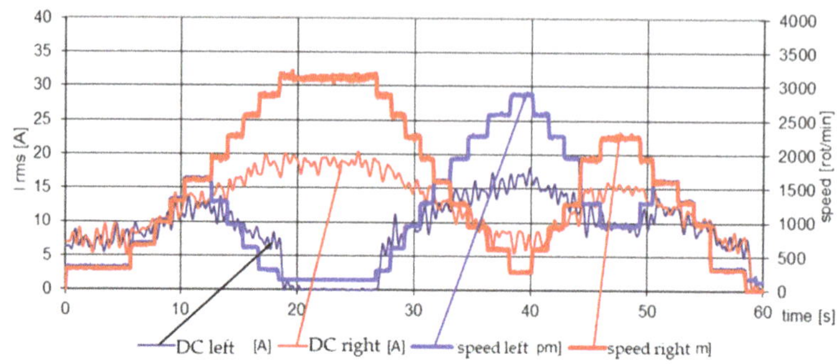

Figure 37. Example of simulating the turn in the suspended test.

The development of experimental research allowed the formulation of the following: the testing program was completed and complied with procedural requirements; when the electric motors are idle, they absorb a current of about 3 A rms; in the case of a rapid change of speed by about 10%, there were short-term increases in the absorbed current of up to 8 A rms; due to the technological and adjustment non-uniformity, the speeds of the two engines differ by about 1–2% when idling; no abnormal noises or overheating were found during the operation of the gearboxes. During the operation of the tracks, the test was suspended: the occurrence of the phenomenon of hooking the tip of the tooth flank of the crawler chain roller was noted, as the phenomenon was found both at the entry into the gear and at the exit of the gear, given that the support branch of the track is a free-forming an arrow of about 35–40 mm, as can be easily seen from Figure 38; the current absorbed by the motors, corresponding to a speed regime close to the maximum speed, by 18–20 A rms.

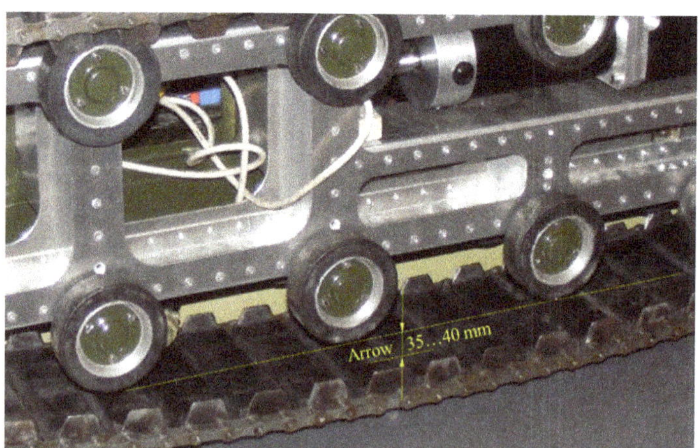

Figure 38. Overview of the mobile platform of the robot with the lower branch of the track hanging, with an arrow within the admissible limits on making the turn.

5.3. *Testing and Evaluation of Rectilinear Displacement Capacity*

Rectilinear gait testing was followed: testing the robot's ability to move forward in a straight line; backwards in a straight line; the ability to accelerate the mobile robot; determination of the maximum speed. A data acquisition system assisted by a portable PC using the specialized program DasyLab was used to perform the measurements.

An example of recording data at rectilinear motion at maximum speed is shown in Figure 39. The maximum speed obtained was higher than 1 m/s.

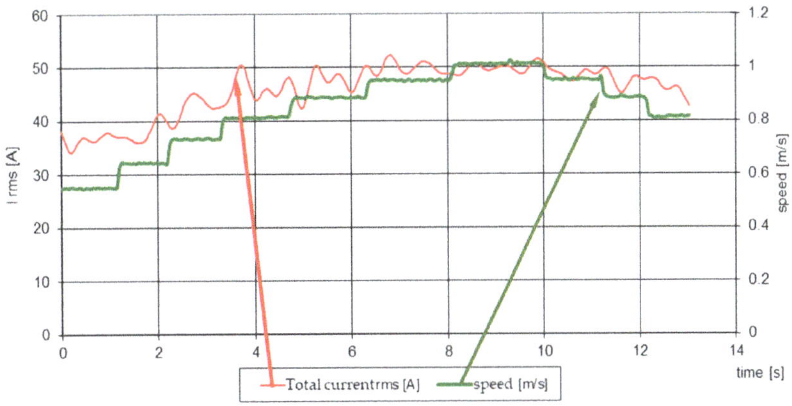

Figure 39. Example of data recorded at rectilinear displacement.

5.4. Testing and Evaluation of the Ability to Make the Turn

Testing of ability to make the turn followed the same things as in Section 5.1.

The same data acquisition systems described above were used to perform the measurements. On-the-go testing at two different engine speeds was performed by analog control, maintaining a relatively long time the command to be able to record data at constant operating speeds. During the suspended test, the following steps were performed: straight forward movement followed by turning right and then left; rectilinear movement backwards followed by the execution of the right and then left turn; execution of the turn with minimum radius only by driving only one track. In the first two stages, the following aspects were pursued: ability to make turns with different radius; the quality of the crawler gear drive; firmness and accuracy of the analog control. In the third stage, the following aspects were further pursued: ability to execute the turn with a minimum radius; smooth, noiseless operation of the tracks; control of the controller in analog mode; efficiency of the track guidance and stretching mechanism. The turning radius was determined:

$$\rho_2 = Bv_2/v_2 - v_1 \ [m]. \tag{46}$$

where B [m]—the gauge of the mobile robot; v [m/s]—left track speeds (index 2) and right track (index 1), respectively.

From the analysis of the graphically represented data, it is found that, at the rectilinear course for which $v_2 = v_1$, the turning radius tends to take an infinite value (Figure 40). During the test when performing the turn by operating a single track, it was found that a turning radius of 0.68 ÷ 0.72 (m) was achieved, depending on the landslide of the immobilized track. During testing the ability to travel for a long time, the robot performed the following travel categories for 80 min: rectilinear movement back and forth; turning left and right; crossing singular obstacles. When operating the robot at maximum speed, the electric motors absorb a current of about 50 A rms, representing less than 50% of the maximum current set for the controller (110 A) and about 35% of the maximum capacity of the controller (140 A), the values recorded differ from those determined by simulation with a maximum of 6.52%, which confirms the validity of the calculation model developed.

Figure 40. Appearance during the test of the ability to move over a rectangular prism.

5.5. Experimental Research on the Ability to Approach Obstacles

Testing the ability of the mobile obstacle robot had the following more important objectives: testing the operation of the propulsion system when crossing obstacles; the control mode of the controller; the operation of the engine at rapid changes in operating mode; the road holding of the mobile robot when approaching obstacles (Figure 40).

The rectilinear gait test aimed at testing the ability to approach the obstacle in terms of energy resource and the characteristics of the track's adhesion to the ground.

An example of data recorded at rectilinear displacement at maximum speed is shown in Figure 41. It is noted that, although the displacement speed was approximately constant (red line), the current shows significant variations due to the corresponding change in the overall resistance coefficient to advance.

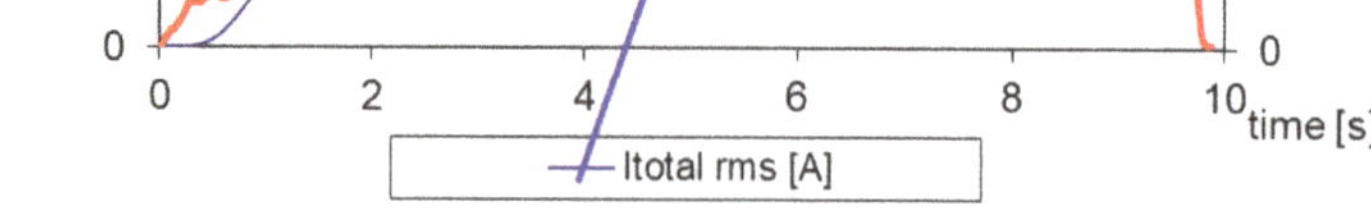

Figure 41. Example of data recorded at rectilinear displacement.

The height of the obstacle was 160 mm and the average speed of approaching the obstacle was 0.363 m/s. During the obstacle approach, the absorbed current had an average value of 57.22 A with a maximum of over 120 A. During the obstacle approach the robot control was firm, and its speed remained at an approximately constant value, the variations being generated by the entry/exit of/in contact with the edges of the obstacle of the support rollers of the track.

6. Conclusions and Discussion

The engine test program was completed in full and complied with the procedural requirements. The test program had a complex character, including the running of motors, reducers and tracks. When the electric motors are idle, they absorb a current of about 3 rms. In the case of a rapid change in speed of about 10%, there were short-term increases in the absorbed current of up to 8 A rms.

When operating during the test, the current absorbed by the motors, corresponding to a speed regime close to the maximum speed, was suspended at 18–20 A rms. Thus, regardless of the speed regimes, when driving in a straight line and/or in a corner, as well as for both directions, no warm-ups of the motors or reducers were found. The temperature of the MOSFET power components in the controller exceeded the ambient temperature by 1–2 °C. The values of the absorbed current represent 12–15% of the maximum current allowed by the controller.

The values obtained differ from those calculated by a simulation with a maximum of 4.7%. Increased current values were recorded at the transient regimes. The tests followed the following functional algorithm:

Through the program of experimental research, the following important issues related to the functional characteristics of the mobile platform were revealed: the constructive characteristics of the ROBTER crawler robot allowed for stable and fully controllable rolling; the ROBTER crawler robot did not lose its grip, and there was no skidding or tilting; the ability to cross the individual obstacles (longitudinal slope) of the ROBTER crawler robot is considered to be in accordance with the formulated requirements;

The main advantages of the studied solution are the possibility of creating a family of intervention robots by replacing the operational platform; simplicity of the calculation logarithm; making modular software; the use of relatively cheap materials and components, the aim being to increase the degree of efficiency and annihilation of potential risks with minimal costs. In addition, this functional model can also be used for research activities in the subject, as well as for doctorates. As the structure is extremely versatile, future users develop it to conduct in-depth research. In further research, we want to focus on describing the influence of the crawler propeller on the ground and on the development of at least two new operational platforms.

The previously-performed and presented study had as main objective the highlighting of the influences that the main geometric parameters exert on the functional characteristics of the mobile robot. The following conclusions regarding the geometric configuration of the mobile robot platform result: the increase in the height at which the center of mass is arranged leads to more favorable conditions for approaching step-type obstacles but generates conditions of loss of stability when sloping and braking, especially when braking when descending the slope; increasing the height of the axis of the drive wheel leads to the corresponding increase of the height of the step-type obstacle that can be approached, but also to the increase of the width of the ditch that can be crossed; the forward movement of the horizontal position of the center of gravity leads to more advantageous conditions for tackling step-type obstacles, but appropriately reduces the width of the ditch that can be crossed and increases the value of the forces required to perform the turn; increasing the adhesion length of the crawler propeller with the ground leads to the corresponding increase in the width of the ditch that can be crossed but also to the increase in the traction forces in turning; the increase of the track width allows to obtain a lower average pressure on the ground, reducing the sinking of the tracked propeller in the deformable ground and, therefore, the resistance to advance; the lower sinking in the deformable soil also leads to the reduction of the lateral excavation effect during the execution of the turn, which leads to the decrease of the resistance moment in the turn.

All these conclusions highlight the contradictory nature of the influence exerted by the main geometric parameters of the crawler engine on the functional characteristics of the mobile robot. This results in the need to optimize the geometrical parameters of the crawler engine according to the characteristics of the specific mission for which the mobile robot is intended. All these considerations were taken into account when making the prototype, which is the subject of this paper.

Author Contributions: Conceptualization, L.Ş.G., I.P., I.O. and D.J.; methodology, I.O.; software, I.P. and D.J.; validation, L.Ş.G.; formal analysis, L.Ş.G. and I.O.; investigation, L.Ş.G. and I.P.; writing—original draft preparation, L.Ş.G.; writing—review and editing, I.O. and D.J.; visualization, I.O. and D.J.; supervision, I.P.; project administration, L.Ş.G., I.P., I.O. and D.J. All authors have read and agreed to the published version of the manuscript.

Funding: This research received no external funding.

Institutional Review Board Statement: Not applicable.

Informed Consent Statement: Not applicable.

Data Availability Statement: The tests were performed with the testing program ROBTER and its components according to the requirements of stage IV of the Robot project to combat terrorist acts, contract 20/2006 within the CEEX program.

Acknowledgments: We thank the Research team from the MOSITEST Research Center of the Military Technical Academy "FERDINAND I", especially Ticuşor Ciobotaru and Valentin Vînturiş. Some of the used experimental results come from the tests performed with the ROBTER test program and its components according to the requirements of stage IV of the Robot for Combating Terrorism, Contract 20/2006 within the CEEX program. ROBTER—Robot for Combating Terrorist Actions CEEX 20/19.10.2006.

Conflicts of Interest: The authors declare no conflict of interest.

References

1. Luxton, A.; Marinov, M. Terrorist Threat Mitigation Strategies for the Railways. *Sustainability* **2020**, *12*, 3408. [CrossRef]
2. Kaewunruen, S.; Alawad, H.; Cotruta, S. A Decision Framework for Managing the Risk of Terrorist Threats at Rail Stations Interconnected with Airports. *Safety* **2018**, *4*, 36. [CrossRef]
3. Kanji, A. Framing Muslims in the "War on Terror": Representations of Ideological Violence by Muslim versus Non-Muslim Perpetrators in Canadian National News Media. *Religions* **2018**, *9*, 274. [CrossRef]
4. Gibbs, J.P. Conceptualization of Terrorism. *Am. Sociol. Rev.* **1989**, *54*, 329–340. [CrossRef]
5. Eikenberry, K.W. *Thoughts on Unconventional Threats and Terrorism*; Working Group on Foreign Policy and Grand Strategy, Hoover Institution, Stanford University: Stanford, CA, USA, 2014.
6. Available online: https://ec.europa.eu/echo/files/about/COMM_PDF_SEC_2010_1626_F_staff_working_document_en.pdf (accessed on 18 February 2021).
7. National Research Council. Technology Development for Army Unmanned Ground Vehicles. ISBN: 0-309-08620-5. Available online: http://www.nap.edu/catalog/10592.html (accessed on 22 February 2021).
8. Available online: https://www.unodc.org/e4j/en/terrorism/module-4/key-issues/defining-terrorism.html (accessed on 17 March 2021).
9. Cassese, A. The Multifaceted Criminal Notion of Terrorism in International Law. *JICJ* **2006**, *4*, 933–958. [CrossRef]
10. Dewulf, S. Human Rights in the Criminal Code: A Critique of the Curious Implementation of the EU and Council of Europe Instruments on Combating and Preventing Terrorism in Belgian Criminal Legislation. *EJC CLCJ* **2014**, *22*, 33–57. [CrossRef]
11. Liem, M.; van Buuren, J.; de Roy van Zuijdewijn, J.; Schönberger, H.; Bakker, E. European Lone Actor Terrorists Versus "Common". *Homicide Stud.* **2018**, *22*, 45–69. [CrossRef]
12. Available online: https://www2.ohchr.org/english/bodies/chr/special/counter-terrorism.htm (accessed on 15 March 2021).
13. Duquet, N.; Goris, K. *Firearms Acquisition by Terrorists in Europe, Research Findings and Policy Recommendations of Project SAFTE*; Flemish Peace Institute: Brussels, Belgium, 2018; ISBN 9789078864912.
14. Available online: https://govinfo.library.unt.edu/911/report/911Report_Exec.htm (accessed on 2 March 2021).
15. Czop, A.; Hacker, K.; Murphy, J.; Zimmerman, T. Low-cost explosive ordnance disposal robot using off-the-shelf parts. In Proceedings of the Unmanned Ground Vehicle Technology VII, Defence and Security, Orlando, FL, USA, 27 May 2005; Volume 5804. [CrossRef]
16. Available online: https://www.un.org/sc/ctc/wp-content/uploads/2017/03/CTED-Trends-Report-March-2017-Final.pdf (accessed on 10 March 2021).
17. Van Moore, A., Jr. Radiological and nuclear terrorism: Are you prepared? *Am. Coll. Radiol.* **2004**, *1*, 54–58. [CrossRef]
18. Available online: https://www.ouvry.com/en/nrbce-terrorist-chemical-weapons-attacks/ (accessed on 2 March 2021).
19. Maynard, R.M.; Tetley, T.D. Bioterrorism: The lung under attack. **2004**, *59*, 188–189. [CrossRef]
20. Available online: https://www.un.org/sc/ctc/wp-content/uploads/2019/01/Compendium_of_Good_Practices_Compressed.pdf (accessed on 2 March 2021).

21. Paulius, D.; Sun, Y. A Survey of Knowledge Representation in Service Robotics. *Robot. Auton. Syst.* **2019**, *118*, 13–30. [CrossRef]
22. Matuszek, C.; Herbst, E.; Zettlemoyer, L. Learning to parse natural language commands to a robot control system. In *Experimental Robotics*; Desai, J., Dudek, G., Khatib, O., Kumar, V., Eds.; Springer: Berlin/Heidelberg, Germany, 2016; Volume 88, pp. 403–415.
23. Liu, R.; Webb, J.; Zhang, X. Natural-Language-Instructed Industrial Task Execution. In Proceedings of the ASME IDETC/CIE, Charlotte, NC, USA, 21–24 August 2016; p. v01bt02a043.
24. Tay, N.N.W.; Saputra, A.A.; Botzheim, J.; Kubota, N. Service robot planning via solving constraint satisfaction problem. *Robomech. J.* **2016**, *3*, 17. [CrossRef]
25. Zhang, S.; Sridharan, M.; Wyatt, J.L. Mixed logical inference and probabilistic planning for robots in unreliable worlds. *IEEE Trans. Robot.* **2015**, *31*, 699–713. [CrossRef]
26. Kress-Gazit, H.; Wongpiromsarn, T.; Topcu, U. Correct, reactive robot control from abstraction and temporal logic specifications. *IEEE Robot. Autom. Mag. Spec. Issue Form. Methods Robot. Autom.* **2011**, *18*, 65–74. [CrossRef]
27. Xie, M. Robot Vision: A Holistic View. In *Climbing and Walking Robots*; Springer: Berlin/Heidelberg, Germany, 2005. [CrossRef]
28. Carey, M.W.; Kurz, E.M.; Matte, J.D.; Perrault, T.D. Novel EOD Robot Design with a Dexterous Gripper and Intituitive Teleoperation. In *Proceedings of the World Automation Congress 2012, Puerto Vallarta, Mexico, 24–28 June 2012*; IEEE: New York, NY, USA, 2012; Volume 1, pp. 463–469. ISBN 9781467344975.
29. Belanche, D.; Casaló, L.V.; Flavián, C.; Schepers, J. Service robot implementation: A theoretical framework and research agenda. *Serv. Ind. J.* **2020**, *40*, 203–225. [CrossRef]
30. Saputra, R.P.; Rijanto, E. Automatic Giuded Vehicles System and Its Coordination Control for Containers Terminal Logistics Application. *Conf. Int. Logist. Semin. Workshop* **2012**, *2*, 4.
31. Grigore, L.Ș.; Priescu, I.; Grecu, D.L. *Inteligența Artificială Aplicată în Sisteme Robotizate Fixe și Mobile*; Editura AGIR: Bucuresti, Romania, 2020; p. 703. ISBN 978-973-720-767-8.
32. Grigore, L.Ș.; Gorgoteanu, D.; Molder, C.; Alexa, O.; Oncioiu, I.; Ștefan, A.; Constantin, D.; Lupoae, M.; Bălașa, R.I. A Dynamic Motion Analysis of a Six-Wheel Ground Vehicle for Emergency Intervention Actions. *Sensors* **2021**, *21*, 1618. [CrossRef]
33. Deng, W.; Zhang, H.; Li, Y.; Gao, F. Research on target recognition and path planning for EOD robot. *Int. J. Comput. Appl. Technol.* **2018**, *57*, 325–333. [CrossRef]
34. Rajan, K.; Saffiotti, A. Towards a science of integrated AI and robotics. *Artif. Intell.* **2017**, *247*, 1–9. [CrossRef]
35. Dudzik, S. Application of the Motion Capture System to Estimate the Accuracy of a Wheeled Mobile Robot Localization. *Energies* **2020**, *13*, 6437. [CrossRef]
36. Russo, M.; Raimondi, L.; Dong, X.; Axinte, D.; Kell, J. Task-oriented optimal dimensional synthesis of robotic manipulators with limited mobility. *Robot. Comput. Integr. Manuf.* **2021**, *69*, 102096. [CrossRef]
37. Sandry, E. *Robots and Communication*, 1st ed.; Palgrave Macmillan: London, UK, 2015; p. 129. ISBN 978-1-137-46837-6. [CrossRef]
38. Zhao, J.; Gao, J.; Zhao, F.; Xu, Z.; Liu, Y. A design Method for Multijoint Explosive-Proof Manipulators by Two Motors. *Appl. Sci.* **2018**, *8*, 712. [CrossRef]
39. Jian-Jun, Z.; Ru-Qing, Y.; Wei-Jun, Z.; Xin-Hua, W.; Jun, Q. Research on Semi-Automatic Bomb Fetching for an EOD Robot. *Int. J. Adv. Robot. Syst.* **2007**, *4*, 27. [CrossRef]
40. Chatziparaschis, D.; Lagoudakis, M.G.; Partsinevelos, P. Aerial and Ground Robot Collaboration for Autonomous Mapping in Search and Rescue Missions. *Drones* **2020**, *4*, 79. [CrossRef]
41. Liao, J.-C.; Chen, S.-H.; Zhuang, Z.-Y.; Wu, B.-W.; Chen, Y.-J. Designing and Manufacturing of Automatic Robotic Lawn Mower. *Processes* **2021**, *9*, 358. [CrossRef]
42. Chen, X.; Huang, F.; Zhang, Y.; Chen, Z.; Liu, S.; Nie, Y.; Tang, J.; Zhu, S. A Novel Virtual-Structure Formation Control Design for Mobile Robots with Obstacle Avoidance. *Appl. Sci.* **2020**, *10*, 5807. [CrossRef]
43. Alexa, O.; Coropețchi, I.; Vasile, A.; Oncioiu, I.; Grigore, L.Ș. Considerations for determining the coefficient of inertia masses for a tracked vehicle. *Sensors* **2020**, *20*, 5587. [CrossRef]
44. Niu, J.; Wang, H.; Shi, H.; Pop, N.; Li, D.; Li, S.; Wu, S. Study on structural modeling and kinematics analysis of a novel wheel-legged rescu robot. *IJARS* **2018**, *I-17*, 17. [CrossRef]
45. Islam, A.J.; Alam, S.S.; Ahammad, K.T.; Nadim, F.K.; Barua, B. Design, kinematic and performance evaluation of a dual arm bomb disposal robot. In Proceedings of the 2017 3rd EICT, Khulna, Bangladesh, 7–9 December 2017; pp. 1–6. [CrossRef]
46. Scalera, L.; Palomba, I.; Wehrle, E.; Gasparetto, A.; Vidoni, R. Natural Motion for Energy Saving in Robotic and Mechatronic Systems. *Appl. Sci.* **2019**, *9*, 3516. [CrossRef]
47. Lei, C.; Chang, F.; Li, S.; Zhang, X. *Control System of the Explosive Ordnance Disposal Robot Based on Active Eye-to-Hand Binocular Vision, Artificial Intelligence and Computational Intelligence*; Springer: Berlin/Heidelberg, Germany, 2011; Volume 7004. ISBN 978-3-642-23895-6.
48. Available online: https://www.nationaldefensemagazine.org/articles/2019/2/15/harris-provides-eod-robots-to-british-army (accessed on 2 March 2021).
49. Yin, Q.; Xiong, Z.; Song, S.; Zhang, G. Modular design method of EOD robot based on genetic algorithm. *Int. J. Wirel. Mob. Comput.* **2018**, *15*. [CrossRef]
50. Available online: https://ceias.nau.edu/capstone/projects/EE/2018/OrdnanceDisposal2/design_subsystems.html (accessed on 2 March 2021).
51. Available online: https://www.militarysystems-tech.com/taxonomy/term/78 (accessed on 2 March 2021).

52. Yu, J.; Jiang, W.; Luo, Z.; Yang, L. Application of a Vision-Based Single Target on Robot Positioning System. *Sensors* **2021**, *21*, 1829. [CrossRef]
53. Ştefan, A.; Constantin, D.; Grigore, L.Ş. Aspects of kinematics and dynamics of a gripping mechanism. In *Proceedings of the 2015 7th ECAI, Bucharest, Romania, 25–27 June 2015*; IEEE: New York, NY, USA, 2015; pp. WF1–WF4. [CrossRef]
54. Available online: https://www.ocean-comms.com.sg/product/portable-xray/ (accessed on 10 March 2021).
55. Ceccarelli, M.; Gasparetto, A. Mechanism Design for Robotics. *Robotics* **2019**, *8*, 30. [CrossRef]
56. Toma, Ş.A.; Sebacher, B.; Focşa, A. On Anomalous Deformation Profile Detection through Supervised and Unsupervised Machine Learning. In Proceedings of the 2019 IEEE International Geoscience and Remote Sensing Symposium, Yokohama, Japan, 28 July–2 August 2019. [CrossRef]
57. Xu, L.; Zhang, L.; Zhao, J.; Kim, K. Cornering Algorithm for a Crawler In-Pipe Inspection Robot. *Symmetry* **2020**, *12*, 2016. [CrossRef]
58. Available online: https://www.antiterrorism.eu/portfolio-posts/expert/ (accessed on 20 March 2021).
59. Grigore, L.Ş.; Ileri, R.; Neculăescu, C.; Soloi, A.; Ciobotaru, T.; Vînturiş, A. A class of autonomous robots prepared for unfriendly sunny environment. In Proceedings of the CAR2011—2011 3rd International Asia Conference on Informatics in Control, Automation and Robotics, Xiamen, China, 25–26 June 2011; Volume 646.
60. Bruch, M.H.; Laird, R.T.; Everett, H.R. *Challenges for Deploying Man-Portable Robots into Hostile Environments*; Space and Naval Warfare Systems Center: San Diego, CA, USA, 2001. [CrossRef]
61. Ogorkiewicy, R.M. *Technology of Tanks*; Jane's Information Group: Coulsdon, UK, 1991; Volume 1–2, p. 424. ISBN 978-0710605955.
62. Haueisen, B. *Mobility Analysis of Small, Lightweight Robotic Vehicles*; Brooke Haueisen: Huntington Beach, CA, USA, 2003; p. 46.
63. Johnson, C.P. Comparative Analysis of Lightweight Robotic Wheeled and Tracked Vehicle. Ph.D. Dissertation, Virginia Polytechnic Institute and State University, Blacksburg, VA, USA, 26 April 2012; p. 155.
64. Gigler, J.K.; Ward, S.M. A simulation model for the prediction of the ground pressure distribution under tracked vehicles. *J. Terramechanics* **1993**, *30*, 461–469. [CrossRef]

applied sciences

MDPI

Article

Design and Modelling of an Amphibious Spherical Robot Attached with Assistant Fins

Xing Chi and Qiang Zhan *

Robotics Institute, School of Mechanical Engineering and Automation, Beihang University, Beijing 100191, China; chixing@buaa.edu.cn
* Correspondence: qzhan@buaa.edu.cn

Abstract: Mobile robots that can survive in unstructured wildernesses is essential in many applications such as environment detecting and security surveillance. In many of these applications, it is highly desirable that the robot can adapt robustly to both terrestrial environment and aquatic environment, and translocate swiftly between various environments. A novel concept of amphibious spherical robot with fins is proposed in this paper, capable of both terrestrial locomotion and aquatic locomotion. Unlike the traditional amphibious robots, whose motions are commonly induced by propellers, legs or snake-like tandem joints, the proposed amphibious spherical robot utilizes the rolling motion of a spherical shell as the principal locomotion mode in the aquatic environment. Moreover, spinning motion of the spherical shell is used to steer the spherical robot efficiently and agilely; several fins are attached to the outer spherical shell as an assistance to the rolling motion. These two motion modes, rolling and spinning, can be used unchangeably in the terrestrial environment, leading to a compact and highly adaptive design of the robot. The work introduced in this paper brings in an innovative solution for the design of an amphibious robot.

Keywords: amphibious robot; spherical robot; assistant fin; buoyancy; hydrodynamic force

Citation: Chi, X.; Zhan, Q. Design and Modelling of an Amphibious Spherical Robot Attached with Assistant Fins. *Appl. Sci.* **2021**, *11*, 3739. https://doi.org/10.3390/app11093739

Academic Editor: Alessandro Gasparetto

Received: 26 March 2021
Accepted: 19 April 2021
Published: 21 April 2021

1. Introduction

Amphibious robots are designed to cope with drudgeries involving miscellaneous terrains, such as lakes, wetlands, shallows and pipelines [1,2]. Compared with a robot moving only on land or underwater, which is easily stuck in the junction area between the land and the water, amphibious robots are capable of both aquatic locomotion and terrestrial locomotion, which are especially suitable for applications in unknown environments, potentially with complex terrains. Tasks calling for this kind of robots include environment detecting, wild exploring, security surveillance as well as scientific inspecting [3].

A variety of amphibious robots have been developed by researchers in recent years. As one of the typical amphibious robots, snake-like robots are inspired by the locomotion mode of a snake as well as a lamprey. A snake-like robot called HELIX is developed by Takayama and Hirose to verify the propulsion principle that helical motion can be created by successive distortions of articulated body segments [4]. The design of one unit of HELIX, including a special spherical mechanism and two servo motors, is introduced in detail by the same authors. As a modified prototype of HELIX, ACM-R5 bears impressive performance in terms of its dexterous locomotion capacity. Underwater swimming fins are attached around the outer surface of each unit to assist the underwater motion, with passive wheels attached on the tip of each fin in order to improve the on-land locomotion performance [2,5]. The AmphiBot robots, a series of amphibious snake-like robots capable of swimming in water and crawling on land, are designed by Ijspeert and Crespi et al. [6]. A central pattern generator (CPG)-based controller is used to generate lateral undulation for this kind of robot [7,8]. Some key parameters that produce fast locomotion gaits, such as the amplitude of oscillation, frequency and wavelength, are identified by a series of experiments. An improved version of AmphiBot, named Salamandra robotica, equipped with a passive tail and four active 3-DOF limbs, is also introduced [9]. To make the snake-like robot to follow a desired path, a straight line path following controller is proposed by Kelasidi et al. [10,11].

Another type of amphibious robots is equipped with flippers that act as paddles in water and as legs on land. Using six paddles for propulsion, AQUA is suitable for navigating in shallow-water environment [12,13]. With the help of an acoustic-based localization system and a vision-based localization system, AQUA obtained the ability to revisit a previously visited site autonomously [14,15]. To enhance its climbing ability, Whegs IV combines two similar body segments with wheel-leg propellers [16,17]. AmphiHex-I obtains versatile gaits with the aid of transformable flipper legs, which makes it flexible in transforming propulsion mechanisms [18]. Its improved version with variable stiffness legs and CPG-based control strategy is also introduced [19]. A quadrupedal micro-robot weighs only several grams is introduced by Cheng et al. [20], which uses passive flaps to swim forward and adjust direction.

Some robots with spherical shape have also drawn researchers' attention. The Groundbot, an amphibious spherical robot that accomplishes its propulsion system by displacing the center of mass, is introduced by Kaznov et al. [21]. Groundbot shows good robustness in the waypoint following mode. An amphibious spherical robot which combines two actuating systems, including the quadruped walking system for terrestrial locomotion and the water-jet propulsion system for underwater locomotion, is designed by Guo et al. [22,23]. A micro-robot which is used as a manipulator and a monitor can be carried and deployed by this amphibious spherical robot. An amphibious spherical robot uses pendulum to move on land and propeller to move in the water is developed by Li et al. [24]. A spherical rotary paddle called Omni-Paddle is proposed to form a robot that mainly moves on the boarder of water and ground [25]. Four Omni-Paddles are disposed in a radical pattern around the outside of the robot, with which the robot is capable of moving towards any direction. The Omni-Paddle works as not only a driving mechanism but also a source of buoyancy in one prototype of the robot.

Each of the amphibious robots mentioned above has its advantages and disadvantages. A snake-like robot adapts well to complex terrains; however, it usually needs many units of bodies in order to achieve a high terrain adaptability, as a result requiring a large amount of motors. This leads to low robustness and high energy consumption, which is the bottleneck in field applications. A robot equipped with water-jet propulsion system can control its position and attitude precisely underwater. However, another actuation system is needed for its locomotion on land, resulting in a complex actuation system and extra weight. A robot with paddles mimics the locomotion modes used by amphibians such as ducks and turtles, which is advantageous in that it uses the same set of actuation system for both on land and in water locomotion; however, it calls for complex algorithms for gait planning in order to reach dynamic stability. In addition, the existence of feet increases risk of being entwined by water plants or shrubs.

A spherical robot bears very few motors that are sufficient for both on land and in water locomotion. Because of its inherent sealing characteristic, good stability and low energy cost, spherical robots have a great potential to be used in amphibious locomotion in field applications. However, the relatively slow locomotion speed on the water due to frictionless spherical shell and the poor maneuverability are main hindrances in using a spherical robot as an amphibian. To overcome these limitations, a novel amphibious spherical robot design is proposed in this paper, by attaching assistant fins to the outside surface of the spherical shell to increase its locomotion speed, and by using a combination of pendulum and rotator as the Inside Drive Unit (IDU) to propel the robot agilely. It is shown that this design strategy significantly improves the mobility capacity of the robot, and enhances its resistance to environment undulation, making it practicable to use a spherical robot both on land and on water.

The remainder of the paper is organized as follows. In Section 2, the mechanical design, the electrical actuation and the control system of the amphibious spherical robot are introduced. In Section 3, the equations of the robot's motion are deduced based on kinematic analyses and force analyses. In Section 4, simulations of the dynamic model are presented, and the experiments to test the performance of the robot are described in detail. Finally, the conclusion is given in Section 5.

2. Mechanical Structure and Electrical Actuation and Control System

2.1. Design Objectives

The design objectives of the amphibious spherical robot project are:

1. the ability to navigate on the water and on the land with exactly the same driving mechanism;
2. a sealed structure, enclosing its vulnerable components to protect them from outside environments, e.g., water, sand and gas, so as to make it capable of surviving harsh environments;
3. this robot can be remotely handled by an operator.

These aims are achieved by the proposed amphibious spherical robot scheme, which consists of a spherical robot and a radio frequency (RF) remote controller.

2.2. Introduction of the Developed Spherical Robot

The spherical robot introduced here is a modified type aiming at amphibious applications, which is originated from traditional spherical robots [26,27]. The spherical robot is mainly composed of two parts: a hollow spherical shell and an IDU, as shown in Figure 1. The spherical shell acts as both a wheel and a shelter; while the IDU, consisting of components inside the spherical shell, acts as the driving unit of the whole robot.

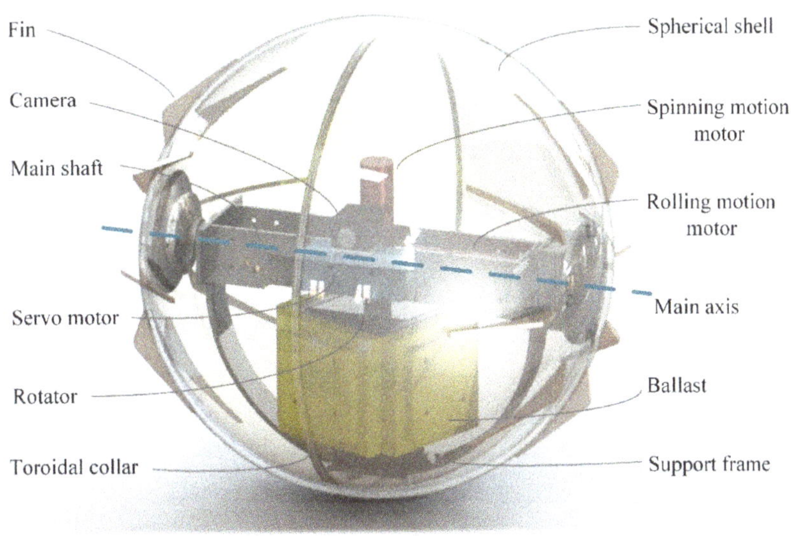

Figure 1. Mechanical structure of the amphibious spherical robot.

The hollow spherical shell consists of two hemispheric shells made of acrylic material, which makes it transparent and with little radio interference. A toroidal collar is attached to the great circle of one hemispherical shell, acting as an installation reference for the other hemispherical shell. Some subtriangular fins are adhered onto the outside surface of the spherical shell, in order to improve the robot's locomotion performance when navigating on the water.

The IDU is mounted completely in the inner space of the hollow spherical shell. Its mechanical structure mainly contains a main shaft, a support frame and a rotator. Two basic motion modes of the spherical robot, i.e., rolling motion and spinning motion, can be actuated by the IDU.

The main shaft lies along the sphere's main axis which is fixed relative to the spherical shell during motion, making it parallel to the ground in a stationary state. Two DC motors are mounted on the main shaft to actuate the IDU, named rolling motion motor and spinning motion motor, respectively. The rolling motion motor is placed along the main axis. When this motor starts to rotate, the IDU will depart from its balanced position and begin to rotate about the main axis. The torque exerted onto the spherical shell will accumulate as a consequence. The spherical shell starts to roll as soon as this torque exceeds the friction resistance, resulting in the linear motion of the whole spherical robot. The spinning motion motor is placed perpendicular to the sphere's main axis, and is along the diameter that passes through the contact point in a stationary state. A rotator is connected to this motor through a coupler. When the angular velocity of the rotator is changed by activating the spinning motion motor, the angular momentum of the rotator succeeds to change. As a result of the conservation of angular momentum principle, the spherical shell will spin in the direction that opposite to the change of angular momentum.

The principle of those two motion modes is illustrated in Figure 2.

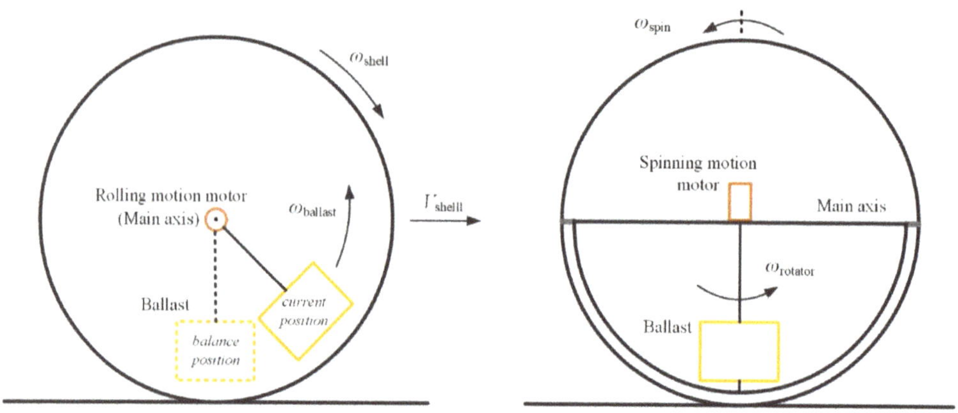

Figure 2. Schematic of rolling motion and spinning motion.

A curved trajectory can be achieved upon combining the rolling motion mode and spinning motion mode, i.e., upon steering the robot when it is moving forward or backward.

The rotator is placed in the interspace between the main shaft and the semi-circular support frame, which is fixed beneath the main shaft. The rotator is composed of a few pieces of ballasts, which are made by stainless steel in order to supply enormous mass and moment of inertial. In order to test the relationship between the depth of immersion and locomotion performance of the spherical robot, the number of ballasts is designed to be adjustable by splitting the ballasts into pieces with different mass. So, the total mass of the rotator can be simply changed by adding or removing some pieces of the ballasts.

Sixteen subtriangular fins, eight on each side, are fixed on the outside surface of the spherical shell to improve the locomotion performance on the water. By applying prudent control strategies, these fins will not have any collision with the ground when the robot is moving on land. By contrast, they will immerge into the water when the robot is navigating on the water. Two factors—the first is the number of fins attached to the spherical shell, and the second is the shape of the fin—are believed to have significant impact on the locomotion performance of the robot on the water.

Apart from those driving facilities mentioned above, the waterproof ability is of great importance for the amphibious spherical robot. Several approaches have been proposed to keep loads inside the spherical shell from water.

2.3. Electrical Actuation and Control System of the Robot

The electrical actuation and control system consist of two DC motors, a servo motor, two electronic speed controls (ESC) working as motor drive units, a signal receiver, a camera and power supply module. It provides the power and impetus needed for the spherical robot. It receives the control signal transmitted by the RF remote controller and the signal emitter, then outputs impetus to the mechanical structure of the IDU and the spherical shell. Its schematic is shown in Figure 3.

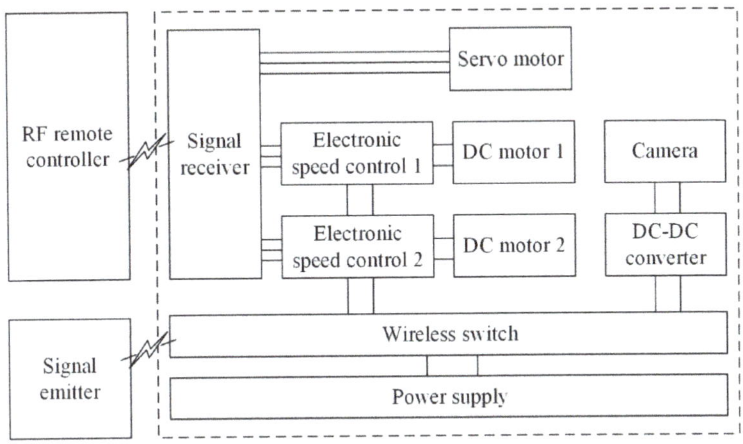

Figure 3. Schematic of electrical actuation and control system.

The signal receiver is used to collect radio frequency signals emitted from the RF remote controller. After some valid signals arrived, it will decode the received signals and then generate control signals for corresponding motors. The signal receiver connects to two DC motors through two ESCs respectively and one servo motor directly. These three motors occupied three of the nine output channels of the signal receiver. The power of the signal receiver is supplied by a battery eliminator circuit (BEC) incorporated in one of the ESCs. The ESC, mainly used for brushed DC motors, acts as motor driver in this application. It converts the digital signals from the signal receiver to voltages exerted upon the motors. Those ESCs are powered by the power supply module. The output voltage of the ESC will change with the variation of the input signal received from the signal receiver. As a result, rotation speed of these DC motors will change at the same time.

To record videos of the robot's surroundings, a digital camera is mounted inside the spherical shell. The servo motor mounted on the main shaft is used for adjusting the vision field of the camera in horizontal plane. This servo motor is connected directly to the signal receiver because it consumes power within the safety threshold of the signal receiver. As a micro wireless signal transmitter is integrated into the camera, videos produced by the camera can be transmitted to the RF remote controller in time.

A 24 V lithium battery is used as the power supply module for the robot, and a wireless switch is placed between the power supply module and those power consuming devices. The on-off state of the power supply is then achieved by sending corresponding signals to the wireless switch from an external signal emitter.

The control block diagram of the robot system is shown in Figure 4. Two PID controllers are implemented in order to control the two system inputs, i.e., τ_l and τ_r, generated by the rolling motion motor and the spinning motion motor, respectively.

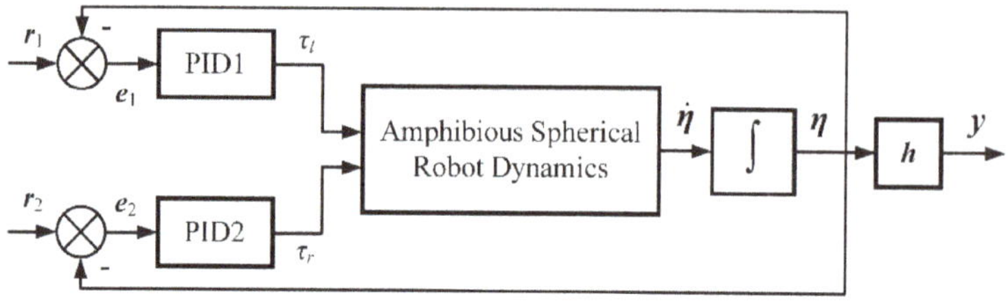

Figure 4. Control block diagram of the robot.

The PID1 controller is used in the basic rolling motion mode, while the PID2 controller is used in the basic spinning motion mode, when the robot needs to adjust its direction. These two controllers work in cooperation when the robot is operating in a curved trajectory.

3. Dynamic Model of the Robot

The amphibious spherical robot is modelled as a system consists of three rigid bodies, the spherical shell with its attached assistant fins, the IDU platform and the rotator. Assumptions about these three components are (1) the spherical shell is a uniform hollow spherical shell whose radius is R; the total mass of the spherical shell and its attached assistant fins is m_s while each fin has mass m_f; (2) the IDU platform is considered as a uniform rectangular bar with mass m_p, length $2R$, neglecting its structural details; it is mounted along the main axis with its center of mass locating at the center of spherical shell; (3) the rotator is a hollow cube with its mass m_r distributed on the four side faces of the cube; the distance between the center of mass of the rotator A and the main axis is l_{OA}.

3.1. Coordinate Frames

Coordinate frames attached to those three rigid bodies are shown in Figure 5. An inertial frame U-XYZ is set with its X axis pointing to the east of the locus, its Y axis pointing to the north and its Z axis pointing vertically upward. The frame O-XYZ is parallel to the inertial frame, with its origin located at the center of spherical shell O. Let O_S-$X_SY_SZ_S$ be the coordinate frame anchored to the spherical shell, whose origin is located at the point O and Y_S axis is always parallel to the main axis. This O_S-$X_SY_SZ_S$ frame is used to describe the configuration of the spherical shell. Let O_D-$X_DY_DZ_D$ be the body frame of the IDU platform, with its origin located also at point O, its Y_D axis coinciding with the Y_S axis and its Z_D axis parallel to the rotator's spinning axis. In addition, the A-$X_AY_AZ_A$ frame, located at the center of mass of the rotator, is used to describe the configuration of the rotator; its Z_A axis lies along the rotator's spinning axis.

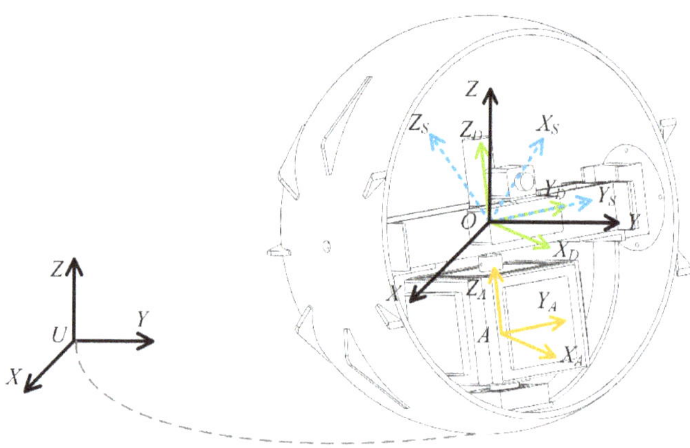

Figure 5. Coordinate frames of the robot.

In general, eight variables $\eta = (x, y, z, \phi, \theta, \psi, \alpha_s, \alpha_r)$ are needed to describe the states of the whole robot, among which, three variables (x, y, z) represent the center of the spherical shell, three Euler angles (ϕ, θ, ψ) represent the orientation of the IDU platform, the angle α_s represents the angle between the body frame of the spherical shell O_S-$X_SY_SZ_S$ and the body frame of the IDU platform O_D-$X_DY_DZ_D$; lastly, the angle α_r is the spinning angle of the rotator. As these eight variables are independent of each other, they are chosen as the generalized coordinates of the robot system.

3.2. Kinematics

Kinematic parameters that are necessary for deducing the dynamic equations, i.e., the translational velocities of the center of mass of each rigid body and the angular velocities with respect to each center of mass, are presented in the following. Let $\boldsymbol{r}_O = (x, y, z)$ be the position vector of the geometric center of the spherical shell O whose velocity is $\dot{\boldsymbol{r}}_O = (\dot{x}, \dot{y}, \dot{z})$, $\boldsymbol{r}_P = (x_P, y_P, z_P)$ be the position vector of the center of mass of the IDU platform whose velocity equals to $\dot{\boldsymbol{r}}_O$, i.e., $\dot{\boldsymbol{r}}_P = \dot{\boldsymbol{r}}_O$, and $\boldsymbol{r}_A = (x_A, y_A, z_A)$ be the position vector of the center of mass of the rotator A. Let ${}^D\boldsymbol{r}_{OA} = (0, 0, -l_{OA})$ denote the position vector of the point A with respect to the O_D-$X_DY_DZ_D$ frame; $\boldsymbol{r}_A$ then is

$$\boldsymbol{r}_A = \boldsymbol{r}_O + {}^U_D\boldsymbol{R}\,{}^D\boldsymbol{r}_{OA}, \tag{1}$$

where ${}^U_D\boldsymbol{R} \in SO(3)$ denotes the rotation matrix between the U-XYZ frame and the O_D-$X_DY_DZ_D$ frame, which can be expressed in terms of the Euler angles, that is,

$${}^U_D\boldsymbol{R} = \begin{bmatrix} \mathrm{c}\psi\mathrm{c}\theta & -\mathrm{s}\psi\mathrm{c}\phi + \mathrm{c}\psi\mathrm{s}\theta\mathrm{s}\phi & \mathrm{s}\psi\mathrm{s}\phi + \mathrm{c}\psi\mathrm{c}\phi\mathrm{s}\theta \\ \mathrm{s}\psi\mathrm{c}\theta & \mathrm{c}\psi\mathrm{c}\phi + \mathrm{s}\psi\mathrm{s}\theta\mathrm{s}\phi & -\mathrm{c}\psi\mathrm{s}\phi + \mathrm{s}\theta\mathrm{s}\psi\mathrm{c}\phi \\ -\mathrm{s}\theta & \mathrm{c}\theta\mathrm{s}\phi & \mathrm{c}\theta\mathrm{c}\phi \end{bmatrix}, \tag{2}$$

where $\mathrm{c}\phi$ is the shorthand for $\cos\phi$, $\mathrm{s}\phi$ for $\sin\phi$, and so on. Substituting ${}^U_D\boldsymbol{R}$ into Equation (1), we get

$$\boldsymbol{r}_A = (x - (\mathrm{s}\psi\mathrm{s}\phi + \mathrm{c}\psi\mathrm{c}\phi\mathrm{s}\theta)l_{OA}, y - (-\mathrm{c}\psi\mathrm{s}\phi + \mathrm{s}\theta\mathrm{s}\psi\mathrm{c}\phi)l_{OA}, z - \mathrm{c}\theta\mathrm{c}\phi l_{OA}). \tag{3}$$

The velocity of the rotator's center of mass $\dot{\boldsymbol{r}}_A$ can be obtained by differentiating $\boldsymbol{r}_A$, its three components are

$$\begin{aligned} \dot{x}_A &= \dot{x} - \left(\mathrm{c}\phi\mathrm{s}\psi\left(\dot{\phi} - \dot{\psi}\mathrm{s}\theta\right) + \mathrm{s}\phi\mathrm{c}\psi\left(\dot{\psi} - \dot{\phi}\mathrm{s}\theta\right) + \dot{\theta}\mathrm{c}\phi\mathrm{c}\psi\mathrm{c}\theta\right)l_{OA} \\ \dot{y}_A &= \dot{y} - \left(\mathrm{s}\phi\mathrm{s}\psi\left(\dot{\psi} - \dot{\phi}\mathrm{s}\theta\right) - \mathrm{c}\phi\mathrm{c}\psi\left(\dot{\phi} - \dot{\psi}\mathrm{s}\theta\right) + \dot{\theta}\mathrm{c}\phi\mathrm{s}\psi\mathrm{c}\theta\right)l_{OA} \\ \dot{z}_A &= \dot{z} + \left(\dot{\theta}\mathrm{c}\phi\mathrm{s}\theta + \dot{\phi}\mathrm{s}\phi\mathrm{c}\theta\right)l_{OA} \end{aligned} \tag{4}$$

As the rotation matrix ${}^U_D\boldsymbol{R}$ is an orthonormal matrix, ${}^U_D\boldsymbol{R}^{-1} = {}^U_D\boldsymbol{R}^{\mathrm{T}}$, so

$${}^U_D\boldsymbol{R}\,{}^U_D\boldsymbol{R}^{\mathrm{T}} = \boldsymbol{I}_{3\times3}. \tag{5}$$

Differentiating both sides of Equation (5), we have

$${}^U_D\dot{\boldsymbol{R}}\,{}^U_D\boldsymbol{R}^{\mathrm{T}} + \left({}^U_D\dot{\boldsymbol{R}}\,{}^U_D\boldsymbol{R}^{\mathrm{T}}\right)^{\mathrm{T}} = 0. \tag{6}$$

If we define $\boldsymbol{S}_p := {}^U_D\dot{\boldsymbol{R}}\,{}^U_D\boldsymbol{R}^{\mathrm{T}}$, then $\boldsymbol{S}_p + \boldsymbol{S}_p^{\mathrm{T}} = 0$. So $\boldsymbol{S}_p \in so(3)$ is an anti-symmetric matrix whose diagonal elements are all zero:

$$\boldsymbol{S}_p = \begin{bmatrix} 0 & -\dot{\psi} + \dot{\phi}\mathrm{s}\theta & \dot{\theta}\mathrm{c}\psi + \dot{\phi}\mathrm{s}\psi\mathrm{c}\theta \\ \dot{\psi} - \dot{\phi}\mathrm{s}\theta & 0 & -\dot{\phi}\mathrm{c}\psi\mathrm{c}\theta + \dot{\theta}\mathrm{s}\psi \\ -\dot{\theta}\mathrm{c}\psi - \dot{\phi}\mathrm{s}\psi\mathrm{c}\theta & \dot{\phi}\mathrm{c}\psi\mathrm{c}\theta - \dot{\theta}\mathrm{s}\psi & 0 \end{bmatrix}. \tag{7}$$

The matrix $\boldsymbol{S}_p$ can be made up from $\boldsymbol{\omega}_p = (\omega_{px}, \omega_{py}, \omega_{pz})$, the angular velocity of the IDU platform with respect to the U-XYZ frame, that is,

$$\boldsymbol{S}_p = \hat{\boldsymbol{\omega}}_p = \begin{bmatrix} 0 & -\omega_{pz} & \omega_{py} \\ \omega_{pz} & 0 & -\omega_{px} \\ -\omega_{py} & \omega_{px} & 0 \end{bmatrix}. \tag{8}$$

The angular velocity $\boldsymbol{\omega}_p$ then can be deduced from $\boldsymbol{S}_p$ reversely:

$$\boldsymbol{\omega}_p = \left(\dot{\phi}\mathrm{c}\psi\mathrm{c}\theta - \dot{\theta}\mathrm{s}\psi, \dot{\theta}\mathrm{c}\psi + \dot{\phi}\mathrm{s}\psi\mathrm{c}\theta, \dot{\psi} - \dot{\phi}\mathrm{s}\theta\right). \tag{9}$$

Note that $\boldsymbol{\omega}_p$ is not a generalized velocity but a quasi-velocity. The angular velocity of the IDU platform with respect to the O_D-$X_DY_DZ_D$ frame can be obtained by coordinate transformation

$${}^{D}\boldsymbol{\omega}_p = {}^{U}_{D}\boldsymbol{R}^{\mathrm{T}}\boldsymbol{\omega}_p = \left(\dot{\phi} - \dot{\psi}\mathrm{s}\theta, \dot{\theta}\mathrm{c}\phi + \dot{\psi}\mathrm{s}\phi\mathrm{c}\theta, -\dot{\theta}\mathrm{s}\phi + \dot{\psi}\mathrm{c}\phi\mathrm{c}\theta\right). \tag{10}$$

To deduce the angular velocity of the spherical shell, let $\boldsymbol{S}_s := {}^{U}_{S}\dot{\boldsymbol{R}}\,{}^{U}_{S}\boldsymbol{R}^{\mathrm{T}}$, where ${}^{U}_{S}\boldsymbol{R}$ is the rotation matrix between the U-XYZ frame and the O_S-$X_SY_SZ_S$ frame. ${}^{U}_{S}\boldsymbol{R}$ can be obtained by matrix multiplication, ${}^{U}_{S}\boldsymbol{R} = {}^{U}_{D}\boldsymbol{R}\,{}^{D}_{S}\boldsymbol{R}$, where ${}^{D}_{S}\boldsymbol{R}$ is the rotation matrix between the O_D-$X_DY_DZ_D$ frame and the O_S-$X_SY_SZ_S$ frame. ${}^{D}_{S}\boldsymbol{R}$ can be expressed as

$${}^{D}_{S}\boldsymbol{R} = \begin{bmatrix} \mathrm{c}\alpha_s & 0 & \mathrm{s}\alpha_s \\ 0 & 1 & 0 \\ -\mathrm{s}\alpha_s & 0 & \mathrm{c}\alpha_s \end{bmatrix}. \tag{11}$$

Substituting ${}^{D}_{S}\boldsymbol{R}$ into the equation of ${}^{U}_{S}\boldsymbol{R}$, ${}^{U}_{S}\boldsymbol{R}$ then is

$${}^{U}_{S}\boldsymbol{R} = \begin{bmatrix} \mathrm{c}\psi\mathrm{c}\theta\mathrm{c}\alpha_s - (\mathrm{s}\psi\mathrm{s}\phi + \mathrm{c}\psi\mathrm{c}\phi\mathrm{s}\theta)\mathrm{s}\alpha_s & \mathrm{c}\psi\mathrm{s}\theta\mathrm{s}\phi - \mathrm{s}\psi\mathrm{c}\phi & \mathrm{c}\psi\mathrm{c}\theta\mathrm{s}\alpha_s + (\mathrm{s}\psi\mathrm{s}\phi + \mathrm{c}\psi\mathrm{c}\phi\mathrm{s}\theta)\mathrm{c}\alpha_s \\ \mathrm{s}\psi\mathrm{c}\theta\mathrm{c}\alpha_s + (\mathrm{c}\psi\mathrm{s}\phi - \mathrm{s}\theta\mathrm{s}\psi\mathrm{c}\phi)\mathrm{s}\alpha_s & \mathrm{c}\psi\mathrm{c}\phi + \mathrm{s}\psi\mathrm{s}\theta\mathrm{s}\phi & \mathrm{s}\psi\mathrm{c}\theta\mathrm{s}\alpha_s + (\mathrm{s}\theta\mathrm{s}\psi\mathrm{c}\phi - \mathrm{c}\psi\mathrm{s}\phi)\mathrm{c}\alpha_s \\ -\mathrm{s}\theta\mathrm{c}\alpha_s - \mathrm{c}\theta\mathrm{c}\phi\mathrm{s}\alpha_s & \mathrm{c}\theta\mathrm{s}\phi & \mathrm{c}\theta\mathrm{c}\phi\mathrm{c}\alpha_s - \mathrm{s}\theta\mathrm{s}\alpha_s \end{bmatrix}. \tag{12}$$

The angular velocity of the spherical shell with respect to the U-XYZ frame, $\boldsymbol{\omega}_s = (\omega_{sx}, \omega_{sy}, \omega_{sz})$, can be deduced from the matrix $\boldsymbol{S}_s$, whose three components are

$$\begin{aligned} \omega_{sx} &= \dot{\phi}\mathrm{c}\psi\mathrm{c}\theta - \dot{\theta}\mathrm{s}\psi + \dot{\alpha}_s(\mathrm{c}\psi\mathrm{s}\theta\mathrm{s}\phi - \mathrm{c}\phi\mathrm{s}\psi) \\ \omega_{sy} &= \dot{\theta}\mathrm{c}\psi + \dot{\phi}\mathrm{s}\psi\mathrm{c}\theta + \dot{\alpha}_s(\mathrm{s}\psi\mathrm{s}\theta\mathrm{s}\phi + \mathrm{c}\phi\mathrm{c}\psi) \\ \omega_{sz} &= \dot{\psi} - \dot{\phi}\mathrm{s}\theta + \dot{\alpha}_s\mathrm{c}\theta\mathrm{s}\phi \end{aligned} \tag{13}$$

Its expression with respect to the O_S-$X_SY_SZ_S$ frame are

$$\begin{aligned} {}^{S}\omega_{sx} &= \dot{\theta}\mathrm{s}\phi\mathrm{s}\alpha_s + \dot{\phi}\mathrm{c}\alpha_s - \dot{\psi}(\mathrm{s}\theta\mathrm{c}\alpha_s + \mathrm{c}\theta\mathrm{c}\phi\mathrm{s}\alpha_s) \\ {}^{S}\omega_{sx} &= \dot{\theta}\mathrm{c}\phi + \dot{\psi}\mathrm{c}\theta\mathrm{s}\phi + \dot{\alpha}_s \\ {}^{S}\omega_{sz} &= \dot{\phi}\mathrm{s}\alpha_s - \dot{\theta}\mathrm{s}\phi\mathrm{c}\alpha_s + \dot{\psi}(\mathrm{c}\theta\mathrm{c}\phi\mathrm{c}\alpha_s - \mathrm{s}\theta\mathrm{s}\alpha_s) \end{aligned} \tag{14}$$

Then, let $\boldsymbol{S}_a := {}^{U}_{A}\dot{\boldsymbol{R}}\,{}^{U}_{A}\boldsymbol{R}^{\mathrm{T}}$, where ${}^{U}_{A}\boldsymbol{R}$ is the rotation matrix between the U-XYZ frame and the A-$X_AY_AZ_A$ frame and ${}^{U}_{A}\boldsymbol{R} = {}^{U}_{D}\boldsymbol{R}\,{}^{U}_{A}\boldsymbol{R}$. ${}^{U}_{D}\boldsymbol{R}$ is the rotation matrix between the O_D-$X_DY_DZ_D$ frame and the A-$X_AY_AZ_A$ frame, which is

$${}^{D}_{A}\boldsymbol{R} = \begin{bmatrix} \mathrm{c}\alpha_r & -\mathrm{s}\alpha_r & 0 \\ \mathrm{s}\alpha_r & \mathrm{c}\alpha_r & 0 \\ 0 & 0 & 1 \end{bmatrix}. \tag{15}$$

Substituting ${}^{D}_{A}\boldsymbol{R}$ into the equation of ${}^{U}_{A}\boldsymbol{R}$, we get

$$
{}^{U}_{A}\boldsymbol{R} = \begin{bmatrix} \mathrm{c}\psi\mathrm{c}\theta\mathrm{c}\alpha_s - (\mathrm{s}\psi\mathrm{s}\phi + \mathrm{c}\psi\mathrm{c}\phi\mathrm{s}\theta)\mathrm{s}\alpha_s & \mathrm{c}\psi\mathrm{s}\theta\mathrm{s}\phi - \mathrm{s}\psi\mathrm{c}\phi & \mathrm{c}\psi\mathrm{c}\theta\mathrm{s}\alpha_s + (\mathrm{s}\psi\mathrm{s}\phi + \mathrm{c}\psi\mathrm{c}\phi\mathrm{s}\theta)\mathrm{c}\alpha_s \\ \mathrm{s}\psi\mathrm{c}\theta\mathrm{c}\alpha_s + (\mathrm{c}\psi\mathrm{s}\phi - \mathrm{s}\theta\mathrm{s}\psi\mathrm{c}\phi)\mathrm{s}\alpha_s & \mathrm{c}\psi\mathrm{c}\phi + \mathrm{s}\psi\mathrm{s}\theta\mathrm{s}\phi & \mathrm{s}\psi\mathrm{c}\theta\mathrm{s}\alpha_s + (\mathrm{s}\theta\mathrm{s}\psi\mathrm{c}\phi - \mathrm{c}\psi\mathrm{s}\phi)\mathrm{c}\alpha_s \\ -\mathrm{s}\theta\mathrm{c}\alpha_s - \mathrm{c}\theta\mathrm{c}\phi\mathrm{s}\alpha_s & \mathrm{c}\theta\mathrm{s}\phi & \mathrm{c}\theta\mathrm{c}\phi\mathrm{c}\alpha_s - \mathrm{s}\theta\mathrm{s}\alpha_s \end{bmatrix}. \tag{16}
$$

The angular velocity of the rotator with respect to the U-XYZ frame, $\boldsymbol{\omega}_r = (\omega_{rx}, \omega_{ry}, \omega_{rz})$, can be deduced from the matrix $\boldsymbol{S}_a$, its three components are

$$
\begin{aligned} \omega_{rx} &= \dot{\phi}\mathrm{c}\psi\mathrm{c}\theta - \dot{\theta}\mathrm{s}\psi + \dot{\alpha}_r(\mathrm{c}\psi\mathrm{s}\theta\mathrm{c}\phi + \mathrm{s}\phi\mathrm{s}\psi) \\ \omega_{ry} &= \dot{\theta}\mathrm{c}\psi + \dot{\phi}\mathrm{s}\psi\mathrm{c}\theta + \dot{\alpha}_r(\mathrm{s}\psi\mathrm{s}\theta\mathrm{c}\phi - \mathrm{s}\phi\mathrm{c}\psi) \\ \omega_{rz} &= \dot{\psi} - \dot{\phi}\mathrm{s}\theta + \dot{\alpha}_r\mathrm{c}\theta\mathrm{c}\phi \end{aligned} \tag{17}
$$

Its expression with respect to the A-$X_AY_AZ_A$ frame are

$$
\begin{aligned} {}^{A}\omega_{rx} &= \dot{\theta}\mathrm{c}\phi\mathrm{s}\alpha_r + \dot{\phi}\mathrm{c}\alpha_r + \dot{\psi}(\mathrm{c}\theta\mathrm{s}\phi\mathrm{s}\alpha_r - \mathrm{s}\theta\mathrm{c}\alpha_r) \\ {}^{A}\omega_{ry} &= \dot{\theta}\mathrm{c}\phi\mathrm{c}\alpha_r - \dot{\phi}\mathrm{s}\alpha_r + \dot{\psi}(\mathrm{c}\theta\mathrm{s}\phi\mathrm{c}\alpha_r + \mathrm{s}\theta\mathrm{s}\alpha_r) \\ {}^{A}\omega_{rz} &= \dot{\psi}\mathrm{c}\theta\mathrm{c}\phi - \dot{\theta}\mathrm{s}\phi + \dot{\alpha}_r \end{aligned} \tag{18}
$$

The above kinematic analyses determine parameters that are important in dynamic modeling, i.e., the velocity $\dot{\boldsymbol{r}}_O, \dot{\boldsymbol{r}}_A$ as well as the angular velocity ${}^{D}\omega_p, {}^{S}\omega_s$ and ${}^{A}\omega_r$.

3.3. Generalized Forces

The equations of motion of the amphibious spherical robot are derived based on force analyses when the robot is navigating on the water, as shown in Figure 6. The coordinate frame O-XYZ is set with its origin located at the center of the spherical shell O and its Y axis pointing outward the paper.

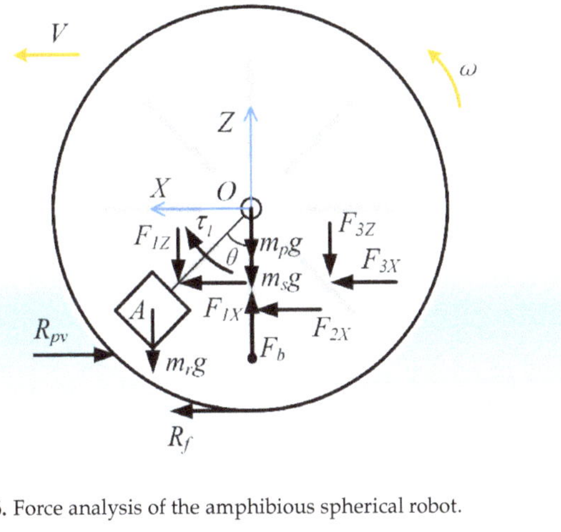

Figure 6. Force analysis of the amphibious spherical robot.

Forces exerted on the spherical shell are mainly due to interactions between the fluid and the spherical shell. It includes the gravity of the spherical shell $m_s\boldsymbol{g}$, the buoyancy $\boldsymbol{F}_b$, the hydrodynamic force $\boldsymbol{F}_{hyd}$, the resultant force exerted on all fins $\boldsymbol{F}_{tf}$ as well as their corresponding torques. $\boldsymbol{F}_{hyd}$ consists of two main components, i.e., the frictional resistance R_f and the pressure resistance R_{pv}. The forces exerted on the IDU platform and the rotator are simply their gravities, $m_p\boldsymbol{g}$ and $m_r\boldsymbol{g}$, respectively. Two input torques τ_l and τ_r are generated from the rolling motion motor and the spinning motion motor.

The buoyancy $\boldsymbol{F}_b$, the frictional resistance R_f, the pressure resistance R_{pv} and the resultant force exerted on all fins $\boldsymbol{F}_{tf}$ are described in detail in the following.

(a) Buoyancy

The buoyancy exerted on the spherical shell equals the gravity of the water which has the same volume as the immerged part of the spherical shell. If we set the generalized coordinate $z = 0$ and let H_0 represents the height from the bottom of the spherical shell to the surface of water when the robot is in static equilibrium on the water, then the buoyancy is

$$F_b = \rho_w g\pi\left(R(H_0 - z)^2 - (H_0 - z)^3/3\right), \tag{19}$$

where ρ_w is the density of water, g is the gravitational acceleration. The center of buoyancy lies vertically below the center of sphere, its z coordinate is

$$z_b = z - \frac{3(H_0 - z - 2R)^2}{4(H_0 - z - 3R)}. \tag{20}$$

(b) Frictional Resistance

In contrast to the pure rolling motion on the ground, the amphibious spherical robot slides relative to the fluid surface when rolling on the water, as a consequence of poor frictional condition at the contact region. The frictional resistance between the spherical shell and the fluid can be derived from the theory adapted to a ship [28], assuming that it is equal to the frictional resistance on a flat plate that is equivalent to the immerged spherical shell. It should be noted that the direction of the frictional resistance on the spherical shell is the same as the robot's moving direction, instead of being in the opposite direction in the case of a ship. As a result, the frictional resistance acts as propulsion for the amphibious spherical robot.

The frictional resistance exerted on the spherical shell R_f can be written as

$$R_f = \left(C_f + \Delta C_f\right)\cdot\frac{1}{2}\rho_w V_{equ}^2 S_w, \tag{21}$$

where C_f is the frictional resistance coefficient of an equivalent flat plate corresponding to the spherical shell; ΔC_f is the roughness coefficient, determined by the roughness of the spherical shell in contact with the water; V_{equ} is the equivalent velocity of the robot relative to the water. S_w is the wetted surface area, which can be written as

$$S_w = \int_0^{H_0} 2\pi\sqrt{R^2 - (h - R)^2}\cdot\sqrt{1 + \frac{(h - R)^2}{R^2 - (h - R)^2}}dh = 2\pi R H_0. \tag{22}$$

The frictional resistance coefficient is determined by the Reynolds number Re, namely,

$$\mathrm{Re} = \frac{V_{equ}\cdot L_{wl}}{\nu}, \tag{23}$$

where ν is the kinematic viscosity of water; L_{wl} represents the waterline length of the spherical shell,

$$L_{wl} = 2 \times \sqrt{2RH_0 - H_0^2}. \tag{24}$$

An estimated value of Re for the amphibious spherical robot described in this paper is about 1.45×10^5 given that $V_{equ} = 0.5$ m/s, $L_{wl} = 0.33$ m, the kinematic viscosity $\nu = 1.139 \times 10^{-6}$ m^2/s, and the temperature of the water is 15 °C. This Re value is less than the critical Reynolds number of a flat plate, which is approximately 1.0×10^6. Therefore, the water flow relative to the spherical shell can be presumed as laminar flow, which is characterized by smooth, constant fluid motion. The frictional resistance coefficient C_f is then given as [29]

$$C_f = 1.328\mathrm{Re}^{-\frac{1}{2}}. \tag{25}$$

As the outside surface of the spherical shell is smooth, the roughness of the spherical shell barely contributes to the total resistance, so the roughness coefficient ΔC_f is set as zero in the calculation.

Solving Equations (21)–(25), we have

$$R_f = 1.328\pi\sqrt{\frac{\nu V_{equ}^3}{L_{wl}}}\rho_w R H_0. \tag{26}$$

Using Equation (26), we can obtain a roughly estimate of the frictional resistance of the spherical shell in terms of its specifications. To get an accurate value of the frictional resistance, experiment or Computational fluid dynamics (CFD) methods are necessary, which, however, are not the emphasis here.

(c) Pressure Resistance

The hydrodynamic force exerted on the spherical shell consists of the frictional resistance and the pressure resistance. The pressure resistance is due to the decrease of pressure between the front of the spherical shell and its rear. It is difficult to separate the frictional resistance and pressure resistance theoretically. However, researchers have carried out CFD simulations and experiments to measure the resultant of those two resistances [30,31]. So, it will be possible to evaluate the pressure resistance approximately by subtracting the frictional resistance from the resultant resistance in a somewhat simple way.

The hydrodynamic force exerted on the spherical shell then can be presented in the O-XYZ frame as ${}^{O}\boldsymbol{F}_{hyd} = (F_{hX}, F_{hY}, F_{hZ})$, its corresponding torque relative to the center of sphere is ${}^{O}\boldsymbol{\tau}_{hyd} = (\tau_{hX}, \tau_{hY}, \tau_{hZ})$.

(d) Resultant Force Exerted on the Fins

According to Healey et al. [32], when a fin moves relative to the water, a drag force and a lift force are exerted on the fin. The drag force D is in line with the inlet water, whose velocity is ${}^{f}V_w$, and the lift force L is perpendicular to the inlet water. They can be calculated as follows [32]

$$L = \frac{1}{2}\rho_w\, {}^{f}V_w^2 S_f C_{L\max}\sin(2\alpha_e), \tag{27}$$

$$D = \frac{1}{2}\rho_w\, {}^{f}V_w^2 S_f C_{D\max}(1-\cos(2\alpha_e)), \tag{28}$$

where S_f is the cross-section area of the fin submerged in the water; the angle of attack α_e is defined as the angle between the inflow water's velocity and the cross-section plane of the fin. The maximum lift coefficient $C_{L\max}$ is determined by the shape of the fin. $C_{D\max}$ is the maximum drag coefficient. Both these two coefficients can be acquired by experiments.

The components of the drag force and the lift force projected in the horizontal direction and vertical direction can be calculated as:

$$F_{iX} = -L\sin(\varphi+\alpha_e) + D\cos(\varphi+\alpha_e), \tag{29}$$

$$F_{iZ} = L\cos(\varphi+\alpha_e) + D\sin(\varphi+\alpha_e), \tag{30}$$

where F_{iX} denotes the components of the ith fin in the X direction, F_{iZ} denotes the components of the ith fin in the Z direction, φ is the phase angle of the fin in its rotation cycle.

The resultant force exerted on all those fins submerged in the water is the sum of forces exerted on each fin, and it can be presented in the O-XYZ frame as ${}^{O}\boldsymbol{F}_{tf} = \left(F_{fX}, F_{fY}, F_{fZ}\right)$; its corresponding torque relative to the center of sphere is ${}^{O}\boldsymbol{\tau}_{tf} = \left(\tau_{fX}, \tau_{fY}, \tau_{fZ}\right)$.

After determining all the applied forces upon the spherical robot, it is necessary to calculate the generalized force due to nonconservative forces before we obtain the equations of motion. This will be accomplished by the principle of virtual work. The virtual work δW_j corresponding to one generalized coordinate can be calculate by multiplying the applied forces with the virtual displacement. The generalized forces can be obtained by dividing the virtual work by its corresponding generalized coordinate, that is,

$$Q_j = \delta W_j / \delta \eta_j. \tag{31}$$

3.4. Equations of Motion

The equations of the spherical robot's motion when it navigates on the surface of water are deduced by Lagrange's methods. When evaluating the forces exerted on the robot's components, forces that can be obtained from a potential function are conservative, and should be formulated into the potential energy. Forces associated with energy dissipation are nonconservative and should be formulated into the generalized forces. First, we need to construct the Lagrangian by calculating the kinetic energy and the potential energy of those three components of the robot.

The kinetic energy of the spherical shell is the sum of the energy due to the translational motion of its center of mass and the energy due to rotation about the center of mass, that is,

$$T_s = \frac{1}{2}m_s\left(\dot{x}^2 + \dot{y}^2 + \dot{z}^2\right) + \frac{1}{2}\left(I_{sXX}\,{}^S\omega_{sx}{}^2 + I_{sYY}\,{}^S\omega_{sy}{}^2 + I_{sZZ}\,{}^S\omega_{sz}{}^2\right), \tag{32}$$

where I_{sXX}, I_{sYY} and I_{sZZ} are moments of inertia of the spherical shell about its three axes of body frame O_S-$X_SY_SZ_S$, respectively. As the spherical shell is symmetrical, those three axes are its principle axes. The total number of fins attached to the spherical shell is $2n$ with n fins on each hemisphere. Each of those attached fins can be reckoned as with mass m_f and radius of gyration $R/2$ approximately, so I_{sXX}, I_{sYY} and I_{sZZ} can be written as

$$I_{sXX} = \frac{2}{3}m_sR^2 - \frac{5}{6}nm_fR^2,\ I_{sYY} = \frac{2}{3}m_sR^2 - \frac{5}{6}nm_fR^2,\ I_{sZZ} = \frac{2}{3}m_sR^2 - \frac{5}{6}nm_fR^2.$$

The gravity and the buoyancy on the spherical shell are related only to the generalized coordinates so that they are conservative, then the potential energy of the spherical shell is

$$V_s = m_sgz + \rho_w g\pi\left(-R(H_0 - z)^3/3 + (H_0 - z)^4/12\right). \tag{33}$$

The first term in the right-hand side of the above equation is potential energy corresponding to the gravity, and the second term is the potential energy arising from the buoyancy.

Similarly, the kinetic energy of the IDU platform is

$$T_p = \frac{1}{2}m_p\left(\dot{x}_P^2 + \dot{y}_P^2 + \dot{z}_P^2\right) + \frac{1}{2}\left(I_{pXX}\,{}^D\omega_{px}{}^2 + I_{pYY}\,{}^D\omega_{py}{}^2 + I_{pZZ}\,{}^D\omega_{pz}{}^2\right), \tag{34}$$

where I_{pXX}, I_{pYY} and I_{pZZ} are moments of inertia of the IDU platform about its three axes of body frame O_D-$X_DY_DZ_D$, respectively. They can be obtained by referring to the cuboid model

$$I_{pXX} = \frac{1}{12}m_p\left(4R^2 + l_p^2\right),\ I_{pYY} = \frac{1}{6}m_pl_p^2,\ I_{pZZ} = \frac{1}{12}m_p\left(4R^2 + l_p^2\right).$$

The potential energy of the IDU platform is

$$V_p = m_pgz_P. \tag{35}$$

The kinetic energy of the rotator is

$$T_r = \frac{1}{2}m_r\left(\dot{x}_A^2 + \dot{y}_A^2 + \dot{z}_A^2\right) + \frac{1}{2}\left(I_{rXX}\,{}^A\omega_{rx}{}^2 + I_{rYY}\,{}^A\omega_{ry}{}^2 + I_{rZZ}\,{}^A\omega_{rz}{}^2\right), \tag{36}$$

where I_{rXX}, I_{rYY} and I_{rZZ} are moments of inertia of the rotator about its three axes of body frame A-$X_AY_AZ_A$, respectively. As the mass of the rotator is assumed to be distributed on the four side faces of the cube, each of which is a square with length l_r, so these moments of inertia are

$$I_{rXX} = \frac{1}{4}m_r l_r^2,\ I_{rYY} = \frac{1}{4}m_r l_r^2,\ I_{rZZ} = \frac{1}{3}m_r l_r^2.$$

The potential energy of the rotator is

$$V_r = m_r g z_A. \tag{37}$$

After determining all those kinetic energies and potential energies associated with the robot's three components, it is ready to construct the Lagrangian of the robot system

$$L = T_s - V_s + T_p - V_p + T_r - V_r. \tag{38}$$

As those eight generalized coordinates describing the states of the robot are independent of each other, the equations of motion of the robot system can be obtained from the Lagrange's equations, that is,

$$\frac{d}{dt}\left(\frac{\partial L}{\partial \dot{\eta}_j}\right) + \frac{\partial \Psi}{\partial \dot{\eta}_j} - \frac{\partial L}{\partial \eta_j} = Q_j, \tag{39}$$

where η_j denotes one of the generalized coordinates, Q_j denotes its corresponding generalized force due to the nonconservative forces and Ψ denotes the viscous damping effect between components of the robot.

The full dynamic equations are lengthy and will not be listed here for simplicity. Instead, its simplifications are presented in the following section in order to reveal its dynamic characteristics.

3.5. Model Simplification

The dynamic equations indicate that the amphibious robot is a highly coupled nonlinear system, which makes it difficult to analyze and control. To make the dynamic model practicable, a method for decoupling is applied, by separating the model into two basic motion modes, i.e., the rolling motion mode and the spinning motion mode.

The dynamic equations, in its original form, represent a highly underactuated system with two inputs and eight state variables. The system has one input τ_l and three state variables (x, θ, α_s) in the basic rolling motion mode, while it bears one input τ_r and two state variables (ψ, α_r) in the spinning motion mode; in a combined motion mode of these two, three extra variables (y, z, ϕ) exist. The focus here is on the linear motion of the robot when navigating on the water. In this case, the dynamic model can be described by just three generalized coordinates $\boldsymbol{\eta}' = (x, \theta, \alpha_s)$ when the others are set to zero. The Lagrangian then can be written as

$$\begin{aligned} L = &\tfrac{1}{2}(m_s + m_p + m_r)\dot{x}^2 + \tfrac{1}{2}(I_{pYY} + I_{rYY} + m_r l_{OA}^2)\dot{\theta}^2 + m_r g l_{OA} c\theta \\ &- m_r l_{OA} c\theta \dot{x}\dot{\theta} + \tfrac{1}{2} I_{sYY}\left(\dot{\theta} + \dot{\alpha}_s\right)^2 + \tfrac{1}{12}\rho_w g \pi (4R - H_0) H_0^3 \end{aligned}. \tag{40}$$

Substituting the above Lagrangian into the Lagrange's equations, and let $\Psi = \zeta \dot{\alpha}_s^2/2$ be the viscous damping effect between the spherical shell and the IDU platform where ζ denotes viscous damping coefficient, we obtain the simplified equations of motion of linear motion,

$$\boldsymbol{M}(\boldsymbol{\eta})\ddot{\boldsymbol{\eta}} + \boldsymbol{C}(\boldsymbol{\eta})\dot{\boldsymbol{\eta}} + \boldsymbol{K}(\boldsymbol{\eta}) = \boldsymbol{\tau}, \tag{41}$$

where

$$M(\eta) = \begin{bmatrix} m_s + m_p + m_r & -m_r g l_{OA} c\theta & 0 \\ -m_r g l_{OA} c\theta & I_{pYY} + I_{rYY} + I_{sYY} + m_r l_{OA}^2 & I_{sYY} \\ 0 & I_{sYY} & I_{sYY} \end{bmatrix}, C(\eta) = \begin{bmatrix} 0 & m_r l_{OA} s\theta\dot{\theta} & 0 \\ 0 & 0 & 0 \\ 0 & 0 & \zeta \end{bmatrix},$$

$$K(\eta) = \begin{bmatrix} 0 \\ m_r l_{OA} g s\theta \\ 0 \end{bmatrix}, \tau = \begin{bmatrix} F_{hX} + F_{fX} \\ \tau_{hY} + \tau_{fY} \\ \tau_{hY} + \tau_{fY} + \tau_l \end{bmatrix}.$$

The generalized forces τ are determined by Equation (31). The elements of τ are corresponding to the generalized coordinates x, θ and α_s, respectively, where F_{hX}, consisting of the frictional resistance and the pressure resistance, is the hydrodynamic force exerted on the spherical shell, F_{fX} is the resultant force exerted on the fins.

It can be concluded from Equation (41) that attaching assistant fins to the outer spherical shell will increase the propulsion force of the amphibious spherical robot. As a result, it will increase the acceleration of the robot, especially in the accelerating process when the robot starts from stationary. This can be validated by experiments presented in the following section. Since the magnitude of the propulsion force induced by one assistant fin correlate positively to the cross-section area of the fin, and the overall effect of all the fins are the summation of propulsion forces exerted on those fins immersed into the water, it will be helpful to use larger fins or more fins in order to improve the performance of the amphibious spherical robot.

It is noteworthy that the forces exerted on the spherical shell by the water are estimated by empirical equations which significantly depends on their corresponding coefficients. Determining the accurate values of these coefficients is important in order to obtain a precise model for the amphibious spherical robot. However, this is not the emphasis here so it is put off for future research.

4. Simulation and Experiments

4.1. Simulation

Simulations were carried out in MATLAB in order to reveal characteristics of the dynamic model. The zero-state response under a prescribed input of the rolling motion motor is provided. Parameters used in the simulation are $m_s = 1.6$ kg, $m_p = 2$ kg, $m_r = 2.2$ kg, $\zeta = 0.02$, $l_{OA} = 0.08$ m, $I_{sYY} = 0.0308$ kg·m^2, $I_{pYY} = 0.00053$ kg·m^2, $I_{rYY} = 0.0045$ kg·m^2, $\tau_l = 0.4$ Nm, $\tau_r = 0$ Nm, $C_{L\max} = 0.92$, $C_{D\max} = 1.12$, respectively. The frictional resistance exerted on the spherical shell is $R_f = 0.04$ N when the equivalent velocity is set as $V_{equ} = 0.4$ m/s.

The time-varying states of those three generalized coordinates are shown in Figure 7.

The simulation results show that the linear velocity of the robot $\dot{x}$, the angle of the IDU platform θ and the angular velocity of the spherical shell $\dot{\alpha}_s$ proceed to a prescribed value in an oscillation manner as time increases, while the angular velocity of the IDU platform $\dot{\theta}$ oscillates and then attenuates to zero; the travel distance x increases in a nearly linear manner, when $\zeta = 0.02$. If $\zeta = 0$, oscillation will be magnified for $\dot{x}$, θ, $\dot{\theta}$ and $\dot{\alpha}_s$, and θ, α_s, $\dot{\alpha}_s$ will have a large amplitude while x keeps the same.

It can be concluded from the simulation results that a steady linear speed can be achieved by increasing the viscous damping between the spherical shell and the IDU platform; moreover, a less oscillating IDU platform can be achieved at the same time, which is favorable in terms of the performance of sensors and cameras mounted on the IDU.

Simulations were carried out to evaluate the effects of two parameters on the robot's steady-state velocity when operating on the water, i.e., m_r and ΔC_f, the adjustable rotator mass and the roughness coefficient of the spherical shell. Figure 8a shows the velocity curve when the rotator mass is set as $0.8m_r$, m_r and $1.2m_r$, respectively, while Figure 8b shows the velocity curve when the roughness coefficient is set as $0R_f$, $0.5R_f$ and $1.5R_f$, respectively, with m_r representing the nominal mass of the rotator and R_f the nominal frictional resistance.

The results indicate that, the reduction of the rotator mass to $0.8m_r$ results in lower steady-state velocity of the robot, while the increase of the rotator mass has no apparent effect. It is observed that, the spherical shell floats upward when the rotator mass is reduced, leading to lower area of fins immersed under water, thereby reducing the forces exerted on the fins, affecting the steady-state velocity.

On the other hand, the spherical shell dives downward when the rotator mass is increased, which leads to larger area of fins immersed under water; however, this effect is found to be less significant, since the water line is approaching the great circle of the spherical shell that parallels to the surface of water. The wetted surface area increases at the same time, leading to increased hydrodynamic forces because it is positively correlated to the wetted surface area.

As a matter of fact, the current value of m_r was chosen such that it roughly maximizes the average area of the fin under water, under the condition that the mass of the robot is kept reasonably low, which is the cause of the behavior observed here.

As for the roughness coefficient, it is found that it has no apparent effect, because the proportion of roughness in the hydrodynamic forces is low.

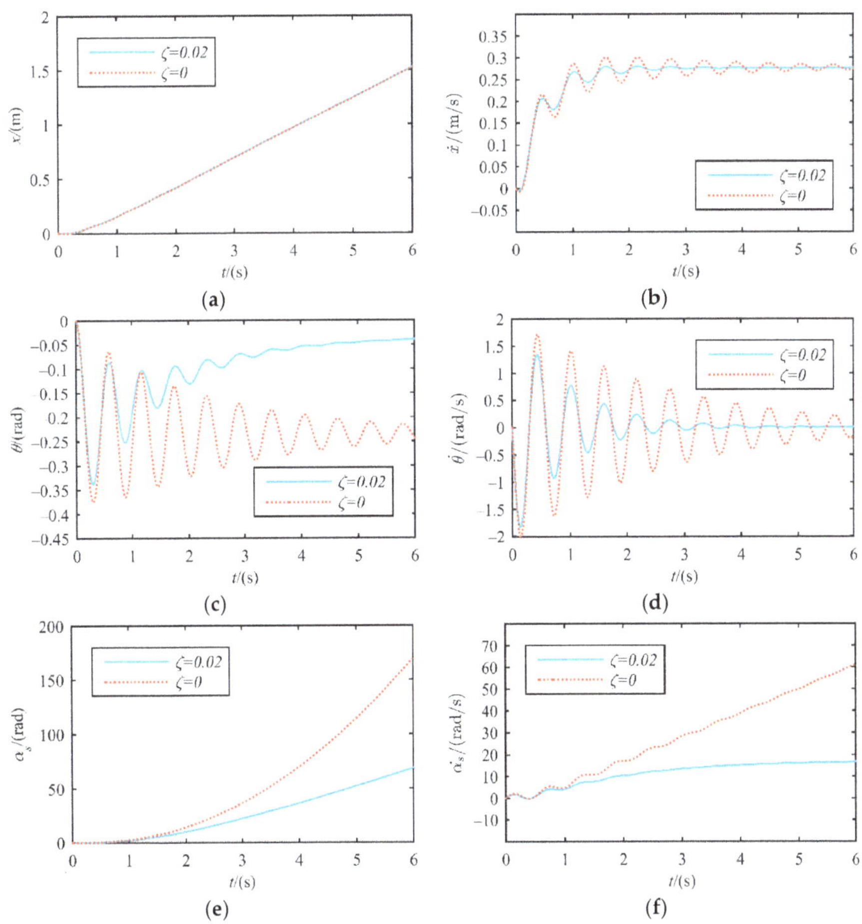

Figure 7. Time-varying states of the generalized coordinates. (**a**) Travel distance of the robot; (**b**) Speed of the robot; (**c**) Angle of the IDU platform; (**d**) Angular velocity of the IDU platform; (**e**) Angle of the spherical shell; (**f**) Angular velocity of the spherical shell.

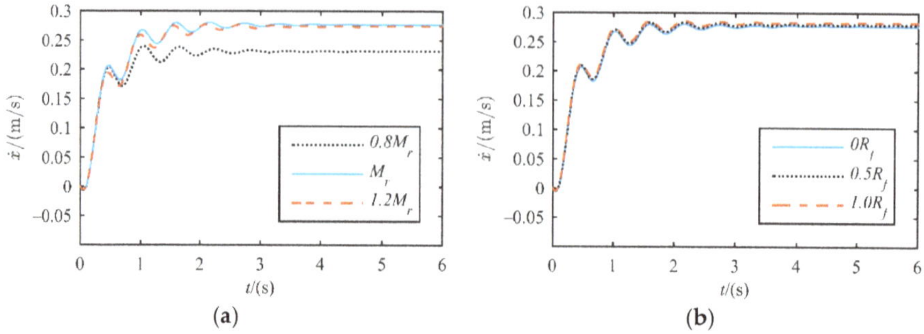

Figure 8. The velocity curve of the robot under different parameters. (**a**) Different rotator masses, (**b**) Different roughness coefficients of the spherical shell.

4.2. Amphibious Spherical Robot Prototype

Following the design considerations, an amphibious spherical robot prototype was constructed, as shown in Figure 9. An RF remote controller (Futaba T9CHP) was used to handle the spherical robot remotely.

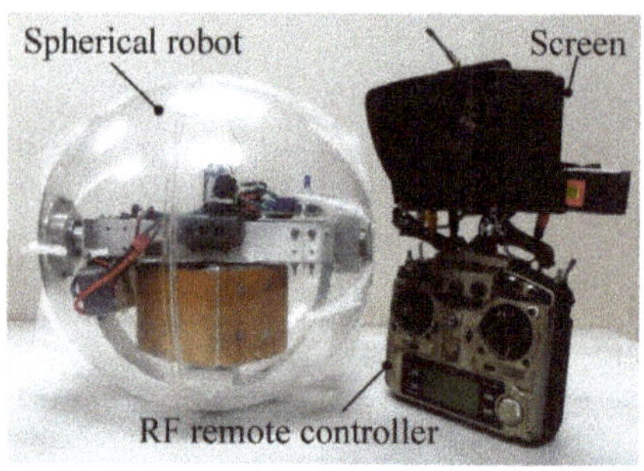

Figure 9. Amphibious spherical robot prototype.

Some specifications of the amphibious spherical robot prototype are given in Table 1.

Table 1. Specifications of the amphibious spherical robot.

Items	Parameters
Diameter	350 mm
Weight	5.8 kg
Power supply voltage	24 V
Duration time	1 h
Speed (on ground)	0.6 m/s
Speed (on the water)	0.4 m/s(16 fins)
Control signal transmitting distance	500 m
Camera signal transmitting distance	60 m

4.3. Movement Experiments

(a) Movement on Land

Curvilinear motion experiments of the robot were carried out in order to validate its on-land motion capacity, especially its turning capacity without stepping over the fins, as shown in Figure 10. The subgraphs were extracted from a recorded video. The time interval between adjacent subgraphs is 1 s. The robot's trajectory is shown in the subgraph numbered 1. During the curvilinear motion, the robot steers by spinning the rotator. It was found that, upon limiting the curvature of the trajectory below a threshold, the robot can avoid stepping over the fins when moving on the land.

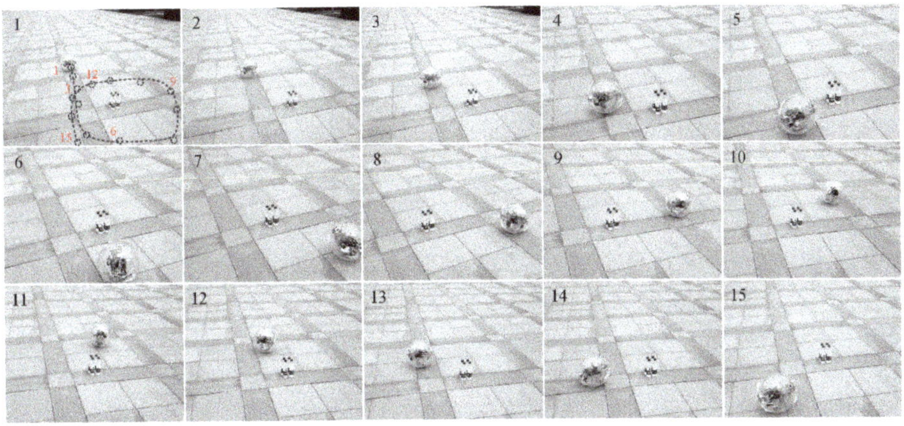

Figure 10. Curvilinear motion on land.

(b) Movement on the Water

Linear motion experiments of the amphibious spherical robot on the water were carried out in order to validate the effectiveness of the assistant fins, as shown in Figure 11. Sixteen assistant fins, eight on each hemisphere shell, were attached to the outside surface of the spherical shell. The amphibious spherical robot with fins was then placed onto the surface of static water out of doors. The rolling motion motor was started at $t = 0$ and the robot's subsequent track of motion was recorded by a video camera. After that, all the assistant fins were removed from the spherical shell, transforming the amphibious spherical robot to a traditional spherical robot. The linear motion experiment was repeated and the track of motion of the robot without fins was also recorded. Comparison of locomotion performances of those two cases are made based on those video records.

Figure 11. Movement on the water. (**a**) Fins attached; (**b**) Fins removed.

The velocities of the spherical robot with fins and without fins during accelerating process are shown in Figure 12. The results show that the spherical robot with fins speeds up more rapidly than that without fins when accelerating. It takes about 5 s for the spherical robot with fins to reach the velocity of 0.4 m/s, while it takes about 12 s for that without fins to reach the same velocity. The final steady velocity of the spherical robot with fins under constant rolling motion motor outputs, however, is only slightly higher than that without fins. The maximum on-water velocity that this amphibious spherical robot can reach is about 0.4 m/s, which is about 70% of the on-land locomotion velocity.

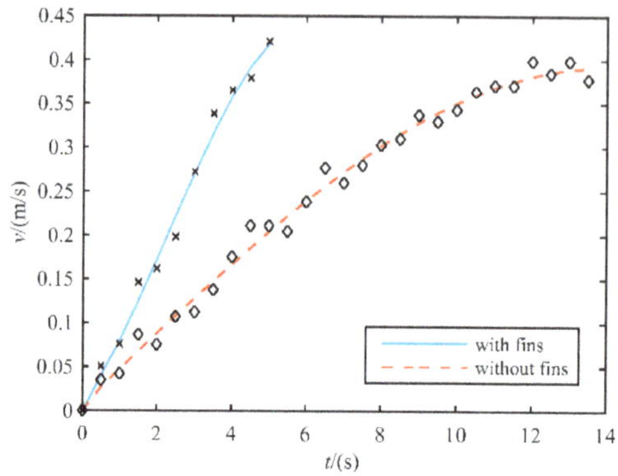

Figure 12. Velocities of the spherical robot with fins and without fins.

5. Conclusions

A novel amphibious spherical robot is here proposed for field applications. This amphibious spherical robot's motion is the composition of the rolling motion based on unbalancing a ballast and the spinning motion based on the principle of conservation of angular momentum. The architecture of the robot is proposed, along with its kinematics and dynamics analyses, which laid out the foundation of manipulation and control of the robot. Moreover, assistant fins are attached on the outside of the spherical shell to increase the propulsion force for the rolling motion, overcoming the disadvantage that a spherical shell slides easily on the surface of water. Results show that the robot with fins speeds up faster than that without fins. With the advantages of high adaptability and robustness, this amphibious spherical robot is suitable for applications in unstructured wild environments.

Future tasks are to carry out experiments to determine coefficients in the empirical equations, so as to obtain a more precise mathematic model for the robot.

Author Contributions: X.C. and Q.Z. provided the research ideas and the theoretical analysis and wrote the paper. Design and experiments, X.C.; analysis, X.C.; supervision, Q.Z. All authors have read and agreed to the published version of the manuscript.

Funding: This research was supported by the National Natural Science Foundation of China under Grant 50705003.

Institutional Review Board Statement: Not applicable.

Informed Consent Statement: Not applicable.

Data Availability Statement: The data presented in this study are available on request from the corresponding author. The data are not publicly available due to restrictions by the affiliation.

Acknowledgments: The authors gratefully acknowledge Tengfei Chai, Yanqiang Zhao, Pengpeng Sun and Yafeng Song from Beihang University for their assistant in constructing the prototype of the amphibious spherical robot.

Conflicts of Interest: The authors claim that there is no conflict of interests regarding the publication of this article.

References

1. Prahacs, C.; Saudners, A.; Smith, M.K.; McMordie, D.; Buehler, M. Towards legged amphibious mobile robotics. In Proceedings of the Canadian Engineering Education Association (CEEA), Montreal, QC, Canada, 29–30 July 2004; pp. 2371–5243.
2. Hirose, S.; Yamada, H. Snake-like robots [tutorial]. *IEEE Rob. Autom. Mag.* **2009**, *16*, 88–98. [CrossRef]
3. Muthugala, M.A.V.J.; Le, A.V.; Cruz, E.S.; Elara, M.R.; Veerajagadheswar, P.; Kumar, M. A self-organizing fuzzy logic classifier for benchmarking robot-aided blasting of ship hulls. *Sensors* **2020**, *20*, 3215. [CrossRef] [PubMed]

4. Takayama, T.; Hirose, S. Amphibious 3D active cord mechanism "HELIX" with helical swimming motion. In Proceedings of the IEEE/RSJ International Conference on Intelligent Robots and Systems, Lausanne, Switzerland, 30 September–4 October 2002; pp. 775–780.
5. Yamada, H.; Hirose, S. Study of a 2-DOF joint for the small active cord mechanism. In Proceedings of the IEEE International Conference on Robotics and Automation (ICRA), Kobe, Japan, 12–17 May 2009; pp. 3827–3832.
6. Crespi, A.; Badertscher, A.; Guignard, A.; Ijspeert, A.J. AmphiBot I: An amphibious snake-like robot. *Robot. Auton. Syst.* **2005**, *50*, 163–175. [CrossRef]
7. Ijspeert, A.J.; Crespi, A.; Ryczko, D.; Cabelguen, J.-M. From swimming to walking with a salamander robot driven by a spinal cord model. *Science* **2007**, *315*, 1416–1420. [CrossRef] [PubMed]
8. Crespi, A.; Ijspeert, A.J. Online optimization of swimming and crawling in an amphibious snake robot. *IEEE Trans. Rob.* **2008**, *24*, 75–87. [CrossRef]
9. Crespi, A.; Karakasiliotis, K.; Guignard, A.; Ijspeert, A.J. Salamandra robotica II: An amphibious robot to study salamander-like swimming and walking gaits. *IEEE Trans. Rob.* **2013**, *29*, 308–320. [CrossRef]
10. Kelasidi, E.; Liljeback, P.; Pettersen, K.Y.; Gravdahl, J.T. Innovation in underwater robots: Biologically inspired swimming snake robots. *IEEE Rob. Autom. Mag.* **2016**, *23*, 44–62. [CrossRef]
11. Kelasidi, E.; Liljeback, P.; Pettersen, K.Y.; Gravdahl, J.T. Integral Line-of-Sight Guidance for Path Following Control of Underwater Snake Robots: Theory and Experiments. *IEEE Trans. Rob.* **2017**, *33*, 610–628. [CrossRef]
12. Georgiades, C.; Nahon, M.; Buehler, M. Simulation of an underwater hexapod robot. *Ocean Eng.* **2009**, *36*, 39–47. [CrossRef]
13. Meger, D.; Higuera, J.C.G.; Xu, A.; Giguere, P.; Dudek, G. Learning legged swimming gaits from experience. In Proceedings of the IEEE International Conference on Robotics and Automation (ICRA), Seattle, WA, USA, 25–30 May 2015; pp. 2332–2338.
14. Dudek, G.; Giguere, P.; Prahacs, C.; Saunderson, S.; Sattar, J.; Torres-Mendez, L.-A.; Jenkin, M.; German, A.; Hogue, A.; Ripsman, A.; et al. AQUA: An Amphibious Autonomous Robot. *Computer* **2007**, *40*, 46–53. [CrossRef]
15. Sattar, J.; Giguère, P.; Dudek, G. Sensor-based behavior control for an autonomous underwater vehicle. *Int. J. Rob. Res.* **2009**, *28*, 701–713. [CrossRef]
16. Boxerbaum, A.S.; Werk, P.; Quinn, R.D.; Vaidyanathan, R. Design of an autonomous amphibious robot for surf zone operation: Part I mechanical design for multi-mode mobility. In Proceedings of the IEEE/ASME International Conference on Advanced Intelligent Mechatronics, Monterey, CA, USA, 24–28 July 2005; pp. 1459–1464.
17. Harkins, R.; Ward, J.; Vaidyanathan, R.; Boxerbaum, A.X.; Quinn, R.D. Design of an autonomous amphibious robot for surf zone operations: Part II-hardware, control implementation and simulation. In Proceedings of the IEEE/ASME International Conference on Advanced Intelligent Mechatronics, Monterey, CA, USA, 24–28 July 2005; pp. 1465–1470.
18. Zhang, S.; Zhou, Y.; Xu, M.; Liang, X.; Liu, J.; Yang, J. AmphiHex-I: Locomotory Performance in Amphibious Environments with Specially Designed Transformable Flipper Legs. *IEEE/ASME Trans. Mechatron.* **2016**, *21*, 1720–1731. [CrossRef]
19. Zhong, B.; Zhang, S.; Xu, M.; Zhou, Y.; Fang, T.; Li, W. On a CPG-based hexapod robot: AmphiHex-II with variable stiffness legs. *IEEE/ASME Trans. Mechatron.* **2018**, *23*, 542–551. [CrossRef]
20. Chen, Y.; Doshi, N.; Goldberg, B.; Wang, H.; Wood, R.J. Controllable water surface to underwater transition through electrowetting in a hybrid terrestrial-aquatic microrobot. *Nat. Commun.* **2018**, *9*, 2495. [CrossRef] [PubMed]
21. Kaznov, V.; Seeman, M. Outdoor navigation with a spherical amphibious robot. In Proceedings of the IEEE/RSJ International Conference on Intelligent Robots and Systems (IROS), Taipei, Taiwan, 18–22 October 2010; pp. 5113–5118.
22. Li, M.; Guo, S.; Hirata, H.; Ishihara, H. Design and performance evaluation of an amphibious spherical robot. *Robot. Auton. Syst.* **2015**, *64*, 21–34. [CrossRef]
23. He, Y.; Shi, L.; Guo, S.; Guo, P.; Xiao, R. Numerical simulation and hydrodynamic analysis of an amphibious spherical robot. In Proceedings of the IEEE International Conference on Mechatronics and Automation, Beijing, China, 2–5 August 2015; pp. 848–853.
24. Li, Y.; Yang, M.; Sun, H.; Liu, Z.; Zhang, Y. A Novel Amphibious Spherical Robot Equipped with Flywheel, Pendulum, and Propeller. *J. Intell. Rob. Syst.* **2018**, *89*, 485–501. [CrossRef]
25. Tadakuma, K.; Tadakuma, R.; Aigo, M.; Shimojo, M.; Higashimori, M.; Kaneko, M. "Omni-Paddle": Amphibious spherical rotary paddle mechanism. In Proceedings of the IEEE International Conference on Robotics and Automation, Shanghai, China, 9–13 May 2011; pp. 5056–5062.
26. Zhan, Q.; Cai, Y.; Yan, C. Design, analysis and experiments of an omni-directional spherical robot. In Proceedings of the IEEE International Conference on Robotics and Automation, Shanghai, China, 9–13 May 2011; pp. 4921–4926.
27. Cai, Y.; Zhan, Q.; Xi, X. Path tracking control of a spherical mobile robot. *Mech. Mach. Theory* **2012**, *51*, 58–73. [CrossRef]
28. Tupper, E.C. *Introduction to Naval Architecture*; Butterworth-Heinemann: Jersey City, NJ, USA, 2013.
29. White, F.M. *Fluid Mechanics*, 4th ed.; McGraw-Hill Higher Education: Boston, MA, USA, 1998.
30. Morsi, S.A.; Alexander, A.J. An investigation of particle trajectories in two-phase flow systems. *J. Fluid Mech.* **1972**, *55*, 193–208. [CrossRef]
31. Cheng, C.M.; Fu, C.L. Characteristic of wind loads on a hemispherical dome in smooth flow and turbulent boundary layer flow. *J. Wind Eng. Ind. Aerodyn.* **2010**, *98*, 328–344. [CrossRef]
32. Healey, A.J.; Rock, S.M.; Cody, S.; Miles, D.; Brown, J.P. Toward an improved understanding of thruster dynamics for underwater vehicles. *IEEE J. Ocean Eng.* **1995**, *20*, 354–361. [CrossRef]

applied sciences

MDPI

Article

Behavior-Based Control Architecture for Legged-and-Climber Robots

Miguel Hernando *, Mercedes Alonso, Carlos Prados and Ernesto Gambao

Centre for Automation and Robotics (UPM-CSIC), Universidad Politécnica de Madrid, 28012 Madrid, Spain; cheripe.caa@gmail.com (M.A.); c.prados@alumnos.upm.es (C.P.); ernesto.gambao@upm.es (E.G.)
* Correspondence: miguel.hernando@upm.es

Featured Application: Control of legged-and-climber robots with at least four legs, under unforeseen failures in one or more legs.

Abstract: In this paper, we present a fully original control architecture for legged-and-climber robots that is level-based, hierarchical, and centralized. The architecture gives the robots the ability to perform self-reconfiguration after unforeseen leg failures, because it can control this kind of robot with different numbers of legs. The results show the capability of performing movements in any direction and inclination planes. The components and functionalities of the developed control architecture for these robots are described, and, the architecture's performance is tested on the ROMHEX robot.

Keywords: behavior-based; climber robot; control; control architecture; fault-tolerant; legged robot; optimization

Citation: Hernando, M.; Alonso, M.; Prados, C.; Gambao, E. Behavior-Based Control Architecture for Legged-and-Climber Robots. *Appl. Sci.* **2021**, *11*, 9547. https://doi.org/10.3390/app11209547

Academic Editors: Alessandro Gasparetto, Stefano Seriani and Lorenzo Scalera

Received: 27 July 2021
Accepted: 9 October 2021
Published: 14 October 2021

Publisher's Note: MDPI stays neutral with regard to jurisdictional claims in published maps and institutional affiliations.

1. Introduction

Due to the increased size and complexity of civil construction, using climbing robots in infrastructure inspection is becoming increasingly relevant. Regular maintenance and surveillance of large complexes are extremely important to guarantee their life-cycle. The European consortium SPARC (euRobotics) carried out an analysis, revealing in the "European Robotics & AI workshop Applied to Inspection and Maintenance" report [1] that in the coming years the task of inspection will be increasingly important, and robots will play the main role in maintenance, inspection and dismantling.

Inspection is a particularly structured repetitive task, that requires permanent attention during the operation, and in many cases, it involves placing the human operator in risk situations. That is why robotics is revealed as a technology of direct application. The number of service robots, mainly driven by drones, has increased by 25% in recent years, and the number of autonomous vehicles in a non-manufacturing environment, has increased by 51% in one year [2].

Many solutions are based on remote visual inspection, but this is not valid for many industrial structures. We may not have direct visual access to the areas to be inspected, and in other cases it is not possible to carry a flying drone because of the narrowness or typology of the environment. In addition, an inspection usually requires much more than seeing. Non-destructive techniques require contact or proximity to the surface that is not possible to achieve while flying. Climbing robots have become increasingly attractive for effective infrastructure inspection due to their ability to overcome these limitations.

A crucial part of these robots is their control. Typically, robots are controlled based on control architectures. The lack of existing control architectures for legged-and-climber robots drives a need to design a new architecture. The state-of-the-art of control architectures for this kind of robot reveal that much work remains, because although plenty of legged-and-climber robots are structurally defined, they lack a defined architecture. Only a few control architectures are found for climber robots. Shen et al. [3] proposed a climber robot for oil tank inspection, whose control architecture only covers the kinematic control of the robot. Several control architectures for legged robots can be found easily, such as [4] or [5], which describe the motion controller of two-legged robots. However, these kinds of legged robots are unable to climb due to their bipedal disposal. Robots with more than two legs, like [6], for example, have a control system that only covers legs management and exclude the high-level control.

A control architecture specifies the organization guidelines of a robot's behavior, establishes the action and movements that the robot must carry out to achieve a goal or a set of goals, according to the robot state. It has the main purpose of establishing a way to organize the system to maintain defined roles, modularize the system components and make the system as fault-tolerant as possible, always trying to keep the system under control over as many situations as possible. The main characteristics of the control architectures are the capacity to (a) face multiple goals simultaneously, (b) integrate data from different sensors, (c) be robust against component failures, (d) be adaptive to new environments, (e) extend and modify its content easily, (f) make its own decisions according to the robot state, and (g) modify the surroundings properly.

Some important definitions should be declared in a control architecture: (a) an agent is a computer system that is capable of autonomous action in its environment to achieve its delegated objectives [7,8], (b) a level is composed of an agent or a group of agents that have the same importance from a determined point of view, (c) hierarchy is how agents of higher levels have a kind of control over the agents of lower levels, (d) a centralized control is characterized by coordinated conduct of the agents, whose decisions depend directly on the state of the other agents, (e) a decentralized control is characterized by freedom in the agent decisions, without directly considering the state of the rest of the agents, and (f) a behavior-based control is a way in which agents are delegated with the main task to achieve a goal through some instructions. Then, it is possible to conclude that a control architecture may be understood as a multi-agent system where the communication rules and protocols are well-defined.

This paper's objective is to design an organic and hierarchical control that allows the safe movement of the legs. Moreover, the control must be generic for a legged robot with any number of legs. In this paper, we test the performance of the control architecture over the legged-and-climber robot ROMHEX. This robot has ben modified; the most remarkable is the change of the initial position of each leg that the robot should adopt to maximize the efficiency of the walking pattern. This maximization has been obtained generically, as explained in Section 3.1. Thus, the initial position directly influences the beginning of the gait, and it is also used as the default position when the robot needs to reconfigure while walking.

We propose a control architecture that fulfills the following requirements: (a) generic for all legged robots, independent of the number of legs, distribution of legs in the body, or the number of joints per leg, (b) agent-based, (c) hierarchical (that is, agents of higher levels should have the control of lower-level agents, and in this way, first agents may disable second agents), and (d) agents of the same level must synchronize their behaviors using synchronization mechanisms.

This paper is organized as follows: In Section 2, we present an overview of the state of the art of control architectures. We describe how different architectures are organized and the levels they use. In Section 3, we describe the hexapod robot used in the tests. Furthermore, we include the developed optimization of the legs' position in the body. In Section 4, we describe the developed control architecture, and we explain its levels, hierarchy and agents. In Section 5, we discuss the results obtained during the tests. Lastly, in Section 6, we detail the obtained conclusions from the results.

2. Related Work

Control architectures are found within autonomous machines, especially in spacecraft. The increasing development of these vehicles has generated generic architectures that may be used in any system. Within the state of the art, we can find Contextual Management of Tasks and Instrumentation (CMTI), which is a mixed architecture between deliberative and reactive architectures [9], originally conceived for an autonomous underwater vehicle (AUV). It is organized into three layers: global supervisory control, local supervisory control and low-level control. CMTI is a well-defined and robust architecture but laborious to implement in short projects. Based on CMTI, Contextual Task Management Architecture (COTAMA) is a control software architecture layered into two levels [10]: decisional level and executive level. The decisional level is in charge of the mission monitoring and decision making according to robot context. The executive level applies these decisions while managing all real-time aspects such as instrumentation conflicts or tasks deadlines. It contemplates the identification of faults to correct them [11]. COTAMA improves the robustness of CMTI while it enlarges the possible applications; however, like CMTI, it is very laborious to implement. Remote Agent architecture (RA) is an autonomous control system capable of closed-loop commanding of spacecraft and other complex systems [12]. It integrates three layers of functionality: a constraint-based planner/scheduler, a reactive executive, and a model identification and recovery system. RA is very useful for spacecraft; however, its usefulness is limited to tasks previously defined in detail. Intelligent Distributed Execution Architecture (IDEA) was created to duplicate RA architecture within a unified agent framework where all the layers have the same structure [12]. IDEA improves the coordination of RA; however, the drawbacks are similar.

Laboratory for Analysis and Architecture of Systems (LAAS) architecture is presented in [13] as an architecture for reflexive autonomous vehicle control. It decomposes the robot software into three main levels, having different temporal constraints and manipulating different data representations. LAAS is thought to improve robustness; however, it is poorly-defined and very open to the developer. Coupled Layer Architecture for Robotic Autonomy (CLARAty) is designed for improving the modularity of system software while more tightly coupling the interaction of autonomy controls [14]. According to the CLARAty developers, typical robot and autonomy architectures comprise three levels: functional, executive and planner. To correct the shortfalls in the three-level architecture, they propose a two-tiered architecture, in which the executive and planner layers are combined. As well as LAAS, CLARAty is very open to the user because it only describes the two main levels. Cooperative Intelligent Real-Time Control Architecture (CIRCA) was designed in [15] to guarantee the control-level goals, but not necessarily the task-level goals. They divide the architecture into three main parts: the real-time subsystem (RTS) that is responsible for implementing the responses, the AI subsystem (AIS) that decomposes task-level goals into plans consisting of several phases, and the scheduler. CIRCA has the limitations that it only works for well-known and defined problems. Open Robot Controller Computer Aided Design (ORCCAD) architecture is an open architecture where qualified users have access to every control level: the application layer is accessed to by the end-user of the system, the control layer is programmed by an automatic control expert, and the lowest one, the system layer, is overseen by a system engineer [16]. ORCCAD has the problem that the system's complexity may increase exponentially with new fault tolerance techniques, while the organization structure may become untenable. The described architectures are generic for any system; however, their application in legged-and-climber robots may be laborious and complex. Thus, the decision to develope a new architecture has been considered a better option than augmenting an existing one.

Control leaders in multiple-legged robots, such as Boston Dynamics, hide their control architectures while other researcher groups show their work. For example, in [17], Jakimovski et al. present an Organic Self-Configuration and Adapting Robot (OSCAR), a hexapod robot that is described through the Organic Robot Control Architecture (ORCA). ORCA [18] proposes creating an entire system out of subsystems, where each of the subsystems is designed for a determined task [19]. More complex subsystems can be generated by combining and cascading smaller subsystems [20]. Each subsystem may be supervised by another subsystem that evaluates its performance and may even change its behavior to optimize the performance of the whole system. As another example, in [21], Pack et al. present Robotic Inspector (ROBIN), a robot designed for climbing infrastructures that uses a behavior-based control architecture arbitration by subsumption [22]. This robot is composed of two vacuum fixtures, so its architecture is completely dependant on the performance of both devices. Lastly, in [23], Ronnau et al. describe LAURON V, a legged robot controlled by its own control architecture, which is a modular and behavior-based design approach. It subdivides the system into understandable hierarchical layers and small individual behaviors. The layers are the hardware architecture, the hardware abstraction layer, and the behavior-based control system. Finally, Fankhauser et al. present Free Gait in [24] a software framework for the task-oriented control of legged robots, which they check over ANYmal [25]. Free Gait consists of a whole-body abstraction layer and several tools designed to interface higher-level motion goals with the lower-level tracking and stabilizing controllers.

Architectures for legged robots exist, but none exist for legged-and-climber robots. Furthermore, these architectures are usually conceived for a defined and not modifiable number of legs. Leg problems are possible, especially in climber robots, due to the harsh conditions they are involved in. OSCAR robots contemplate the situation of leg amputation; however, the visible face of its architecture does not allow to define clearly the behavior of a new robot.

3. The Climber Hexapod Robot ROMHEX

The ROMHEX robot is a commercial platform called *xyzrobot bolide crawler Y-01* with some modifications. The robot is a hexapod with three degrees of freedom in each leg. The reference systems of each leg according to the robot body are referred to as shown in Figure 1a, while the axes of the leg joints are illustrated in Figure 1b. Mainly, the robot is composed of an electronic board called *MCU board Y-01* and motors called *xyzrobot smart servo A1-16*.

The development kit *Intel Euclid* has been added to the robot through a plastic piece that locates it in a proper position to take advantage of all its features. This device provides a motion camera (not used, so external obstacles are not considered), a computer processing unit and a depth camera. Furthermore, suction cups have been added to the legs extremes in order to hold on to any surface and allow the robot to climb. Every suction cup is equipped with its own centrifugal impeller and motor that creates and maintains the vacuum even on porous surfaces, extracting the internal air [26]. The complete griping system consists of (a) an electronic circuit inside the cup that sensorizes the system and measures the pressure and the distance to the support surface, (b) a turbine motor with its variator, (c) an electronic board that acts as a link between sensorization circuits and the control system of the suction cups and the microcontroller, and (d) a mechanical system with three rotary degrees of freedom to properly align with the surface.

Lastly, to increase the work-space of the legs, the configuration of the motors has been modified, changing the position of the second motor. In this way, the center of mass is lowered, increasing the robot stability.

A critical aspect of controlling the robot while climbing is to ensure the normal and shear forces at the suction cup do not exceed certain limits during movement, creating the risk of loss of grip [27]. The hexapod robot is a hyper-static system whose elastic model is too complex to be included in a control loop. Given the hyperstatic nature of the problem, a simplified dynamic model is calculated in [28] and included in the control.

Climber robots are deployed in dangerous situations, where the power consumption must be optimized to guarantee the finalization of the task. Possible solutions are found in weight reduction, calculation of the best path, or optimization of the walking patterns.

The leg position is a critical aspect that determines the distance traveled in a given period of time. For this reason, it is highly desired to optimize the robot's leg positions.

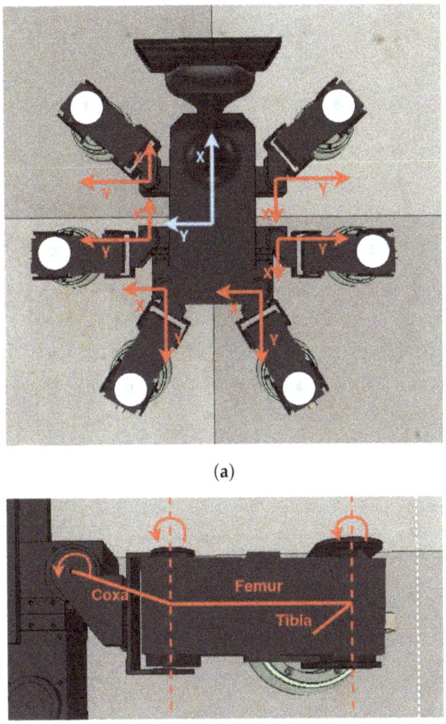

(**a**)

(**b**)

Figure 1. Joint axis and reference systems of the legs. (**a**) ROMHEX with the reference system of the body and legs, and leg identifier. (**b**) Axis of ROMHEX joints.

3.1. Optimization of the Leg Positions

Making use of genetic algorithms (in this case, MatLab's *ga* function), it is possible to find the best position of the robot's legs' initial configuration according to a criterion. The algorithm uses a combination of the distance the robot can walk and the forces produced at the legs as a cost function. Because the objective is to obtain the optimal initial position of the legs to walk, the "genes" or decision variables are the initial positions of the legs (X and Y for each leg and a global Z with respect to the center of the robot, this is, 13 positions in total).

An analysis has been carried out on the center of mass and how its position affects the forces produced in the different legs to improve the results. The objective of the cost function is to obtain a genetic individual (legs position configuration) that achieves the greatest distance while walking with a given number of movements following a predefined pattern, keeping the robot safe. It considers the distance that the robot moves, as well as the maximum permissible forces in the legs, as indicated in (1), where C is the cost function, D is the distance traveled, and F is a matrix where each row corresponds to a moment in the execution of the walking movement and each element of the row corresponds to each leg. The cost function is negative because it is required to minimize the value. Both $f(x)$ and

D are positive values, being a single S calculated in (2), where S represents the vector of forces applied over a leg, and $S(2)$ is the force applied over the z axis.

$$C = -min(f(max(F)), f(min(F))) \cdot D \quad (1)$$
$$S = norm(S) \cdot sign(S(2)) \quad (2)$$

Function $f(x)$ is distance scaling and piecewise defined, as shown in Figure 2. The objective of scaling the distance obtained by the factor produced by the function is to penalize the individuals that produce the highest forces, even if they manage to move a greater distance. The function takes into account the sign on the z axis, to differentiate the dangerous forces. Both signs are considered to eliminate individuals that make the suction cups detach and reduce extreme forces on a single leg to reduce unequal wearing.

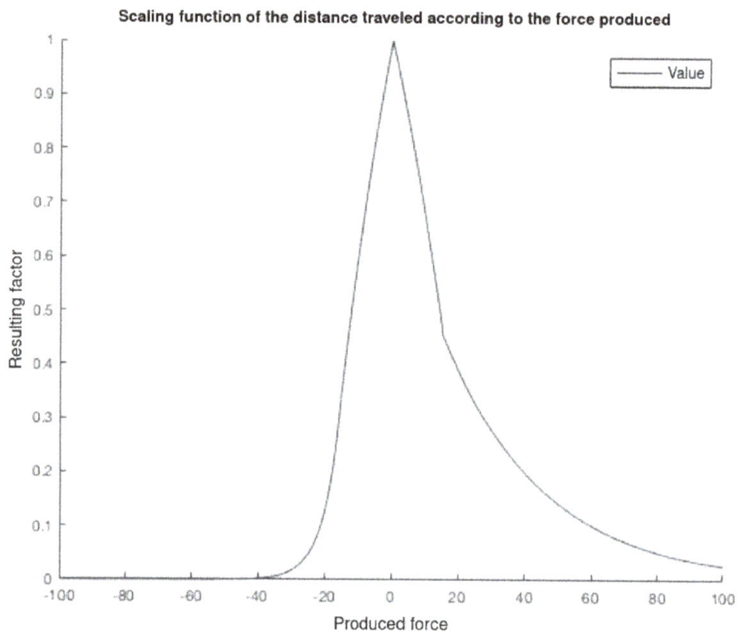

Figure 2. Distance scaling function.

To obtain the value of *dist*, the developed walking movement has been simulated in the following way: First, it is checked that the individual is valid, this is, (a) the position of all the legs is reachable with the inverse kinematics, (b) the position of the motors is within the specified ranges, and (c) there is no collision between legs. Second, the cost function value is obtained.

The results of the genetic algorithm are an increase of 107% in the distance traveled (from 355 mm to 735 mm) and a decrease of 10% in the force. Figure 3 shows a representation of the optimized version over the previous one. As illustrated in that picture, the position of the legs has undergone a slight variation to achieve an initial position that optimizes the evaluation criteria. Table 1 denotes the joint initial position increment between before and after the optimization, with the references in the motor encoder origins. Furthermore, both tables show the end-effector positions (feet) when the motors are in the given initial position.

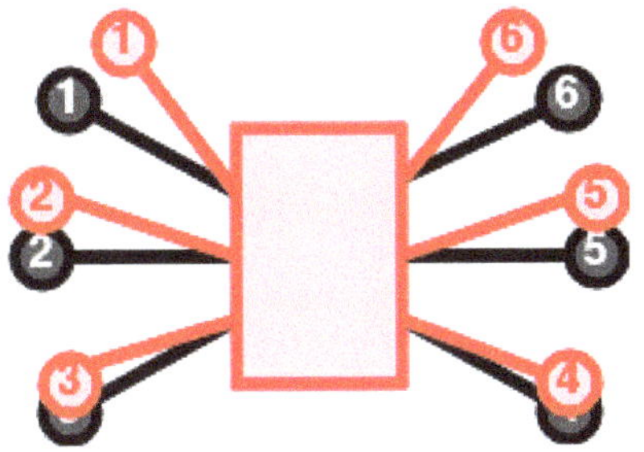

Figure 3. Comparison between the position of the legs before (gray) and after (red) the optimization through the genetic algorithm. Positions specified in Table 1.

Table 1. Variation of the position of each joint and suction cup after the optimization.

Leg	Joint Angles (rad)			Feet Position (mm)		
	q_0	q_1	q_2	x	y	z
1	−0.1	−0.13	0.33	28	6	−3
2	−0.1	−0.18	0.49	22	35	−3
3	0.36	−0.36	−1.15	79	−129	−3
4	−0.66	0.15	−0.75	−17	127	−3
5	−0.11	−0.08	0.19	−21	−11	−3
6	0.11	−0.19	0.49	36	−11	−3

4. Control Architecture

A new control architecture that considers safety under unforeseen circumstances is needed to guide legged-and-climber robots. The proposed control architecture is characterized as a behavior-based control, hierarchical and centralized. As shown in Figure 4, the architecture is split in the Executive, the Planner and the User Interface. The Planner is divided into three main levels, which make use of complementary modules located in the Executive. The architecture includes a User interface, with which the user may control the behavior of the robot and observe the state of the robot and the legs. Each level of the Planner has a set of critical and given objectives:

1. **Level 1**: Corresponds to the nominal and continuous behavior without checking the safety at any moment. This level is responsible for the body movement in the desired direction, through the performance of the robot legs.
2. **Level 2**: Corresponds to behaviors about movements under expected situations, having considered basic safety issues. It is responsible for determining if a movement may still be developed.
3. **Level 3**: Corresponds to the critical safety checks to ensure that the robot is not in a hazardous situation. This level is vitally important in robots like the one in question here, where the goal is to allow it to walk safely on the wall and ceiling.

There is a hierarchical relationship between the different levels in that the higher level is able to disable the lower level. Dependencies occur from top to bottom; in other words, what happens at the upper level is unknown by lower levels. The agents of the same level are in a situation of equality, so they need a synchronization mechanism in case two behaviors are mutually exclusive. A token synchronization has been used to do this: the agent with the token is the one that can be executed. When it stops executing, it will drop the token and other behavior will be free to catch it.

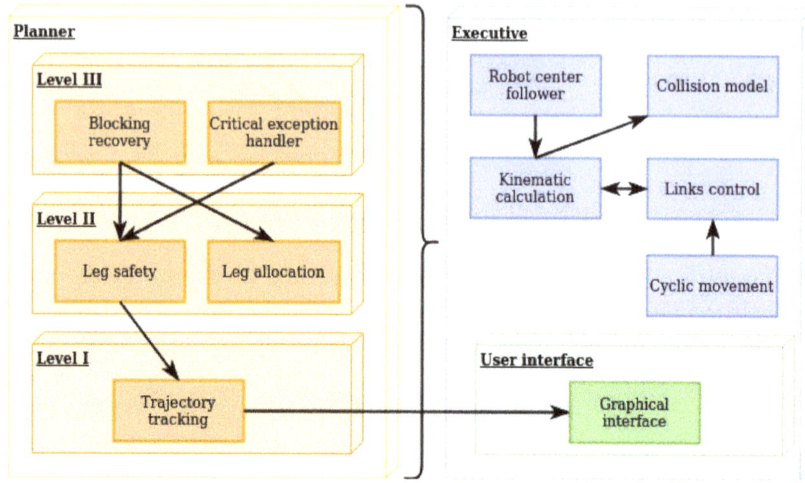

Figure 4. Control architecture.

Each behavior has its own functionalities, inputs, outputs, and implementation features. Architecture modularity allows developers to add more, increasing control capabilities.

4.1. Level 1: Nominal Movement of the Body

Trajectory Tracking

The objective of this behavior is that the robot center follows a trajectory in terms of different global positions and orientations, without explicit information about velocities. Basically, it carries out the inverse kinematics of the robot, where the input is the robot center trajectory, and the output is the position of the leg extremity. The output is generated dynamically through a close chain. However, this agent neither checks the stability nor sends commands if the inverse kinematics can not obtain a required point. Interpolation is needed if two consecutive poses are too far apart to obtain as many intermediate ones as needed. In this case, the point is divided into position and orientation, where spherical linear interpolation (SLERP) [29] is used for the orientation interpolation, to obtain the maximum precision, while the position interpolation is linearly done.

Because several legs are attached to the ground to move the center of the body in the world coordinates, the legs will move opposite to the body robot coordinates. When P_n is the position of the center of the robot, R_n is its orientation, v_n is the vector that describes the position of one of the leg extremities, v_{n+1} is the vector in the position to be achieved, (P_n, R_n) denotes the robot pose, and (P_{n+1}, R_{n+1}) is the pose to be achieved. Then, the position of the leg extremity in both references (3) is obtained, while v_{n+1} is calculated in (4). For better understanding, Figure 5 shows a comparison of the movement in the robot's and world coordinates.

$$P_n + R_n \cdot v_n = P_{n+1} + R_{n+1} \cdot v_{n+1} \tag{3}$$

$$v_{n+1} = (R_{n+1})^t \cdot (P_n - P_{n+1} + R_n \cdot v_n) \tag{4}$$

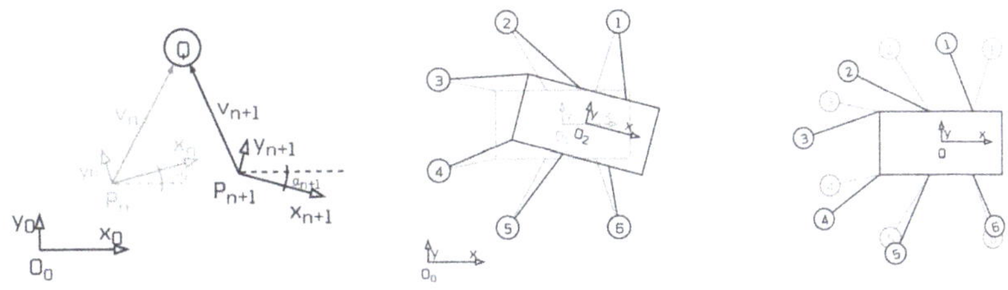

(a) (b) (c)

Figure 5. Comparison of the movement in the robot's and world coordinates. (**a**) Representation of the reference change of the point Q between (P_n, R_n) and (P_{n+1}, R_{n+1}), where α_k represent the angle for the rotation matrix R_k. (**b**) Comparison between the initial (light color) and final position in global coordinates. (**c**) Comparison between the initial (light color) and final position in the robot's coordinates.

4.2. *Level 2: Expected Situations and Leg Allocation*

4.2.1. Leg Safety

The objective of this behavior is to predict when a leg will find an instability or blocking situation and move it to avoid this state. It takes as input the current pose of each leg and the current motion tendency, among others. The main output is which leg is required to relocate to where to ensure the robot's stability; that is, move a leg to a new position.

To implement this behavior, a metric about how urgently each leg should be relocated is obtained, in this case: how close the joints are to their limits, from 0.6 rad (not urgent) to 0 rad (critical); and how close each foot is to the center of mass (COM) of the robot, from 20 cm (not urgent) to 5 cm (critical). Furthermore, it checks that, in the future position, each leg's kinematics will allow lifting them in case a reallocation is needed in that state.

Two limits have been set, a motion limit and a danger limit. They represent the limit within which a leg can move and the limit within which the leg can move without entering a dangerous situation, respectively. If the robot enters a dangerous situation, the behavior is blocked until the legs in hazard are reallocated.

Whenever a leg is expected to move, it moves the maximum possible distance without colliding with other legs in the direction of movement, minimizing the number of movements needed. To move a leg, it must have enough space to move and lift without creating an unstable situation, considering how the rest of the legs are positioned. For that, the force model generated in [28] is needed.

During the behavior execution, it is possible to enter in a blocked state, in which there is no leg to move. In this case, it informs the upper level to take control to return the robot to a safe state. When this behavior is disabled, the trajectory control is disabled to avoid hazardous situations. Whenever it is enabled again, it enables the trajectory control to continue with the previous execution.

4.2.2. Leg Allocation

This behavior is responsible for safely moving all the legs, or a subset of them, to a given position. The input is the desired position, and the output is a movement of each leg, in the proper order and moment. The order is defined as a function of the safety in which the movement may be done (state while a leg is lifted). Leg allocation checks that the forces of lifting the legs do not produce any hazard and the possibility of reaching the desired position without colliding with other legs exits.

This behavior can not work at the same time as the "leg safety" behavior due to the possibility of conflicts. Thus, a timing mechanism of a *token* is used, because there is no hierarchy of any kind between them.

4.3. Level 3: Critical Exceptions and Blocking Situations

4.3.1. Blocking Recovery

This behavior is thought of as an external observer. It is responsible for detecting when there is a blocking state to unlock it. The input is the system state, and the output is the position in which the legs should move to solve the blocking state.

The "leg safety" behavior informs when it detects that no legs can move because it is not possible to lift a leg or move enough in the required direction. When the advice is received, this behavior blocks the advice emitter to avoid any leg movement during the reallocation. It asks the "leg allocation" behavior to move all legs to a default safety position. When this process finishes, it enables the "leg safety" behavior to continue with the nominal execution.

4.3.2. Critical Exception Handler

This behavior is responsible for ensuring that the robot is not in danger. The input is the system state with the information from all sensors. The output is the fact of blocking the motion system, keeping the robot completely still, and informing the user about the critical error. Then, the user should solve the problem using the graphical user interface to check forces, leg positions, and other factors.

4.4. Complementary Modules

These modules are needed to make the behaviors work. They include different geometric calculations and control over the links at a low level.

4.4.1. Robot Center Follower

This module obtains the movement of the robot center. Its input is the position of the legs, and its output is the position of the robot center in the world coordinates. To observe how the robot follows the trajectory from an initial state (P_n, R_n) to a target state (P^*_{n+1}, R^*_{n+1}), the real state of the center of the robot (P, R) is calculated in an instant between t_n and t_{n+1} in absolute coordinates, making use of the relative position vectors of each of the different legs attached to the ground. u^k are the vectors of the relative position of the leg k-th respecting the robot center in the current time (with reference (P, R)), u^k_n are the vectors of the relative position at the t_n time respecting the reference (P_n, R_n), and K is the number of legs attached to the ground. A system of $3 \cdot K$ equations is obtained (5), where the unknown values of P and R are obtained through numeric solvers due to the non-linearity of the problem. For that, the gradient descent method has been used.

$$P + Ru_k = P_n + R_n u^k_n, \quad k = 1, \ldots, K \tag{5}$$

4.4.2. Collision Model

This module describes a simplified collision model with which is possible to calculate if a leg collides with another leg in a given configuration. The simplification consists of a 2D model of the ROMHEX robot, where each leg is represented as a linear segment, and each suction cup is represented as a circle. The module checks if there is a collision of the type (a) between two circles, (b) between a circle and a segment, or (c) between two segments.

4.4.3. Kinematics Calculation

This module obtains the direct and inverse kinematic of a leg, with the reference system in the robot center. It is completely dependent on the robot, so this module must be changed if another robot is used. Using the tests with ROMHEX, we present the forward and inverse kinematics of this robot, obtaining the algebraic solution.

Following Figure 1, forward kinematics is calculated in (6)–(8), where A_{coxa} is the angle between the first motor origin and the femur. P_x, P_y and P_z denote the position of the end-effector with respect to the leg coordinate system, L_{coxa}, L_{femur}, and L_{tibia} denote the link lengths, while q_1, q_2 and q_3 denote the joint angles. Furthermore, thanks to Figure 1a it is possible to obtain the transformation matrix between the body center and each leg origin. These transformation matrices must be applied over the body and legs reference systems.

$$P_x = L_{coxa} \cdot cos(A_{coxa} + q_1) + L_{femur} \cdot cos(q_2) \cdot sin(q_1) + L_{tibia} \cdot cos(q_2 + q_3) \cdot sin(q_1) \quad (6)$$

$$P_y = L_{coxa} \cdot sin(A_{coxa} + q_1) + L_{femur} \cdot cos(q_2) \cdot cos(q_1) + L_{tibia} \cdot cos(q_2 + q_3) \cdot cos(q_1) \quad (7)$$

$$P_z = L_{coxa,z} + L_{femur} \cdot sin(q_2) + L_{tibia} \cdot sin(q_2 + q_3) \quad (8)$$

With respect the inverse kinematic, the solution is found in (9)–(11), where variables B, A_1, etc. are calculated in (12)–(16).

$$q_3 = \begin{cases} B - \pi & if\ third\ link\ up \\ \pi - B & if\ third\ link\ down \end{cases} \quad (9)$$

$$q_2 = \begin{cases} (A_1 + A_2) - \frac{\pi}{2} & if\ third\ link\ up \\ \frac{\pi}{2} - (A_1 + A_2) & if\ third\ link\ down \end{cases} \quad (10)$$

$$q_1 = \frac{P_x}{P_y} - \frac{L_{coxa,x}}{LP + L_{coxa,y}} \quad (11)$$

$$B = arccos(\frac{HF^2 - L_{femur}^2 - L_{tibia}^2}{-2 \cdot L_{femur} \cdot L_{tibia}}) \quad (12)$$

$$HF = \sqrt{LP^2 + (P_z - L_{coxa,z})^2} \quad (13)$$

$$LP = \sqrt{P_x^2 + P_y^2 - L_{coxa,x}^2} - L_{coxa,y} \quad (14)$$

$$A_1 = arctan(\frac{LP}{|P_z - P_{coxa,z}|}) \quad (15)$$

$$A_2 = arccos(\frac{L_{tibia}^2 - L_{femur}^2 - HF^2}{-2 \cdot L_{femur} \cdot HF}) \quad (16)$$

4.4.4. Center of Mass Calculation

This module calculates the center of mass with respect to the robot center. It is implemented with the knowledge of the joints' state, the links' mass and the links' shape.

4.4.5. Links Control

This module is responsible for unifying all movements for the different legs, which are sent from the different behaviors to be executed at the link level. Whenever a behavior desires to move a leg, it sends a command with the leg identifier, where to move it (in cartesian or articular coordinates), and the priority of the movement. Behaviors that send movements must be aware that some movements may be overwritten and partially executed because a higher priority message is received.

4.4.6. Cyclic Movement

This module aims to generate a default cyclic movement that allows the robot to walk forward. Without inputs, the output is the positions that the legs should reach in each moment. The module checks the performance of the control architecture, so a simple walking process has been developed. It consists of moving the legs individually from the back to the front after moving the whole body forward. Its behavior may be replaced by changing and complex patterns.

4.5. Graphical User Interface

A graphical user interface (GUI) for a generic legged robot has been developed to make interacting with the robot easier. The GUI is shown in Figure 6, and it includes (a) information about the robot status, (b) plots of the positions of the motors along a given period of time, and (c) a graphical representation of the position of the legs and the robot trajectory. It is also possible to create trajectories and send them to the robot.

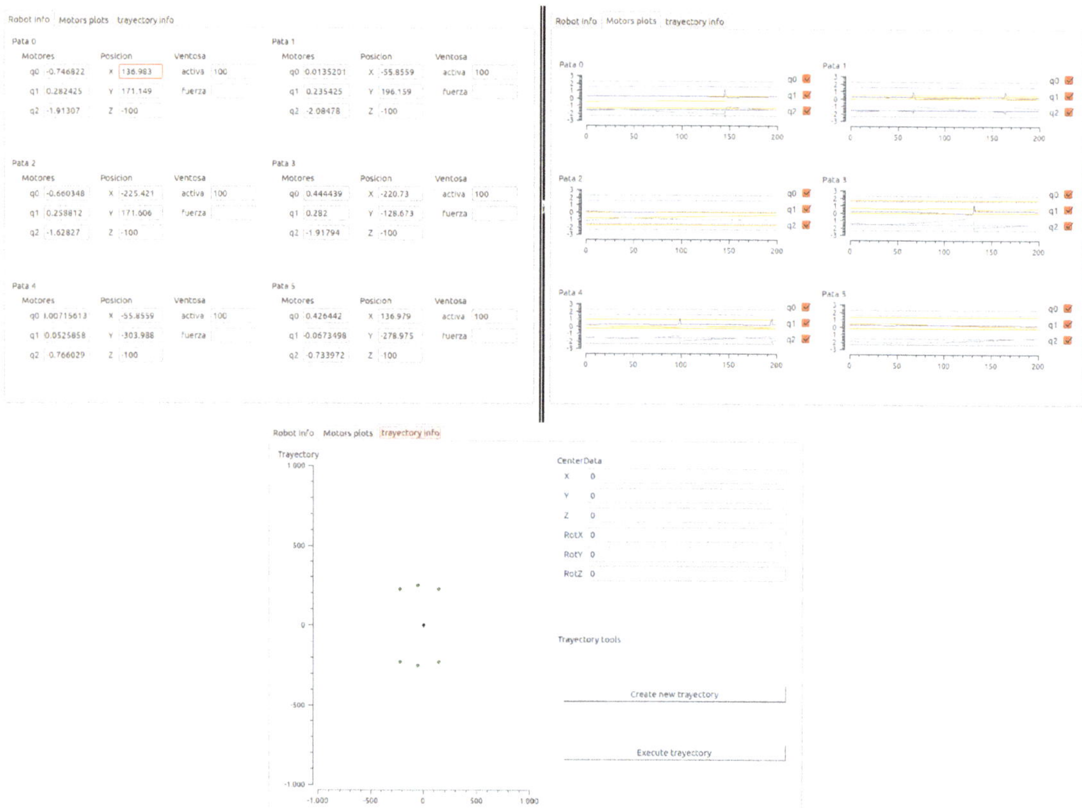

Figure 6. Graphical User Interface. The first tab shows information about the robot, including the motors and suction cup. The second tab shows the motors position during a given period. The third tab shows the whole robot trajectory, and it allows setting the goal position to execute a new trajectory.

5. Experimental Results

The performance of the simulated robot while walking on a flat plane was tested in CoppeliaSim. The system was studied while moving forward, laterally, diagonally, rotatory, and walking with a combination of movements. After validating the performance in these conditions, the system was tested on sloping (45°) and vertical walls with successful results. However, the presence of gravity in different axes required adjusting the controller gains.

To test the generalization of the control, the performance was checked when the number of legs was changed. The behavior when two legs are removed is also valid, even though the control code is not modified. Figure 7 shows the generated walking pattern when the robot detects six legs and when it detects a malfunction in two of its legs. As it is observed, the walking pattern when the robot has six legs is periodic, because the tolerances specified for a leg to move were adjusted with a hexapod robot. In the case of four legs, the walking pattern is not periodic. In this case, legs are moved a non-defined distance when the space in front of them is higher than a threshold.

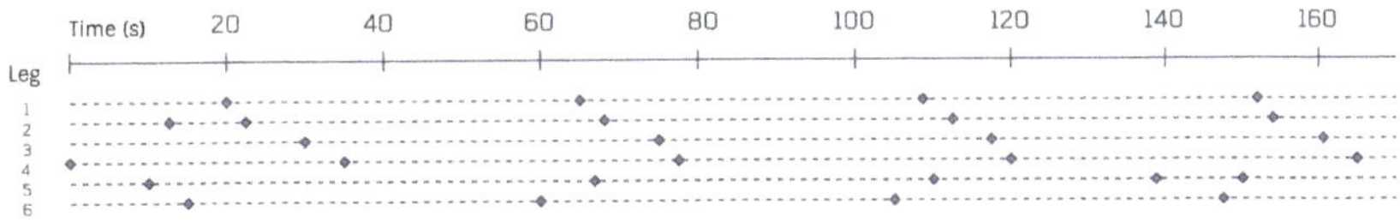

(**a**)

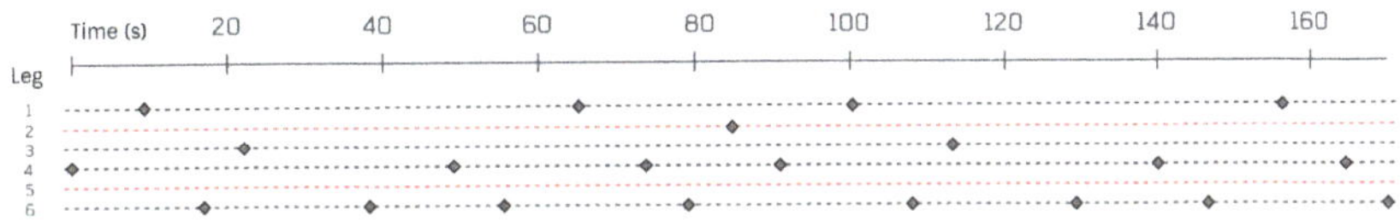

(**b**)

Figure 7. Walking patterns automatically generated for different numbers of legs detected. Each point represents the reallocation of a leg, that is, the turn of a leg to move. For example, in the upper walking pattern, first the fourth leg moves, second the fifth leg, third the second leg, etc. (**a**) Six legs detected. (**b**) Four legs detected. Red lines represent the disabling of a leg.

Figure 8 shows the motion sequences that the robot follows during a walking pattern. Furthermore, we tested how capable the robot is of moving its center to desired positions and orientations. A video summary of the robot's movements during this simulation is found in https://youtu.be/ex1Dj-uwluE, accessed on 12 October 2021.

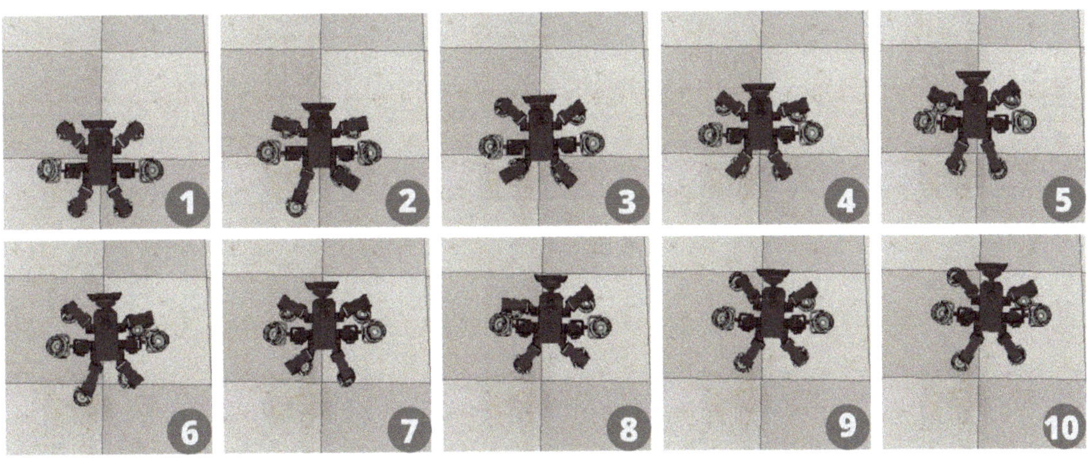

Figure 8. Motion sequences during a walking pattern. The two center legs are disabled, pointing the suction cup up.

In the tests with the real robot, it is crucial to determine the time the suction cups spend to be attached to the ground or wall and the amount of time they spend to be detached. These times are 0.5 s and 1.5 s, respectively. The tests were carried out in the ROMHEX robot, to check the feasibility of our approach for its implementation in the ROMERIN robot (a modular climber robot for infrastructure inspection) [28]. The tests reveal a good performance during the movements in the horizontal plane. However, the tests on the sloping wall reveal hardware problems. The first problem is related to the suction cups, which have three free joints. These joints make the suction cup focus down instead of against the wall, spoiling the correct pulling force. As a result, one of the free joints has been removed, while another has been limited in movement. Once the first problem was solved, the second problem involved the grip force of the suction cups. The maximum inclination that the robot can manage to hold by itself is 60°. However, in this situation, a small perturbation may make the robot fall. The walking pattern during the tests with ROMHEX is shown in Figure 9 with successful results. The video of the robot moving can be found in https://youtu.be/-ASO8B4THEU, accessed on 12 October 2021.

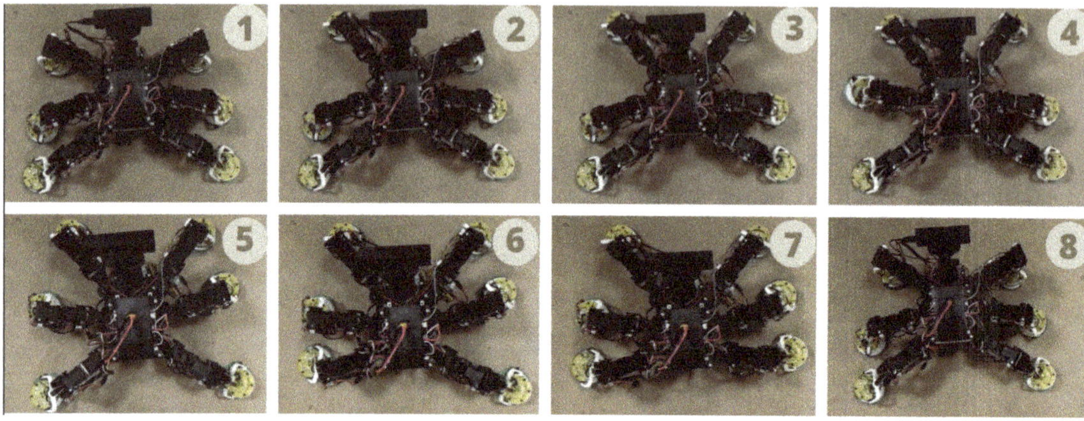

Figure 9. Motion sequences during a walking pattern with ROMHEX.

Finally, the control architecture has been tested and found to work when the robot loses more legs than allowed. For example, if the hexapod robot loses three legs, it is statically unstable, but it can stay still with three legs supporting its weight.

6. Conclusions

First of all, implementing the described control has completed the task of making the robot capable of walking in any direction while maintaining safety. Thanks to behavior-based control, it has been possible to divide the global problem into smaller and more encompassing parts, obtaining a more modular control. This structure also allows adding new functionality in a simple way, by adding layers in the control without changing the current control. The generality of the system allows using a large part of the control with any legged robot typically between four to eight legs, because the majority of legged-and-climber robots dispose of these number of legs. However, the control architecture could be used for a legged robot of more than eight legs, because there is no upper limit.

We achieve a generic control for a robot with an unpredefined number of legs. A cyclic walking pattern has been tested in the hexapod ROMHEX robot with successful results, even when the robot suffers a malfunction of two legs. Taking advantage of the agent-based structure, the system may be improved with the easy addition of new agents over the used standard framework ROS.

Optimizing the initial position of the legs allows increasing the mobility of the robot and obtaining a better understanding of how the forces are distributed when walking. As the movement is generated dynamically, it sometimes reaches a configuration where it cannot easily move. In this case, all legs are reconfigured to this optimized initial position, which allows the robot to continue moving easily. The tests carried out with the real robot demonstrate its potential for climbing, although the hardware may undergo some modifications. Each iteration carried out on the robot has improved its ability to walk, and increase knowledge about the effects of gravity.

All results and changes made with the current robot, as well as improving its ability to move and climb correctly, serve as inspiration for designing future robots. It is important to consider all the details in which ROMHEX fails to obtain a more complete and robust platform in these designs.

Contrasting with state of art, this paper presents a new architecture especially created for legged-and-climber robots, where the number of layers is reduced from the typical three-layer architecture [30] to only two layers, as done previously in CLARAty and COTAMA. Unlike CLARAty, where the internal behaviors are open to the developer, we define specific behaviors for legged-and-climber. Unlike COTAMA architecture, we dispense with the supervisors and scheduler, to particularize our problem.

Author Contributions: Conceptualization, M.H., M.A., C.P. and E.G.; methodology, M.H. and M.A.; software, M.A.; validation, M.A.; formal analysis, M.H. and M.A.; investigation, M.H. and M.A.; resources, M.H.; data curation, M.A.; writing—original draft preparation, C.P.; writing—review and editing, C.P. and E.G.; visualization, M.A. and C.P; supervision, M.H.; project administration, M.H. and E.G.; funding acquisition, M.H. and E.G. All authors have read and agreed to the published version of the manuscript.

Funding: This research is part of The ROMERIN project (DPI2017-85738-R) funded by the Spanish Ministry of Science and Innovation (RETOS research and innovation program).

Institutional Review Board Statement: Not applicable.

Informed Consent Statement: Not applicable.

Data Availability Statement: Not applicable.

Conflicts of Interest: The authors declare no conflict of interest.

Abbreviations

The following abbreviations are used in this manuscript:

ROMHEX	Romerin Hexapod
SLERP	Spherical linear interpolation
COM	Center of mass
GUI	Graphical user interface
ROMERIN	Modular Climber Robot for Infrastructure Inspection
ROS	Robot Operating System

References

1. SPARC. The Partnership for Robotics in Europe. Robotics 2020 Multi-Annual Roadmap for Robotics in Europe. Horizon 2020. Available online: https://www.eu-robotics.net/sparc/about/roadmap (accessed on 27 August 2021).
2. Sostero, M. *Automation and Robots in Services: Review of Data and Taxonomy*; European Commission, Joint Research Centre (JRC): Seville, Spain, 2020.
3. Shen, W.; Gu, J.; Shen, Y. Proposed wall climbing robot with permanent magnetic tracks for inspecting oil tanks. In Proceedings of the IEEE International Conference Mechatronics and Automation, Niagara Falls, ON, Canada, 29 July–1 August 2005; Volume 4, pp. 2072–2077.
4. Yi, S.J.; Lee, D.D. Dynamic heel-strike toe-off walking controller for full-size modular humanoid robots. In Proceedings of the 2016 IEEE-RAS 16th International Conference on Humanoid Robots (Humanoids), Cancun, Mexico, 15–17 November 2016; pp. 395–400.
5. Park, I.W.; Kim, J.Y.; Oh, J.H. Online biped walking pattern generation for humanoid robot khr-3 (kaist humanoid robot-3: Hubo). In Proceedings of the 2006 6th IEEE-RAS International Conference on Humanoid Robots, Genova, Italy, 4–6 December 2006; pp. 398–403.
6. Vladareanu, V.; Boscoianu, C.; Munteanu, R.I.; Yu, H.; Vladareanu, L. Dynamic control of a walking robot using the versatile intelligent portable robot platform. In Proceedings of the IEEE 2015 20th International Conference on Control Systems and Computer Science, Bucharest, Romania, 27–29 May 2015; pp. 38–45.
7. Wooldridge, M. Intelligent agents. *Multiagent Syst.* **1999**, *6*, 3–50.
8. Jamroga, W.; Ågotnes, T. What agents can achieve under incomplete information. In Proceedings of the Fifth International Joint Conference On Autonomous Agents and Multiagent Systems, Hakodate, Japan, 8–12 May 2006; pp. 232–234.
9. El Jalaoui, A.; Andreu, D.; Jouvencel, B. Contextual Management of Tasks and Instrumentation within an AUV control software architecture. In Proceedings of the IEEE International Workshop on Intelligent Robots and Systems (IROS), Beijing, China, 9–15 October 2006; pp. 3761–3766.
10. El Jalaoui, A.; Andreu, D.; Jouvencel, B. A control architecture for contextual tasks management: Application to the AUV Taipan. In Proceedings of the IEEE Europe Oceans 2005, Brest, France, 20–23 June 2005; Volume 2, pp. 752–757.

11. Durand, B.; Godary-Dejean, K.; Lapierre, L.; Passama, R.; Crestani, D. Fault tolerance enhancement using autonomy adaptation for autonomous mobile robots. In Proceedings of the IEEE 2010 Conference on Control and Fault-Tolerant Systems (SysTol), Nice, France, 6–8 October 2010; pp. 24–29.
12. Muscettola, N.; Dorais, G.A.; Fry, C.; Levinson, R.; Plaunt, C.; Clancy, D. Idea: Planning at the Core of Autonomous Reactive Agents. In Proceedings of the 3rd International NASA Workshop on Planning and Scheduling for Space, Houston, TX, USA, 27–29 October 2002.
13. Alami, R.; Chatila, R.; Fleury, S.; Ghallab, M.; Ingrand, F. An architecture for autonomy. *Int. J. Robot. Res.* **1998**, *17*, 315–337. [CrossRef]
14. Volpe, R.; Nesnas, I.; Estlin, T.; Mutz, D.; Petras, R.; Das, H. The CLARAty architecture for robotic autonomy. In Proceedings of the 2001 IEEE Aerospace Conference Proceedings (Cat. No. 01TH8542), Big Sky, MT, USA, 10–17 March 2001; Volume 1, pp. 1–121.
15. Musliner, D.J.; Durfee, E.H.; Shin, K.G. CIRCA: A cooperative intelligent real-time control architecture. *IEEE Trans. Syst. Man Cybern.* **1993**, *23*, 1561–1574. [CrossRef]
16. Simon, D.; Espiau, B.; Castillo, E.; Kapellos, K. Computer-aided design of a generic robot controller handling reactivity and real-time control issues. *IEEE Trans. Control Syst. Technol.* **1993**, *1*, 213–229. [CrossRef]
17. Jakimovski, B.; Meyer, B.; Maehle, E. Self-reconfiguring hexapod robot OSCAR using organically inspired approaches and innovative robot leg amputation mechanism. In Proceedings of the International Conference on Automation, Robotics and Control Systems (ARCS-09), Orlando, FL, USA, 13–16 July 2009.
18. Maehle, E.; Brockmann, W.; Grosspietsch, K.E.; Auf, A.E.S.; Jakimovski, B.; Krannich, S.; Litza, M.; Maas, R.; Al-Homsy, A. Application of the organic robot control architecture ORCA to the six-legged walking robot OSCAR. In *Organic Computing—A Paradigm Shift for Complex Systems*; Springer: Basel, Switzerland, 2011; pp. 517–530.
19. Brockmann, W.; Maehle, E.; Grosspietsch, K.E.; Rosemann, N.; Jakimovski, B. ORCA: An organic robot control architecture. In *Organic Computing—A Paradigm Shift for Complex Systems*; Springer: Basel, Switzerland, 2011; pp. 385–398.
20. Maas, R.; Maehle, E.; Großpietsch, K.E. Applying the organic robot control architecture ORCA to cyber-physical systems. In Proceedings of the IEEE 2012 38th Euromicro Conference on Software Engineering and Advanced Applications, Cesme, Turkey, 5–8 September 2012; pp. 250–257.
21. Pack, R.T.; Christopher, J.L.; Kawamura, K. A rubbertuator-based structure-climbing inspection robot. In Proceedings of the IEEE International Conference on Robotics and Automation, Albuquerque, NM, USA, 25 April 1997; Volume 3, pp. 1869–1874.
22. Brooks, R. A robust layered control system for a mobile robot. *IEEE J. Robot. Autom.* **1986**, *2*, 14–23. [CrossRef]
23. Rönnau, A.; Heppner, G.; Nowicki, M.; Dillmann, R. LAURON V: A versatile six-legged walking robot with advanced maneuverability. In Proceedings of the 2014 IEEE/ASME International Conference on Advanced Intelligent Mechatronics, Besacon, France, 8–11 July 2014; pp. 82–87.
24. Fankhauser, P.; Bellicoso, C.D.; Gehring, C.; Dubé, R.; Gawel, A.; Hutter, M. Free gait—An architecture for the versatile control of legged robots. In Proceedings of the 2016 IEEE-RAS 16th International Conference on Humanoid Robots (Humanoids), Cancun, Mexico, 15–17 November 2016; pp. 1052–1058.
25. Hutter, M.; Gehring, C.; Jud, D.; Lauber, A.; Bellicoso, C.D.; Tsounis, V.; Hwangbo, J.; Bodie, K.; Fankhauser, P.; Bloesch, M.; et al. Anymal—A highly mobile and dynamic quadrupedal robot. In Proceedings of the 2016 IEEE/RSJ International Conference on Intelligent Robots and Systems (IROS), Daejeon, Korea, 9–14 October 2016; pp. 38–44.
26. Hernando, M.; Gómez, V.; Brunete, A.; Gambao, E. CFD Modelling and Optimization Procedure of an Adhesive System for a Modular Climbing Robot. *Sensors* **2021**, *21*,1117. [CrossRef] [PubMed]
27. Ko, H.; Yi, H.; Jeong, H.E. Wall and ceiling climbing quadruped robot with superior water repellency manufactured using 3D printing (UNIclimb). *Int. J. Precis. Eng. Manuf.-Green Technol.* **2017**, *4*, 273–280. [CrossRef]
28. Hernando, M.; Brunete, A.; Gambao, E. ROMERIN: A Modular Climber Robot for Infrastructure Inspection. *IFAC-PapersOnLine* **2019**, *52*, 424–429. [CrossRef]
29. Barrera, T.; Hast, A.; Bengtsson, E. Incremental spherical linear interpolation. In *The Annual SIGRAD Conference. Special Theme-Environmental Visualization*; Linköping University Electronic Press: Linköping, Sweden, 2004; pp. 7–10.
30. Gat, E.; Bonnasso, R.P.; Murphy, R. On three-layer architectures. *Artif. Intell. Mob. Robot.* **1998**, *195*, 210.

applied sciences

MDPI

Article

Data-Driven Control Algorithm for Snake Manipulator

Kai Hu [1,2,*], Lang Tian [1,3], Chenghang Weng [1], Liguo Weng [1,2], Qiang Zang [1,2], Min Xia [1,2] and Guodong Qin [4]

1 School of Automation, Nanjing University of Information Science & Technology, Nanjing 210044, China; 20181223062@nuist.edu.cn (L.T.); 20201249145@nuist.edu.cn (C.W.); 002311@nuist.edu.cn (L.W.); zangq@nuist.edu.cn (Q.Z.); xiamin@nuist.edu.cn (M.X.)
2 Jiangsu Collaborative Innovation Center of Atmospheric Environment and Equipment Technology (CICAEET), Nanjing University of Information Science & Technology, Nanjing 210044, China
3 China Telecom Stocks Co., Ltd., Zhangjiagang Branch, Zhangjiagang 215600, China
4 College of Mechanical & Electrical Engineering, Nanjing University of Aeronautics and Astronautics, Nanjing 210001, China; guodongqin@nuaa.edu.cn
* Correspondence: 001600@nuist.edu.cn

Abstract: In some environments where manual work cannot be carried out, snake manipulators are instead used to improve the level of automatic work and ensure personal safety. However, the structure of the snake manipulator is diverse, which renders it difficult to establish an environmental model of the control system. It is difficult to obtain an ideal control effect by using the traditional manipulator control method. In view of this, this paper proposes a data-driven snake manipulator control algorithm. After collecting data, the algorithm uses the strong learning and decision-making ability of the deep deterministic strategy gradient to learn these system data. A data-driven controller based on the deep deterministic policy gradient was trained in order to solve the manipulator system control problem when the control system environment model is uncertain or even unknown. The data of simulation experiments show that the control algorithm has good stability and accuracy in the case of model uncertainty.

Keywords: deep deterministic policy gradients; snake manipulator; data-driven; accuracy

Citation: Hu, K.; Tian, L.; Weng, C.; Weng, L.; Zang, Q.; Xia, M.; Qin, G. Data-Driven Control Algorithm for Snake Manipulator. *Appl. Sci.* **2021**, *11*, 8146. https://doi.org/10.3390/app11178146

Academic Editor: Alessandro Gasparetto, Stefano Seriani and Lorenzo Scalera

Received: 6 August 2021
Accepted: 30 August 2021
Published: 2 September 2021

1. Introduction

Existing manipulator control theory can be divided into three categories: (1) Accurate mathematical models are required, such as optimal control strategies, linear or nonlinear control strategies, and pole assignment methods. Some (2) mathematical models are known, such as sliding mode variable structure control, fuzzy control, adaptive control, and intelligent control. (3) The mathematical model is unknown, or it is difficult to establish a mathematical model, such as iterative learning control, model-free adaptive control, and other data-driven control strategies [1]. Currently, the commonly adopted control strategies of manipulators include PID control, fuzzy control, adaptive control, and hybrid control strategy. Mendes designed an adaptive fuzzy controller to solve the contact problem of a manipulator [2]. However, with continuous improvement of the control requirements of manipulators, the scale of control systems is increasing, there are a large number of coupling phenomena between the systems, and the traditional manipulator control strategy has been unable to meet the control requirements. Due to their strong self-learning ability and nonlinear system mapping ability, neural networks have been introduced into manipulator control to compensate for the uncertainty of manipulator models. Aiming at the trajectory tracking control problem caused by uncertainty and disturbance of the manipulator, Vu proposed a robust adaptive control strategy based on a fuzzy wavelet neural network system with dynamic structure. The control strategy can effectively reduce system error and can improve the control accuracy of the manipulator system [3,4]. Concerning the problem of the unknown dynamic model of a manipulator, Yu proposed an adaptive neural network tracking control strategy based on a disturbance observer, which is used to compensate for the unknown disturbance of the system [5]. Jung proposed an improved sliding mode control method based on an RBF neural network to solve the problem of nonlinear function gain selection of a sliding mode controller and the uncertainty of the three-link manipulator model [6].

The hybrid control strategy of neural networks [7] and classical strategy can improve the control performance of the manipulator and can improve its application in many fields, such as stirring, welding, polishing, and assembly. However, with continuous improvement in industrial production accuracy, the neural network model's uncertainty compensation has been unable to meet control accuracy requirements. Therefore, the deep neural network algorithm was introduced into manipulator control. Due to its strong perception and decision-making ability, deep reinforcement learning can perceive the response of the environment to change and improve the accuracy of the behavior of the agent. Therefore, deep reinforcement learning is more widely used in deep neural network algorithms [8,9].

Deep Reinforcement Learning (DRL) is an artificial intelligence method that combines deep learning with a perceptual ability and reinforcement learning with a decision-making ability. DRL can be divided into two categories: value-based function and strategy-based gradient. The value-based learning algorithm is mainly an approximate representation of the value function. The representative algorithms are the Deep Q Network (DQN) algorithm, Nature DQN algorithm, Double DQN algorithm [10], prioritized replay DQN algorithm [11], and Dueling DQN algorithm [12]. The representative algorithm based on strategy learning is the Policy Gradient algorithm [13]. The algorithms that combines strategy and value are the Actor-Critic algorithm [14], the Deep Deterministic Policy Gradient (DDPG) algorithm, and the Asynchronous Advantage Actor-Critic (A3C) [15]. The algorithms are summarized in Table 1 [16].

Table 1. Classical Deep reinforcement learning algorithms.

Classification Algorithm	Algorithm Name	Algorithm Improvement
Value-based reinforcement learning (DQN)	Nature DQN	Two identical Q network structures.
	Double DQN	A choice of action between decoupling target Q value and the calculation of target Q value .
	Prioritized replay DQN	The sample is prioritized.
	Dueling DQN	The value function of the Q network is divided into two parts: value function and advantage function.
Strategy-based reinforcement learning method	Policy gradient	Value-based methods are replaced by policy-based methods.
Hybrid algorithm actor-critic	Actor-critic	The two methods, namely policy-based and value-based, are combined.
	Asynchronous advantage actor-critic (A3C)	Asynchronous training framework, network structure optimization, and evaluation point optimization. A general asynchronous concurrent reinforcement learning framework
	Deep deterministic policy gradient (DDPG)	Two actor networks and two critic networks, a total of four neural networks, are used to update the model parameters iteratively.

DRL has been proven to be effective in solving complex control problems of manipulators in OpenAI, such as operating [17], grasping [18,19], and mobile tasks [20,21]. Luo applied the DRL strategy to the assembly task of a manipulator. The device completed a task that could not be realized using the traditional control strategy [22]. Through the priority division of the DRL network, Wu realized high-precision millimeter-scale automatic assembly technology [23]. Wen designed an obstacle avoidance algorithm based on DDPG, which solved the convergence problem of the obstacle avoidance motion of the manipulator and ensured its continuity and stability [24].

The contribution of this paper is to propose a data-driven snake manipulator control strategy to solve the control problem of snake manipulators in some complex environments. The main work is as follows:

(1) Based on the model control method, a control system of a two-link model snake manipulator based on DDPG was designed. First, according to the structure of the snake manipulator and the Lagrangian dynamic equation, the dynamic model of the two-link snake manipulator was established. Second, the Q network model and action network model of the DDPG agent were designed. Finally, the simulation results show that the control strategy based on DDPG has good convergence and strong anti-jamming ability.

(2) In order to solve the problem of model uncertainty, a data-driven control method of the snake manipulator is proposed. When the number of connecting rods of the serpentine manipulator increases and the environment becomes complex, it is difficult to establish the model of an integral snake manipulator that renders the control effect of the model-based control strategy poor. In view of this, a data-driven control method based on DDPG is proposed. First, the data-driven control method requires a large number of input and output data, and the data set was established by using the results of the traditional control method as the training sample of this method. Second, the DDPG agent model was designed according to the input and output parameters. Finally, based on simulation and comparative results analysis, the feasibility and superiority of the control method was verified.

The main sections of this paper are as follows. The Section 1 mainly introduces the control methods currently applied to the manipulator. It briefly combs through the control method of the manipulator, focusing on the DRL algorithm and its application in the manipulator. Based on this, aiming at the problems in the application of the snake manipulator, the main innovation of this paper is pointed out. Combined with previous research work on DRL in manipulators, Section 2 outlines the chosen DRL algorithm and DDPG algorithm used in this study. The main structure, workflow, and related calculation methods of DDPG are described in detail. In the Section 3, the design of the DDPG control method simulation of the two-link snake manipulator is explained. The method of establishing the environment object and agent of the DDPG control system is introduced in detail. This mainly includes the two-link dynamic model of the serpentine manipulator, Q network design, and action network design. The simulation results show that the DDPG algorithm is effective and superior in the control of snake manipulators. Section 4, based on the research outlined in the Section 3, reports that it is difficult to establish the entire snake manipulator model, with significant error in the accuracy of the model, which reduces the stability and accuracy of the control system. In view of this, a data-driven control method of the snake manipulator is proposed in this paper. The simulation experiment of this method was designed, and the simulation results verify the feasibility of the method. In addition, this method not only avoids the complicated task of establishing the manipulator model but also improves the stability and accuracy of the control system.The data-driven control algorithm uses the existing motion data of the serpentine robot. All the motion data of the serpentine robot are integrated, and its motion database is established. The DDPG is used to train using the data in the database, and the current best motion path is obtained. With the increasing amount of data in the database, the accuracy and stability of robot control continue to improve. Section 5 summarizes the work of this paper.

2. Deep Deterministic Policy Gradient

In the study of motion control of a two-link manipulator, Jianping Wang compared 3 kinds of DRL algorithms: A3C, DPPO, and DDPG. He found that the DDPG algorithm has the best convergence effect on the control system. The convergence reward value of the algorithm is the most stable [25]. Since the DDPG algorithm has a better control effect on the manipulator [26] and the motion of the snake manipulator is mainly continuous, the application of the DDPG algorithm was studied in this work. DDPG adds a deterministic strategy network on the basis of the DQN to output action values; thus, compared with the DQN, it only needs to learn the Q network, while DDPG also needs to learn the strategy network. The network structure of DDPG is shown in Figure 1.

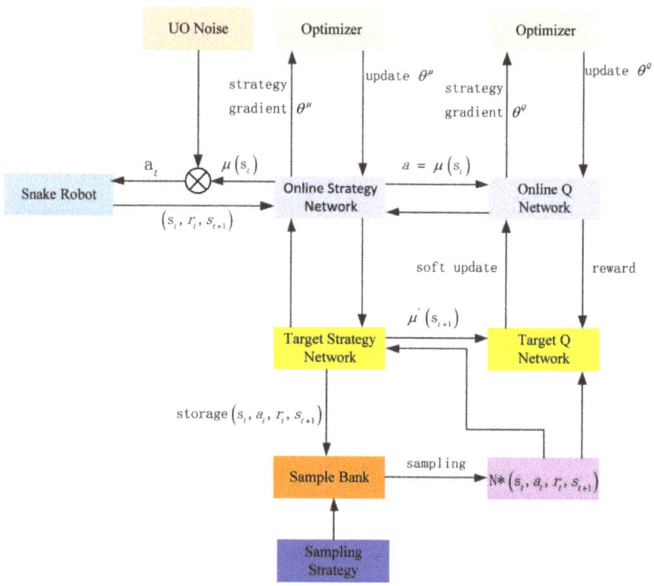

Figure 1. DDPG network structure.

From the figure, we can observe that DDPG has four networks: the current action network, the target action network, the current Q network, and the target Q network. The current action network is mainly used to update the policy network parameter δ^{μ}. The network selects the current action A through the current state S. By utilizing target action network sampling, one state S' selects the optimal next action, and its network parameter $\delta^{\mu'}$ duplicates the parameter value of δ^{μ}. The current Q network is mainly used to update the value network parameter θ^{Q} and to calculate the current Q value. The target Q network is used to calculate the Q' part of the Q value, and the network parameter $\delta^{Q'}$ copies the parameter θ^{Q}. The two processes of updating network parameters are different. The current network uses the SGD algorithm to update network parameters. The target network uses the soft update algorithm to update the network parameters, as shown in Equation (1). By using the soft update algorithm, the change in the target network parameters is small, and the training is easy to converge, but the learning speed is slow.

$$\begin{aligned} \theta^{Q'} &\leftarrow \delta\theta^{Q} + (1-\tau)\theta^{Q'} \\ \theta^{\mu'} &\leftarrow \delta\theta^{\mu} + (1-\tau)\theta^{\mu'} \end{aligned} \quad (1)$$

In Equation (1), $\theta^{Q'}$ denotes the target Q network parameters. θ^{Q} denotes the current Q network parameters. The current network parameters θ^{μ} and target network parameters $\theta^{\mu'}$ are constructed by policy μ. δ is the update coefficient, and the value is often 0.1 or 0.01.

For the definition of the network loss function, the current Q network loss function is similar to that in the DQN, mainly using mean square error, as shown in Formula (2).

$$Loss = \sum_{i=1}^{m} (r_i - Q(s_i, a_i, \omega))^2 \tag{2}$$

In Formula (2), Loss denotes the loss function. r_i denotes the reward value of i moment. $Q(s_i, a_i, \omega)$ denotes the current Q network Q value calculation function. Since the deterministic strategy is adopted in the current action network, its loss function is different from that of PG and A3C.

$$\nabla_J(\theta) = \sum_{i}^{m} [\nabla_a Q(s_i, a_i, \omega) \mid_{s=s_i, a=\pi_\theta(s)} * \nabla_{\theta * \pi_{theta}}(s) \mid_{s=s_i}] \tag{3}$$

In Formula (3), ∇ denotes the gradient decreases. $\pi_\theta(s)$ denotes the action selection strategy. For the loss function, the greater the Q value of the target action, the smaller the Loss. The smaller the Q value, the greater the Loss; thus, the status network returns to a negative Q value.

$$J(\theta) = -\sum_{i}^{m} Q(s_i, a_i, \omega) \tag{4}$$

In Formula (4), s_i denotes the state of the manipulator at i moment, including joint angle and angular velocity. a_i denotes the moment value of the manipulator at time i, which is transformed from the state s_i to s_{i+1} through the dynamic model.

$$r_{Time} = \sum_{i=t}^{Time} \gamma^{i-t} * r(s_i, a_i) \tag{5}$$

In Formula (5), γ denotes the weighted value of the reward, the range of the value is $[0 \quad 1]$, and *Time* denotes the total time. $r(s_i, a_i)$ is used to calculate the single-step reward value obtained by the dynamic model after performing the behavior a_i in the state s_i. r_{Time} denotes the weighted total value of all single-step awards r from the current state to a certain state throughout the process.

3. Simulation of DDPG Control Based on 2-Link Model

In order to verify whether the DDPG algorithm can be feasibly and effectively applied to the space manipulator, a DDPG control simulation based on two connecting rods was designed, as outlined in this section. In addition, a 2-link model lays the foundation for the establishment of the whole model.

3.1. Overall Design Scheme of Control System

The DDPG-based control system is shown in Figure 2. According to the given parameters of the target trajectory point, the angular value, angular velocity value, and angular acceleration value of each joint moving to the target point are calculated by using the ridge method. These parameters are used as input to the DDPG controller. The DDPG agent is trained, saving the agent that meets the conditions, and finally the state value with the lowest reward value is the output. The position of the actual trajectory point is obtained by the forward kinematics of the manipulator. The control precision value is obtained from the error between the actual position and the target position.

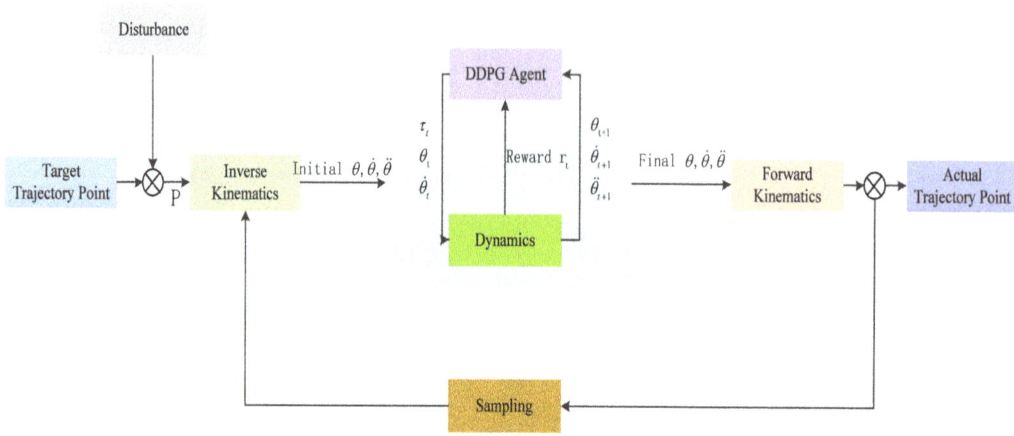

Figure 2. Control block diagram based on DDPG control system.

3.2. Environmental Object Establishment

A modular design method was adopted for the snake manipulator in this study to facilitate equipment maintenance. In order to improve the flexibility of the manipulator, a 1-joint 2-degrees-of-freedom structure was adopted to enable the manipulator to move in both the yaw and pitch directions, and the three-dimensional motion of the manipulator was realized. A structure diagram of the snake manipulator model is shown in Figure 3. Due to the large number of structural joints, a 2-joint structure was used as an example in this study to establish the dynamic model.

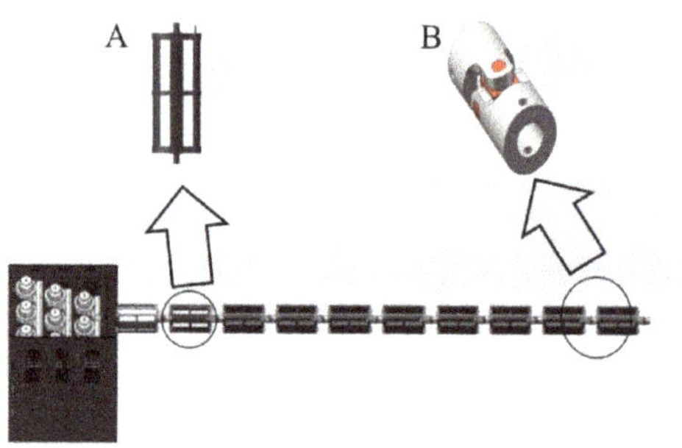

Figure 3. Structural model diagram of snake manipulator. (**A**) is the connecting rod structure and (**B**) is the universal joint structure.

(1) Definition of input signal: In this experiment, the joint torque of the 2-link manipulator was used as the input signal of the environment object.

(2) Establishing a dynamic equation to evaluate the environment: The dynamic model of the two connecting rods was established according to the Lagrangian dynamic method. The joint coordinate system of the manipulator is shown in Figure 4. The yaw angle is α. The pitch angle is β.

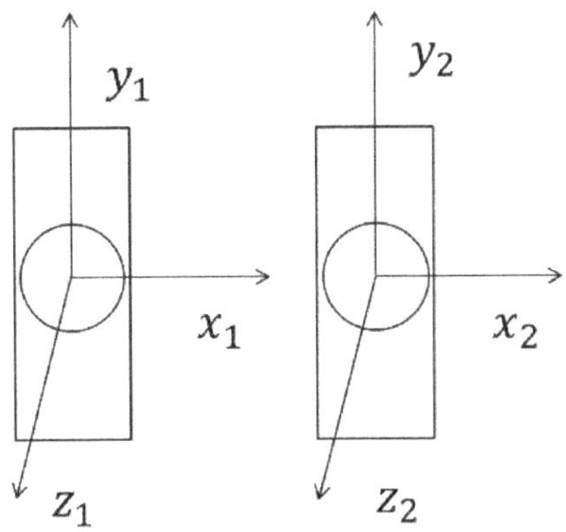

Figure 4. Joint coordinate system.

Suppose that the mass of the connecting rod is concentrated at the center of mass; the inertia tensor of the connecting rod is set to 0. Therefore, according to the Lagrange formula, the kinetic energy and potential energy of each connecting rod are expressed as follows.

$$E_{k1} = \frac{1}{2} m_1 \dot{P}_{c1}^T \dot{P}_{c1} \tag{6}$$

$$E_{p1} = m_1 g L_{c1} sin(\beta_1) \tag{7}$$

$$E_{k2} = \frac{1}{2} m_2 \dot{P}_{c2}^T \dot{P}_{c2} \tag{8}$$

$$E_{p2} = m_2 g(asin(\beta_1) + L_{c2} sin(\beta_1 + \beta_1)) \tag{9}$$

In Formulas (6)–(9), E_{k1} denotes the kinetic energy of the first connecting rod; E_{p1} denotes the potential energy of the first connecting rod; $\dot{P_{c1}}$ denotes the velocity at the center of mass of the first connecting rod; "g" denotes gravitational acceleration; "a" represents the length of the connecting rod; and L_{c1} denotes the length of the center of mass of the first connecting rod, etc. The definitions are consistent for all formulas in this paper.

$$P_{c1} = \begin{bmatrix} L_{c1} cos(\alpha_1) cos(\beta_1) \\ L_{c1} sin(\beta_1) \\ -L_{c1} sin(\alpha_1) cos(\beta_1) \end{bmatrix} \tag{10}$$

$$P_{c1} = \begin{bmatrix} a * cos(\alpha_1) cos(\beta_1) + L_{c2} cos(\alpha_1 + \alpha_2) cos(\beta_1 + \beta_2) \\ a * sin(\beta_1) + L_{c2} sin(\beta_1 + \beta_2) \\ -a * sin(\alpha_1) cos(\beta_1) - L_{c2} sin(\alpha_1 + \alpha_2) cos(\beta_1 + \beta_2) \end{bmatrix} \tag{11}$$

$\theta_1 = \begin{bmatrix} 0 & \alpha_1 & \beta_1 \end{bmatrix}^T$ denotes the rotation angle of joint 1. $\dot{\theta_1} = \begin{bmatrix} 0 & \dot{\alpha_1} & \dot{\beta_1} \end{bmatrix}^T$ denotes the angular velocity of joint 1. $\ddot{\theta_1} = \begin{bmatrix} 0 & \ddot{\alpha_1} & \ddot{\beta_1} \end{bmatrix}^T$ denotes the angular acceleration of joint 1. The angular value, angular velocity, and angular acceleration of joint 2 are the same as those of joint 1. Set the center of mass of the connecting rod to the center of the connecting rod, such that $L_c = \frac{1}{2} a$.

According to the position of the centroid point, the expressions of Formulas (6) and (8) can be obtained, which are $\dot{P}_{c1}^T \dot{P}_{c1}$ and $\dot{P}_{c2}^T \dot{P}_{c2}$ of Formulas (6) and (8), as shown in Formulas (12) and (13).

$$\dot{P}_{c1}^T \dot{P}_{c1} = L_{c1}^2 cos(\beta_1)^2 \dot{\alpha_1}^2 + L_{c1}^2 \dot{\beta_1}^2 \tag{12}$$

$$
\begin{aligned}
\dot{P}_{c2}^{T}\dot{P}_{c2} &= a^2cos(\beta_1)^2\dot{\alpha_1}^2 + a^2\dot{\beta_1}^2 + L_{c2}^2(\dot{\beta_1}+\dot{\beta_2})^2 \\
&+ L_{c2}^2(cos(\beta_1+\beta 2))^2(\dot{\alpha_1}+\dot{\alpha_2})^2 \\
&+ 2aL_{c2}cos(\beta_1)cos(\alpha_2)cos(\beta_1+beta_2)(\dot{\alpha_1}^2+\dot{\alpha_2}) \\
&+ 2aL_{c2}sin(\beta_1)cos(\alpha_2)sin(\beta_1+beta_2)(\dot{\beta_1}^2+\dot{\beta_2}) \\
&+ 2aL_{c2}cos(\beta_1)cos(\beta_1+\beta_2)(\dot{\beta_1}^2+\dot{\beta_2})
\end{aligned}
\tag{13}
$$

L_{c2} in Formulas (12) and (13) denotes the centroid length of connecting rod 2, which is only half of the length of a connecting rod. Formula (12) is substituted into Formula (6). Formula (13) is substituted into Formula (8). The total kinetic energy of connecting rod 1 and connecting rod 2 can be calculated.

$$
\begin{aligned}
E_k &= E_{k1} + E_{k2} \\
&= \frac{1}{2}\dot{\theta}^T H(\theta)\dot{\theta}
\end{aligned}
\tag{14}
$$

The Lagrangian dynamic formula deduces the dynamic equation by making use of the difference between the kinetic energy and the potential energy of the mechanical system of the manipulator. The second kind of Lagrange equation is shown in Formula (15). The dynamic equation of the manipulator is shown in Formula (16).

$$
\frac{d}{d_t}\frac{\partial L}{\partial\dot{\theta}} - \frac{\partial L}{\partial\theta} = \tau \tag{15}
$$

$$
H(\theta)\ddot{\theta} + C(\theta,\dot{\theta}))\dot{\theta} + G(\theta) = \tau \tag{16}
$$

In Formula (16), $H(\theta)$ denotes the equivalent inertia matrix. $C(\theta,\dot{\theta})$ denotes a Coriolis matrix. $G(\theta)$ denotes the gravity matrix. τ denotes the moment. The complete dynamic expression equation Formula (17) can be deduced according to Formulas (15) and (16).

$$
\ddot{\theta} = [H(\theta)]^{-1}(\tau - C(\theta,\dot{\theta})\dot{\theta} - G(\theta)) \tag{17}
$$

(3) Calculation and updated observations: According to the dynamic equation, the angular acceleration of the joint at the next moment is calculated. The joint angular acceleration integral operation is used to calculate the corresponding angle value and angular velocity value. It is outputted as an observation of the agent.

(4) Calculation of the reward value: The reward value of each sampling is calculated according to the angle value and angular velocity value. The reward value for each process is the sum of all sample reward values. The simulation time for each episode is set to 10 s, and the sampling time is set to 0.05 s. Each episode is sampled 200 times. The process reward value function is set to the following.

$$
r_t = -\sum_{i}^{200} 7(\theta_{i1}+\theta_{i2})^2 + (\dot{\theta_{i1}}+\dot{\theta_{i2}})^2 \tag{18}
$$

In Formula (18), r_t denotes the sum of the reward values of the t process. θ_{i1} denotes the angle value of joint 1 during the i time sampling. $\dot{\theta_{i1}}$ denotes the angle velocity of joint 1 during i time sampling. Joint 2 is defined in the same manner.

3.3. DDPG Agent Establishment

(1) Q network design. The Q network uses a deep convolution neural network with three inputs and one output. The three inputs are the observed value (joint angle value $[\theta_1 \quad \theta_2]$; the angular velocity value $[\dot{\theta_1} \quad \dot{\theta_2}]$), and the action value (torque value $[\tau_1 \quad \tau_2]$), and the output is the evaluation reward value. The input parameter of the observed value input is eight. The network is set to the first full connection layer to set the number of nodes as 128. The activation function is ReLU. The number of nodes in the second layer of the

full connection layer is 200. The input parameter of the action value is four. The number of nodes in the fully connected layer is 200. The final output parameters of the full connection layer of the three inputs are added through the ReLU activation function. The setting parameter of the third layer is one. The output value can be obtained. The Q network configuration is shown in Figure 5. The network optimizer is set to the adaptive matrix estimation (adam) optimizer. The learning rate is set to 0.0005.

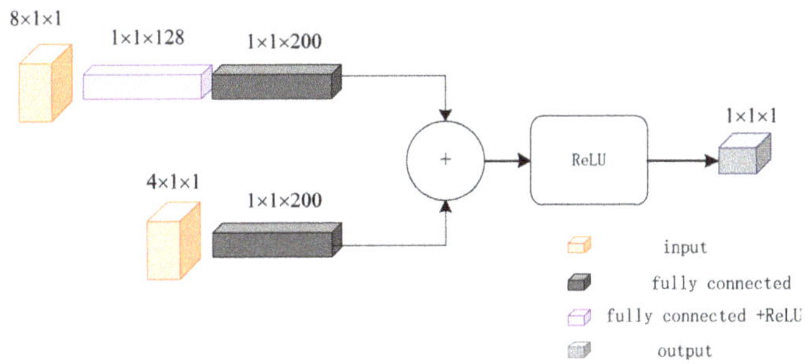

Figure 5. Q network structure design.

(2) Action network design. The action network adopts a deep convolution network with one input and one output. The input is the observed value. The output end is the torque value $[\tau_1 \quad \tau_2]$. The input parameter is eight. The network is set to the first full connection layer to set the number of nodes to 128. The activation function is ReLU. The second full connection layer sets the number of nodes 200. The activation function is ReLU. The number of nodes in the full connection layer of the third layer is set to two. The activation function is tanh. The magnification of the scale layer is five times. The action network configuration is shown in Figure 6. The network optimizer is set to the adaptive matrix estimation (adam) optimizer. The learning rate is set to 0.001.

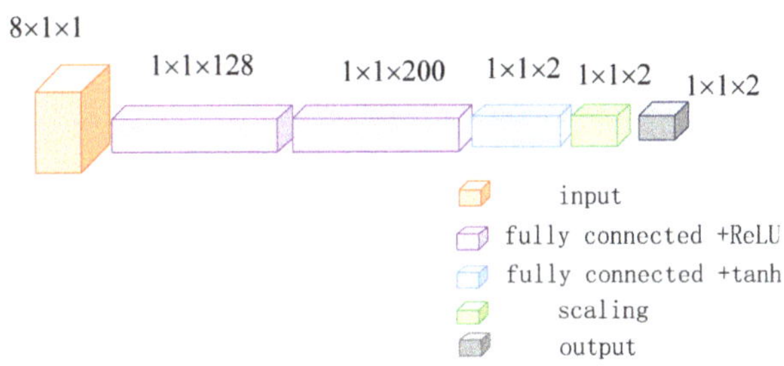

Figure 6. Action network structure design.

(3) Parameter setting of agent training: Training is set at 1800 episodes with 200 samples per episode. The training end mark is the point at which the total value of a certain episode reward is zero or the required number of training episodes is completed.

3.4. Analysis of Experimental Results

This simulation design is a single-point control precision simulation experiment. The sampling time is 0.02 s. Single episode time is 10 s, with a total of 1800 episodes. Two simulation experiments were designed. They involve the additive torque disturbance value $\begin{bmatrix} 1 & 1 \\ 1 & 1 \end{bmatrix}$ and undisturbed value. The training progress is shown in Figures 7 and 8.

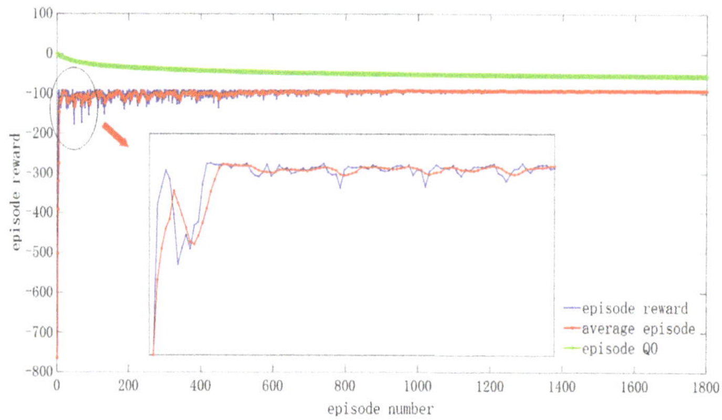

Figure 7. Undisturbed training chart.

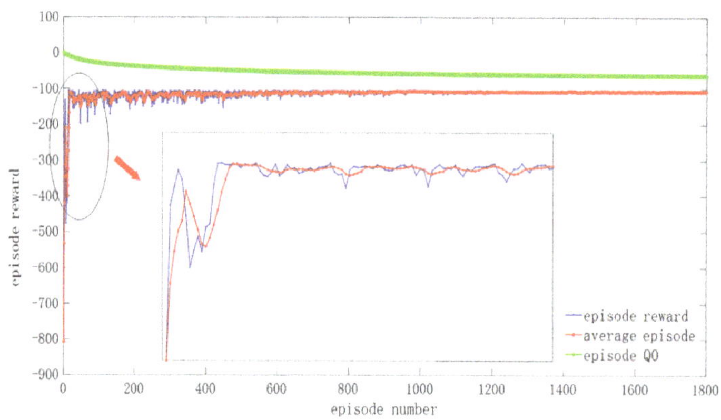

Figure 8. Diagram of disturbance training.

In Figures 7 and 8, the blue curve represents the change in the total reward value for each episode of training. The red curve represents the change curve of the average value of the total episode reward. The cyan curve is the Q value change curve for each episode of the evaluation network. With the increase in the number of training episodes, the fluctuation of the curve tends to be stable, the network finally converges, and the agent reaches the optimal value in this state. Overall, there is no difference between the two situations. Due to the anti-disturbance adjustment of the controller, the curve fluctuation is different in early training. It is proven that the DDPG controller has a strong anti-disturbance ability and strong stability.

Through the 2-joint experimental simulation of the snake manipulator, it is proven that the DDPG algorithm can improve the anti-interference and stability of the manipulator control system. Through many repetitions of training and learning, the position control accuracy of the end of the manipulator can be improved. However, it is difficult to establish the overall model of the snake manipulator. Thus, it is difficult to realize the model-based DDPG control method.

4. Control Simulation of Snake Manipulator Based on Data-Drive

According to the 2-link model, the dynamic model of the snake manipulator is complex and difficult to establish. The modeling of the whole snake manipulator not only is difficult but also involves a large number of parameters. According to the model-free data-driven control method, a data-driven control method based on DDPG is proposed for which an accurate snake manipulator and environment model are not required. This method avoids the influence of model uncertainty on the control system to improve the safety and reliability of the system.

4.1. Data Set Establishment

Since the snake manipulator mainly adopts the joint modular design method, as shown in Figure 3, according to the 2-link model of the snake manipulator, the overall 10-link snake manipulator model can be established. By using the traditional motion control method of the snake manipulator, the motion parameters of the snake manipulator are obtained. The data in the database are mainly based on the snake manipulator using the firework-optimized BP neural network PID (FWA-BPNN-PID) control method to perform varied trajectory motion. Part of the trajectory is shown in Figure 9.

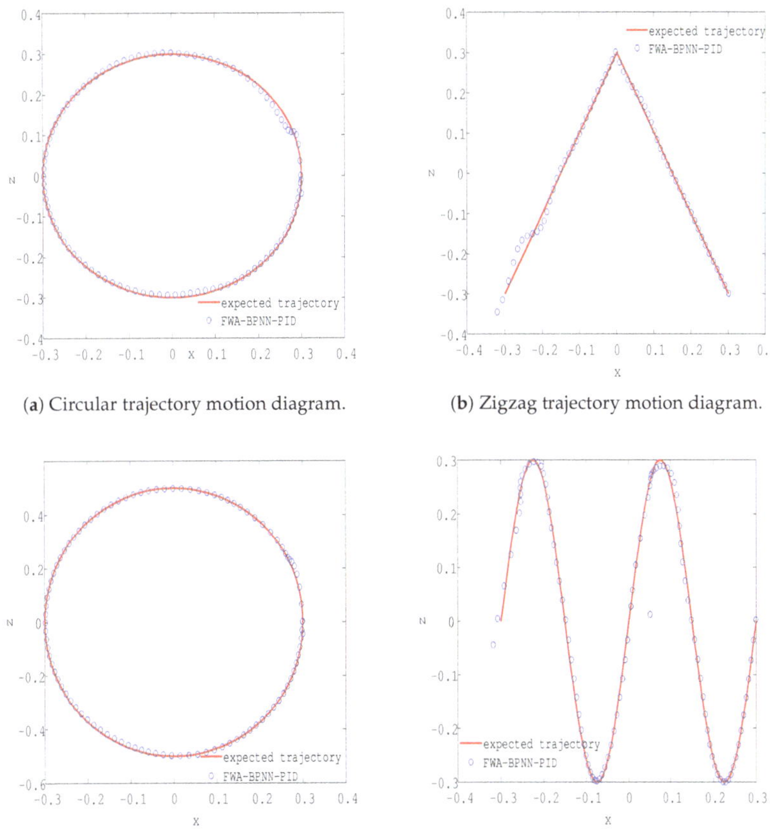

(**a**) Circular trajectory motion diagram. (**b**) Zigzag trajectory motion diagram.

(**c**) Ellipse trajectory motion diagram. (**d**) Sinusoidal trajectory motion diagram.

Figure 9. Partial trajectory motion diagram of snake arm in the database.

The data set of this simulation takes the origin as the center and 0.3 as the radius. The endpoint trajectory moves counterclockwise from the point (0.3), as shown in Figure 9a. Since the trajectory of the endpoint in the figure is designed on the plane of y = 0.65, the coordinate value y of the endpoint of the track is expected to remain the same. The coordinate values of the output result of the FWA-BPNN-PID method in Figure 9a are used as the input data set, and the sampling data are shown in Figure 10b. The desired track position coordinate values in Figure 9a are taken as the output data set, and the sampling data are shown in Figure 10b. The three curves in Figure 10 represent the values of the coordinate location in the data set.

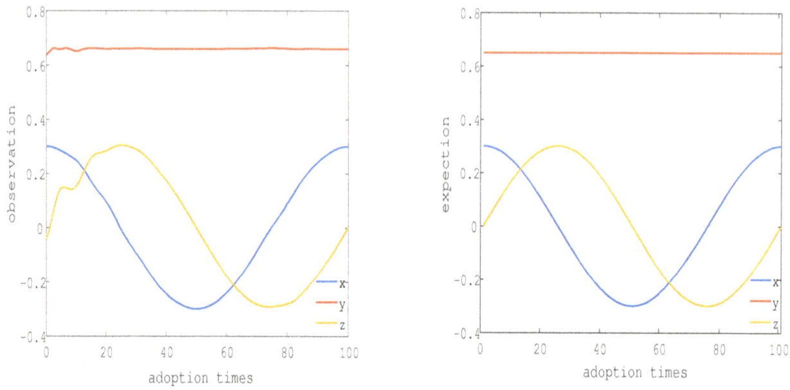

(**a**) Input data set. (**b**) Output data set.

Figure 10. Input and output data set.

4.2. Analysis of Simulation Results

A new DDPG agent was built according to the method described in Section 3.3. The observed value of the agent is the coordinate position $\begin{bmatrix} x_o & y_o & z_o \end{bmatrix}^T$. The output action value is the coordinate location point $\begin{bmatrix} x_a & y_a & z_a \end{bmatrix}^T$. The reward value function is set to the error between the action value and the expected value. The sampling time of the training network is 1 s, the simulation time is 100 s, and the number of samples for a single episode is 100. The number of sample training episodes is set to 3000. The 3000 training results of the agent are shown in Figure 11. In Figure 11, it can be found that with the increase in training times, the reward value of the sample gradually tends toward zero, and the gap between the sample Q value and the sample Q value decreases and tends to be stable. When the sample reward value tends to be stable, it means that the system has converged and the network training has reached the current best state. The closer the sample reward value is to zero, the smaller the system error value is and the higher the system accuracy is.

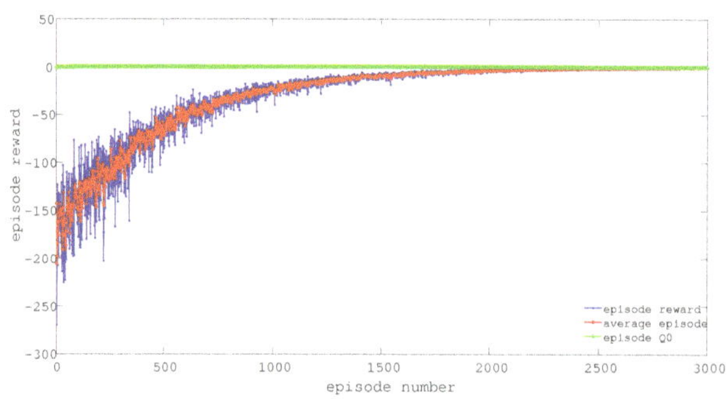

Figure 11. One hundred samples for a single episode and 3000 training maps for agents.

The 12th, 1000th, and 3000 training states were randomly selected, and the error values are shown in Table 2. The output results and error changes of the control system during training are shown in Figure 12.

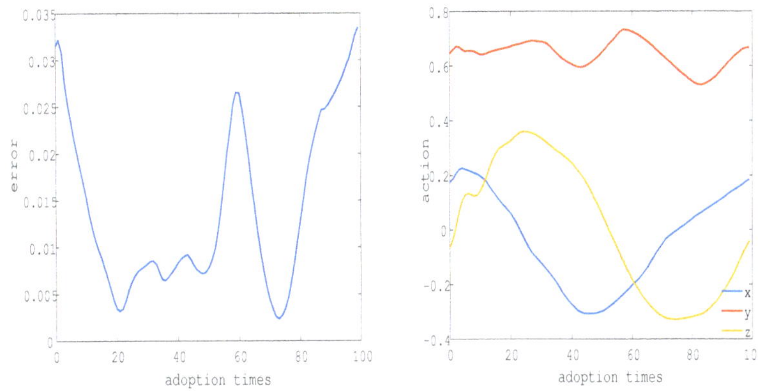

(**a**) 12th training state.

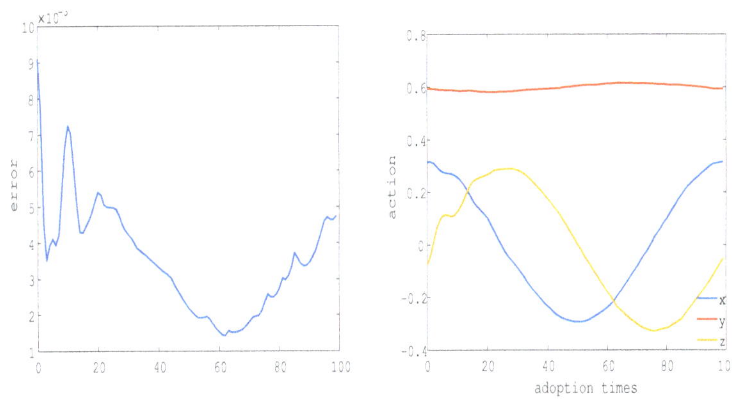

(**b**) 1000th training state.

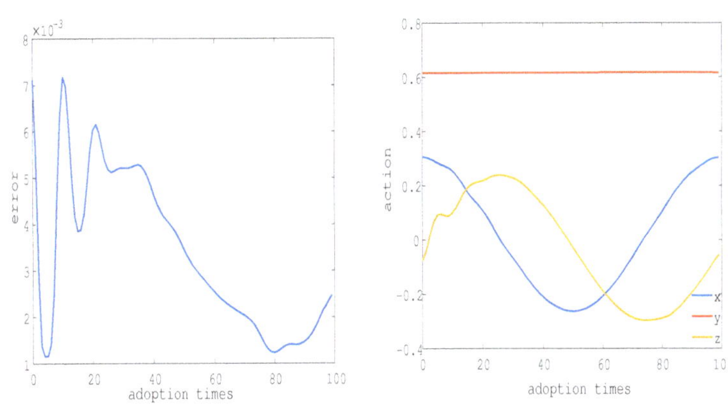

(**c**) 3000th training state.

Figure 12. Sampling of the system training process: (**left**) error curve; (**right**) output value.

Figure 12 shows the 12th training state in the initial stage, the 1000th in the middle of the training, and the 3000th at the end of the training. The three pictures on the left are error curves, which represent the data output error values for the database. The average errors of the 12th, 1000th, and 3000th training states are 0.0142, 0.006, and 0.0034, respectively. The specific data are shown in Table 2. The three pictures on the right side of Figure 12 are the output action graphs of the three training states, which clearly show that the curves become smoother and that the fluctuations are reduced. The fluctuation of the output curve represents the stability and accuracy of the output of the control system. With the increase in the number of training episodes, the stability of the system increases.

Table 2. Training error data of data-driven control method based on DDPG.

	FWA-BPNN-PID	12th	1000th	3000th
Average	0.0119	0.1420	0.0060	0.0034
MSE (10^{-5})	2.543	8.119	0.265	0.284

It can be found from Table 2 that, with the increase in training times, the control effect of the snake manipulator control system is better, and the control accuracy is continuously improved. In addition, compared with the traditional control methods, the control accuracy and stability of the snake manipulator control system are significantly improved, which verifies the effectiveness and superiority of the data-driven control algorithm based on DDPG.

4.3. Comparative Experimental Analysis

In order to test the influence of intelligent agent single-episode training times and agent network setting on the control system, a comparative experiment was designed. The first group of simulations is the original network, but the number of single-process training episodes is 10. The simulation results are shown in Figure 13. In the second group of simulations, the number of single training episodes is set to 10, and the agent action network adds a hidden layer. The hidden layer is set to the full connection layer, and the number of network nodes is set to 150. The simulation results are shown in Figure 14. The third group of simulations is based on the second group of experiments. The number of single training episodes is set to 100. The simulation results are shown in Figure 15.

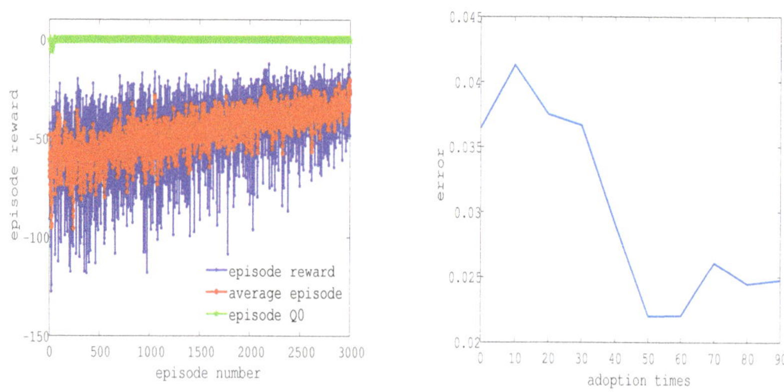

(**a**) Training chart. (**b**) Sampling error curve.

Figure 13. The first group of simulations.

As observed in Figure 13a, the training simulation chart converges slowly and fluctuates greatly compared with Figure 11. The average error of the sampling sample is 0.0301, and the Mean Square Error (MSE) is 4.7225×10^{-5}. Compared with the simulation of 100 samples, the average error and MSE are larger. This shows that the number of training samples is too small, which results in deterioration of the accuracy and stability of the control system.

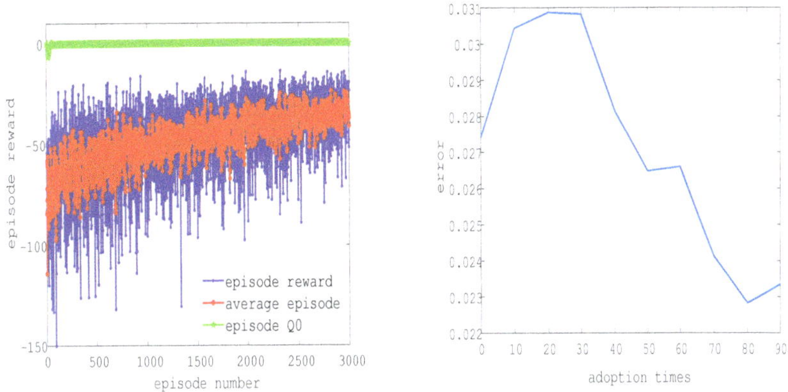

(**a**) Training chart. (**b**) Sampling error curve.

Figure 14. The second group of simulations.

When comparing the training chart of Figure 14a to the training chart of Figure 13a , we can observe that, after adding the network, the fluctuation value and convergence speed of the network do not change much. However, after increasing the network, the average error is 0.0271, and the MSE is 8.2381×10^{-6}. From numerical analysis, the increase in the number of network layers improves the accuracy and stability of the system.

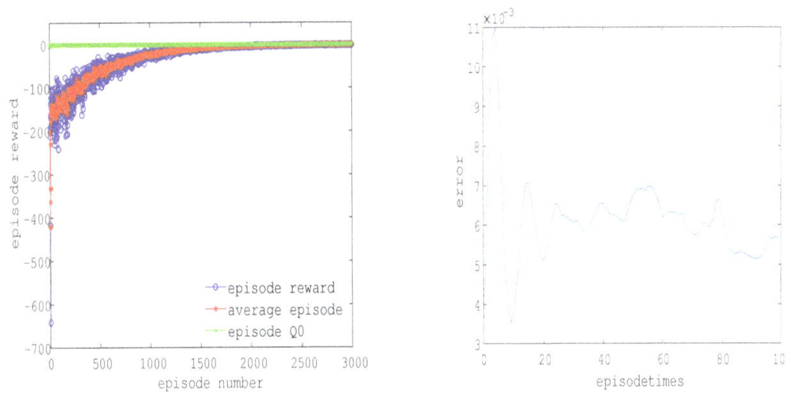

(**a**) Training chart. (**b**) Sampling error curve.

Figure 15. The third group of simulations.

By comparing the training chart of Figure 15a with that of Figure 11, it can be concluded that the convergence speed of the network is faster when the number of samples is the same. However, there are large fluctuations in the early stage, and the overall training time is longer. The average error value of 0.0061, and the MSE value of 1.1585×10^{-6} can be obtained from the sampling error curve. Compared with the original network, the control accuracy of the simulation training system is reduced, but the stability is improved. The simulation results are shown in Table 3.

Table 3. Comparison of simulation training error.

	Original	1st Group	2nd Group	3rd Group
Average	0.0034	0.0301	0.0271	0.0061
MSE (10^{-5})	0.284	4.7225	0.8238	0.1156

It can be observed from Table 3 that the number of samples and the number of network layers have an impact on the stability and accuracy of the system. Too few samples result in deterioration of the accuracy and stability of the control system. The control performance of the system can be improved by increasing the number of samples and training network layers. However, when the number of samples reaches a certain value, increasing the number of network layers can continue to improve the stability of the system. This, however, results in a decline in the accuracy of the system. In addition, increasing the number of samples and the number of network layers results in a longer training time of the system. In practical application, according to the needs of control, the number of samples of training or the number of network layers can be increased. This adjusts the manipulator control speed, stability, and accuracy to achieve a relative balance.

5. Summary and Prospects

In this study, the DRL algorithm was applied to a snake manipulator to verify whether it can improve its control accuracy. Through the study of the theory of DRL and the continuity of the action of the snake manipulator, the DDPG method is proposed as a means of designing the controller of the manipulator's joint. A model-based DDPG control method is proposed to verify the effectiveness and superiority of the DDPG controller through the control of two joints, the simulation object environment and the network structure of the agent are established. The agent was trained, and the simulation without disturbance and torque disturbance was carried out. The simulation results verify the feasibility of the method. However, through the modeling of the 2-link snake manipulator, it was found that it is difficult to establish the entire manipulator model. It is difficult to realize the model-based DDPG control method for the whole manipulator control system. In order to solve the control problem of an unknown model of the snake manipulator, a data-driven control method based on DDPG is proposed. It does not need to establish an accurate manipulator model and avoids the influence of the accuracy error of the manipulator model. The training data were established according to the expected trajectory and the actual output trajectory of the traditional control method. Compared with the traditional control algorithm, it is proven that the data-driven control system based on DDPG has higher accuracy and better stability. At present, the control system of the snake manipulator is only based on data theory. In order to establish a complete snake manipulator motion library for training, the actual manipulator motion data and a large number of trajectory data are needed. The trained control system has a good ability to control known movements. For unknown actions, it also achieves a better control effect. By utilizing continuous learning, the strong adaptability of the snake manipulator control system can be realized.

Author Contributions: Conceptualization, K.H. and L.T.; methodology, K.H.; software, L.T.; validation, L.T., C.W., L.W.and Q.Z.; formal analysis, L.T., L.W., and M.X.; investigation, L.T. and G.Q.; resources, K.H., Q.Z.; data curation, K.H.; writing—original draft preparation, L.T., C.W., and L.W.; writing—review and editing, L.T., C.W., and L.W.; visualization, L.T., C.W., and L.W.; supervision, K.H.; project administration, K.H.; funding acquisition, K.H. All authors have read and agreed to the published version of the manuscript.

Funding: Research in this article is supported by the National Natural Science Foundation of Chinese (42075130, 61773219, and 61701244), the key special project of the National Key R&D Program (2018YFC1405703), and I would like to express my heartfelt thanks. I would like to express my heartfelt thanks to the reviewers who submitted valuable revisions to this article.

Institutional Review Board Statement: Ethical review and approval were waived for this study due to the data being provided publicly.

Informed Consent Statement: Not applicable.

Data Availability Statement: The data and code used to support the findings of this study are available from the corresponding author upon request (001600@nuist.edu.cn).

Conflicts of Interest: The authors declare no conflicts of interest.

References

1. Hou, Z.; Xu, J. Review and prospect of data-driven control theory and method. *J. Autom.* **2009**, *35*, 650–667.
2. Mendes, N.; Neto, P. Indirect adaptive fuzzy control for industrial robots: A solution for contact applications. *Expert Syst. Appl.* **2015**, *42*, 8929–8935. [CrossRef]
3. Yen, V.; Nan, W.; Cuong, P.; Quynh, N.; Thich, V. Robust adaptive sliding mode control for industrial robot manipulator using fuzzy wavelet neural networks. *Int. J. Control* **2017**,*15*, 2930–2941. [CrossRef]
4. Yen, V.; Nan, W.; Cuong, P. Recurrent fuzzy wavelet neural networks based on robust adaptive sliding mode control for industrial robot manipulators. *Neural Comput. Appl.* **2019**, *31*, 6945–6958. [CrossRef]
5. Yu, X.; He, W. Research on adaptive neural network tracking control of manipulator based on disturbance observer. *J. Autom.* **2019**, *45*, 1307–1324.
6. Jung, S. Improvement of tracking control of a sliding mode controller for robot manipulators by a neural network. *Int. J. Control Autom. Syst.* **2018**, *16*, 937–943. [CrossRef]
7. Xia, M.; Zhang, X.; Weng, L.; Xu, Y. Multi-stage Feature Constraints Learning for Age Estimation. *IEEE Trans. Inf. Forensics Secur.* **2020**, *15*, 2417–2428. [CrossRef]
8. Xia, M.; Wang, K.; Song, W.; Chen, C.; Li, Y. Non-intrusive load disaggregation based on composite deep long short-term memory network. *Expert Syst. Appl.* **2020**, *160*, 113669. [CrossRef]
9. Xia, M.; Wang, K.; Zhang, X.; Xu, Y. Non-intrusive load disaggregation based on deep dilated residual network. *Electr. Power Syst. Res.* **2019**, *170*, 277–285. [CrossRef]
10. Hasselt, H.; Guez, A.; Silver, D. Deep reinforcement learning with double q-learning. In Proceedings of the AAAI Conference on Artificial Intelligence, Phoenix, AZ, USA, 12–17 February 2016; pp. 1–13.
11. Horgan, D.; Quan, J.; Budden, D.; Barth-Maron, G.; Hessel, M.; Hasselt, H. Distributed prioritized experience replay. 2018. Available online: https://arxiv.org/abs/1803.00933 (accessed on 6 August 2021).
12. Wang, Z.; Schaul, T.; Hessel, M.; Hasselt, H.; Lanctot, M.; Freitas, N. Dueling network architectures for deep reinforcement learning. In Proceedings of the International Conference on Machine Learning, New York, NY, USA, 19–24 June 2016; pp. 1995–2003.
13. Sutton, R.; Mcallester, D.; Singh, S.; Mansour, Y. Policy gradient methods for reinforcement learning with function approximation. *Submitt. Adv. Neural Inf. Process. Syst.* **1999**, *99*, 1057–1063.
14. Barto, A.; Sutton, R.; Anderson, C. Neuron like elements that can solve difficult learning control problems. *IEEE Trans. Syst. Man Cybern.* **1970**, *13*, 834–846.
15. Mnih, V.; Badia, A.; Mirza, M.; Graves, A.; Lillicrap, T.; Harley, T.; Kavukcuoglu, K. Asynchronous methods for deep reinforcement learning. In Proceedings of the International Conference on Machine Learning, New York, NY, USA, 19–24 June 2016; pp. 1928–1937.
16. Weng, L.; Tian, L.; Hu, K.; Zang, Q.; Chen, X. Overview of robot force control algorithms based on neural network. In Proceedings of the 2020 Chinese Automation Congress (CAC), Shanghai, China, 6–8 November 2020; pp. 6800–6803.
17. Levine, S.; Finn, C.; Darrell, T. End-to-end training of deep visuomotor policies. *J. Mach. Learn. Res.* **2016**, *17*, 1334–1373.
18. Pinto, L.; Gupta, A. Supersizing self-supervision: Learning to grasp from 50k tries and 700 robot hours. In Proceedings of the 2016 IEEE International Conference on Robotics and Automation(ICRA), Stockholm, Sweden, 16–21 May 2016; pp. 3406–3413.
19. Levine, S.; Pastor, P.; Krizhevsky, A.; Ibarz, J.; Quillen, D. Learning hand-eye coordination for robotic grasping with deep learning and large-scale data collection. *Int. J. Robot. Res.* **2018**, *37*, 421–436. [CrossRef]
20. Lillicrap, T.; Hunt, J.; Pritzel, A.; Heess, N.; Erez, T.; Tassa, Y.; Wierstra, D. Continuous control with deep reinforcement learning. *Comput. Ence* **2015**, *8*, 1–10.
21. Schulman, J.; Levine, S.; Abbeel, P.; Jordan, M.; Moritz, P. Trust region policy optimization. In Proceedings of the International Conference on Machine Learning, Lille, France, 6–11 June 2015; pp. 1889–1897.
22. Luo, J.; Solowjow, E.; Wen, C.; Ojea, J. A.; Agogino, A. Deep reinforcement learning for robotic assembly of mixed deformable and rigid objects. In Proceedings of the 2018 IEEE/RSJ International Conference on Intelligent Robots and Systems (IROS), Madrid, Spain, 1–5 October 2018; pp. 1–5.
23. Wu, X.; Zhang, D.; Qin, F.; Xu, D. Deep reinforcement learning of robotic precision insertion skill accelerated by demonstrations. In Proceedings of the 2019 IEEE 15th International Conference on Automation Science and Engineering (CASE), Vancouver, BC, Canada, 25–28 August 2019; pp. 1651–1656.
24. Wen, S.; Chen, J.; Wang, S.; Zhang, H.; Hu, X. Path planning of humanoid arm based on deep deterministic policy gradient. In Proceedings of the 2018 IEEE International Conference on Robotics and Biomimetics (ROBIO), Kuala Lumpur, Malaysia, 12–15 December 2018; pp. 1755–1760.

25. Wang, J.; Wang, G.; Mao, X.; Ma, Q. Motion control method of two-link manipulator based on deep reinforcement learning. *Comput. Appl.* **2021**, *41*, 1799–1804. (In Chinese with English Abstract)
26. Li, Z.; Ma, H.; Ding, Y.; Wang, C.; Jin, Y. Motion Planning of Six-DOF Arm Robot Based on Improved DDPG Algorithm. In Proceedings of the 2020 39th Chinese Control Conference (CCC), Shenyang, China, 27–29 July 2020; pp. 3954–3959.

applied sciences

Article

Online Foot Location Planning for Gait Transitioning Using Model Predictive Control

Xiangming Liu [1], Hongxu Ma [1], Lin Lang [1,2,*] and Honglei An [1]

[1] College of Intelligence Science and Technology, National University of Defense Technology, Changsha 410000, China; XiangmingL2021@163.com (X.L.); mhx1966@163.com (H.M.); eric_nudt@163.com (H.A.)
[2] College of Information Technology and Management, Hunan University of Finance and Economics, Changsha 410000, China
* Correspondence: langlin_8502@nudt.edu.cn

Abstract: This paper proposes an online uniform foot location planning method (UPMPC) based on model predictive control (MPC) for solving the problem of large posture changes during gait transitioning. This method converts the foot location planning into a discrete-time MPC problem. The core part of the method is to complete the planning of the foot location based on the linear inverted pendulum (LIP) model and the simplified robot dynamics model. By unifying the input foot location at each time step, the solution time is shortened. The final simulation experiment compares the results of using the UPMPC and foot location planning method with heuristic function (HF) for gait transitioning, respectively. This result demonstrates that the UPMPC can complete the gait transitioning task and adapt to large changes in posture during gait transitioning. In addition, the results also show the good performance of UPMPC in fixed gait.

Keywords: MPC; foot location; motion planning; gait transitioning; legged robots

Citation: Liu, X.; Ma, H.; Lang, L.; An, H. Online Foot Location Planning for Gait Transitioning Using Model Predictive Control. *Appl. Sci.* **2021**, *11*, 7866. https://doi.org/10.3390/app11177866

Academic Editors: Alessandro Gasparetto, Stefano Seriani and Lorenzo Scalera

Received: 19 July 2021
Accepted: 23 August 2021
Published: 26 August 2021

1. Conventions

To illustrate the content of this paper, the following abbreviations listed in Table 1 are used.

Table 1. The abbreviations are used in this manuscript.

Abbreviations	Meaning
MPC	Model Predictive Control
LIP	Linear Inverted Pendulum
COM	Center of Mass
UPMPC	A foot location planning method proposed in this paper. It contains LIP model, dynamics model of a quadruped robot and MPC method.
HF	A foot location planning method used heuristic function
GRFs	Ground reaction forces
COT	Cost of Transport
QP	Quadratic Programming
FL	Front Left Leg
FR	Front Right Leg
HL	Hind Left Leg
HR	Hind Right Leg

To illustrate the formulas in subsequent sections, the mathematical notation listed in Table 2 is used. Matrices and vectors are upright and bold. Scalars are italicized.

Table 2. Main Symbol Conventions in this paper.

Variables	Meaning
x	The position of the COM on the x-axis in the world coordinate system
y	The position of the COM on the y-axis in the world coordinate system
z	The position of the COM on the z-axis in the world coordinate system
$\mathbf{p}$	The position of the COM in the world coordinate system
p_x	The position of supporting foot in the LIP model on the x-axis in the world coordinate system
p_y	The position of supporting foot in the LIP model on the y-axis in the world coordinate system
p_z	The position of supporting foot in the LIP model on the z-axis in the world coordinate system
$\mathbf{x}$	Robot state variables without angular velocity
$\mathbf{p_f}$	The position of the robot's foot in the world coordinate system
$\mathbf{f}$	Ground reaction force
$\mathbf{I}$	The robot's inertia tensor
ω	The robot's angular velocity
$\mathbf{X}$	Robot state variables

2. Introduction

Legged robots have greater environmental adaptability than wheeled and tracked robots. In order to meet the needs of different tasks in various environments, users are no longer satisfied with quadruped robots moving in a fixed gait and expect legged robots to be able to switch stably between multiple gaits during moving.

Quadruped robots usually perform gait transitioning to reduce energy consumption. Horses [1] have varying energy expenditure at different gaits and speeds. In order to maximize energy efficiency, the horse will choose the gait with the least energy consumption at each speed. Quadruped robots designed with reference to quadrupeds naturally have variable energy consumption at different gaits. The energy consumption of a quadruped robot at various gaits and speeds is explored in detail in [2]. Each gait has a significant energy gap at the same speed. It is also found that the long-period, large duty cycle gait and the short-period, low duty cycle gait consume less energy at low and high speeds, respectively. Therefore, it is of utmost importance to switch from a high-energy-consuming gait to a gait with relatively low energy consumption.

However, in actual use, the appropriate gait is also selected according to the task needs. When traversing uneven terrain, it is expected that the impact on the ground is small and the height of the quadruped robot is basically constant while maintaining a certain speed. Thus, the gait chosen at this point may not be the less energy intensive one. The choice of gait is the result of many factors. Therefore, the quadruped robot needs to successfully switch between different gaits at a certain speed, not just from a high-energy-consuming gait to a low-energy-consuming gait.

The successful locomotion of the robot cannot be separated from the choice of the foot location. A lot of researches have been done on foot location planning. HF [3] is a commonly used method. By adjusting the gain parameters, better results can be obtained. The method has been applied in several MIT's robots [4,5]. Combining HF and MPC [4], MIT's Mini Cheetah implements multiple gaits. However, the fixed-parameter HF is difficult to cope with transitioning between various gaits with different periods. The properties of gaits are distinct, in the case of different periods, leading to the inability of the fixed-parameter HF to adapt to the large changes of posture after the gait transitioning. The capture point technique which is frequently used in bipedal robots is also a common method for foot location planning. Pratt et al. simplify the bipedal robot into a LIP model with a flywheel, and determine the range of foot locations and the foot location point of the bipedal robot by establishing the capture domain and the capture point where the robot can stop in one step [6]. Simões et al. find the capture domain that can two-step stop based on the capture points and N-Step Capturability theories, and successfully verify the effect of the algorithm on simulation [7]. Seyde et al. [8] and Kojio et al. [9] both utilize capture point theory for foot location planning and achieve good results on bipedal robots. However, the capture point method requires setting the number of steps to stop, which limits the performance of the robot. MPC is capable of generating behavior with a wide range of adaptations based on real-world situations [10]. Saraf et al. use the MPC to complete a terrain adaptive gait transitioning between bounding and trotting [11]. MPC-based foot location planning has also been widely studied. A number of researchers achieve better robustness and interference resistance on quadruped robots based on the LIP model for foot location planning using MPC [12–14]. Xin et al. design LIP-based foot location planning using MPC [15]. The method applies the LIP model to predict the COM position and velocity. Furthermore, the prediction result is used as the initial state of MPC for planning the foot location. It is successfully applied not only to quadruped robots but also to bipedal robots, and has good anti-interference capability. However, the calculating time grows rapidly with increasing prediction length.

This paper aims at addressing the transitioning of walking, trotting and flight trotting with different gait periods at the same speed. Walking (four beat or two beat) has a duty cycle greater than 50%. The duty cycle is the proportion of the support period to the entire gait period. Furthermore, the four-legged support period can exist in this gait. The two diagonally opposite legs move together for trotting. The duty cycle of trotting is equal to 50%. The two diagonal opposite leg of the flight trotting gait also moves simultaneously, but differs from trotting in that the duty cycle is less than 50%. There is a period of quadrupedal vacancy. The gait cycles for three gaits are shown in Figure 1. Walking is a common gait and employed by all terrestrial vertebrates [16]. Walking is stable with little change in height and posture. And the quadruped robot has little contact force with the ground. The results in [1,2] indicate that either two-beat walking or four-beat walking has less energy expenditure at low speeds than the other gaits. Trotting is widely adaptable to a greater range of speeds and has high energy efficiency when the speed is below 1.2 m/s [4]. Trotting with a vacant period (flight trotting) has a better performance in terms of energy utilization at high speeds [2]. Furthermore, these three gaits are representative on quadruped robots that can adapt to a variety of environments and task needs. During the gait transitioning process, the quadruped robot posture and speed will experience large changes. By choosing a suitable foot location, the effect of the foot location on posture and speed can be reduced. The quadruped robot can return to the stable position more quickly. The three gaits which are used for gait transitioning, with widely varying characteristics, have different gait periods. The HF gain parameters required for each gait are also different. There is no fixed order for transitioning. The difference in gait characteristics leads to large changes in posture after transitioning, making it difficult for the fixed parameter HF to cope with. The MPC-based foot location planning method can select a foot location that best meets the desired future state based on the actual situation during the transitioning process and the prediction results. Thus, the quadruped robot can better complete the gait transitioning and quickly correct the state deviation caused by the transitioning, with better robustness and adaptability. Therefore, it was decided to use UPMPC as the foot location planning method for quadruped robots. During the motion of the quadruped robot, the position of the foot in the world coordinate system remains unchanged. By unifying the foot positions at each time step, the final optimization variables are ensured to be independent of the predicted horizon length. The matrix dimensions to be computed are greatly reduced, which can significantly improve the computational speed of UPMPC to satisfy the demand of high-frequency foot location planning. The gait transitioning results given at the end of the article well demonstrate that UPMPC has better practical performance than HF and can be applied to the foot location planning for fixed gait.

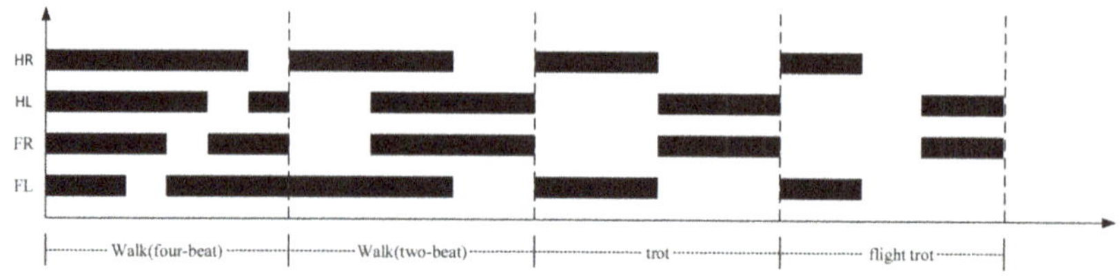

Figure 1. The black block indicates that the single leg of the quadruped robot is in the support phase. FL: Front Left Leg. FR: Front Right Leg. HL: Hind Left Leg. HR: Hind Right Leg.

The main contributions of this paper are: (a) An MPC-based method for selecting the foot location is proposed based on the LIP model and the dynamics model of the quadruped robot, which solves the problem of large posture oscillation during gait transitioning in different gait periods. (b) UPMPC is able to complete some of gait transitioning that cannot be done by the HF. (c) UPMPC selects the foot location based on the predicted future state, which improves the overall adaptability of the quadruped robot.

The paper is organized as follows. Section 3 presents the overall control architecture of the quadruped robot. The composition of MPC and the functions of the UPMPC are briefly demonstrated. A detailed mathematical description of UPMPC is illustrated in Section 4. A comparison of UPMPC with HF in a simulation environment is given in Section 5. Section 6 is a summary of the whole text.

3. Control Architecture

The control method used in this paper is MIT's Convex MPC [4]. The overall control block diagram [4] is shown in Figure 2.

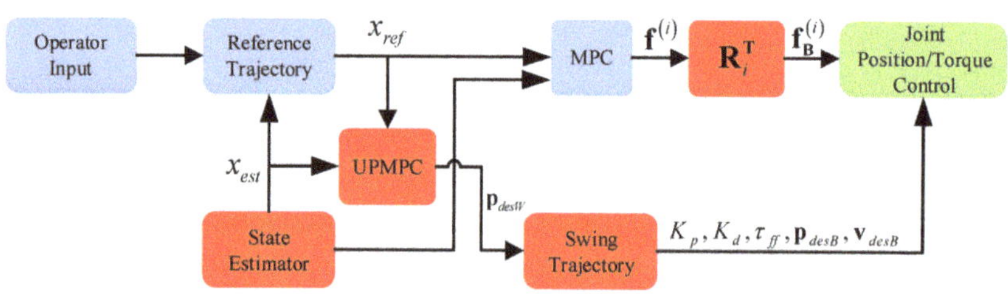

Figure 2. The red block indicates the control frequency at 0.5 kHz, the blue block indicates the control frequency at 30 Hz, and the green block indicates the control frequency at 4.5 kHz [4]. $x_{ref}$is the reference state of the robot. x_{est} is the estimated state of the robot. $\mathbf{p}_{desW}$ is the desired foot location in the world coordinate system. $\mathbf{f}^{(i)}$ is the GRFs in the world coordinate system and $\mathbf{f}_B^{(i)}$ is the GRFs in the ontology coordinate system. $K_p, K_d, \tau_{ff}, \mathbf{p}_{desB}, \mathbf{v}_{desB}$ denotes the PD control parameters of the joint, the feedforward force, the desired foot location under the body coordinate system, and the desired foot velocity in the body coordinate system, respectively. The swing leg is position controlled and the control effect of the joint is adjusted according to the PD control gain K_p, K_d.

Based on Convex MPC, the UPMPC for selecting the foot location is added. Operator Input is the user input for commands such as gait, speed and direction. Reference Trajectory generates a reference state for the robot based on user input, such as the robot's pose and speed. State Estimator estimates the robot's state based on the sensor data. Swing Trajectory is used in three main purposes: selection of swing legs, determination of swing time and generation of swing trajectory. Swing Trajectory generates the desired swing trajectory for the swing leg to track based on the reference foot location input by the UPMPC. MPC outputs the desired GRFs based on the output of Reference Trajectory. The UPMPC selects the foot location of the swing leg.

The UPMPC input variables are the desired state, the current estimated state and the GRFs planned by MPC when switching from the swing phase to the support phase. The output variables are the desired foot location. Using the LIP model and dynamics model of the robot, the selection of the foot location is simplified to a QP problem. The desired result is obtained from the solution of the QP problem. The detailed mathematical derivation and introduction are in shown in Section 4.

4. Foot Location Selection Based on MPC

The expression for the HF is shown in Equation (1).

$$\mathbf{p}_{\mathrm{f}}^{\mathrm{des}} = \mathbf{p}_{\mathrm{h}}^{\mathrm{ref}} + \mathbf{v}^{\mathrm{COM}}\Delta t/2 + k_d\left(\mathbf{v}^{\mathrm{COM}} - \mathbf{v}^{\mathrm{des}}\right) \tag{1}$$

$\mathbf{p}_{\mathrm{f}}^{\mathrm{des}}$ is the desired foot location of the quadruped robot. $\mathbf{p}_{\mathrm{h}}^{\mathrm{ref}}$ is the location of the hip projected onto the ground. $\mathbf{v}^{\mathrm{COM}}$ is the current speed of the COM. Δt is the time the foot will spend on the ground. $\mathbf{v}^{\mathrm{des}}$ is the desired speed of the COM. All the variables above are in the world coordinate system. k_d is the gain of the difference between $\mathbf{v}^{\mathrm{COM}}$ and $\mathbf{v}^{\mathrm{des}}$. Due to the characteristics of different gaits, the posture of the quadruped robot after gait transitioning will have a large offset from the steady state. In this case, HF may choose a foot location that is not conducive to the stability of the quadruped robot's posture in order to correct the velocity offset. If it is desired to do a better job of transitioning between several different gaits, a heuristic gain value needs to be determined for each gait. Therefore, a new way of foot location planning is needed to enable gait transitioning of quadruped robots in multiple gaits with a single set of parameters. The variables used in the equations below are shown in Figure 3.

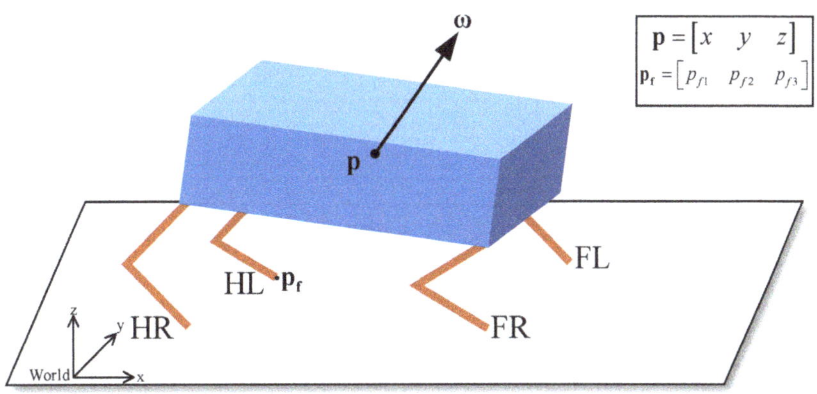

Figure 3. A model of the quadurped robot is shown in the figure. The *World* in the bottom left corner of the diagram indicates the defined world coordinate system. The black frame in the top right corner indicates the composition of the variables labelled in the model. $\mathbf{p}_{\mathrm{f}}$ is the position of the robot's foot end. $\mathbf{p}$ is the position of the robot's COM. $\mathbf{f}$ is the ground reaction force. All the variables mentioned are in the world coordinate system.

The LIP model has been widely used in bipedal robot motion [17,18]. The simplified LIP model allows better planning of the bipedal robot's foot location. LIP can also be applied to quadrupedal robots [19,20]. With the virtual leg concept, a quadrupedal robot can be equated to a bipedal robot. At this point the quadruped robot can also apply the LIP model to plan the foot location. Therefore, the LIP model can be used to predict the velocity of the quadruped robot. In a single-point support, the COM is pushed by the support foot.

The forward acceleration $\ddot{x}$ is driven by the difference between the support foot position p_x and the COM position x.

$$\ddot{x} = \frac{g}{z_0}(x - p_x) \tag{2}$$

where p_x represents the position of the supporting foot in the x-axis direction in the world coordinate system, and x, $\ddot{x}$ and z_0 represent the position of the COM, the acceleration in the x-axis and the height in the z-axis in the world coordinate system, respectively. g is the acceleration of gravity in the z-axis in the world coordinate system. With the height of COM z_0 constant, the value of $\ddot{x}$ is determined only by x and p_x.

The lateral acceleration $\ddot{y}$ is driven by the difference between the position of support foot p_y and the position y of the COM.

$$\ddot{y} = \frac{g}{z_0}(y - p_y) \tag{3}$$

where p_y represents the position of the supporting foot in the y-axis direction in the world coordinate system, and y and $\ddot{y}$ represent the position of the COM and the acceleration in the y-axis under the world coordinate system, respectively. Let the COM state be $\mathbf{x} = [x, \dot{x}]^{\mathbf{T}}$, then the equation of state is shown in Equation (4).

$$\dot{\mathbf{x}} = \mathbf{A}\mathbf{x_0} + \mathbf{B}p_x \tag{4}$$

$$\mathbf{A} = \begin{bmatrix} 0 & 1 \\ \frac{g}{z_0} & 0 \end{bmatrix}, \mathbf{B} = \begin{bmatrix} 0 \\ -\frac{g}{z_0} \end{bmatrix}, \mathbf{x_0} = \begin{bmatrix} x_0 \\ \dot{x}_0 \end{bmatrix} \tag{5}$$

$\mathbf{x_0}$ is the robot's current state. Extending the COM state to $\mathbf{x} = [\mathbf{p}, \dot{\mathbf{p}}]^{\mathbf{T}}$, $\mathbf{p} \in \mathbf{R}^{3\times1}$ is the position of the COM in the world coordinate system. $\dot{\mathbf{p}} \in \mathbf{R}^{3\times1}$ is the velocity of the COM in the world coordinate system. The position of the supportting foot is extended to the coordinates of the foot in the world coordinate system during the motion of the quadruped robot. The equation of state is shown in Equation (6).

$$\dot{\mathbf{x}} = \mathbf{A}\mathbf{x} + \mathbf{B}\mathbf{p_f} \tag{6}$$

$$\mathbf{A} = \begin{bmatrix} \mathbf{0}^{3\times3} & \mathbf{1}^{3\times3} \\ \mathbf{w}^{3\times3} & \mathbf{0}^{3\times3} \end{bmatrix} \tag{7}$$

$$\mathbf{B} = \begin{bmatrix} \mathbf{0}^{3\times3} & \mathbf{0}^{3\times3} & \mathbf{0}^{3\times3} & \mathbf{0}^{3\times3} \\ \frac{-\varepsilon(|\mathbf{f}^{(1)}|)\cdot\mathbf{w}^{3\times3}}{2} & \frac{-\varepsilon(|\mathbf{f}^{(2)}|)\cdot\mathbf{w}^{3\times3}}{2} & \frac{-\varepsilon(|\mathbf{f}^{(3)}|)\cdot\mathbf{w}^{3\times3}}{2} & \frac{-\varepsilon(|\mathbf{f}^{(4)}|)\cdot\mathbf{w}^{3\times3}}{2} \end{bmatrix} \tag{8}$$

$$\mathbf{w}^{3\times3} = \begin{bmatrix} g/z_0 & 0 & 0 \\ 0 & g/z_0 & 0 \\ 0 & 0 & 0 \end{bmatrix} \tag{9}$$

$$\varepsilon\left(\left|\mathbf{f}^{(i)}\right|\right) = \begin{cases} 0 & \left|\mathbf{f}^{(i)}\right| > 0 \\ 1 & \left|\mathbf{f}^{(i)}\right| = 0 \end{cases} \tag{10}$$

where $\mathbf{p_f} = \begin{bmatrix} \mathbf{p_f}^{(1)} & \mathbf{p_f}^{(2)} & \mathbf{p_f}^{(3)} & \mathbf{p_f}^{(4)} \end{bmatrix}^{\mathbf{T}}$ is the position of the foot in the world coordinate system. $\mathbf{p_f}^{(i)} = \begin{bmatrix} p_{f1}^{(i)} & p_{f2}^{(i)} & p_{f3}^{(i)} \end{bmatrix}^{\mathbf{T}} \in \mathbf{R}^{3\times1}$ and $\mathbf{f}^{(i)} = \begin{bmatrix} f_1^{(i)} & f_2^{(i)} & f_3^{(i)} \end{bmatrix} \in \mathbf{R}^{3\times1}$ are the coordinates of the foot i and the GRFs in the world coordinate system, respectively. In the LIP model, the choice of the foot location affects the velocity of the COM, but not the height of the COM. Equation (10) is used to determine the swing leg.

The presence of GRFs makes the location of the foot affect the COM posture. Therefore, the choice of foot location needs to consider the effect on the velocity and the posture.

The dynamics of a quadruped robot rotating around the COM in the world coordinate system is shown in Equation (11).

$$\frac{\mathrm{d}}{\mathrm{dt}}(\mathbf{I}\omega) = \sum_{i=1}^{n} \mathbf{r}^{(i)} \times \mathbf{f}^{(i)} \tag{11}$$

where $\mathbf{I}$ is the robot's inertia tensor, $\omega \in \mathbf{R}^{3\times1}$ is the angular velocity of the robot. $\mathbf{r}$ is the vector between the position of the foot end and the COM. When ω is small, the left side of Equation (11) can be approximated with:

$$\frac{\mathrm{d}}{\mathrm{dt}}(\mathbf{I}\omega) = \mathbf{I}\dot{\omega} + \omega \times (\mathbf{I}\omega) \approx \mathbf{I}\dot{\omega} \tag{12}$$

Converting the right-hand side of Equation (11) to a representation of the COM position and the foot position in the world coordinate system.

$$\sum_{i=1}^{n} \mathbf{r}^{(i)} \times \mathbf{f}^{(i)} = \sum_{i=1}^{n} \left(\mathbf{p_f}^{(i)} - \mathbf{p}\right) \times \mathbf{f}^{(i)} = \sum_{i=1}^{n} \mathbf{f}^{(i)} \times \left(\mathbf{p} - \mathbf{p_f}^{(i)}\right) \tag{13}$$

Then Equation (11) can be converted to:

$$\begin{aligned} \dot{\omega} = {} & \begin{bmatrix} \mathbf{I}^{-1}\left[\sum_{i=1}^{n} \varepsilon\left(\left|\mathbf{f}^{(i)}\right|\right)\mathbf{f}^{(i)}\right] & 0 \end{bmatrix} \cdot \begin{bmatrix} \mathbf{p} \\ \omega \end{bmatrix} \\ & + \begin{bmatrix} \mathbf{I}^{-1}\left[-\mathbf{f}^{(1)}\right] & \mathbf{I}^{-1}\left[-\mathbf{f}^{(2)}\right] & \mathbf{I}^{-1}\left[-\mathbf{f}^{(3)}\right] & \mathbf{I}^{-1}\left[-\mathbf{f}^{(4)}\right] \end{bmatrix} \begin{bmatrix} \mathbf{p_f}^{(1)} \\ \mathbf{p_f}^{(2)} \\ \mathbf{p_f}^{(3)} \\ \mathbf{p_f}^{(4)} \end{bmatrix} \end{aligned} \tag{14}$$

where $\left[-\mathbf{f}^{(i)}\right] = \begin{bmatrix} 0 & \mathbf{f}_3^{(i)} & -\mathbf{f}_2^{(i)} \\ -\mathbf{f}_3^{(i)} & 0 & \mathbf{f}_1^{(i)} \\ \mathbf{f}_2^{(i)} & -\mathbf{f}_1^{(i)} & 0 \end{bmatrix}$ is the GRFs of MPC planning when switching from the swing phase to the support phase in the previous period. The equation of state of the system is

$$\dot{\mathbf{X}} = \mathbf{A}\mathbf{X} + \mathbf{B}\mathbf{p_f} \tag{15}$$

$$\mathbf{A} = \begin{bmatrix} \mathbf{0}^{3\times3} & \mathbf{1}^{3\times3} & \mathbf{0}^{3\times3} \\ \mathbf{w}^{3\times3} & \mathbf{0}^{3\times3} & \mathbf{0}^{3\times3} \\ \mathbf{I}^{-1}\left[\sum_{i=1}^{n} \varepsilon\left(\left|\mathbf{f}^{(i)}\right|\right)\mathbf{f}^{(i)}\right] & \mathbf{0}^{3\times3} & \mathbf{0}^{3\times3} \end{bmatrix} \tag{16}$$

$$\mathbf{B} = \begin{bmatrix} \mathbf{0}^{3\times3} & \mathbf{0}^{3\times3} & \mathbf{0}^{3\times3} & \mathbf{0}^{3\times3} \\ \frac{-\varepsilon\left(\left|\mathbf{f}^{(1)}\right|\right)\cdot\mathbf{w}^{3\times3}}{2} & \frac{-\varepsilon\left(\left|\mathbf{f}^{(2)}\right|\right)\cdot\mathbf{w}^{3\times3}}{2} & \frac{-\varepsilon\left(\left|\mathbf{f}^{(3)}\right|\right)\cdot\mathbf{w}^{3\times3}}{2} & \frac{-\varepsilon\left(\left|\mathbf{f}^{(4)}\right|\right)\cdot\mathbf{w}^{3\times3}}{2} \\ \mathbf{I}^{-1}\left[-\mathbf{f}^{(1)}\right] & \mathbf{I}^{-1}\left[-\mathbf{f}^{(2)}\right] & \mathbf{I}^{-1}\left[-\mathbf{f}^{(3)}\right] & \mathbf{I}^{-1}\left[-\mathbf{f}^{(4)}\right] \end{bmatrix} \tag{17}$$

where $\mathbf{X} = \begin{bmatrix} \mathbf{p} \\ \dot{\mathbf{p}} \\ \omega \end{bmatrix}$, $\mathbf{1}^{3\times3} = \begin{bmatrix} 1 & 0 & 0 \\ 0 & 1 & 0 \\ 0 & 0 & 1 \end{bmatrix}$. Discretize Equation (15) using a zero-order retainer:

$$\mathbf{X}[n+1] = \hat{\mathbf{A}}\mathbf{X}[n] + \hat{\mathbf{B}}\mathbf{p_f} \tag{18}$$

Combining the current COM state $\mathbf{X}_0$, the GRFs planned by the MPC when the foot touched the ground in the previous period and the reference state of the robot, the location of the foot is determined according to Equation (18). Let the desired foot location be $\mathbf{p_f}$. After the foot touches the ground without slipping, the position in the world coordinate system does not change. The equation for the change of the state for the next N steps can be obtained.

$$
\begin{aligned}
&\mathbf{X}_1 = \hat{\mathbf{A}}\mathbf{X}_0 + \hat{\mathbf{B}}\mathbf{p}_\mathrm{f} \\
&\mathbf{X}_2 = \hat{\mathbf{A}}\mathbf{X}_1 + \hat{\mathbf{B}}\mathbf{p}_\mathrm{f} = \hat{\mathbf{A}}^2\mathbf{X}_0 + (\hat{\mathbf{A}}\hat{\mathbf{B}} + \hat{\mathbf{B}})\mathbf{p}_\mathrm{f} \\
&\mathbf{X}_3 = \hat{\mathbf{A}}\mathbf{X}_2 + \hat{\mathbf{B}}\mathbf{p}_\mathrm{f} = \hat{\mathbf{A}}^3\mathbf{X}_0 + \left(\hat{\mathbf{A}}^2 \cdot \hat{\mathbf{B}} + \hat{\mathbf{A}} \cdot \hat{\mathbf{B}} + \hat{\mathbf{B}}\right)\mathbf{p}_\mathrm{f} \\
&\ldots \\
&\mathbf{X}_N = \hat{\mathbf{A}}\mathbf{X}_{N-1} + \hat{\mathbf{B}}\mathbf{p}_\mathrm{f} = \hat{\mathbf{A}}^N\mathbf{X}_0 + \left(\hat{\mathbf{A}}^{N-1} \cdot \hat{\mathbf{B}} + \hat{\mathbf{A}}^{N-2} \cdot \hat{\mathbf{B}} + \cdots + \hat{\mathbf{B}}\right)\mathbf{p}_\mathrm{f} = \bar{\mathbf{A}}\mathbf{X}_0 + \bar{\mathbf{B}}\mathbf{p}_\mathrm{f} \\
&\bar{\mathbf{A}} = \hat{\mathbf{A}}^N, \bar{\mathbf{B}} = \hat{\mathbf{A}}^{N-1} \cdot \hat{\mathbf{B}} + \hat{\mathbf{A}}^{N-2} \cdot \hat{\mathbf{B}} + \cdots + \hat{\mathbf{B}}
\end{aligned} \tag{19}
$$

The control problem can be transformed into the selection of the foot location that the robot will achieve the desired state at step N.

$$\min J = \sum_{i=1}^{N} \left(\mathbf{X}_{(i)} - \mathbf{X}_{(i,ref)}\right)^\mathrm{T} \mathbf{L}\left(\mathbf{X}_{(i)} - \mathbf{X}_{(i,ref)}\right) + (\mathbf{p}_\mathrm{f} - \mathbf{p})^\mathrm{T}\mathbf{K}(\mathbf{p}_\mathrm{f} - \mathbf{p}) \tag{20}$$

$$s.\,\mathrm{t.}\ \mathbf{X}[n+1] = \hat{\mathbf{A}}\mathbf{X}[n] + \hat{\mathbf{B}}\mathbf{p}_\mathrm{f} \tag{21}$$

$$\mathbf{p}_{\mathbf{f}\mathbf{min}} \le \mathbf{p}_\mathrm{f} \le \mathbf{p}_{\mathbf{f}\mathbf{max}} \tag{22}$$

$$\mathbf{D}\mathbf{p}_\mathrm{f} = \mathbf{0} \tag{23}$$

Both **L** and **K** are adjustable weight matrices and semi-positive definite. Equation (23) selects the foot that is currently in the swing period. By using Equations (19) and (20) can be converted into the following quadratic optimization form.

$$\min_{\mathbf{x}} \frac{\mathbf{1}}{\mathbf{2}}\mathbf{p}_\mathrm{f}^\mathrm{T}\mathbf{H}\mathbf{p}_\mathrm{f} + \mathbf{p}_\mathrm{f}^\mathrm{T}\mathbf{g} \tag{24}$$

where $\mathbf{H} = 2\left(\bar{\mathbf{B}}^\mathrm{T}\mathbf{L}\bar{\mathbf{B}} + \mathbf{K}\right)$, $g = 2\bar{\mathbf{B}}^\mathrm{T}\mathbf{L}(\bar{\mathbf{A}}\mathbf{X}_0 - \mathbf{X}_d) - 2\mathbf{K}\mathbf{p}$. By limiting the foot location of each time step to be equal, the dimension of the matrix to be computed is greatly reduced and the efficiency of the computation is improved. The dimensionality of the final matrices **H** and vector **g** is independent of the prediction length and is only related to the number of variables. Therefore, it can greatly reduce the amount of computation to improve the computation speed and increase the frequency of prediction.

$$\ddot{\mathbf{p}} = \frac{\sum_{i=1}^{4} \mathbf{f}^{(i)}}{m} - \mathbf{g} \tag{25}$$

$\mathbf{g} \in \mathbf{R}^{3\times 1}$ is the acceleration of gravity. Equations (11) and (25) form the complete dynamics model of the quadruped robot. In Equation (11), angular acceleration is determined by the foot location and GRFs. The GRFs also determines the linear acceleration in Equation (25). The user not only wants the quadruped robot to be able to keep up with the desired speed, but also to ensure that the posture is stable around the desired value. This places demands on both linear and angular acceleration. The MPC controller will select the GRFs based on Equations (11) and (25) and the foot location. Therefore, the choice of foot location will directly influence the outcome of the GRFs.

The HF selects the foot location only according to the current state so that the gain value greatly affects the result. The UPMPC considers not only the current state but also the influence of the foot location on the future state. It improves adaptability by selecting a foot location where the future state is more in line with the desired one, rather than just based on the current state. Therefore, the UPMPC can adapt to multiple gaits and complete gait transitioning between different periods.

5. Results

5.1. Simulation Setup

The proposed UPMPC is validated based on the simulation platform of MIT [21]. The simulation environment is shown in Figure 4 below. It is based on Ubuntu 18.04. MIT's Mini Cheetah is used [5] as the verification model for the algorithm. In which the robot with 12 degrees of freedom weighs 9 kg [5], the friction between the foot and the ground is 0.4. All of the joints are motor driven and torque controlled. LCM is used for communication between the simulation environment and the control interface. LCM is lightweight real-time communication software [22]. In this paper, qpOASES [23] is used as a solver for the QP problem. qpOASES is specifically used to solve the MPC quadratic optimization problem. The walking gait chosen in this paper is two-beat walking, so the diagonal leg of walking, trotting and flight trotting can be equated to a virtual leg. At this point, the quadruped robot is equivalent to a bipedal robot. Furthermore, the LIP model can be applied. The control frequency of UPMPC is 0.5 kHz.

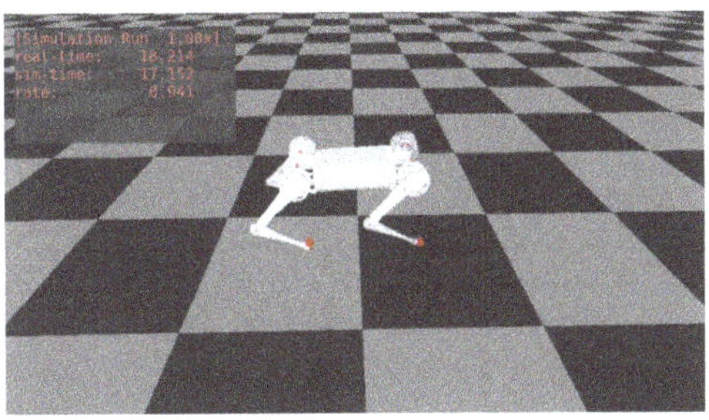

Figure 4. The model in simulation environment is the Mini Cheetah. The real-time in the upper left corner indicates the time of the real world. The sim-time represents the time in the simulation environment. The rate denotes the simulation rate.

During the simulation experiments, the control parameters of MPC are kept constant. The prediction length of MPC method is 10 steps in all three gaits. The time length of each step is 0.03 s. The prediction time is 0.3 s and the control frequency is 33 Hz. The prediction length of UPMPC is three steps. The time length of each step is 0.03 s. None of the parameters of the HF or UPMPC are changed during the motion. The periods of walking, trotting and flight trotting are 0.6 s, 0.42 s and 0.3 s, respectively. The duty cycle of walking, trotting and flight trotting are 0.6, 0.5 and 0.3, respectively. Each gait transitioning interval is 3 s. The swing leg is planned with a Bessel curve. Figure 5 shows the COT [24] of three gaits. COT is used to indicate the energy consumption level of the quadruped robot.

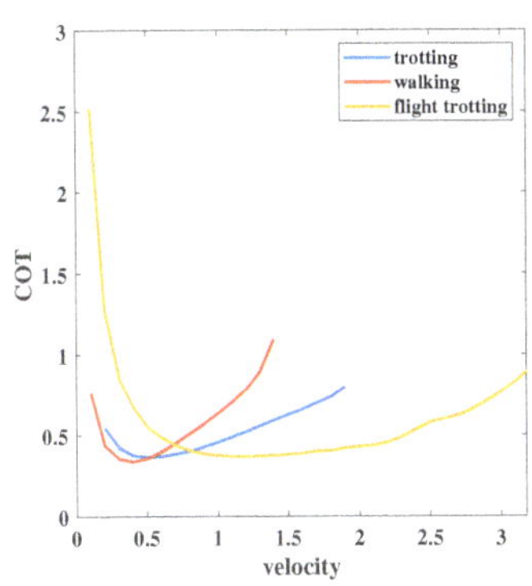

Figure 5. COT of walking, trotting and flight trotting at different speeds, respectively.

The three curves in Figure 5 represent the different gaits. The horizontal axis coordinate corresponding to the rightmost end of each curve indicates the maximum speed that can be achieved at the current gait. According to the curves in the image, the energy consumption at various speeds for different gaits is variable. To reduce energy consumption or to meet task requirements, the robot needs to transitioning between multiple gaits. Because gait transitioning is performed at multiple speeds, the experiment is divided into five parts to verify the performance of UPMPC:

- Transitioning between three gaits at the speed of 0.4 m/s;
- Transitioning between three gaits at the speed of 0.8 m/s;
- Transitioning between trotting and flight trotting at the speed of 1.2 m/s;
- Transitioning between trotting and flight trotting at the speed of 1.6 m/s;
- Movement of trotting gait using the UPMPC

At 0.4 m/s, walking has lower energy consumption. Higher than this speed, the energy consumption of trotting is lower than walking. Transitioning between trotting and walking at this speed can complete the transitioning of energy consumption smoothly. At this time, flight trotting energy consumption is higher, but can have greater acceleration. Flight trotting is indispensable when the robot wants to accelerate quickly from low speed to high speed. Therefore, it is of great importance to perform the transitioning between the three gaits.

At 0.8 m/s, it is the approximate speed at which the trotting and flight trotting COT curves appear to intersect. Greater than this speed, flight trotting energy consumption is less than trotting energy consumption. Flight trotting can accelerate to very fast speed in the short time while walking can ensure posture stability at a certain speed. Therefore, flight trotting and walking transitioning at this speed are equally relevant for applications.

For walking, 1.2 m/s and 1.6 m/s are already quite large speed of movement, very close to the maximum speed in that gait, which itself is already not moving stably enough. Feet in swing may collide with feet lifting off the ground sometimes. Furthermore, the COT in this gait is already much larger than trotting and flight trotting. Therefore, transitioning between trotting or flight trotting and walking at this speed lacks practical meaning. Only the gait transitioning between trotting and flight trotting is considered.

To eliminate the effect of incidental factors, multiple gait transitioning is performed at a given speed during the transitioning process. During the motion of a quadruped robot, the posture change is closely related to the stability of the control. With small variations in posture, the quadruped robot performs better in actual motion. Ideally, a quadruped robot is most stable when the posture is at the desired value and does not change during motion. Therefore, comparing the change in posture at the time of gait transitioning and throughout the gait transitioning process gives a good indication of how well the foot location planning method works. With the control algorithm unchanged, the choice of foot location affects the planning result of MPC in gait transitioning, which in turn influences the posture control of the quadruped robot. The posture is the result of the combined effect of force and foot location. The advantages and disadvantages of the two methods can be determined by posture changes and foot-end forces. Because the selected gaits are all symmetrical, the force change at the foot end of one leg can better reflect the planning result of MPC. The effectiveness and superiority of the UPMPC can be better demonstrated by integrating the results of force planning and posture change. Finally, the posture changes in trotting are shown to demonstrate that UPMPC can also get good results in a single gait.

5.2. *UPMPC vs. HF*

In this section, the peak value and the mean of absolute value are used to compare the advantages and disadvantages of the two methods. The peak value is able to reflect the stability of the posture after transitioning. Smaller peak indicates less posture oscillation and better stability. The mean of absolute value gives a better evaluation to the change in posture throughout the process and eliminates the effect of positive and negative values. A smaller mean of absolute value of the posture indicates that the quadruped robot recovers from postural oscillations more quickly after a gait transitioning or the change in posture is relatively small over the whole time. If the mean of absolute value is large, the quadruped robot is unable to return to a stable state more quickly after a gait transitioning, or has a large oscillation over the whole time.

The numbers marked in the figures below are the times at which the gait transitioning takes place. The peak value used is the maximum value of the posture for the duration after the gait transitioning has occurred and before the next starts. There is a time delay between the peak values and the transitioning point, as the quadruped robot needs a very short amount of time after the gait transitioning to reach its most unstable moment. This part of the time is not negligible. Therefore, it will be seen that there is a lag between the time of the transitioning point and the time of peak values.

Figure 6 is the gait transitioning between trotting and walking at 0.4 m/s. The gait transitioning using the HF in Figure 6a fails directly and using the UPMPC in Figure 6b succeeds despite facing a large posture oscillation. In Figure 6a, a large posture oscillation has happened during the gait transitioning but the foot location of the HF planning cannot cope with this situation. It makes the roll and yaw deviate more and more from the stable position, leading to the posture divergence and gait transitioning failure. The posture change is not so clear in Figure 6a,b. In Figure 6e, it becomes clear. In this graph the peak of the posture after the gait transitioning is shown. The numbers 1, 2 and 3 stand for roll, pitch and yaw, respectively. It is clear from Figure 6e that the gait transitioning using the HF faces a large posture peak after the transitioning and far exceeds that of the UPMPC. This is more than the controller can steadily control, making the system falter and the transitioning fail. In Figure 6b,e, with the controller unchanged, the UPMPC planned foot location effectively reduces the peak values of the posture comparing with the HF, allowing the posture to return to near the stable value. Even if the transitioning results in a large posture change, the UPMPC is able to help keep the robot moving stably by planning a good foot location. Figure 6c,d shows the foot-end force planned by MPC for the left front leg foot of the quadruped robot during the transitioning process. Figure 6f shows the comparison of the peak force at the foot end after transitioning. The difference in peak force is not so clear from Figure 6c,d. However, in Figure 6f, a huge difference is shown. The huge force may cause unnecessary vibration and instability. It is not desirable that the peak of the foot-end force is too large if the gait transitioning can be done. This is because it will have a large impact on the posture and affects the stability of the quadruped robot after gait transitioning. As can be seen from Figure 6f, the forces required by the robot using UPMPC are much smaller during gait transitioning. The quadruped robot regains stability in its posture by the combined effect of force and foot-end position. In Figure 6f, the HF uses a greater force to try to restore stability, but the gait transitioning still fails. In contrast, the UPMPC successfully completes the gait transitioning with less force.

These would suggest that the foot location chosen by the UPMPC is far superior to the HF. Gait transitioning is accomplished with less foot-end force and less posture oscillation. UPMPC shows a great advantage over HF in completing the gait transitioning between trotting and walking.

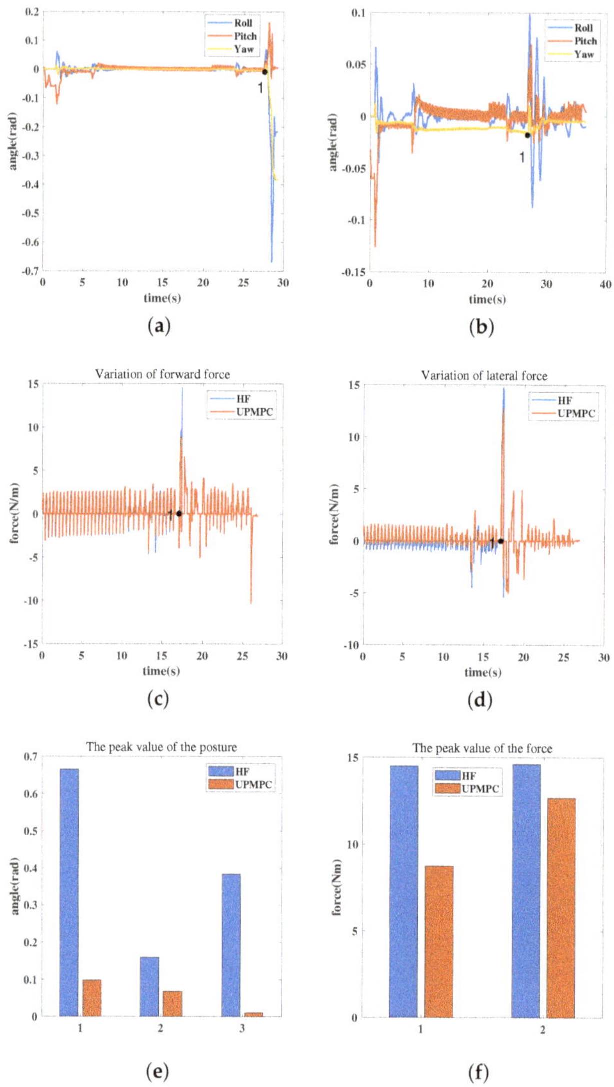

Figure 6. The figure shows the posture and GRFs of trotting in a gait transitioning to walking at 0.4 m/s. The number 1 in (**a**–**d**) indicate gait transitioning occuring. The horizontal coordinate of the number 1 in the diagram represents the exact time of the transitioning. The transitioning process is: trotting (0–27.6 s), walking (27.6-final). (**c**,**d**) Only capture the GRFs near the transitioning point to facilitate analysis of the forces, so the transitioning time is not consistent with (**a**,**b**). All data in the plots are from the same dataset. (**a**) Posture change using the HF at 0.4 m/s gait transitioning. (**b**) Posture change using the UPMPC at 0.4 m/s gait transitioning. (**c**) Change in forward force at 0.4 m/s during gait transitioning. (**d**) Change in lateral force at 0.4 m/s during gait transitioning. (**e**) Comparison of the peak values of roll, pitch and yaw for the two methods after transitioning. The numbers 1, 2 and 3 on the horizontal axis in the graph represent roll, pitch and yaw, respectively. (**f**) Comparison of peak values after transitioning for forward and lateral forces under both methods. The numbers 1 and 2 on the horizontal axis in the diagram represent forward and lateral forces, respectively.

Figure 7 is the posture change during gait transitioning between flight trotting and walking at 0.8 m/s. The transitioning points are marked by the numbers 1, 2, 3, 4 and 5 in the figure. In Figure 7a, the change of roll using UPMPC is basically consistent with HF. All have oscillation after the transitioning. In Figure 7b, the numbers 1, 2, 3, 4, 5 show the peak of the posture after the gait transitioning. A distinct difference between the two methods can be seen in in Figure 7b. In this figure, the peak of roll using the UPMPC is smaller than that of the HF, except for the fourth gait transitioning. The location of the foot chosen by the UPMPC enables a smaller roll peak after transitioning. This indicates that the quadruped robot with the UPMPC is more stable after switching. The number 6 represents the mean of absolute value of the roll. Smaller mean of the absolute values can indicate that the quadruped robot is more stable throughout the whole transitioning process. In Figure 7b, the mean of absolute value of roll using the HF is significantly larger than that using the UPMPC. The roll with the less oscillations over the whole time has the small mean of the absolute values. Combining the mean of absolute value and the peak values, the UPMPC has better control over roll. In Figure 7c, the curves of the two methods overlap highly. In Figure 7d, both the peak values and the mean of absolute value of pitch at the gait transitioning using UPMPC are smaller. Therefore, the UPMPC has better control of the pitch. In Figure 7e, yaw drifts regardless of the foot location planning method. As the number of transitioning accumulates, the yaw drift of UPMPC is smaller than that of the HF in Figure 7e. UPMPC has better control of yaw during gait transitioning. Comparing Figure 7b,d, the good control of yaw by UPMPC does not come at the expense of roll and pitch. Figure 8 is the foot-end force change during gait transitioning between flight trotting and walking at 0.8 m/s. Figure 8a,c show the force planned by MPC for the foot-end of the left front leg during the transitioning process. In Figure 8b, the force peaks of the UPMPC are significantly smaller in the second, third and fourth gait transitioning. The HF has a slightly smaller force peak in the first and fifth gait transitioning. It is not possible to distinguish exactly which method has the smaller mean of absolute values. The forward forces required by the robot are nearly the same for both methods over the whole time. In Figure 8d, the peak using the HF is smaller than that of UPMPC over the whole time. However, there is no difference in the mean of absolute value between the two methods. This means that both methods require essentially the same size force over the whole time. There is no considerable difference between the two methods in regards to forward forces or lateral forces. Combined with the change of posture, the UPMPC uses the same force to achieve smaller posture peak and mean of absolute values of posture over the whole time. Therefore, the UPMPC enables better control of posture. The UPMPC helps the quadruped robot to perform better in gait transitioning tasks at 0.8 m/s. The overall oscillations in posture are also smaller. However, it still cannot avoid the oscillation of posture at the moment of transitioning. The better effect of UPMPC has been demonstrated.

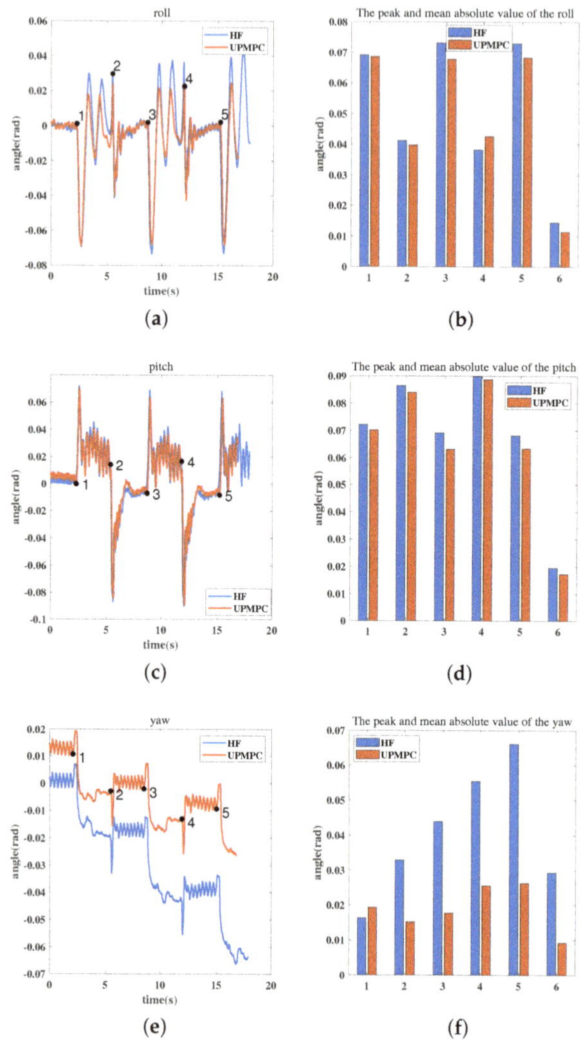

Figure 7. The figure shows the posture changes in gait transitioning between flight trotting and walking at 0.8 m/s. The numbers 1, 2, 3, 4 and 5 in (**a**,**c**,**e**) indicate gait transitioning occuring. The horizontal coordinate of these numbers in the diagram represents the exact time of the transitioning. The transitioning process is: flight trotting (0–2.27 s (number 1)), walking (2.27–5.27 s (number 2)), flight trotting (5.27–8.77 s (number 3)), walking (8.77–11.81 s (number 4)), flight trotting (11.81–15.20 s (number 5)) and walking (15.20-final). The numbers 1, 2, 3, 4 and 5 in (**b**,**d**,**f**) denote the peak values of the corresponding posture (roll, pitch or yaw) after the gait transitioning. Furthermore, the number 6 indicates the mean of the absolute value of the corresponding posture over the whole time. (**a**) Changes in roll during gait transitioning at 0.8 m/s. (**b**) Comparison of the peak values of roll after transitioning and mean of the absolute values of roll in the whole time for the two methods at 0.8 m/s. (**c**) Changes in pitch during gait transitioning at 0.8 m/s. (**d**) Comparison of the peak values of pitch after transitioning and mean of the absolute values of pitch in the whole time for the two methods at 0.8 m/s. (**e**) Changes in yaw during gait transitioning at 0.8 m/s. (**f**) Comparison of the peak values of yaw after transitioning and mean of the absolute values of yaw in the whole time for the two methods at 0.8 m/s.

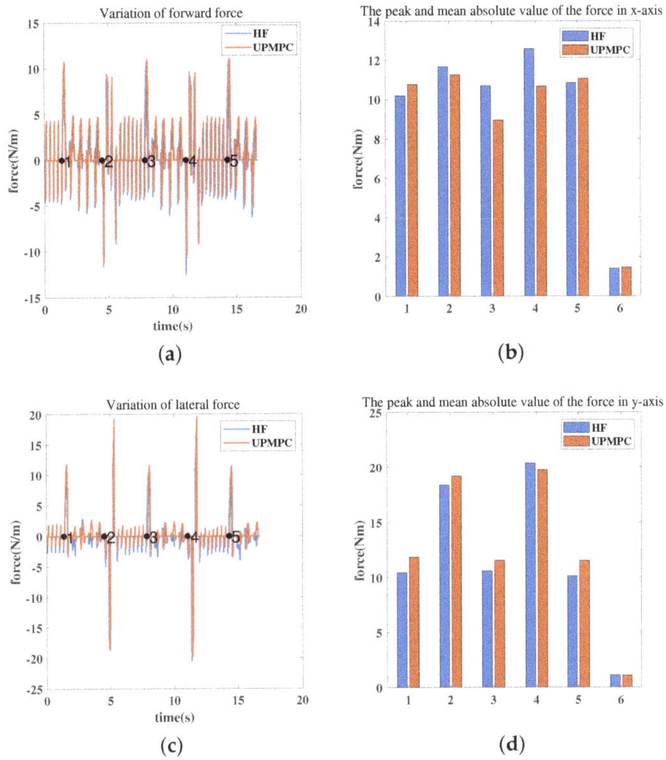

Figure 8. The figure shows the force changes in gait transitioning between flight trotting and walking at 0.8 m/s. The numbers 1, 2, 3, 4 and 5 in (**a**,**c**) indicate gait transitioning occuring. The horizontal coordinate of these numbers in the diagram represents the exact time of the transitioning. The transitioning process is: flight trotting (0–1.37 s (number 1)), walking (1.37–4.54 s (number 2)), flight trotting (4.54–7.86 s (number 3)), walking (7.86–11.05 s (number 4)), flight trotting (11.05–14.33 s (number 5)) and walking (14.33-final). The numbers 1, 2, 3, 4 and 5 in (**b**,**d**) denote the peak values of the corresponding forward forces or lateral forces after the gait transitioning. Furthermore, the number 6 indicates the mean of the absolute value of the corresponding forces over the whole time. (**a**) Change in forward force at 0.8 m/s. (**b**) Comparison of the peak values of forward force after transitioning and mean of the absolute values of forward force in the whole time for the two methods at 0.8 m/s. (**c**) Change in lateral force at 0.8 m/s. (**d**) Comparison of the peak values of lateral force after transitioning and mean of the absolute values of lateral force in the whole time for the two methods at 0.8 m/s.

Figure 9 is the posture change during gait transitioning between flight trotting and trotting at 1.2 m/s. The transitioning points are marked by the numbers 1, 2, 3, 4 and 5 in the figure. In Figure 9a, many parts are overlapped. At three transitioning points, the peak of the roll using UPMPC is slightly larger than that of HF in Figure 9c. However, the HF does not show a clear advantage over the UPMPC in reducing the peaks values. The mean of absolute value of the two methods is basically the same. For a slightly larger peak values than the HF, the UPMPC has the same mean of absolute value. It shows that the UPMPC allows the quadruped robot to recover stable from the postural oscillation of the transitioning more quickly. Despite the small peak values, the HF requires a longer time of posture oscillation to remove the effects of the transitioning. Therefore, it cannot be stated which method is really better for roll control. In Figure 9e, both methods have a large change in pitch at the instant of transitioning. The pitch using UPMPC has a smaller peak. It can be clearly seen from Figure 9b. At the same time, the mean of absolute values of pitch using the UPMPC is also smaller than that using the HF. Therefore, the UPMPC has better control for pitch. In Figure 9d, the yaw of both methods has some variation. In contrast to the UPMPC, HF does not control yaw well. Yaw gradually deviates from the stable value as gait transitioning occurs. The yaw using UPMPC also deviates, but its deviation under multiple gait transitioning is much smaller than HF in Figure 9f. Thus, the UPMPC has better control for yaw. Stability of posture is a prerequisite for the stable movement of a quadruped robot. Figure 10a,b shows the foot-end force of the left front leg planned by the MPC during the transitioning process. In Figure 10a, the vast majority of the curves from both methods overlap. A clear feedback is also obtained in Figure 10c. It was not possible to clearly determine which method had a smaller peak value of the forward force after gait transitioning. The mean of absolute values are also essentially the same. The forces are essentially the same under both methods. From the peak values, the lateral forces planned by UPMPC become significantly smaller than those of HF in Figure 10d. During the whole transitioning process, the lateral forces of HF are significantly greater than those of the UPMPC. Comparing to HF, the robot with the UPMPC has less posture change during gait transitioning, along with same forward forces and less lateral forces over the whole time at the speed of 1.2 m/s. The gait transitioning task is done better by the UPMPC. This indicates that the UPMPC selects a better location for the landing point. Combining the forces and posture, UPMPC's results are much better than HF's.

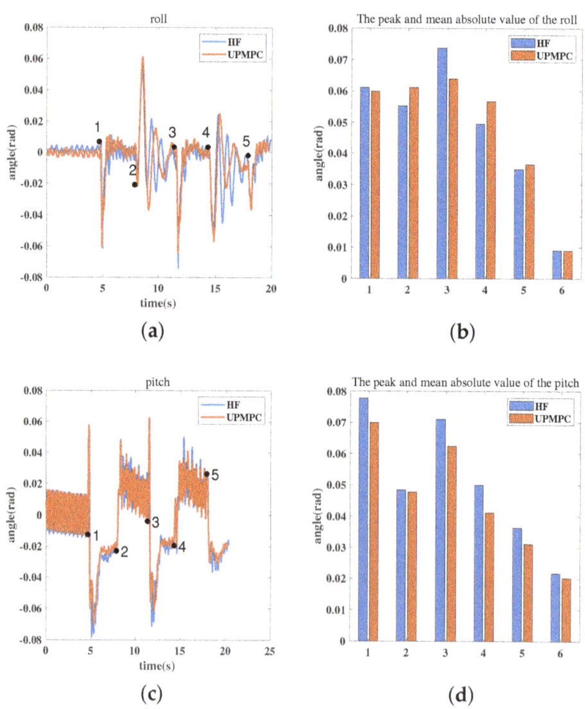

Figure 9. *Cont.*

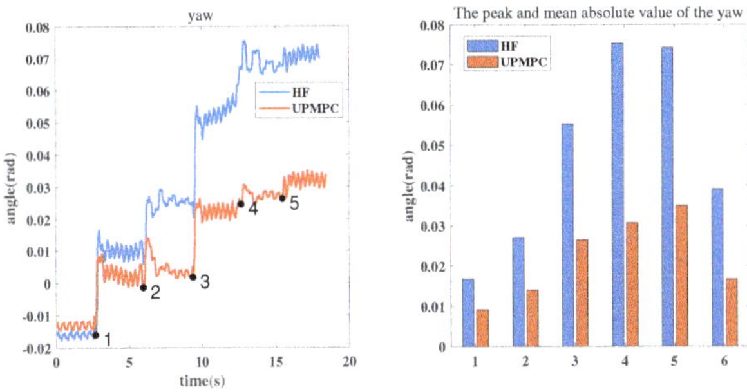

(e) (f)

Figure 9. The figure shows the posture changes in gait transitioning between trotting and flight trotting at 1.2 m/s. The numbers 1, 2, 3, 4 and 5 in (**a**,**c**,**e**) indicate gait transitioning occuring. The horizontal coordinate of these numbers in the diagram represents the exact time of the transitioning. The transitioning process is: trotting (0–4.66 s (number 1)), flight trotting (4.66–7.87 s (number 2)), trotting (7.87–11.36 s (number 3)), flight trotting (11.36–14.38 s (number 4)), trotting (14.38–17.88 s (number 5)) and flight trotting (17.88-final). The numbers 1, 2, 3, 4 and 5 in (**b**,**d**,**f**) denote the peak values of the corresponding posture (roll, pitch or yaw) after the gait transitioning. Furthermore, the number 6 indicates the mean of the absolute value of the corresponding posture over the whole time. (**a**) Changes in roll during gait transitioning at 1.2 m/s. (**b**) Comparison of the peak values of roll after transitioning and mean of the absolute values of roll in the whole time for the two methods at 1.2 m/s. (**c**) Changes in pitch during gait transitioning at 1.2 m/s. (**d**) Comparison of the peak values of pitch after transitioning and mean of the absolute values of pitch in the whole time for the two methods at 1.2 m/s. (**e**) Changes in yaw during gait transitioning at 1.2 m/s. (**f**) Comparison of the peak values of yaw after transitioning and mean of the absolute values of yaw in the whole time for the two methods at 1.2 m/s.

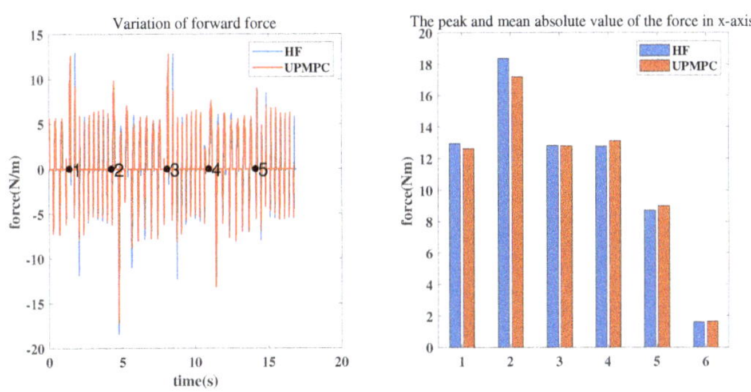

(a) (b)

Figure 10. *Cont.*

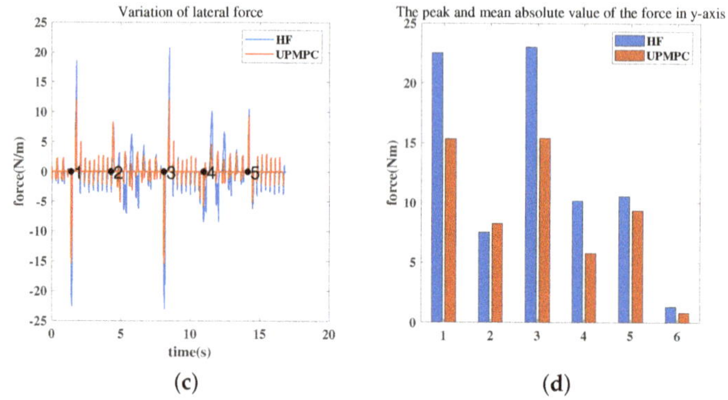

Figure 10. The figure shows the force changes in gait transitioning between flight trotting and trotting at 1.2 m/s. The numbers 1, 2, 3, 4 and 5 in (**a**,**c**), indicate gait transitioning occuring. The horizontal coordinate of these numbers in the diagram represents the exact time of the transitioning. The transitioning process is: trotting (0–1.4 s (number 1)), flight trotting (1.4–4.26 s (number 2)), trotting (4.26–8.08 s (number 3)), flight trotting (8.08–10.94 s (number 4)), trotting (10.94–14.11 s (number 5)) and flight trotting (14.11-final). The numbers 1, 2, 3, 4 and 5 in (**b**,**d**) denote the peak values of the corresponding forward forces or lateral forces after the gait transitioning. Furthermore, the number 6 indicates the mean of the absolute value of the corresponding forces over the whole time. (**a**) Change in forward force at 1.2 m/s gait transitioning. (**b**) Comparison of the peak values of forward force after transitioning and mean of the absolute values of forward force in the whole time for the two methods at 1.2 m/s. (**c**) Change in lateral force at 1.2 m/s gait transitioning. (**d**) Comparison of the peak values of lateral force after transitioning and mean of the absolute values of lateral force in the whole time for the two methods at 1.2 m/s.

5.3. Walk with a Fixed Gait

The UPMPC can also be used for walking with a fixed gait. Let the gait be trotting and the foot location planning method be UPMPC. In this gait the quadruped robot accelerates from stepping in place to 1.6 m/s. Each speed increase is 0.1 m/s with time interval of 30 gait periods (12.6 s). The roll, pitch and yaw changes during this process are shown in Figure 11.

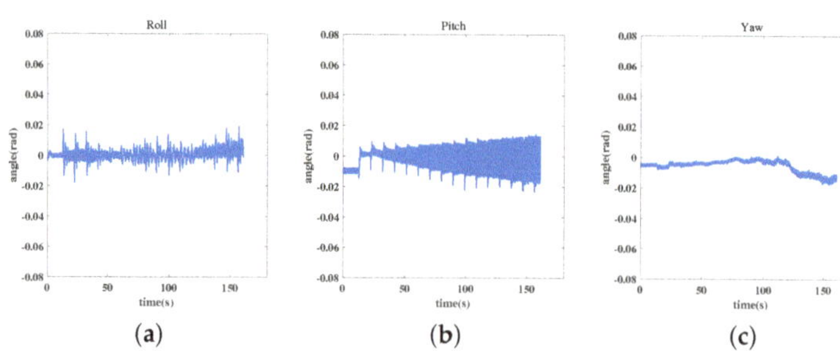

Figure 11. (**a**) The change of roll in trottting. (**b**) The change of pitch in trottting. (**c**) The change of yaw in trottting.

The quadruped robot has a small spike at each acceleration and the amplitude of the oscillation increases as the speed increases. Yaw has deviated from the stable value. However, it basically oscillates back and forth around the stable value. The peak values of posture has not yet exceeded ±0.03 rad (1.72 deg) which is so small that it can be ignored. With the same control method, the use of UPMPC also allows the quadruped robot to advance at a uniform speed and accelerate steadily. Therefore, UPMPC can be applied not only in the process of gait transitioning, but also in the planning of foot location under fixed gait.

5.4. Discussion

Figure 12 is the gait transitioning between trotting and walking at 0.8 m/s. This figure is different from Figure 6a. The gait transitioning using the HF in Figure 12a fails. The three postures after the transitioning are a straight line with no change, indicating that the quadruped robot is at stationary position. There is no command given after transitioning in the simulation. Thus, the gait transitioning has failed and the robot falls to the ground. It fails because of the large posture changes generated after the gait transitioning and the foot-end forces do not provide good control of the posture. The gait transitioning using the UPMPC in Figure 12b succeeds. However, the posture still face a huge oscillation. The control method for the quadruped robot is MPC all the time. Despite the greater foot end forces, the UPMPC planned foot location can perform gait transitioning tasks together with the controller. Although the force is less, the gait transitioning using the HF cannot be completed. It indicates that the foot location chosen by HF at this point is not reasonable. The foot location and foot-end forces under the HF can not correct for postural deviations. The HF cannot cope with this gait transitioning. Therefore the UPMPC's choice of foot location is still preferable to HF. The rest of the simulation results are in the Appendix A.

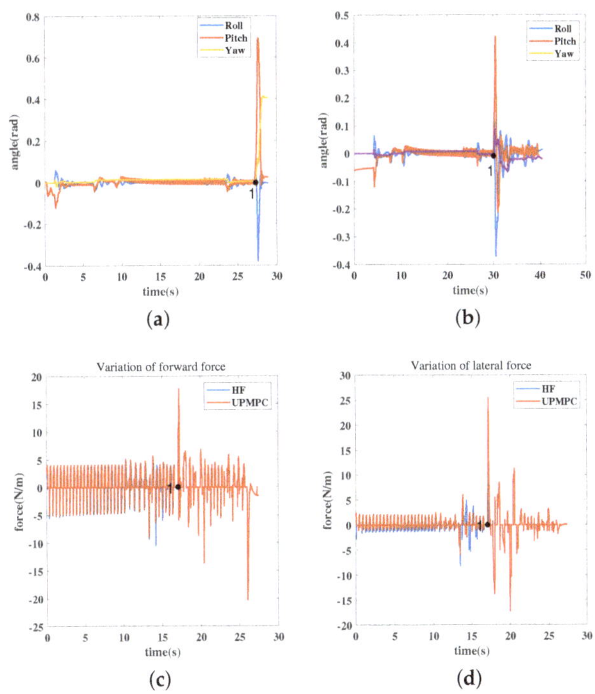

Figure 12. *Cont.*

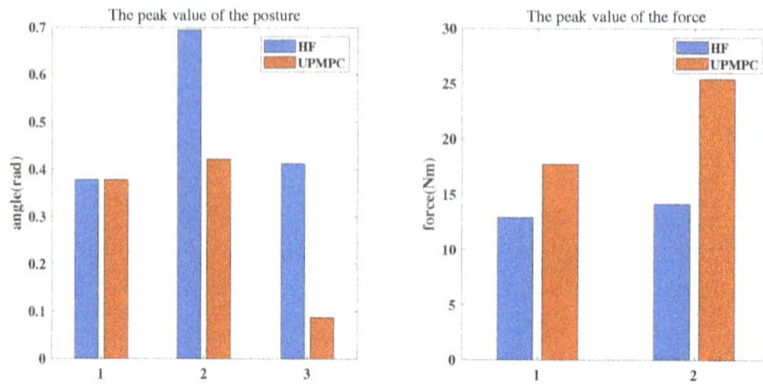

(e) (f)

Figure 12. The figure shows the posture and GRFs of trotting in a gait transitioning to walking at 0.8 m/s. The number 1 in (**a**–**d**) indicate gait transitioning occuring. The horizontal coordinate of the number 1 in the diagram represents the exact time of the transitioning. The transitioning process is: trotting (0–28.3 s), walking (28.3-final). (**c**,**d**) only capture the GRFs near the transitioning point to facilitate analysis of the forces, so the transitioning time is not consistent with (**a**,**b**). All data in the plots are from the same dataset. (**a**) Posture change using the HF at 0.8 m/s gait transitioning. (**b**) Posture change using the UPMPC at 0.8 m/s gait transitioning. (**c**) Change in forward force at 0.8 m/s during gait transitioning. (**d**) Change in lateral force at 0.8 m/s during gait transitioning. (**e**) Comparison of the peak values of roll, pitch and yaw for the two methods after transitioning. The numbers 1, 2 and 3 on the horizontal axis in the graph represent roll, pitch and yaw, respectively. (**f**) Comparison of peak values after transitioning for forward and lateral forces under both methods. The numbers 1 and 2 on the horizontal axis in the diagram represent forward and lateral forces, respectively.

6. Conclusions

Combining multiple gaits transitioning at various speeds, UPMPC performs better than the HF in all cases. In particular, the gait transitioning between trotting and walking cannot be done by the HF but can be done by UPMPC. Furthermore, it has much better control over yaw and pitch than HF. At most gait transitioning, the mean of absolute values of roll under UPMPC is smaller. UPMPC with fixed parameters is able to accommodate gait transitioning between multiple gaits at a variety of speeds.

Despite the use of a simplified model to describe the kinematics and dynamics of the quadruped robot, the UPMPC-planned foot location still meets the requirements for gait transitioning and is better than HF. The use of the uniform foot location not only speeds up the solution, but also converts the problem to be solved into a QP form with a smaller matrix. The coordinates of the foot in the world coordinate system are used as the optimization solution variable more in line with the actual situation. Due to the good migration of the used platform, it is believed that UPMPC can be applied in real systems to meet the need of gait transitioning on real world.

Author Contributions: Conceptualization, methodology and writing—original draft preparation, X.L.; writing—review and editing, L.L., H.A. and H.M. All authors have read and agreed to the published version of the manuscript.

Funding: This work was funded by National Science Foundation of China (No. 61903131) and Outstanding Youth of Department of Education of Hunan Province (No. 20B096) and the China Postdoctoral Science Foundation (No. 2020M683715).

Institutional Review Board Statement: Not applicable.

Informed Consent Statement: Not applicable.

Data Availability Statement: The simulation code can be downloaded in https://github.com/lxm-1/Mini-Cheetah.git accessed on 23 August 2021.

Conflicts of Interest: The authors declared no potential conflict of interest with respect to the research, authorship, and/or publication of this article.

Appendix A

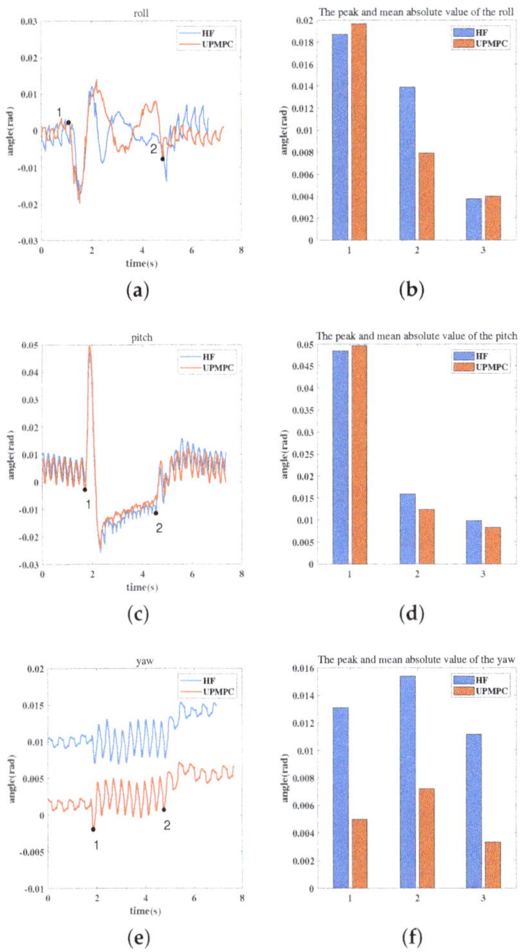

Figure A1. The figure shows the posture changes in gait transitioning between trotting and flight trotting at 0.4 m/s. The numbers 1 and 2 in (**a**,**c**,**e**) indicate gait transitioning occuring. The horizontal coordinate of these numbers in the diagram represents the exact time of the transitioning. The transitioning process is: trotting (0–1.80 s (number 1)), flight trotting (1.80–4.76 s (number 2)) and trotting (4.76-final). The numbers 1 and 2 in (**b**,**d**,**f**) denote the peak values of the corresponding posture (roll, pitch or yaw) after the gait transitioning. Furthermore, the number 3 indicates the mean of the absolute values of the corresponding posture over the whole time. (**a**) Changes in roll during gait transitioning at 0.4 m/s. (**b**) Comparison of the peak values of roll after transitioning and mean of the absolute values of roll in the whole time for the two methods at 0.4 m/s. (**c**) Changes in pitch during gait transitioning at 0.4 m/s. (**d**) Comparison of the peak values of pitch after transitioning and mean of the absolute values of pitch in the whole time for the two methods at 0.4 m/s. (**e**) Changes in yaw during gait transitioning at 0.4 m/s. (**f**) Comparison of the peak values of yaw after transitioning and mean of the absolute values of yaw in the whole time for the two methods at 0.4 m/s.

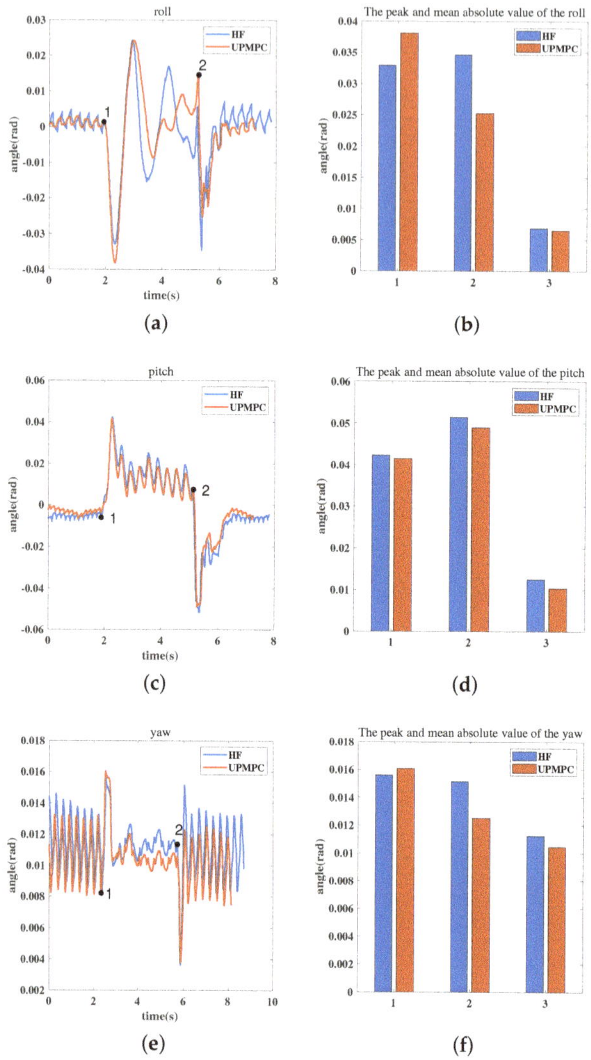

Figure A2. The figure shows the posture changes in gait transitioning between flight trotting and walking at 0.4 m/s. The numbers 1 and 2 in (**a**,**c**,**e**) indicate gait transitioning occuring. The horizontal coordinate of these numbers in the diagram represents the exact time of the transitioning. The transitioning process is: flight trotting (0–1.88 s (number 1)), walking (1.88–5.12 s (number 2)) and flight trotting (5.12-final). The numbers 1 and 2 in (**b**,**d**,**f**) denote the peak values of the corresponding posture (roll, pitch or yaw) after the gait transitioning. Furthermore, the number 3 indicates the mean of the absolute value of the corresponding posture over the whole time. (**a**) Changes in roll during gait transitioning at 0.4 m/s. (**b**) Comparison of the peak values of roll after transitioning and mean of the absolute values of roll in the whole time for the two methods at 0.4 m/s. (**c**) Changes in pitch during gait transitioning at 0.4 m/s. (**d**) Comparison of the peak values of pitch after transitioning and mean of the absolute values of pitch in the whole time for the two methods at 0.4 m/s. (**e**) Changes in yaw during gait transitioning at 0.4 m/s. (**f**) Comparison of the peak values of yaw after transitioning and mean of the absolute values of yaw in the whole time for the two methods at 0.4 m/s.

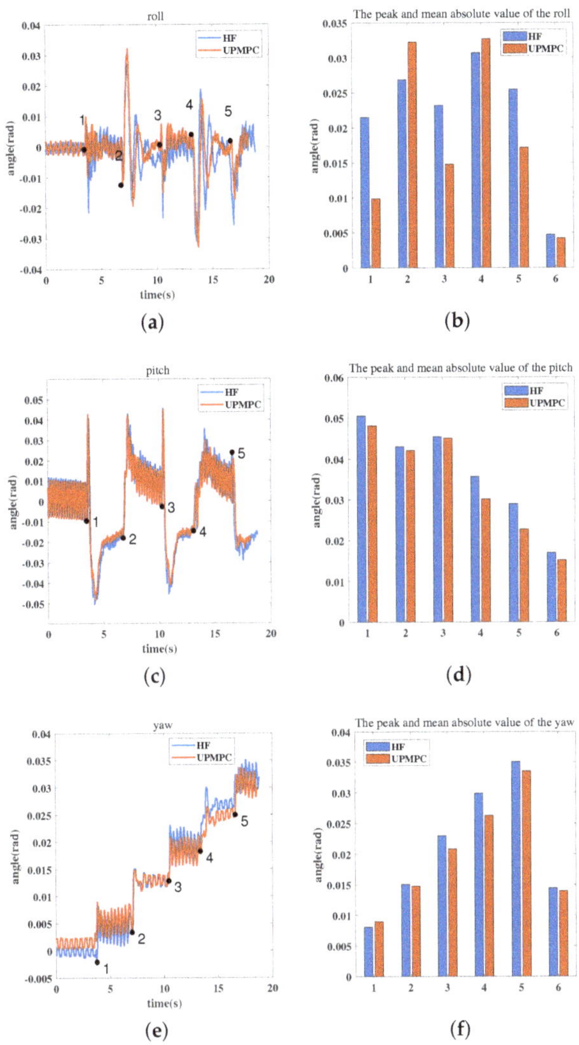

Figure A3. The figure shows the posture changes in gait transitioning between trotting and flight trotting at 0.8 m/s. The numbers 1, 2, 3, 4 and 5 in (**a**,**c**,**e**) indicate gait transitioning occuring. The horizontal coordinate of these numbers in the diagram represents the exact time of the transitioning. The transitioning process is: trotting (0–3.48 s (number 1)), flight trotting (3.48–6.77 s (number 2)), trotting (6.77–10.28 s (number 3)), flight trotting (10.28–13.10 s (number 4)), trotting (13.10–16.61 s (number 5)) and flight trotting (16.61-final). The numbers 1, 2, 3, 4 and 5 in (**b**,**d**) denote the peak values of the corresponding posture (roll, pitch or yaw) after the gait transitioning. Furthermore, the number 6 indicates the mean of the absolute value of the corresponding posture over the whole time. (**a**) Changes in roll during gait transitioning at 0.8 m/s. (**b**) Comparison of the peak values of roll after transitioning and mean of the absolute values of roll in the whole time for the two methods at 0.8 m/s. (**c**) Changes in pitch during gait transitioning at 0.8 m/s. (**d**) Comparison of the peak values of pitch after transitioning and mean of the absolute values of pitch in the whole time for the two methods at 0.8 m/s. (**e**) Changes in yaw during gait transitioning at 0.8 m/s. (**f**) Comparison of the peak values of yaw after transitioning and mean of the absolute values of yaw in the whole time for the two methods at 0.8 m/s.

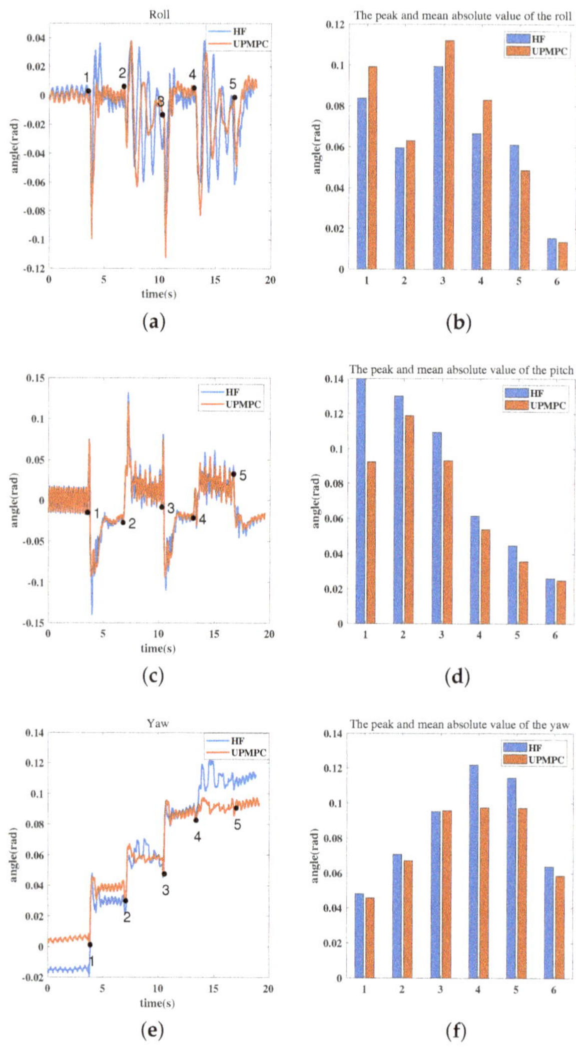

Figure A4. The figure shows the posture changes in gait transitioning between trotting and flight trotting at 1.6 m/s. The numbers 1, 2, 3, 4 and 5 in (**a**,**c**,**e**) indicate gait transitioning occuring. The horizontal coordinate of these numbers in the diagram represents the exact time of the transitioning. The transitioning process is: trotting (0–3.57 s (number 1)), flight trotting (3.57–6.79 s (number 2)), trotting (6.79–10.27 s (number 3)), flight trotting (10.27–13.09 s (number 4)), trotting (13.09–16.72 s (number 5)) and flight trotting (16.72-final). The numbers 1, 2, 3, 4 and 5 in (**b**,**d**) denote the peak values of the corresponding posture (roll, pitch or yaw) after the gait transitioning. Furthermore, the number 6 indicates the mean of the absolute value of the corresponding posture over the whole time. (**a**) Changes in roll during gait transitioning at 1.6 m/s. (**b**) Comparison of the peak values of roll after transitioning and mean of the absolute values of roll in the whole time for the two methods at 1.6 m/s. (**c**) Changes in pitch during gait transitioning at 1.6 m/s. (**d**) Comparison of the peak values of pitch after transitioning and mean of the absolute values of pitch in the whole time for the two methods at 1.6 m/s. (**e**) Changes in yaw during gait transitioning at 1.6 m/s. (**f**) Comparison of the peak values of yaw after transitioning and mean of the absolute values of yaw in the whole time for the two methods at 1.6 m/s.

References

1. Hoyt, D.F.; Taylor, C.R. Gait and the energetics of locomotion in horses. *Nature* **1981**, *292*, 239–240. [CrossRef]
2. Xi, W.; Yesilevskiy, Y.; Remy, C.D. Selecting gaits for economical locomotion of legged robots. *Int. J. Robot. Res.* **2016**, *35*, 1140–1154. [CrossRef]
3. Raibert, M.H. *Legged Robots That Balance*; MIT Press: Cambridge, MA, USA, 1986.
4. Di Carlo, J.; Wensing, P.M.; Katz, B.; Bledt, G.; Kim, S. Dynamic locomotion in the mit cheetah 3 through convex model-predictive control. In Proceedings of the 2018 IEEE/RSJ International Conference on Intelligent Robots and Systems (IROS), Madrid, Spain, 1–5 October 2018; pp. 1–9.
5. Katz, B.; Di Carlo, J.; Kim, S. Mini cheetah: A platform for pushing the limits of dynamic quadruped control. In Proceedings of the 2019 International Conference on Robotics and Automation (ICRA), Montreal, QC, Canada, 20–24 May 2019; pp. 6295–6301.
6. Pratt, J.; Carff, J.; Drakunov, S.; Goswami, A. Capture point: A step toward humanoid push recovery. In Proceedings of the 2006 6th IEEE-RAS International Conference on Humanoid Robots, Genova, Italy, 4–6 December 2006; pp. 200–207.
7. Simões, I.F.E.; Cordero, A.F. Walking in the 2-Step Capture Region; pushes, ramps and speed modulation. In Proceedings of the 2019 19th International Conference on Advanced Robotics (ICAR), Belo Horizonte, Brazil, 2–6 December 2019; pp. 449–455.
8. Seyde, T.; Shrivastava, A.; Englsberger, J.; Bertrand, S.; Pratt, J.; Griffin, R.J. Inclusion of angular momentum during planning for capture point based walking. In Proceedings of the 2018 IEEE International Conference on Robotics and Automation (ICRA), Brisbane, Australia, 21–25 May 2018; pp. 1791–1798.
9. Kojio, Y.; Ishiguro, Y.; Sugai, F.; Kakiuchi, Y.; Okada, K.; Inaba, M. Unified balance control for biped robots including modification of footsteps with angular momentum and falling detection based on capturability. In Proceedings of the 2019 IEEE/RSJ International Conference on Intelligent Robots and Systems (IROS), Macau, China, 4–8 November 2019; pp. 497–504.
10. Wieber, P.B. Trajectory free linear model predictive control for stable walking in the presence of strong perturbations. In Proceedings of the 2006 6th IEEE-RAS International Conference on Humanoid Robots, Genova, Italy, 4–6 December 2006; pp. 137–142.
11. Saraf, P.; Sarkar, A.; Javed, A. Terrain Adaptive Gait Transitioning for a Quadruped Robot using Model Predictive Control. *arXiv* **2021**, arXiv:2106.03307.
12. Faraji, S.; Pouya, S.; Atkeson, C.G.; Ijspeert, A.J. Versatile and robust 3d walking with a simulated humanoid robot (atlas): A model predictive control approach. In Proceedings of the 2014 IEEE International Conference on Robotics and Automation (ICRA), Hong Kong, China, 31 May–7 June 2014; pp. 1943–1950.
13. Faraji, S.; Pouya, S.; Ijspeert, A. Robust and agile 3d biped walking with steering capability using a footstep predictive approach. In Proceedings of the Robotics Science and Systems (RSS), Berkeley, CA, USA, 12–16 July 2014.
14. Feng, S.; Xinjilefu, X.; Atkeson, C.G.; Kim, J. Robust dynamic walking using online foot step optimization. In Proceedings of the 2016 IEEE/RSJ International Conference on Intelligent Robots and Systems (IROS), Daejeon, Korea, 9–14 October 2016; pp. 5373–5378.
15. Xin, S.; Orsolino, R.; Tsagarakis, N. Online relative footstep optimization for legged robots dynamic walking using discrete-time model predictive control. In Proceedings of the 2019 IEEE/RSJ International Conference on Intelligent Robots and Systems (IROS), Macau, China, 4–8 November 2019; pp. 513–520.
16. Lewis, S.B. *Animals in Motion*; Eadweard Muybridge: New York, NY, USA, 1975.
17. Kajita, S.; Kanehiro, F.; Kaneko, K.; Fujiwara, K.; Harada, K.; Yokoi, K.; Hirukawa, H. Biped walking pattern generation by using preview control of zero-moment point. In Proceedings of the 2003 IEEE International Conference on Robotics and Automation (Cat. No. 03CH37422), Taipei, Taiwan, 14–19 September 2003; Volume 2, pp. 1620–1626.
18. Dini, N.; Majd, V.J. An MPC-based two-dimensional push recovery of a quadruped robot in trotting gait using its reduced virtual model. *Mech. Mach. Theory* **2020**, *146*, 103737. [CrossRef]
19. Kajita, S.; Tani, K. Study of dynamic biped locomotion on rugged terrain-derivation and application of the linear inverted pendulum mode. In Proceedings of the 1991 IEEE International Conference on Robotics and Automation, Sacramento, CA, USA, 9–11 April 1991; pp. 1405–1406.
20. Alexander, R.M. The gaits of bipedal and quadrupedal animals. *Int. J. Robot. Res.* **1984**, *3*, 49–59. [CrossRef]
21. Di Carlo, J. Software and Control Design for the MIT Cheetah Quadruped Robots. Ph.D. Thesis, Massachusetts Institute of Technology, Cambridge, MA, USA, 2020.
22. Huang, A.S.; Olson, E.; Moore, D.C. LCM: Lightweight communications and marshalling. In Proceedings of the 2010 IEEE/RSJ International Conference on Intelligent Robots and Systems, Taipei, Taiwan, 18–22 October 2010; pp. 4057–4062.
23. Ferreau, H.J.; Kirches, C.; Potschka, A.; Bock, H.G.; Diehl, M. qpOASES: A parametric active-set algorithm for quadratic programming. *Math. Program. Comput.* **2014**, *6*, 327–363. [CrossRef]
24. Collins, S.; Ruina, A.; Tedrake, R.; Wisse, M. Efficient bipedal robots based on passive-dynamic walkers. *Science* **2005**, *307*, 1082–1085. [CrossRef]

applied sciences

MDPI

Article

Design of JET Humanoid Robot with Compliant Modular Actuators for Industrial and Service Applications

Jaehoon Sim [1], Seungyeon Kim [1], Suhan Park [1], Sanghyun Kim [2], Mingon Kim [1] and Jaeheung Park [1,3,*]

1 Graduate School of Convergence Science and Technology, Seoul National University, Seoul 08826, Korea; simjeh@snu.ac.kr (J.S.); ksy0711@snu.ac.kr (S.K.); psh117@snu.ac.kr (S.P.); mingonkim@snu.ac.kr (M.K.)
2 Korea Institute of Machinery and Materials (KIMM), Daejeon 34103, Korea; kim87@kimm.re.kr
3 Advanced Institute of Convergence Technology (AICT), Suwon 16229, Korea
* Correspondence: park73@snu.ac.kr; Tel.: +82-31-888-9146

Abstract: This paper presents the development of the JET humanoid robot, which is based on the existing THORMANG platform developed in 2015. Application in the industrial and service fields was targeted, and three design concepts were determined for the humanoid robot. First, low stiffness of the actuator modules was utilized for compliance with external environments. Second, to maximize the robot whole-body motion capability, the overall height was increased. However, the weight was reduced to satisfy power requirements. The workspace was also increased to enable various postures, by increasing the range of motion of each joint and extending the links. Compared to the original THORMANG platform, the lower limb length increased by approximately 20%, and the hip range of motion increased by 39.3%. Third, the maintenance process was simplified through modularization of the electronics and frame design for improved accessibility. Several experiments, including stair climbing and egress from a car, were performed to verify that the JET humanoid robot performance enhancements reflected the design concepts.

Keywords: humanoid robots; robot design; legged robots

Citation: Sim, J.; Kim, S.; Park, S.; Kim, S.; Kim, M.; Park, J. Design of JET Humanoid Robot with Compliant Modular Actuators for Industrial and Service Applications. *Appl. Sci.* **2021**, *11*, 6152. https://doi.org/10.3390/app11136152

Academic Editors: Alessandro Gasparetto, Stefano Seriani, Lorenzo Scalera and Manuel Armada

Received: 31 May 2021
Accepted: 30 June 2021
Published: 2 July 2021

1. Introduction

Robots have many increasingly different applications. Recently, their feasibility in various service and industry settings, where interaction with humans or environments is required, has been explored. Among the many possible robot forms for these applications, humanoids can be very effective as they have similar kinematic structures to humans. That is, utilizing two arms and two legs, humanoid robots can perform various tasks in a human-centered environment. In particular, the DARPA Robotics Challenge (DRC) Finals in 2015 demonstrated the potential of humanoid robots for use in disaster-response scenarios [1]. Furthermore, the use of humanoid robots in industrial applications, such as airplane assembly and nuclear power plant cleaning, is now being actively investigated [2].

As humanoid robots are required to interact with people as well as the environment in industry and service applications, both high performance and safety are important. Thus, several design features should be considered: compliance with unexpected external forces, whole-body motion capability through an increased workspace, and straightforward maintenance.

Compliance reduces the physical damage or instability from unexpected collisions between the robot and external objects and is an essential design consideration for a humanoid robot. There are two approaches to achieving compliance: hardware methods using low-stiffness actuators and software approaches such as torque-based controllers and impedance controllers [3,4]. Recently, a method for controlling compliant contact with objects has been proposed through an optimization-based approach [5]. In addition, artificial intelligence-based methods of controlling desired compliance have also been introduced for many applications [6]. However, software methods are limited by their dependence on the performance of a sensory system, with their compliance bandwidth limited by those of the sensors and control systems. Furthermore, if a sensor malfunction or software failure occurs, robot compliance cannot be guaranteed. Therefore, humanoid robots with low-stiffness actuators have recently been developed [7,8].

Second, whole-body motion capability is fundamental for various tasks, and a large robot workspace enhances this capability. In the DRC Finals, most teams reported whole-body motion planning as a main factor determining manipulation capability in human-centered environments [9]. For instance, the robot should have the capability to perform complex tasks such as stair climbing, driving, and exiting a vehicle.

Finally, and most importantly, industrial and service humanoids should be easy to maintain. As humanoid robots frequently interact with humans or objects, they require periodic maintenance to avoid malfunction. Modularization is a dominant approach to easy maintenance and is adopted for many robots [10].

In this study, the JET humanoid platform shown in Figure 1 is introduced to implement the three features listed above. JET was developed on the basis of THORMANG, which was developed by ROBOTIS and was modified by TEAM SNU for the DRC Finals [11]. Through the new design proposed herein, JET overcomes the THORMANG platform limitations while conserving its advantages. The main aim of the new design is to acquire more practical capabilities suited to industrial and service applications.

Figure 1. Humanoid robot JET.

The remainder of this paper is organized as follows. Section 2 describes the THORMANG platform specifications and our design goals, while Section 3 presents the newly developed JET humanoid robot design. Section 4 reports experimental validation of the developed robot and Section 5 concludes the paper.

2. Design Goal Selection

This section presents the specifications of our previous humanoid robot platform, THORMANG, and the design goals of our newly proposed humanoid robot for maximized performance.

2.1. Limitation of THORMANG Platform

THORMANG, which was developed by ROBOTIS, is an open source platform humanoid. This robot has two representative characteristics. First, THORMANG is modular and can, therefore, be easily modified according to user aims. For example, although four teams used the THORMANG platform at the DRC Finals (TEAM SNU, TEAM THOR, TEAM, Hector, and TEAM Robotis), the robot shapes, sizes, and weights varied, as shown in Figure 2 [11].

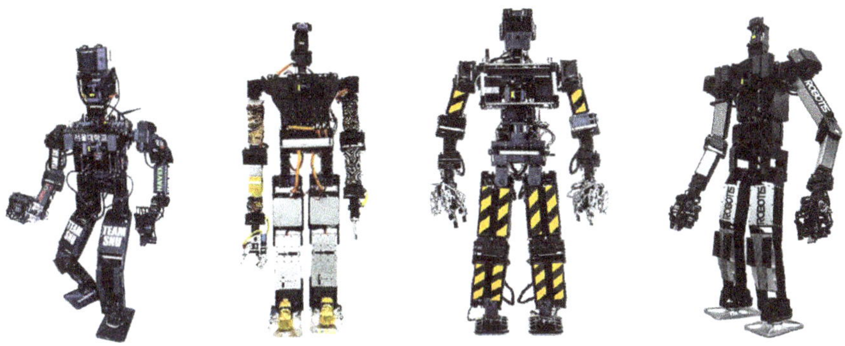

(a) (b) (c) (d)

Figure 2. THORMANG platform at DRC Finals in 2015: (**a**) THORMANG, TEAM SNU (1.47 m, 60 kg), (**b**) THOR-RD, TEAM THOR (1.50 m, 54 kg), (**c**) Johnny05, TEAM Hector (1.47 m, 55 kg), (**d**) THORMANG2, TEAM ROBOTIS (1.60 m, 60 kg).

Second, THORMANG employs Dynamixel Pro, which is an actuator module developed as a commercial product by ROBOTIS. Dynamixel Pro has low stiffness, which is more comparable to a series elastic actuator (SEA) than to other conventional electric actuators with harmonic drives. The SHG-17-100-2SO harmonic drive, which has a similar torque capacity to Dynamixel Pro, has a stiffness of 10,000 Nm/rad or higher [12,13]. In contrast, SEAs for humanoids have a lower stiffness of 500–10,000 Nm/rad [14,15]. Notably, the H54-200-S500-R Dynamixel Pro stiffness is 900 Nm/rad, on a similar scale as the COMAN and WALKMAN SEAs [16]. As mentioned above, mechanical compliance from low-stiffness actuators aids robot control in complex multiple-contact situations, as it can provide robustness to the disturbance caused by unexpected contact with external objects.

However, the THORMANG platform has several limitations in terms of robot performance. First, the robot kinematic structure is not suited to various tasks in human-centered environments such as industrial and service sites. Humanoids require whole-body motion capability to perform diverse tasks, but the THORMANG leg length is insufficient and the joint ranges are limited. In the DRC Finals, all THORMANG-based robots failed the egress task, as detailed in Table 1, because the robot could not reach the floor while sitting in the vehicle. Further, the stair task, which required the robot to climb a staircase, was one of the most difficult tasks for the THORMANG-based designs because of their short legs and narrow workspace [11]. To statically climb the stairs, the knees required sufficient bending for lateral movement. To move the center of mass (COM) over the feet on the stairs, it was necessary for the leg joints to rotate by more than the shorter length between the hip and foot. On a step with a 24 cm height, self-collision of the hip and knee joints occurred because of the short legs of the THORMANG design.

Second, unmeasurable deformation occurred at the hip joints because of the high lower body weight and actuator module compliance. The lower body of the original THORMANG platform consists of two lithium polymer (LiPo) batteries (22,000 mAh, 22.2 V) and heavy links [17]. Therefore, when the robot walks, the pelvis tilt is so extensive that control of the swing foot is very difficult. During the DRC competition, most THORMANG-based robots, excluding that of TEAM SNU, fell when walking.

Table 1. Scores of THORMANG-based teams in DRC Finals. Team SNU successfully performed 4 missions (**O**), but failed 4 missions (**X**), including the egress task.

Mission	Team SNU	Team ROBOTIS	Team THOR	Team HECTOR
Driving	O	O	O	O
Egress	X	X	X	X
Door	O	O	O	X
Valve	O	O	O	X
Wall	O	X	X	X
Surprise	X	X	X	X
Rubble	X	X	X	X
Stairs	X	X	X	X
Score	4	3	3	1

Therefore, we concentrated on developing our hardware to upgrade the performance and overcome the above limitations.

2.2. Design Goals

In addition to overcoming the above THORMANG limitations, to achieve compliance, whole-body motion capability, and easy maintenance, as mentioned in Section 1, the following three philosophies were adopted as the JET design goals.

First, increasing the lower body workspace allows a robot to fit various situations. As main structures such as stairs, seats, and ladders are designed for humans, a proper leg length is advantageous. Furthermore, the robot range of motion (ROM) can be increased by avoiding self-collision. Hence, the designed robot can perform various motions such as full squats.

Second, robot weight reduction is necessary for increased energy efficiency and to alleviate actuator loads. However, increasing the link length to enhance the robot workspace can overload the actuators. To solve this problem, we considered better weight distribution as well as weight reduction for JET.

Third, easy maintenance is important for increased economic feasibility. Thus, the JET links are composed of many small parts. We designed most link parts to have a flat shape for easy laser cutting, which is cheaper and quicker than milling. Additionally, the frames containing actuators and circuits were designed to be easily removable for easy disassembly and assembly.

3. JET Design

As detailed in Table 2, the height, weight, and wingspan of the developed robot are 1.63 m, 48 kg, and 2.12 m, respectively. Other robots, including TALOS, HRP-5P, and former THORMANG platforms, and human characteristics are also shown in Table 2. Similar to the THORMANG platforms, TALOS is a commercial off-the-shelf research platform which is developed by PAL-Robotics [18]. HRP-5P is an adult-sized humanoid developed to realize the use of humanoids in large-scale assembly industries. Compared with the THORMANG of Team SNU, JET is taller and lighter. However, JET is heavier than THORMANG3, the latest released version from ROBOTIS because THORMANG3 is the shortest platform among all of the THORMANG platforms. Both THORMANG3 and JET are developed on the basis of THORMANG, but THORMANG3 is designed to focus on increasing speed by reducing size. On the other hand, JET is designed to focus on the workspace to conduct difficult tasks, such as egress from the vehicle and climbing stairs in the DRC Finals, for THORMANG due to its small size. With a similar leg length to JET, TALOS is taller but much heavier due to stronger actuators and torque sensors. Developed to perform a variety of industrial tasks with a wide joint range, HRP-5P has 23 mm longer legs, and it is taller than JET. However, HRP-5P is much heavier due to its larger degrees of freedom (DOF) and stronger actuators. Notably, the JET height, weight, and leg length are similar to the averages for Korean young women aged 20–24 [19]. Due to the torque capacity of the hip pitch joint, the length of each link in the shin and thigh was adjusted. Therefore, the JET hardware specifications are suitable for performance of various complex tasks in a human-centered environment.

Table 2. Comparison of JET, THORMANG (Team SNU), THORMANG3, TALOS, HRP-5P, and average Korean young female (aged 20–24) characteristics.

	JET	THORMANG (Team SNU)	THORMANG3	TALOS [20]	HRP-5P [21]	Korean Young Female
Height (m)	1.63	1.47	1.38	1.75	1.83	1.61
Weight (kg)	48	60	42	95	101	55
Wingspan (m)	2.12	1.95	1.69	2.1	-	1.62
Leg length (mm)	839	696	716	837	862	799
Thigh length (mm)	370	300	300	380	380	384
Shin length (mm)	373	300	300	325	380	352
Foot height (mm)	96	96	116	132	102	63
Degree of freedom	32	32	32	31	37	-

Figure 3 shows the JET joint configuration, which consists of a total of 32 DOF, excluding the LiDAR actuator. There are eight joints in each arm with the end-effectors, six in each leg, two in the waist, and two in the head. In the waist, we used two DOF joints in the yaw and roll directions to increase not only the bipedal walking stability but also the robot workspace [22].

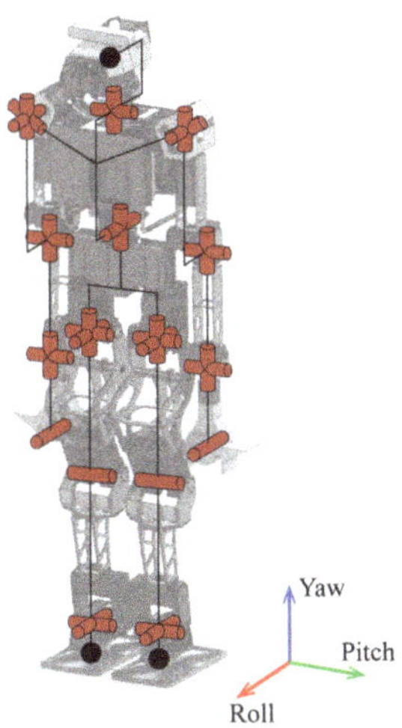

(a) Joint configurations

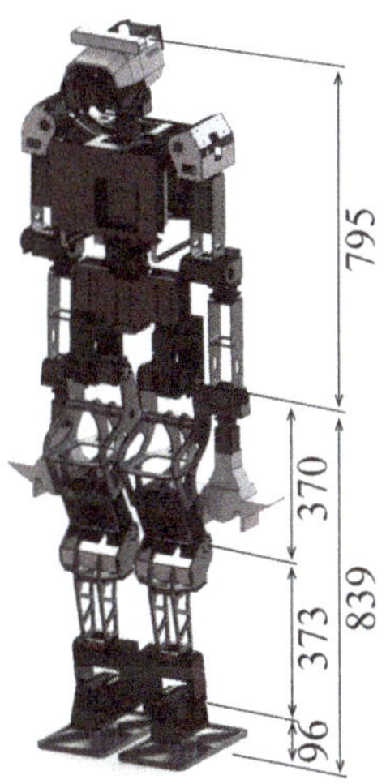

(b) Link length

Figure 3. Kinematic structure of JET.

The following subsections present details of the proposed hardware, i.e., the electrical system and lower and upper body designs.

3.1. Electrical System Design

Figure 4 shows the overall electrical system of JET. A mini-ITX PC is used as the computing system, with the computer board being placed at the front of the robot's chest. The computer CPU is an Intel Core i7-4790 Processor (3.6 GHz) to provide sufficient computation performance for robot control. The computer is equipped with a SENSORAY 826 DAQ board for acquiring analog sensor data and a DIAMOND DS-MPE-OPT4485 4-Port RS-485 MiniCard Module for actuator communication. Using a PCIe-based communication interface, JET can be controlled with a 200 Hz command rate. In contrast, the THORMANG command rate is 100 Hz [11].

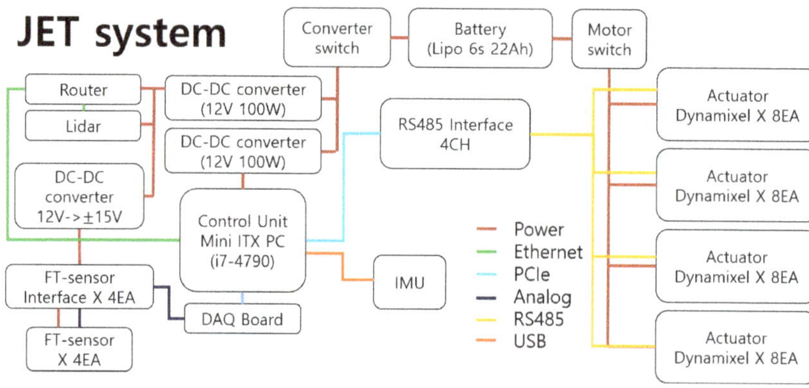

Figure 4. Electrical system diagram of JET.

To supply power to the robot, we designed a power system including regulators and a battery. To connect the battery, electronics, and regulators, wire connectors were unified with the same actuator connectors. Parts unification is a well-known method of reducing both manufacturing cost and maintenance. In the JET system, one LiPo (22,000 mAh, 22.2 V) battery supplies power to all actuators and voltage transformer modules. A voltage transformer module is composed of three DC–DC converters. To match the computer ground level with the actuators, two 12 V converters are nonisolated. However, the converter for the FT sensor interface is isolated to reduce the FT sensor power source noise. The sensor ground level is matched with the main ground via the DAQ board. The LiPo battery has excellent discharge performance and small voltage fluctuation, even under high load. On standby, JET consumes 5 A (0.2 C). The peak current while walking is 9 A (0.4 C); this is a small current compared to the LiPo battery discharge rate. Thus, the battery voltage fluctuations due to the discharge rate are sufficiently small to integrate the actuator power source with the computer source. With a fully charged LiPo battery, JET can stand for 4 h.

The sensory system required for humanoids includes an environment perception sensor and a measurement sensor of the robot state. Three sensor types are used for the perception system: one LiDAR, one IMU, and four FT sensors. A HOKUYO UTM-30LX-EW sensor, which is a 2D LiDAR, is mounted at the head to acquire 3D point cloud data of the robot surroundings; this is achieved by continuously rotating the sensor about its vertical axis. To facilitate continuous LiDAR rotation, a slip ring is used to connect the electric wires. A MicroStrain 3DM-GX4-25 IMU is placed at the robot coordinate origin in the pelvis. Finally, ATI Mini 58 and Mini 45 FT sensors are mounted on both feet and hands to measure the forces acting on the feet and wrists. Further system integration and software development details are given in our previous paper [23]. Using a robot operating system (ROS)-based system framework, JET can be controlled stably with many sensors and actuators.

3.2. Lower Body Design

The lower body design goal was to achieve a wide ROM. To accommodate various environments and obstacles tailored to an adult body, a wide lower body workspace is required. The foot workspace can be extended by increasing the leg length or by increasing the joint ROM. However, in the case of JET, the lower body length is limited by the robot actuator. Therefore, to secure a large working space, the ROM of each joint must be increased.

JET has the same hip joint structure as THORMANG. The hip joint consists of two actuators in roll-pitch order, as shown in Figure 5a, and each axis of rotation is orthogonal at the initial position. To increase the ROM, an S-shape thigh frame was devised, as shown in Figure 5b. This thigh frame reduces interference between the motor and thigh frames, thereby increasing the hip joint ROM, as shown in Figure 5c. Figure 6 compares the JET, THORMANG, THORMANG3, and TALOS ROMs. Because of self-collision, the ROM of one joint is not independent of another joint position. Thus, thigh frame was designed for ROM expansion when both hip joints bend. Hence, the ROM of the two hip joints was expanded by 39.3% and 36.7%, respectively, compared to THORMANG and THORMANG3. The hip joint ROM of TALOS, which is introduced in [18], is also smaller than the ROM of JET. The authors of [21] developed HRP-5P, which has 14.8% larger hip roll ROM than that of JET when the pitch angle is 110°. JAXON3-P, which has a 1.7 m height and 70 kg weight, also has 14.8% larger hip roll ROM than that of JET when the hip pitch is 110° [24]. To increase the knee ROM, the curves of the back of the thigh and shin frame were designed similarly, and the knee actuator was relocated from the shin to the thigh. These changes allow JET to perform more varied postures than THORMANG.

THORMANG has a larger upper-to-lower body height ratio than that of a typical young female with similar height, at 1:0.79 compared to 1:1, respectively. The large head, which includes many cameras, is one of the reasons for this difference. Generally, a humanoid robot with a shorter pelvis and lower shoulder height than humans has performance limitations for various tasks involving human life tools such as ladders and stairs.

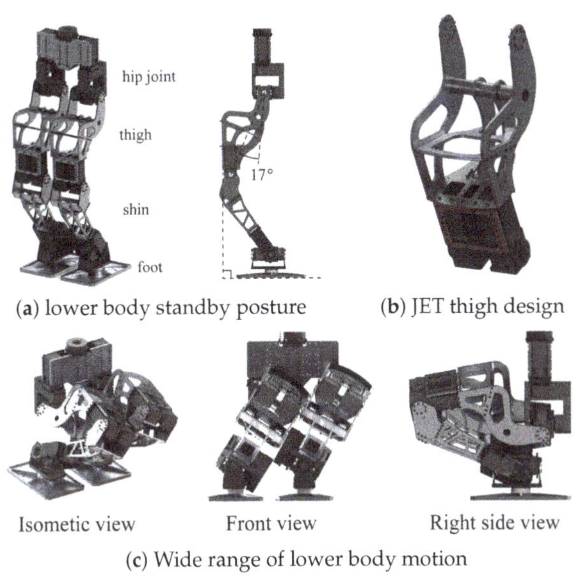

Figure 5. Lower limb design: (**a**) lower body standby posture, (**b**) thigh design, (**c**) lower body squatting pose. The lower body frames were designed to avoid self-collisions. In particular, the thigh frames were designed such that the link front surfaces are perpendicular to the ground.

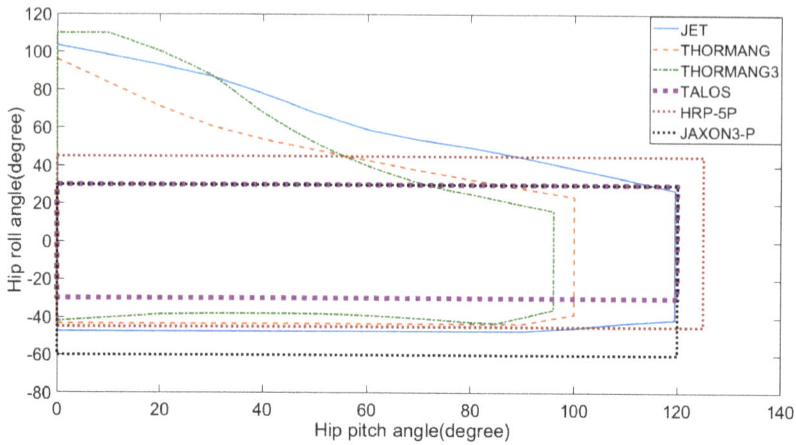

Figure 6. ROM comparison for JET, THORMANG, THORMANG3, TALOS, HRP-5P, and JAXON3-P hip joints.

The JET lower body design is based on the average body proportions of a young female for more natural behavior in industry and service applications. Thus, the lower body is lengthened from 696 to 839 mm compared to THORMANG. The the thigh length to shin length to ankle joint height ratio is 1:1:0.26, which is adjusted to the average ratio for a young female. Figure 3b shows the length of each part. With this leg length increase, the upper-to-lower body height ratio is almost 1:1.

In addition to the functional advantages, the S-shape thigh frame has aesthetic benefits, having a fuller shape than that of the shin and human-like body proportions. As shown in Figure 5c, the S-shape thigh frame seems to be stretching the knees in the standby pose, because the thigh front surface is perpendicular to the ground.

3.3. Upper Body Design

JET features a yaw and roll combination in the waist for the following reasons. First, downward tilting of the upper body in the forward direction is difficult, because the actuator has limited maximum torque. However, the waist pitch motion can be generated by the two hip pitch actuators. Second, the upper body roll motion is useful to control the center of mass in the lateral direction during walking [22,25]. This also extends the arm reach in the vertical direction. Finally, the yaw motion is useful to reverse the upper body orientation to avoid collision between obstacles and the shin, which protrudes during knee bending [11,26].

For easy maintenance, most electronic components are located on the open side of the torso. The advantage of this is that most electrical components can be accessed without disassembling the entire frame. The part where the circuit is fixed is modularized for easy assembly and disassembly. Figure 7 shows the torso and modularized computer design, which allows easy removal of the computer module from the robot.

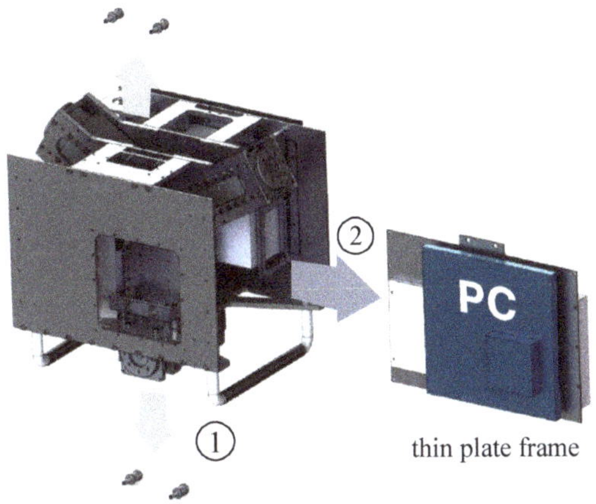

Figure 7. Exploded view of computer module. The modularized computer board can be sideways-disassembled by removing four bolts.

3.4. Design for Load Alleviation

Greater leg length increases the moment of inertia as well as the gravity load of the same joint space configuration. As JET employs the same actuators as THORMANG, it was essential to alleviate the load in accordance with the actuator capacity. The frame weight was reduced by adding polygon holes. Through finite element analysis (FEA), a lightweight link was created that could maintain the strength to withstand an applied load without excessive deformation.

Figure 8 shows the overall weight proportions of each JET component. The actuators occupy more than 50% of the total robot weight. As the ratio of the actuators to the total weight is significant, the actuator arrangement played an important role in load reduction. In various studies, four bar linkages or ball screws have been used to position the actuators close to the first limb joint [27,28]. Although transmission helps reduce inertia, these approaches increase the overall weight and mechanical element number. Such changes would complicate achievement of the design goals in this case.

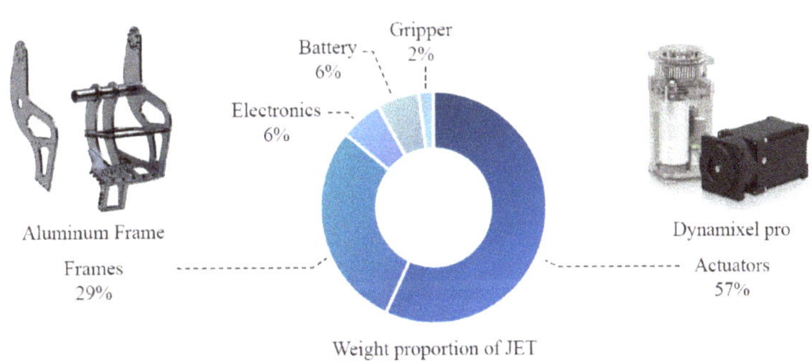

Figure 8. JET weight proportions.

Therefore, the JET actuators were placed as close as possible to the parent joint. However, a simple structure is maintained. As shown in Figure 9, various actuator arrangements are possible. The JET actuator positions are colored blue, and the worst cases, with maximum inertia, are colored orange. The inertia values at the hip and shoulder calculated for each case are listed in Table 3. Note that all values were calculated from the initial pose, with full extension of the robot legs and arms.

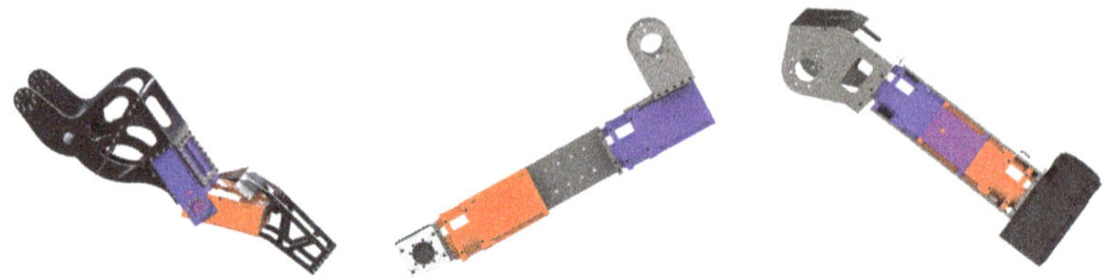

(a) (b) (c)

Figure 9. Possible actuator arrangements, blue for JET and orange for bad case: (**a**) Knee actuator arrangements, (**b**) Forearm actuator arrangements, (**c**) Arm actuator arrangements.

For example, the total inertia of the entire leg at the hip roll joint is 2.53 kgm^2 for the blue case, with a knee-actuator inertia of 0.22 kgm^2. For the orange case, however, the knee actuator contributes 0.34 kgm^2 to the total inertia with respect to the hip roll joint. Therefore, the inertia due to the knee actuator increases by 50%, and the total inertia increases by 5%. The foot has a major effect on the total inertia. When the robot bends its knees, the proportion of inertia due to the knee actuator increases. Note that the total inertia decreases when the knees bend, but the inertia due to the knee actuator is independent of the joint configuration. Therefore, the proposed design has a greater impact in the walking configuration. Similar to the knee actuator, the forearm and arm actuator positions affect the inertia at the shoulder joint. Without weight reduction, inertia reduction is achieved through actuator rearrangement.

Table 3. Comparison of inertia values for various actuators and arrangements.

	Joint Inertia (kg m^2) for Blue Case (Total/Actuator)	**Joint Inertia (kg m^2) for Orange Case (Total/Actuator)**
Hip roll (Knee actuator)	2.53/0.22	2.65/0.34
Hip pitch (Knee actuator)	2.53/0.22	2.65/0.34
Shoulder pitch (forearm actuator)	1.13/0.08	1.26/0.20
Shoulder pitch (arm actuator)	1.13/0.02	1.26/0.03
Shoulder roll (forearm actuator)	1.15/0.08	1.29/0.20
Shoulder roll (arm actuator)	1.15/0.02	1.29/0.03

Although each JET link was lengthened, the entire link weight is 3 kg less than that of THORMANG. We designed a frame with polygonal-shaped holes, not only for weight reduction but also to enhance the actuator connector accessibility. The actuator modules were utilized as the frame itself to reduce the weight. Thus, in the final design, most actuator modules are exposed. Hence, most actuators can be easily disassembled and replaced. For safety, every frame was assessed via FEA to confirm a yield strength with a gravity load.

Most frames were designed to have flat plate shapes for cheap fabrication. These flat frames were manufactured by laser cutting, which is faster and cheaper than CNC milling. Sanding is essential for laser cut frames to remove the roughness of the laser cut surface. However, both rough and sanded surfaces are too inaccurate for assembly. Therefore, we designed flat frames that do not require high edge tolerance for assembly. This design approach allowed us to manufacture complex links, such as the thigh links in Figure 5b, at low cost.

To reduce the total weight, the battery capacity was reduced from 44,000 to 22,000 mAh. THORMANG utilizes two low-performance computers. However, JET features one high-performance computer. Additionally, for weight reduction, JET has no bumpers. Therefore, the total JET weight is 12 kg lower than that of THORMANG.

3.5. Design for Maintenance

Sections 3.1, 3.3 and 3.4 describe the electric connector frame design and unification. In addition to the above design approaches, the number of bolt types used for assembly was also limited. Dynamixel Pro requires M3 bolts. Thus, many of the bolts for the frames were selected to have an M3 size. Overall, more than 95% of the bolts in the finalized design are M3. Further, more than 90% of the bolts are the same size, with one of three different lengths: 6, 8 and, 12 mm. Thus, only one tool is required to disassemble most of the actuators.

4. Performance Demonstrations

This section reports and discusses several experiments performed to demonstrate the performance of the developed hardware. In particular, to assess the compliance and whole-body motion capability, various tests involving locomotion, stair climbing, exiting a vehicle, etc., were performed.

As mentioned in Section 1, compliance reduces physical damage or instability from unexpected contacts. Therefore, compliance helps increase robot stability and balance during contact situations. However, compliance from low-stiffness actuators in the supporting limb often causes problems during humanoid bipedal walking, which degrades the performance and stability. To overcome this problem, in a previous work, we presented the actuator elasticity compensator [16].

To enhance the swing leg compliance and overcome the negative features of the supporting leg, the following studies were conducted. External joint encoders were installed to measure the actuator deformation [23]. A linear quadratic regulator (LQR)-based trajectory tracking controller and disturbance observer (DOB)-based compliance controller, which were presented in our previous work [29], were also employed, along with a walking pattern generator with improved stability proposed by Kim et al. [30]. Using these new algorithms, JET can walk faster and with better stability than THORMANG. To minimize the COM velocity fluctuation, model predictive control-based pattern generation was developed and verified for JET [31].

Whole-body motion capability was verified through the following experiments. Stair climbing and egress experiments were chosen among the DRC Finals missions. Both tasks were the hardest tasks in the DRC Finals because THORMANG has short legs for these missions. First, the lower body workspaces were validated through stair climbing experiments on 23 cm high stairs, which were the same height as the stairs used in the DRC Finals. As shown in Figure 10, JET walked up the stairs without any of the THORMANG strategies applied at the DRC. That is, THORMANG climbed the stairs without bumping into the stairs themselves by grabbing the rails and walking backwards [11]. The required foot workspaces of the JET for climbing stairs are ± 30 cm for x-axis, ± 17 cm for y-axis, and 23 cm for z-axis as shown in Figure 10. The x and z axis workspaces are determined by the dimensions of the stairs. Figure 11 shows the total foot workspace of JET, THORMANG, and HRP-5P and the required workspaces of the robots. In Figure 11d, JET has a 32 cm z-axis workspace from -75 cm to -43 cm. In the same way, HRP-5P has a 34 cm z-direction workspace from -77 cm to -43 cm, but THORMANG has only a 17 cm z-direction workspace from -59 cm to -42 cm. In this experiment, JET used the same walking pattern generation algorithm which was used at the DRC Finals [11].

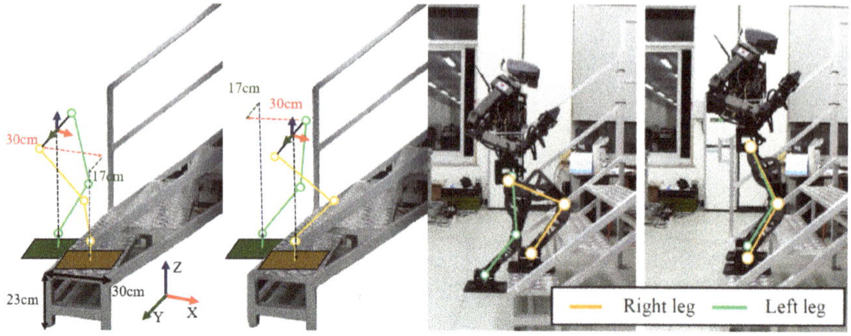

Figure 10. Snapshots of stair climbing test.

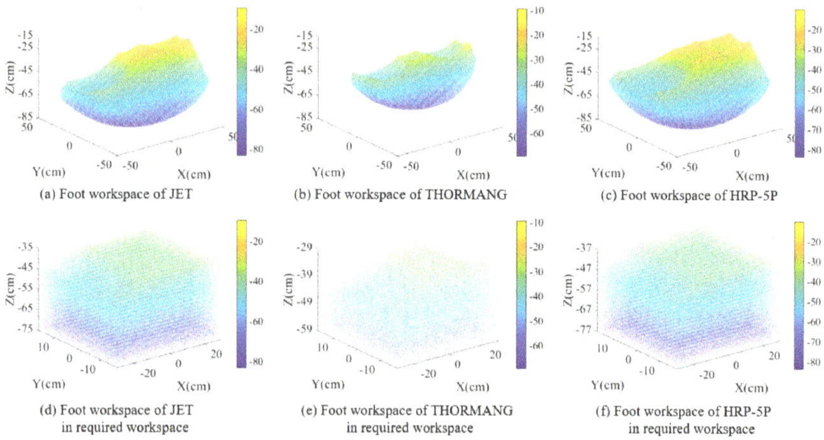

Figure 11. Comparison of foot workspace among JET, THORMANG, and HRP-5P. (**a–c**) are the total foot workspaces of JET, THORMANG, and HRP-5P. (**d–f**) are the required workspaces with 60 cm × 34 cm on x- and y-axis.

Next, we confirmed the robot could rise from a 64 cm high vehicle seat using whole-body motion with multiple contacts. Figure 12 shows snapshots of the robot egress from the vehicle. In this experiment, the robot performed whole-body motion with contacts at the gripper, pelvis, and two feet. The gripper holds the hand rail to stabilize the body during the JET stand-up. This demonstrated not only the improved robot workspaces, but also the robot's capability to execute compliant motion in multiple-contact situations. In this experiment, the personal mobility vehicle which was designed for adults is used. The height of the vehicle seat is 56 mm, which is shorter than the fully stretched leg length of THORMANG.

In Figure 13, JET manipulated objects with the gripper developed by Kim and Park and teleoperation systems [32]. Videos of experiments on manipulation and locomotion can be viewed at https://youtu.be/jWbCwT1IYJ4 (accessed on 1 July 2021), https://youtu.be/T48uXPhgeoU (accessed on 1 July 2021), and https://youtu.be/ZfyvAklrUBQ (accessed on 1 July 2021).

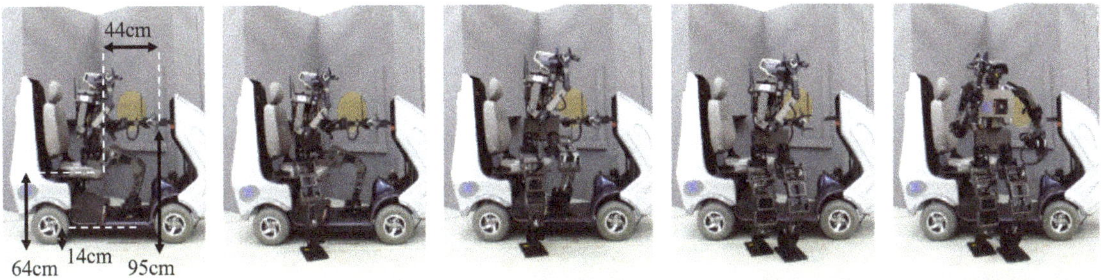

Figure 12. Snapshots of egress test with personal mobility. Videos can be seen at https://youtu.be/6NWdPNTjnjs (accessed on 1 July 2021).

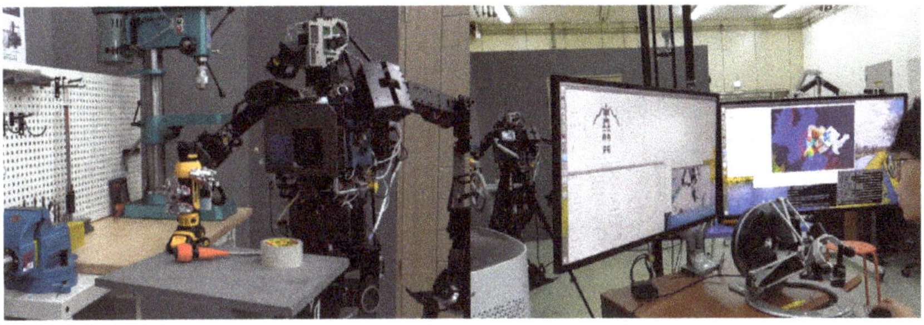

Figure 13. Manipulating objects with teleoperation.

5. Conclusions

We developed the JET humanoid robot, which is based on the THORMANG platform. To develop JET, the limitations of THORMANG were considered, as revealed by our experiences at the DRC. However, THORMANG has the advantage of easy maintenance as it employs the Dynamixel commercial modular actuator. To retain the advantages of THORMANG but overcome its platform limitations, three design goals were identified with the aim of effective application to the industrial and service fields: compliance, whole-body motion capability, and easy maintenance. To achieve compliance, the Dynamixel low stiffness was utilized to achieve compliance with external environments.

To overcome the small workspaces of THORMANG, the overall height was increased. However, the weight and inertia were reduced to satisfy the power requirement. Notably, load alleviation is essential for a taller robot with the same actuators. The leg-joint ROM was also increased to achieve a larger lower limb workspace.

Both manufacturing and maintenance methods were considered for the JET design. The frame was designed to provide easy access to the electronics, including the computer, power modules, and actuators. With this exposed design, the modular actuators can be replaced easily. Further, by limiting the number of electric connector and bolt types, the number of parts required for actuator replacement was reduced. The number of links requiring manufacture with expensive CNC milling was reduced through the use of flat-plate-shaped links, which can be manufactured through low-cost laser cutting.

Several experiments were conducted to verify the enhancements of the JET humanoid robot based on the above three design concepts. Compliance and enhanced whole-body motion capability were observed for stair climbing and egress from a personal mobile vehicle. In addition, the entire system was verified through a teleoperation test. These experiments validated the JET performance.

As a future study, an optimization-based control and path generation algorithm to improve walking speed and stability will be developed based on the JET proposed in this paper.

Author Contributions: Conceptualization, methodology, formal analysis, J.S. and S.K. (Seungyeon Kim); software, S.P., M.K. and S.K. (Sanghyun Kim); validation, M.K. and S.K. (Sanghyun Kim); investigation, J.S., S.K. (Seungyeon Kim) and M.K.; resources, J.S.; data curation, J.S.; writing—original draft preparation, J.S., S.K. (Seungyeon Kim) and S.K. (Sanghyun Kim); writing—review and editing, J.S.; visualization, S.K. (Seungyeon Kim); supervision, J.P.; project administration, J.P.; funding acquisition, J.P. All authors have read and agreed to the published version of the manuscript.

Funding: This work was supported by a National Research Foundation of Korea (NRF) grant funded by the Korean government (MSIT) (No. 2021R1A2C3005914).

Conflicts of Interest: The authors declare no conflict of interest.

Abbreviations

The following abbreviations are used in this manuscript:

ROM	Range of Motion
DRC	DARPA Robotics Challenge
SEA	Series Elastic Actuator
DOF	Degree of Freedom
FEA	Finite Element Analysis
CNC	Computerized Numerical Control

References

1. Krotkov, E.; Hackett, D.; Jackel, L.; Perschbacher, M.; Pippine, J.; Strauss, J.; Pratt, G.; Orlowski, C. The DARPA Robotics Challenge Finals: Results and Perspectives. *J. Field Robot.* **2016**, *34*, 229–240. [CrossRef]
2. Caron, S.; Kheddar, A. Multi-contact walking pattern generation based on model preview control of 3D COM accelerations. In Proceedings of the 16th International Conference on Humanoid Robots (Humanoids) IEEE-RAS, Cancun, Mexico, 15–17 November 2016; pp. 550–557.
3. Negrello, F.; Garabini, M.; Catalano, M.G.; Kryczka, P.; Choi, W.; Caldwell, D.G.; Bicchi, A.; Tsagarakis, N.G. WALK-MAN humanoid lower body design optimization for enhanced physical performance. In Proceedings of the International Conference on Robotics and Automation (ICRA), Stockholm, Sweden, 16–21 May 2016; pp. 1817–1824.
4. Dietrich, A.; Wimböck, T.; Albu-Schäffer, A. Dynamic whole-body mobile manipulation with a torque controlled humanoid robot via impedance control laws. In Proceedings of the 2011 IEEE/RSJ International Conference on Intelligent Robots and Systems, San Francisco, CA, USA, 25–30 September 2011; pp. 3199–3206.
5. Sands, T. Optimization Provenance of Whiplash Compensation for Flexible Space Robotics. *Aerospace* **2019**, *6*, 93. [CrossRef]
6. Ren, T.; Dong, Y.; Wu, D.; Chen, K. Learning-based variable compliance control for robotic assembly. *J. Mech. Robot.* **2018**, *10*. [CrossRef]
7. Paine, N.; Mehling, J.S.; Holley, J.; Radford, N.A.; Johnson, G.; Fok, C.L.; Sentis, L. Actuator Control for the NASA-JSC Valkyrie Humanoid Robot: A Decoupled Dynamics Approach for Torque Control of Series Elastic Robots. *J. Field Robot.* **2015**, *32*, 378–396. [CrossRef]
8. Knabe, C.; Griffin, R.; Burton, J.; Cantor-Cooke, G.; Dantanarayana, L.; Day, G.; Ebeling-Koning, O.; Hahn, E.; Hopkins, M.; Neal, J.; et al. Team valor's escher: A novel electromechanical biped for the darpa robotics challenge. *J. Field Robot.* **2017**, *34*, 912–939. [CrossRef]
9. DeDonato, M.; Polido, F.; Knoedler, K.; Babu, B.P.; Banerjee, N.; Bove, C.P.; Cui, X.; Du, R.; Franklin, P.; Graff, J.P.; et al. Team WPI-CMU: Achieving Reliable Humanoid Behavior in the DARPA Robotics Challenge. *J. Field Robot.* **2017**, *34*, 381–399. [CrossRef]
10. Stentz, A.; Herman, H.; Kelly, A.; Meyhofer, E.; Haynes, G.C.; Stager, D.; Zajac, B.; Bagnell, J.A.; Brindza, J.; Dellin, C.; et al. Chimp, the cmu highly intelligent mobile platform. *J. Field Robot.* **2015**, *32*, 209–228. [CrossRef]
11. Kim, S.; Kim, M.; Lee, J.; Hwang, S.; Chae, J.; Park, B.; Cho, H.; Sim, J.; Jung, J.; Lee, H.; et al. Team SNU's Control Strategies for Enhancing a Robot's Capability: Lessons from the 2015 DARPA Robotics Challenge Finals. *J. Field Robot.* **2016**, *34*, 359–380. [CrossRef]
12. SHG-17-100-2SO | Harmonic Drive. Available online: https://www.harmonicdrive.net/products/gear-units/simplicity-gear-units/shg-2so/shg-17-100-2so (accessed on 11 June 2021).
13. Robotis e-Manual. Available online: https://emanual.robotis.com/docs/en/dxl/pro/h54-200-s500-ra/ (accessed on 11 June 2021).
14. Tsagarakis, N.G.; Caldwell, D.G.; Negrello, F.; Choi, W.; Baccelliere, L.; Loc, V.G.; Noorden, J.; Muratore, L.; Margan, A.; Cardellino, A.; et al. Walk-man: A high-performance humanoid platform for realistic environments. *J. Field Robot.* **2017**, *34*, 1225–1259. [CrossRef]
15. Tsagarakis, N.G.; Morfey, S.; Cerda, G.M.; Zhibin, L.; Caldwell, D.G. Compliant humanoid coman: Optimal joint stiffness tuning for modal frequency control. In Proceedings of the 2013 IEEE International Conference on Robotics and Automation (ICRA), Karlsruhe, Germany, 6–10 May 2013; pp. 673–678.

16. Kim, J.; Kim, M.; Park, J. Improvement of humanoid walking control by compensating actuator elasticity. In Proceedings of the 2016 IEEE-RAS 16th International Conference on Humanoid Robots (Humanoids), Cancun, Mexico, 15–17 November 2016; pp. 29–34.
17. Kim, S.; Kim, M.; Lee, J.; Hwang, S.; Chae, J.; Park, B.; Cho, H.; Sim, J.; Jung, J.; Lee, H.; et al. Approach of team snu to the darpa robotics challenge finals. In Proceedings of the 2015 IEEE-RAS 15th International Conference on Humanoid Robots (Humanoids), Seoul, Korea, 3–5 November 2015; pp. 777–784.
18. Stasse, O.; Flayols, T.; Budhiraja, R.; Giraud-Esclasse, K.; Carpentier, J.; Mirabel, J.; Del Prete, A.; Souères, P.; Mansard, N.; Lamiraux, F.; et al. TALOS: A new humanoid research platform targeted for industrial applications. In Proceedings of the 2017 IEEE-RAS 17th International Conference on Humanoid Robotics (Humanoids), Birmingham, UK, 15–17 November 2017; pp. 689–695.
19. Korean Size Report. Available online: Https://sizekorea.kr/ (accessed on 7 May 2021).
20. Talos Description. Available online: https://github.com/pal-robotics/talos_robottree/kinetic-devel/talos_description (accessed on 11 June 2021).
21. Kaneko, K.; Kaminaga, H.; Sakaguchi, T.; Kajita, S.; Morisawa, M.; Kumagai, I.; Kanehiro, F. Humanoid robot HRP-5P: An electrically actuated humanoid robot with high-power and wide-range joints. *IEEE Robot. Autom. Lett.* **2019**, *4*, 1431–1438. [CrossRef]
22. Ogura, Y.; Aikawa, H.; Shimomura, K.; Morishima, A.; Lim, H.O.; Takanishi, A. Development of a new humanoid robot WABIAN-2. In Proceedings of the 2006 IEEE International Conference on Robotics and Automation, ICRA 2006, Orlando, FL, USA, 15–19 May 2006; pp. 76–81.
23. Park, S.; Sim, J.; Park, J. System design of humanoid robot dyros-jet. In Proceedings of the 2019 IEEE/SICE International Symposium on System Integration (SII), Paris, France, 14–16 January 2019; pp. 746–750.
24. Kojima, K.; Kojio, Y.; Ishikawa, T.; Sugai, F.; Kakiuchi, Y.; Okada, K.; Inaba, M. A robot design method for weight saving aimed at dynamic motions: Design of humanoid JAXON3-P and realization of jump motions. In Proceedings of the 2019 IEEE-RAS 19th International Conference on Humanoid Robots (Humanoids), Toronto, ON, Canada, 15–17 October 2019; pp. 586–593.
25. Ogura, Y.; Lim, H.O.; Takanishi, A. Stretch walking pattern generation for a biped humanoid robot. In Proceedings of the 2003 IEEE/RSJ International Conference on Intelligent Robots and Systems, 2003 (IROS 2003), Las Vegas, NV, USA, 27–31 October 2003; Volume 1, pp. 352–357.
26. Lim, J.; Lee, I.; Shim, I.; Jung, H.; Joe, H.M.; Bae, H.; Sim, O.; Oh, J.; Jung, T.; Shin, S.; et al. Robot System of DRC-HUBO+ and Control Strategy of Team KAIST in DARPA Robotics Challenge Finals. *J. Field Robot.* **2017**, *34*, 802–829. [CrossRef]
27. Lohmeier, S.; Buschmann, T.; Schwienbacher, M.; Ulbrich, H.; Pfeiffer, F. Leg design for a humanoid walking robot. In Proceedings of the 2006 6th IEEE-RAS International Conference on Humanoid Robots, Genova, Italy, 4–6 December 2006; pp. 536–541.
28. Lahr, D.F. Design and Control of a Bipedal Robot. Ph.D. Thesis, Virginia Polytechnic Institute and State University, Blacksburg, VA, USA, 2014.
29. Kim, M.; Kim, J.H.; Kim, S.; Sim, J.; Park, J. Disturbance observer based linear feedback controller for compliant motion of humanoid robot. In Proceedings of the 2018 IEEE International Conference on Robotics and Automation (ICRA), Brisbane, QLD, Australia, 21–25 May 2018; pp. 403–410.
30. Kim, M.; Lim, D.; Park, J. Online walking pattern generation for humanoid robot with compliant motion control. In Proceedings of the 2019 International Conference on Robotics and Automation (ICRA), Montreal, QC, Canada, 20–24 May 2019; pp. 1417–1422.
31. Park, B.; Park, J. Walking Pattern Generation using MPC with minimization of COM Velocity Fluctuation. In Proceedings of the 2020 20th International Conference on Control, Automation and Systems (ICCAS), Busan, Korea, 13–16 October 2020; pp. 268–273.
32. Kim, S.; Park, J. Gripper with Thumb Adduction/Abduction Joint for Enhanced In-hand Orientation Manipulation. In Proceedings of the 2018 IEEE-RAS 18th International Conference on Humanoid Robots (Humanoids), Beijing, China, 6–9 November 2018; pp. 1–9.

applied sciences

MDPI

Article

A Workspace-Analysis-Based Genetic Algorithm for Solving Inverse Kinematics of a Multi-Fingered Anthropomorphic Hand

Chun-Tse Lee and Jen-Yuan (James) Chang *

Department of Power Mechanical Engineering, National Tsing Hua University, Hsinchu 30013, Taiwan; ctlee@gapp.nthu.edu.tw

* Correspondence: jychang@pme.nthu.edu.tw; Tel.: +886-3-574-2498

Abstract: Although the solution of inverse kinematics for a serial redundant manipulator has been widely researched, many algorithms still seem limited in dealing with complex geometries, including multi-finger anthropomorphic hands. In this paper, the inverse kinematic problems of multiple fingers are an aggregate problem when the target points of fingers are given. The fingers are concatenated to the same wrist and the objective is to find a solution for the wrist and two fingers simultaneously. To achieve this goal, a modified immigration genetic algorithm based on workspace analysis is developed and validated. To reduce unnecessary computation of the immigration genetic algorithm, which arises from an inappropriate inverse kinematic request, a database of the two fingers' workspace is generated using the Monte Carlo method to examine the feasibility of inverse kinematic request. Furthermore, the estimation algorithm provides an optimal set of wrist angles for the immigration genetic algorithm to complete the remaining computation. The results reveal that the algorithm can be terminated immediately even when the inverse kinematic request is out of the workspace. In addition, a distribution of population in each generation illustrates that the optimized wrist angles provide a better initial condition, which significantly improves the convergence of the immigration genetic algorithm.

Keywords: inverse kinematics; genetic algorithm; workspace analysis; multi-fingered anthropomorphic hand

Citation: Lee, C.-T.; Chang, J.-Y. A Workspace-Analysis-Based Genetic Algorithm for Solving Inverse Kinematics of a Multi-Fingered Anthropomorphic Hand. *Appl. Sci.* **2021**, *11*, 2668. https://doi.org/10.3390/app11062668

Academic Editor: Alessandro Gasparetto

Received: 24 February 2021
Accepted: 15 March 2021
Published: 17 March 2021

1. Introduction

The solution of inverse kinematics (IK) is one of the most critical and elemental issues in robotics. Many related research areas such as motion planning, robotic grasping, manipulation, and manufacturing are all involved in IK. The IK of a manipulator is used to find the map between the joint coordinate (θ) and the Cartesian coordinates (x, y, z), where θ represents the joint angles of each joint in the manipulator and (x, y, z) represent the position of the manipulator's end-effector. While most IK solutions are performed on simple manipulators, several studies focus on redundant manipulators. For a redundant robot, in addition to the basic accurate positioning, different requirements such as computation time, robustness, and minimum displacement are considered suboptimal conditions. One of these complex redundant manipulators is the multi-fingered anthropomorphic robotic hand, whose joints of fingers move simultaneously and two wrist joints are deemed as the base of the hand. To obtain the IK solution of the anthropomorphic hand, a robust algorithm is essential to solve this multiple end-effector problem.

Traditionally, there are three kinds of methods used to acquire the IK solutions of a redundant manipulator, namely, the algebraic [1,2], geometric [3], and iterative methods [4–7]. Each of them has its own advantages and disadvantages. For example, the algebraic method can derive a closed-form solution from a Denavit and Hatenberg (D–H) table [8] and, thus, offers efficiency in computation. However, the algebraic method cannot guarantee a closed-form solution for all configurations of manipulators. On the other hand, the geometric method can obtain a closed-form solution, while it is usually limited in specific configurations such as sphere–revolute–sphere manipulators. Likewise, the iterative algorithm utilizes local linearization to get a linear trajectory, and it is suitable for most manipulators. Nevertheless, the iterative method may encounter a singularity problem in which it can accidentally converge to the same solution if the same initial point is given. All of these traditional methods have one point in common, i.e., high computational efficiency, but they all inevitably suffer from certain restrictions. Alternatively, thanks to the advancement of computer hardware, compute-intensive algorithms can come in handy. For these reasons, although the computation is intensive, researchers have focused on using biologically inspired optimization to solve the IK problem.

Recently, many papers associated with neural networks [9–13] and evolutionary algorithms [14–18] have been published to allow obtaining the IK solution of redundant manipulators with some convincing results. KöKer [19] presented a hybrid approach using neural networks and the genetic algorithm (GA) to solve the IK problem under the consideration of error minimization in the end-effector. Three Elman neural networks were applied to create the floating-point initial population for the GA. The simulation experiments were performed on a six-axis serial robot and showed a critical high precision. Tabandeh et al. [20] presented an adaptive niching strategy to acquire several possible solutions to IK problem. By using the niching genetic algorithm accompanied by the quasi-Newton algorithm, the precision and resolution of the simulation results were improved dramatically. A sequential mutation genetic algorithm combined with an extreme-learning machine was presented by Zhou et al. [21]. In their study, the extreme-learning machine first computed the preliminary IK solution. With the use of the simple GA, the refined solution was optimized. The grasping experiments were then conducted on the 6 degree of freedoms (DOFs) Stanford MT-ARM robotic manipulator, from which an improvement in the computational time without reducing the accuracy of IK solutions was demonstrated. Momani et al. [17] provided a continuous genetic algorithm to solve the IK problems of a 3R planar manipulator. The continuous genetic algorithm operators used in the initialization phase, crossover, and mutation were designed to smoothen the joint space while the Cartesian path was kept accurate.

Although the aforementioned algorithms made a great contribution to solving the IK of redundant robots, they focused only on single kinematic chain redundant manipulators which are widely used in industry. For instance, the most common applications using GA to solve the IK problem are 6-DOFs manipulators, although there are closed-form solutions already; this can also be found in 7-DOFs manipulators, for which GA is quite applicable because it includes one redundant DOF in the joint space when the end-effector moves in the Cartesian coordinate system. Given that previous research did not consider the inverse kinematic problem of multiple manipulators such as robotic hands and dual-arm robots, further investigation on the IK problem of multiple kinematic chain manipulators is crucial. The other drawback of GA is its ignorance of an unreachable target. Specifically, GA is implemented regardless of feasibility, which is a fatal flow in solving the IK problem of a dual-arm robot. Without checking the viability in advance, it is frequently found that, when two target points are given to each arm, only one arm can reach the target point while the other one fails because of the limited range of motion. Therefore, feasibility analysis beforehand is indispensable.

In this paper, a workspace-analysis-based immigration genetic algorithm (IGA) is proposed to solve the inverse kinematics of a multi-fingered anthropomorphic hand. The architecture of the proposed algorithm is divided into two stages, namely, the offline stage and the online stage. In the offline stage, workspaces of the exemplary index finger and the thumb are established by the Monte Carlo method and stored in a database for the feasibility check. In the online stage, a feasibility estimation function examines whether the IK request is out of the workspace or not. If the request is unreachable, the algorithm is suspended immediately. Conversely, with a reachable request, the feasibility function provides a better initial condition for the IGA to get high-precision IK solutions. This paper aims to solve the IK problem of multiple chain manipulators and to provide a mechanism that can avoid unnecessary computation if the target point is inappropriate. Meanwhile, the proposed mechanism can offer a superior range to variables, which improves the convergence of the IGA. The performance of this algorithm is validated through simulations, which include both reachable and unreachable IK requests. A population distribution during the evolutionary process is conducive to the understanding of the improvement in convergence.

The remainder of this paper is organized as follows: in Section 2, the configuration of the anthropomorphic hand is described, and the IK problem is defined. In Section 3, the establishment of the workspace and mechanism to check the feasibility is explained. Section 4 introduces the architecture of the proposed algorithm and illustrates details of each operator in IGA, while Section 5 presents the simulation results of the proposed algorithm applied to several IK requests. Lastly, the contributions of this paper are summarized in the conclusion.

2. Problem Formulation

In this section, the geometric description of the anthropomorphic hand and the scenario of the motion task are discussed. The anthropomorphic hand is designed according to the human hand anatomy. In this anthropomorphic hand, the wrist mechanism is designed to possess two degrees of freedom in order to mimic the motion of a human's wrist. For fingers, the thumb and index finger each has three joints, namely, the metacarpophalangeal (MCP), the proximal interphalangeal (PIP), and the distal interphalangeal (DIP) joints, as shown in Figure 1. Among these three, the PIP and DIP joints of a human finger are anatomically coupled by a tendon. As a result, to mimic a human finger's anatomy, the design of a four-bar linkage mechanism to create coupled motion by a ratio of 1:1 is adopted in this research, whereby θ_{t3} is equal to θ_{t4} and θ_{i3} is equal to θ_{i4}, where subscript t stands for thumb and i stands for index finger. As many dexterous gestures and tasks are executed by the thumb and index finger, such as picking up small items and handling a key, the kinematics of the thumb and index finger of the anthropomorphic hand is the main focus of this paper. The other fingers, i.e., the middle finger, ring finger, and little finger, could also be taken into consideration for the IK problem by using the same procedure presented in this study. However, to fix ideas and to convey the concept clearly, the remainder of this paper uses only the thumb and index finger as examples for simplification.

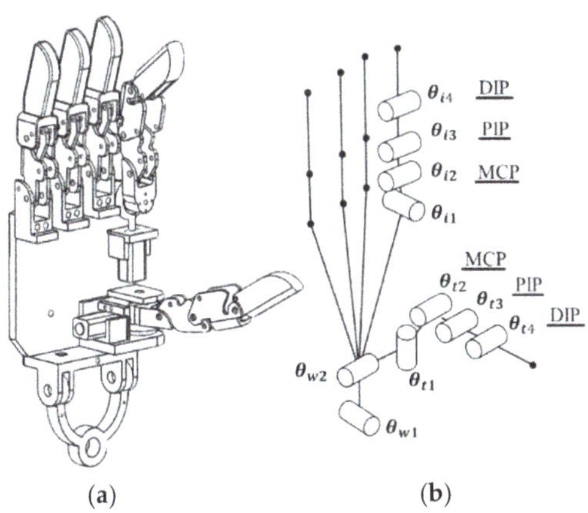

Figure 1. Schematics of (**a**) the design and (**b**) the kinematic structure of the model anthropomorphic hand.

In this study, the target points of the index finger and the thumb are given at the same time as the inverse kinematic requests and the associated task space is restricted only to a translation of posture. Because both the index finger and the thumb are connected to the wrist, it is evident that the robotic hand has multiple serial kinematic chains, and each of them is dependent. In other words, the tips of both the index finger and the thumb comprise the same angle θ_{w1} and θ_{w2}, where the subscript w represents the wrist. Mechanically, the inverse kinematics of the two fingers are highly coupled.

With the problem described according to Cartesian coordinates, an IGA is then used to solve this highly nonlinear and coupled inverse kinematics. However, from theory, an IGA always returns a minimum value that it has computed even if the target point is out of the workspace. To avoid this ambiguous situation, a procedure to generate the workspace is proposed, and a technique to check the feasibility to assist the computation of the inverse kinematics is discussed in the next section.

3. Workspace Analysis

3.1. Forward Kinematics

Through kinematic analysis, the relationship between the position of the end-effector and the angles of each joint can be easily defined for a robot. The kinematics can be conducted using two analyses, namely, forward kinematics analysis and inverse kinematics analysis. The forward kinematics analysis allows one to obtain orientation and translation of the end-effector on the basis of the given angles of joints. Therefore, the workspace of the fingers is easily acquired using the forward kinematics analysis.

Before analyzing kinematics, the problem must be formulated using a mathematic tool. Denavit and Hatenberg in 1995 introduced the convention for two spatial coordinates [8]. From this convention, a matrix, which contains the information on the translation and rotation, can be obtained. To access the values of the D–H parameters, the coordinate of each joint must be defined first. The long axis of each cylinder, as shown in Figure 1, represents the rotational direction of each joint for the index finger and the thumb. Tables 1 and 2 give the geometric parameters of the index finger and the thumb, respectively. The forward kinematic model of the index finger or thumb is given through D–H parameter convention, and the transformation between adjacent i-th and $(i-1)$-th joint coordinate systems can be written in the form of the following homogeneous transformation matrix:

$$^{i-1}T_i(\theta_i) = \begin{bmatrix} cos\theta_i & -sin\theta_i cos\alpha_i & sin\theta_i sin\alpha_i & a_i cos\theta_i \\ sin\theta_i & cos\theta_i cos\alpha_i & -cos\theta_i sin\alpha_i & a_i sin\theta_i \\ 0 & sin\alpha_i & cos\alpha_i & d_i \\ 0 & 0 & 0 & 1 \end{bmatrix}, \quad (1)$$

where α_i, a_i, θ_i, and θ_i are D–H parameters. From Tables 1 and 2, one can observe that each finger has six joints from the base to its fingertip. Starting from joint 0 to joint 5, the homogeneous transformation matrix $^0T_5$ can be written as

$$^0T_5 = {}^0T_1 \cdot {}^1T_2 \cdot {}^2T_3 \cdot {}^3T_4 \cdot {}^4T_5 = \begin{bmatrix} n_x & o_x & a_x & P_x \\ n_y & o_y & a_y & P_y \\ n_z & o_z & a_z & P_z \\ 0 & 0 & 0 & 1 \end{bmatrix}, \quad (2)$$

where $(P_x\ P_y\ P_z)^T$ is the position vector of the fingertip measured from the base coordinate system. By giving many sets of joint angles, the workspace of the finger can be collected through the forward kinematics analysis.

3.2. Database Collection

The workspace of the robot mathematically represents all possible solutions of locations that the robot's end-effector can reach in space. To ensure that the finger of the model anthropomorphic hand can reach the target points, careful analysis of its workspace is crucial.

Table 1. Denavit and Hatenberg (D–H) table of the index finger.

Joint i	α_i (°)	a_i (mm)	θ_i	d_i (mm)	Range
0	90°	33	θ_{w1}	0	−60° to +60°
1	−90°	93	θ_{w2}	38	−60° to +90°
2	90°	15.5	θ_{i1}	−4	−45° to 0°
3	0	20	θ_{i2}	0	0° to +90°
4	0	25	θ_{i3}	0	0° to +90°
5	0	36	θ_{i4}	0	0° to +90°

Table 2. D–H table of the thumb.

Joint i	α_i (°)	a_i (mm)	θ_i	d_i (mm)	Range
0	90°	33	θ_{w1}	0	−60° to +60°
1	90°	1.8	θ_{w2}	38	−60° to +90°
2	−90°	21	θ_{t1}	31	−30° to +90°
3	0	20	θ_{t2}	0	−90° to 0°
4	0	25	θ_{t3}	0	−90° to 0°
5	0	36	θ_{t4}	0	−90° to 0°

The conventional workspace analysis is to find the range of the locations according to all joints. For example, the dual-arm robot investigated by Kang et al. [22] has two workspaces to indicate the motion range of each arm. In this dual-arm robot, the two arms are connected to the same fixed frame, and the joints of the left arm are independent of the right arm's joints. Consequently, we can simply use the workspace to check whether one arm can reach the target point.

Unlike the dual-arm robot, the index finger and the thumb of the anthropomorphic hand are connected to a moving palm. As they share the same degrees of freedom of the wrist, the workspaces of two fingers are definitely dependent and coupled. Even if the two target points of the index finger and the thumb are all in their workspaces, with the traditional method, the calculated angle of the wrist may show different value at each point, which implies that there may exist mechanical interference and that the index finger and thumb cannot reach their target points at the same time. To resolve this problem, the workspace analysis below is proposed to make the workspace of the thumb independent of the workspace of the index finger in this research.

Unlike the traditional method to build a workspace, we propose the workspace analysis under the condition that the angle of the wrist is separated from the angle of the fingers. The angle of the wrist is denoted as $\vec{q_w} = [\theta_{w1}, \theta_{w2}]^T$. The angles of the two fingers are denoted as $\vec{q_{f1}} = [\theta_{t1}, \theta_{t2}, \theta_{t3}, \theta_{t4}]^T$ and $\vec{q_{f2}} = [\theta_{i1}, \theta_{i2}, \theta_{i3}, \theta_{i4}]^T$, with the subscript f standing for finger. Once the angle of the wrist is decided, the workspaces of the index finger and the thumb are determined by substituting sets of $\vec{q_{f1}}$ and $\vec{q_{f2}}$ into Equation (2) to generate the cloud of points, as shown in Figure 2a, in which the red dots represent the possible positions of the fingertip of the index finger. The workspace of the index finger for a given set of wrist angles is denoted as $I\left(\vec{q_w}\right)$ and that of the thumb is denoted as $T\left(\vec{q_w}\right)$.

The Monte Carlo method, a statistical method to solve problems by random sampling, was adopted to analyze the workspace. A considerable number of particles were used to approach the real shape of the workspace. Algorithm 1 shows the details for generating the workspace of the index finger and the thumb for a given set of wrist angle. Figure 2b,c illustrate the point cloud of the index finger (red dots) and the thumb (green dots) under different conditions of wrist angle.

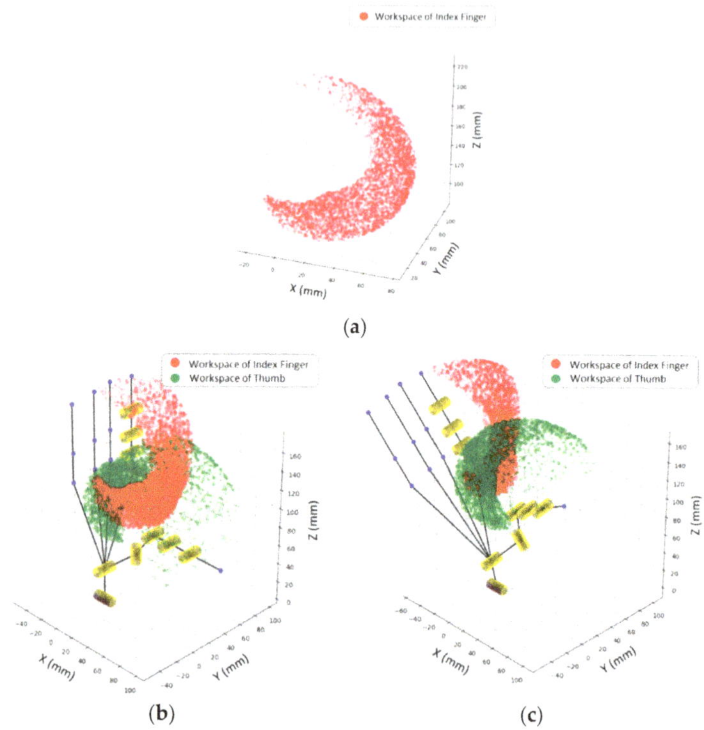

Figure 2. Representation of three-dimensional (3D) finger workspace: (**a**) point cloud of the index finger's tip generated by the Monte Carlo Method; (**b**) workspace of the index finger's tip and the thumb's tip at wrist angle $(\theta_{w1}, \theta_{w2}) = (0^{\circ}, 0^{\circ})$; (**c**) workspace of the index finger's tip and the thumb's tip at the wrist angle $(\theta_{w1}, \theta_{w2}) = (12^{\circ}, 30^{\circ})$.

Since the purpose of workspace analysis was to roughly examine whether the target point is near the finger's workspace or not, the number of sampling points of each finger in a certain configuration of wrist was intentionally set to be low. As a result, the sub-workspaces $I\left(\vec{q_w}\right)$ and $T\left(\vec{q_w}\right)$ were established every 10 degrees in wrist angle to cover the whole workspace. Using this algorithm, the database for two fingers was established and the data were sorted by the angle of the wrist. With this database, a further estimation function was created to check the feasibility of the IK request.

Algorithm 1. Algorithm for constructing the workspace of fingers

Purpose: To obtain the workspace of the fingers at certain wrist angle
Input: $\vec{q}_w, \vec{q}_{f,range}$
Output: S

1: $S = \varnothing, I(\vec{q_w}) = \varnothing, T(\vec{q_w}) = \varnothing$
2: Choose sample size N_{sample}
3: for $i \in \{1, 2, \ldots, N_{sample}\}$ do
4: Obtain $\vec{q_{f1}}$ from randomly sampling $\vec{q}_{f,range}$
5: Obtain $\vec{q_{f2}}$ from randomly sampling $\vec{q}_{f,range}$
6: $P_T = {}^0T_5(\vec{q_w}, \vec{q_{f1}})$
7: $P_I = {}^0T_5(\vec{q_w}, \vec{q_{f2}})$
8: *add* P_T *to* $T(\vec{q_w})$
9: *add* P_I *to* $I(\vec{q_w})$
10: end for
11: *add* $T(\vec{q_w}), I(\vec{q_w})$ *to* S
12: return S

3.3. Feasibility Check

Now, we have two databases for the index finger and the thumb, and each of them has many point clouds to indicate the workspace of the finger's tip according to the angle of the wrist. The feasibility of the motion command can be determined by computing the nearest distance between the target point and the point in the workspace. This distance is defined as

$$\kappa(S, p) = \min_{x_i \in S} \|p - x_i\|, \tag{3}$$

where S is the point cloud of the finger's tip which is stored in the database, and p is the target point. This function can be achieved by many algorithms such as the k-nearest neighbor algorithm which returns the nearest point and the distance. Because the point cloud is not concentrated enough, a threshold value is needed as the basis for proximity to the workspace. Using Equation (3) and the database of reachable area, the coupling phenomenon can be solved by the following equation:

$$Dist = \min\left[\kappa\left(T\left(\vec{q_w}\right), p_1\right) + \kappa\left(I\left(\vec{q_w}\right), p_2\right)\right], \tag{4}$$

where $\kappa(a, b)$ denotes the operation with the k-nearest neighbor algorithm for entities a and b, and p_1 and p_2 are the target points for the thumb and the index finger, respectively. Please note that the angle set of the wrist $\vec{q_w}$ is intentionally set to be the same in the database of the thumb and the index finger so that the inconsistency between the wrist angles of two fingers can be avoided. The result of Equation (4) is used as a basis for performing the IGA to prevent unnecessary computations. Moreover, the database is sorted in the order of the angle of the wrist for quick search and access to the corresponding optimized angles of the wrist. This optimized angle set of the wrist $\vec{q_w}^*$ is further provided to the IGA through the estimation function below as the initial guess to accelerate the computation and to offer better convergence. Equation (5) defines the estimation function.

$$\theta_{w1}^*, \theta_{w2}^* = \arg\min_{\vec{q_w}}\left[\kappa\left(T\left(\vec{q_w}\right), p_1\right) + \kappa\left(I\left(\vec{q_w}\right), p_2\right)\right]. \tag{5}$$

4. Algorithm

The proposed hybrid genetic algorithm mainly includes two steps:

1. Examining the feasibility of the motion command on the basis of the workspace and providing the angle set of the wrist if it is reachable.
2. Using IGA to solve the inverse kinematics and finding joint values of two fingers.

A schematic illustration of the proposed algorithm is depicted in Figure 3, and a flowchart of the proposed hybrid GA is shown in Figure 4. A detailed explanation of each step is addressed in the sections below.

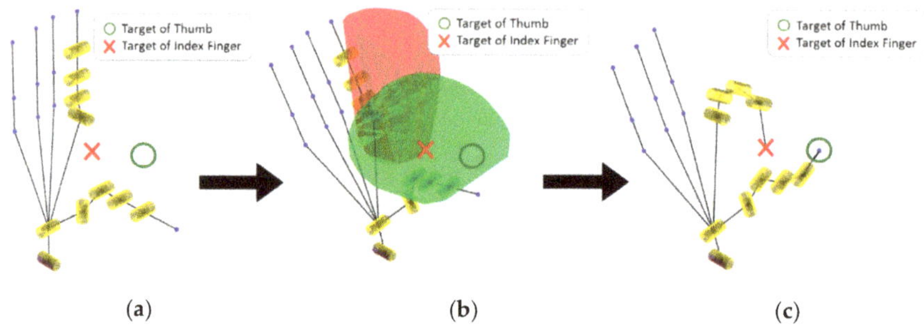

Figure 3. Process of the proposed algorithm: (**a**) assign inverse kinematics (IK) request; (**b**) feasibility check and generate wrist angle range; (**c**) perform immigration genetic algorithm (IGA) to obtain precise solution.

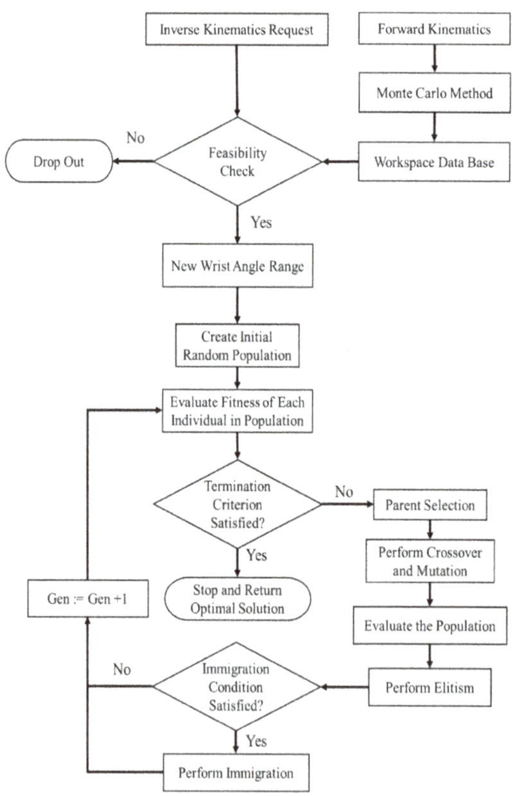

Figure 4. Flowchart of the proposed algorithm.

4.1. Feasibility Check and Feedforward of Wrist Angle

Utilizing the techniques elaborated in Section 3, in which collection and establishment of the database are conducted, the feasibility check is performed when the inverse kinematics request is assigned. The value obtained from Equation (4) is compared to the heuristic threshold value to decide whether the motion command should be dropped out or not. The heuristic threshold value and the fitness value of termination are the sum of the distances between target points and end-effectors of the two fingers. Basically, the heuristic threshold value is set to be a bit bigger than the fitness value of termination. Provided that the heuristic threshold value is as low as termination value, the feasibility check will frequently fail. On condition that the heuristic threshold value is much larger than the fitness value of termination, the fitness value will not converge to termination 1. In this research, the fitness value of termination is set to be 1 and the heuristic threshold value is set to be 5.

On the assumption that the motion command passes the feasibility check, Equation (5) yields an optimal wrist angle set $\vec{q_w}^*$ according to the aforementioned workspace database. The new wrist angle range provided to the IGA is, therefore, generated in the form of $\vec{q_w}^* \pm \theta_w$. The value θ_w is associated with the intervals of the wrist angle used in creating the workspace database. A wider interval denotes a larger value of θ_w. In this paper, θ_w was set to be 10° to cover the adjacent workspace. For example, the original range of θ_{w1} was $-60°$ to +60° and that of θ_{w2} was $-60°$ to +90°. Supposing that the $\vec{q_w}^*$ provided by the feasibility estimation function is [0°, 12°], then the new range of θ_{w1} becomes $-10°$ to +10° and that of θ_{w2} becomes +2° to 22°. Note that, if the new angle range exceeds the original angle limit, the boundary shall be set to the original one.

4.2. Chromosome Coding

In an IGA, a chromosome is the individual in a population, and it consists of many genes to show the individual differences. In this case, the chromosome represents a solution to the IK problem, while a series of genes constitutes a chromosome. The population contains many chromosomes and acts as a set of numerous possible IK solutions. To begin with the IGA, it is necessary to encode the joint angle configurations as the genes for each chromosome. The genetic encoding format is a binary string representation for each joint. By concatenating the binary string of each joint, the chromosome is built and provided for further operations. Figure 5 shows the binary encoding structure of the chromosome.

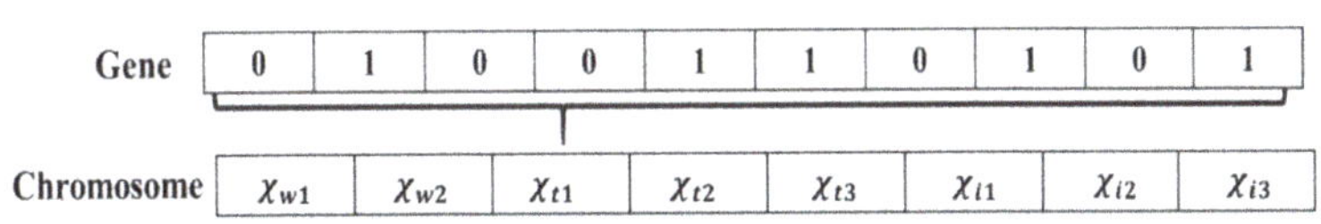

Figure 5. Binary encoding structure.

The decoding process can help a joint angle recover from a chromosome. Using binary–decimal conversion, the bit value is used to calculate the exact joint angle and to ensure that the joint variable is within the bounded joint space, which is mathematically written as

$$\theta_i = \theta_{i_{min}} + \frac{\left(\sum_{j=1}^{n} \chi_i[j] * 2^j\right)}{2^n} * \left(\theta_{i_{max}} - \theta_{i_{min}}\right), \tag{6}$$

where $\theta_{i_{min}}$ is the lowest value for a certain joint, and $\theta_{i_{max}}$ is the highest one. It is hard for iterative algorithms based on the Jacobian matrix to avoid exceeding the joint limits. Through this encoding mechanism, the resulting angles computed by the IGA always fall in the range of the joint limit and, thus, prevent unpredictable errors.

4.3. Evaluation and Selection

Better chromosomes are chosen from the population in order to reproduce the offspring during the process of the IGA. To evaluate a chromosome, a fitness function, which is a quantitative measure, is adopted to measure how satisfactory a chromosome is. In this study, the task-space is only within the position. As a result, the fitness function is defined as

$$Error_p = \|p_1 - x_t\| + \|p_2 - x_i\|, \tag{7}$$

$$Error_d = \max_{k=1,\ldots,8} \theta_{k,i} - \theta_{k,f}, \tag{8}$$

$$Fitness = Error_p + K \cdot Error_d, \tag{9}$$

where p_1 is the target point for the thumb, x_t is the tip of the thumb, p_2 is the target point for the index finger, x_t is the tip of the index finger, $\theta_{k,i}$ is the initial angle of each joint before computing the IGA, $\theta_{k,f}$ is the angle of the joint contained in the chromosome, and K is the weight of the second criterion (infinity-norm term).

The position of x_t is calculated using the forward kinematics according to the angle set contained in the chromosome. Equation (7) computes the distance in Cartesian space to indicate how far the tips of fingers are from the target points and, thus, sums the distances to be a criterion for the fitness function. Equation (8) takes the maximum absolute value within the angle differences between the initial angle and the angle generated by the IGA. Equation (8) is derived from the concept of the infinity norm and is used to guarantee the smoothness of the motion. The fitness function is minimized by the IGA, which means that the tips of two fingers are set to reach the target points as closely as possible, with each joint having about the same angular difference.

On the other hand, to maintain the quality and diversity of the population, the selection operator is performed to select superior chromosomes for breeding of the next population. Supposing that the chosen chromosomes are always the best individuals, the diversity of the IGA will be narrow. Hence, the k-tournament selection operator is used to choose which chromosomes should reproduce new offspring. The k-tournament selection randomly withdraws k chromosomes from the population and picks the top two chromosomes for the crossover operator. The advantage of the k-tournament selection is that it is unnecessary to sort the entire population to get the top two chromosomes, thereby reducing computational resources. In addition, the top two chromosomes may be poor ones, which keeps the population diverse and eliminates a local minimum.

4.4. Crossover and Mutation

In this step, the new chromosomes are produced by genetic operations, such as a crossover and a mutation. The crossover operation is adopted to create two offspring from two parent chromosomes. The parents swap with the partial section of genes and then generate two children. Here, k-point crossover is used. The k-point crossover first selects k crossover points to make the k + 1 interval in one chromosome and then changes the section of genes every two intervals to form two offspring.

The crossover operation is followed by the mutation operation. Mutation is utilized to change some genes of the children chromosomes. The mutation rate is defined as the reciprocal of the chromosome's length. The mutation operation is implemented on each gene and flips it with the probability of the mutation rate. Because of these two genetic operations, the diversity of the IGA is highly promoted; simultaneously, it becomes more possible to get close to the global minimum.

4.5. Elitism

Elitism is a mechanism that guarantees the best chromosome will survive in every generation. Elitism can accelerate the convergence of the IGA but slightly decrease the diversity of the population. In this study, the feasibility check function was used to provide an optimal wrist angle set that narrows down the search space of the IGA. Therefore, the possibility of being trapped in local minimum was lower, thus validating elitism as a good strategy to be adopted.

4.6. Immigration

An immigration strategy can be performed periodically to maintain the diversity level of the genetic algorithm [23]. The best individuals immigrate to the new population to replace the worst ones at defined intervals, which is called "structured memory immigration operator." In this research, the immigration operator was set to occur every three generations in order to obtain better optimal solutions. Half of the population was replaced with the best individuals of the previous generation in the same quantity. As mentioned, we can ensure that the solution to the IK problem is near the best individuals of the population through the feasibility check. In this case, the introduced immigration operator is more suitable for the population than a random immigration operator. As a result, the tournament selection can have a better chance of selecting better chromosomes and maintaining the diversity.

5. Results and Discussion

In this section, the performance of the proposed hybrid GA is validated by examination through parametric numerical simulations. The simple GA and the IGA are also performed to compare the results. To highlight the capability of determining the feasibility, both appropriate and improper IK requests are given to different algorithms for computing IK solutions. The detailed parameters used in the illustrated cases through numerical simulations with the proposed algorithm are shown in Table 3.

Table 3. Parameters used in the proposed hybrid genetic algorithm. GA, genetic algorithm.

Parameter	Value
GA type	Generational
Representation	Binary
Chromosome length	80
Population number	500
Selection	10-tournament
Crossover	2-point
Crossover rate	0.8
Mutation	Bit-flip
Mutation rate	1/L
Survivor	1000
Termination	3 s or fitness lower than 0.6

Among the parameters in Table 3, the fitness value of termination dominates the computation time. As exhibited in Table 4, when the termination value increases, it takes less time to compute, while the accuracy of the solution is sacrificed. It also compares the average execution time of the simple genetic algorithm and the performance of the proposed hybrid GA, which shows that the proposed hybrid GA is two times faster than the simple GA. Additionally, it is known that the selection of the fitness value of termination depends on the application scenario. Thus, the termination condition applied in different tasks was varied accordingly.

Table 4. Effect of termination value on the convergence speed of GA.

	Termination Fitness Value	Average Number of Generation	Average Execution Time (s)
Simple GA Proposed hybrid GA	1	52 24	4.41 2.06
Simple GA Proposed hybrid GA	5	18 10	1.68 0.72
Simple GA Proposed hybrid GA	10	7 4	0.63 0.35

In calculating the inverse kinematics, two points were assigned as the target points of the index finger and thumb. Different algorithms were performed to solve the inverse kinematics, and the best fitness values were recorded for the graphical representation to analyze the performance of each algorithm. Table 5 lists the conditions and results for three illustrated cases. The illustrated case #1 and case #2 were those with reachable target points, whereas an unreachable target point was assigned in the illustrated case #3. In the illustrated case #1, the target fingertip point sets were (50, 0, 130) and (75, 30, 125) for the index finger and the thumb, respectively. As shown in Figure 6a, the two target points were in the workspace of two fingers and they passed the feasibility check. Additionally, the corresponding angle set of the wrist $(\theta_{w1}, \theta_{w2}) = (12°, 15°)$ was provided to the hybrid GA as the initial condition. By using the solutions computed by the GA, the tips of the index finger and the thumb reached their target points. The simple GA and IGA were also performed to solve the same IK problem as a comparison. As shown in Figure 6b, the proposed hybrid GA was observed to have a lower fitness value than the other GAs in the beginning. The proposed hybrid GA was found to obtain the solution for the inverse kinematic problem faster than the other GAs with fewer generations.

Table 5. Results of illustrated cases performed by the proposed algorithm.

Case	Target Points (mm)	Feasibility	Feedforward Wrist Angle	Solutions of the Proposed Algorithm		Error Sum of Two Fingers (mm)
#1	Index finger: (50, 0, 130) Thumb: (75, 30, 125)	Pass	$\theta_{w1}: 12°$ $\theta_{w2}: -15°$	θ_{w1} θ_{w2} θ_{t1} θ_{t2} θ_{t3} θ_{i1} θ_{i2} θ_{i3}	14.4981° −5.0000° 14.9817° −8.4396° −21.9780° −45.0001° 55.2968° 45.0110°	0.5795
#2	Index finger: (45, 52, 172) Thumb: (81, 60, 111)	Pass	$\theta_{w1}: 0°$ $\theta_{w2}: 0°$	θ_{w1} θ_{w2} θ_{t1} θ_{t2} θ_{t3} θ_{i1} θ_{i2} θ_{i3}	0.2515° −4.9987° 15.0110° −19.6923° −8.8571° −19.6374° 14.5494° 41.4726°	0.5875
#3	Index finger: (50, 0, 300) Thumb: (60, 50, 125)	Drop out	Not Computed	Not Computed		Not Computed

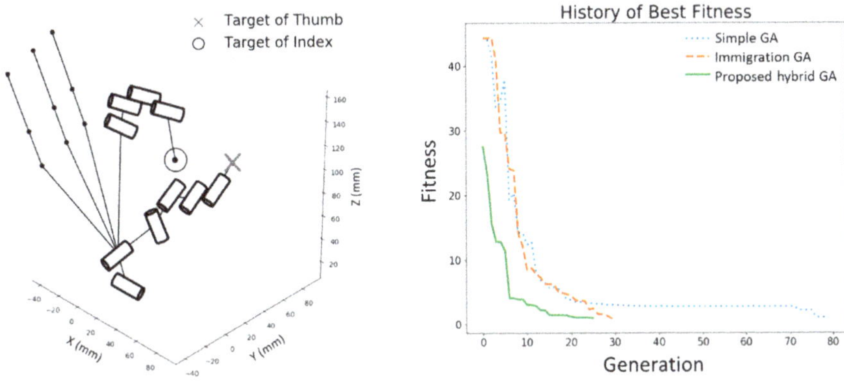

(a) (b)

Figure 6. (**a**) Schematic illustration showing the solution resulting from an appropriate motion command and (**b**) the corresponding fitness performance of the illustrated case #1.

In the illustrated case #3, the target fingertip point sets were (50, 0, 300) and (60, 50, 125) for the index finger and thumb, respectively. As shown in Figure 7a, the target point of the index finger was obviously out of the workspace. In this circumstance, as shown in Figure 7b, the proposed hybrid GA immediately ceased the inappropriate motion command to prevent unnecessary computation. However, the other GAs still conducted computation until meeting certain termination criteria. As a result, Figure 7b shows high fitness values of the simple GA and IGA, while the proposed hybrid GA successfully stopped computing. The hybrid GA, thus, was successful in avoiding the unnecessary computation for a given inappropriate IK request.

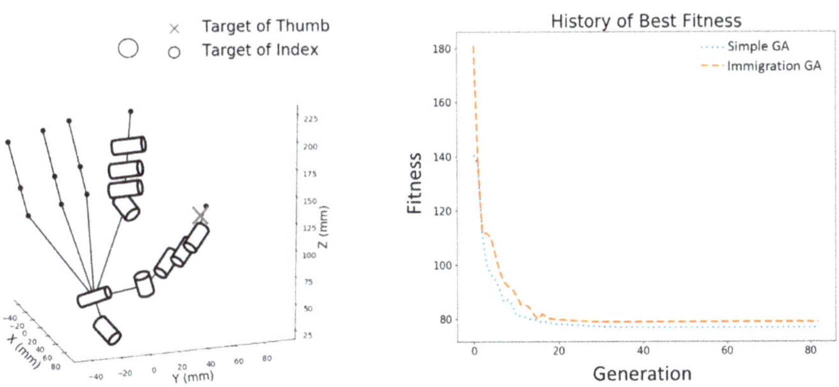

(a) (b)

Figure 7. (**a**) Schematic illustration showing the solution resulting from an inappropriate motion command and (**b**) the corresponding fitness performance of the illustrated case #3.

Results in Table 6 demonstrate the distribution of population by exemplary generations during the evolution process in the illustrated case #1 for the simple GA and the proposed hybrid GA. The blue dots represent all the thumbs' tips of the population in one generation, whereas the red dots serve as those of the index finger. The corresponding cross marks stand for the target point or the answer of the IK problem. In the initial generation, the population distribution of the simple GA was scattered more widely than that of the proposed hybrid GA. Because of the feedforward wrist angle, the proposed hybrid GA started with a smaller searching scope. When the generation increased, the proposed hybrid GA had more chances to get close to the target points and, therefore, converged faster than the simple GA.

To surpass the performance of the other GA, the workspace of two fingers plays an important role in the proposed hybrid GA. Initially, the workspace analysis can provide the nearest distance between the target point and the workspace. As a result, the feasibility of the IK request can be checked accordingly. Second, the workspace was pre-established and stored in the order of the wrist angle. Consequently, the workspace analysis gives a feedforward wrist angle to the hybrid GA. The hybrid GA then reduces the variable range of the wrist angle to offer a superior initial condition. In general, the proposed hybrid GA has better convergence than the other GA and can effectively terminate the computation when an improper IK request is given.

Table 6. Comparison of population distribution at different generations in the illustrated case #1.

	Proposed Hybrid GA	**Simple GA**
Generation No. 1	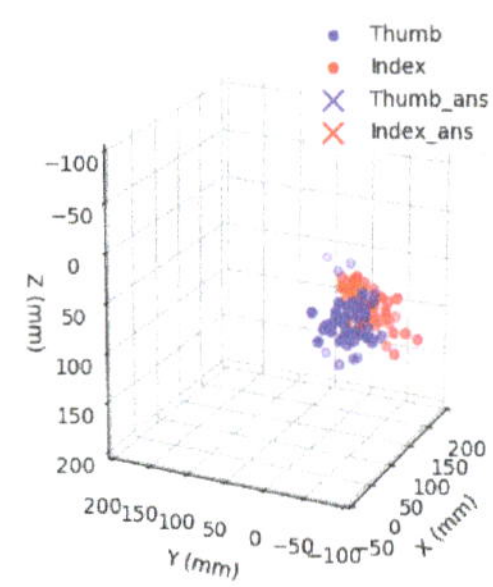	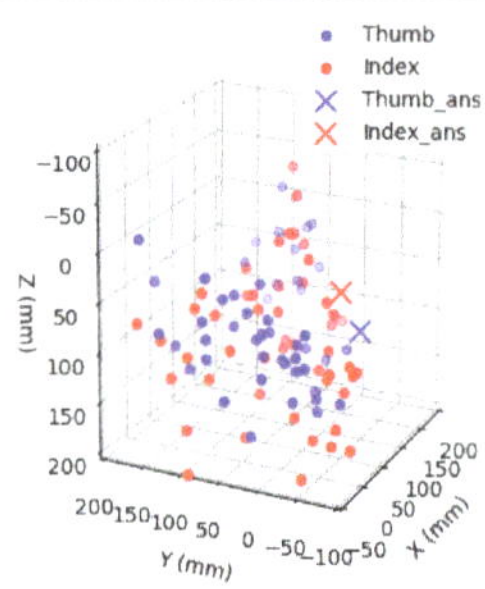
Generation No. 2	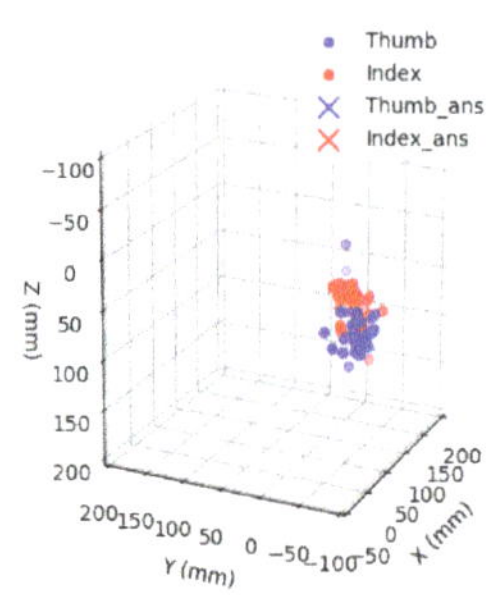	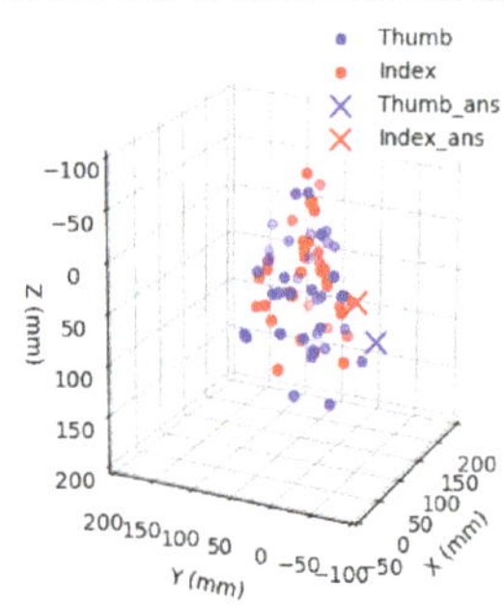
Generation No. 5	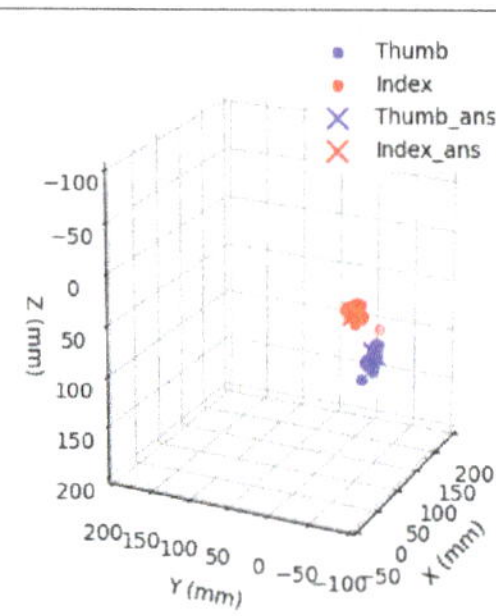	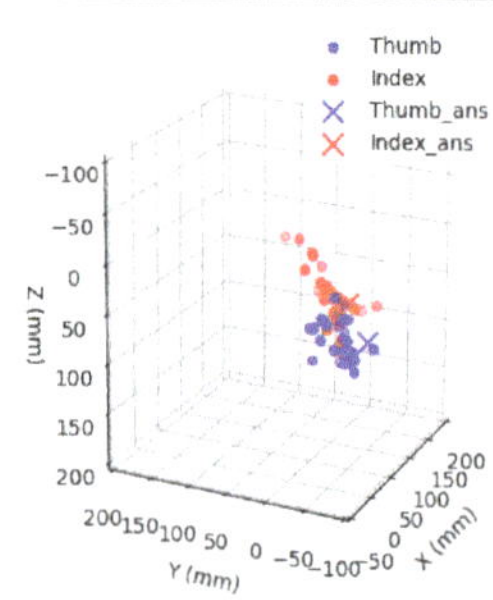

Table 6. *Cont.*

	Proposed Hybrid GA	Simple GA
Generation No. 20	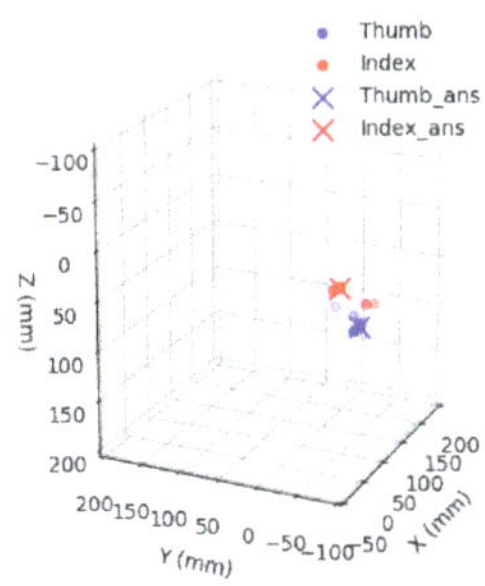	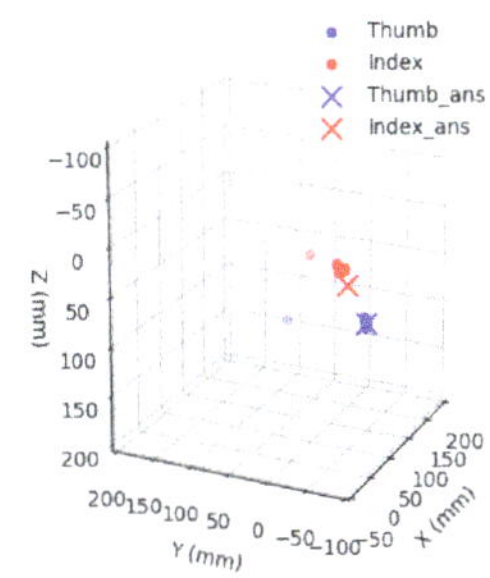

In addition to the interaction between a thumb and an index finger, the proposed algorithm can solve the IK problem for a thumb and the other three fingers. Numerical simulations were performed to solve the IK under all three scenarios, namely, the IK of the thumb and middle finger, the IK of the thumb and ring finger, and the IK of the thumb and little finger. Similarly, the same procedures were adopted in these IK problems, including building the database of the workspace, checking the feasibility, and providing the feedforward wrist angle. Figure 8 shows a schematic illustration of the IK solutions for the thumb and the other three fingers. The diagram of the best fitness value over generations indicates that the proposed algorithm outperformed the simple GA and the immigration GA. Through these cases, it was confirmed that the proposed algorithm is capable of solving the IK problem of a multi-fingered robotic hand.

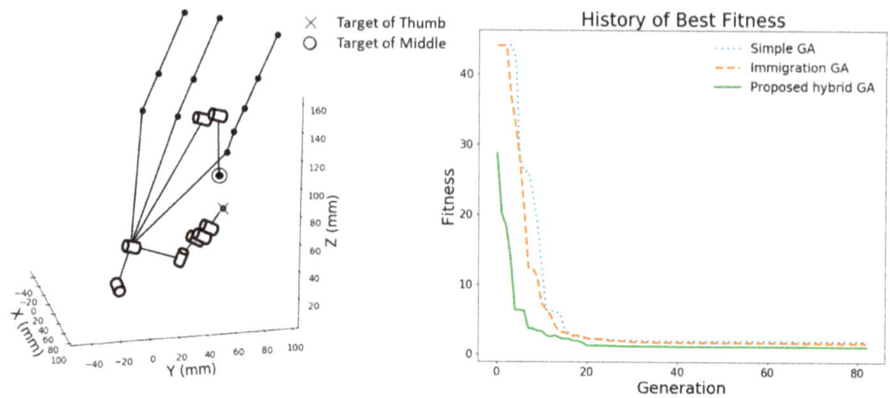

(**a**)

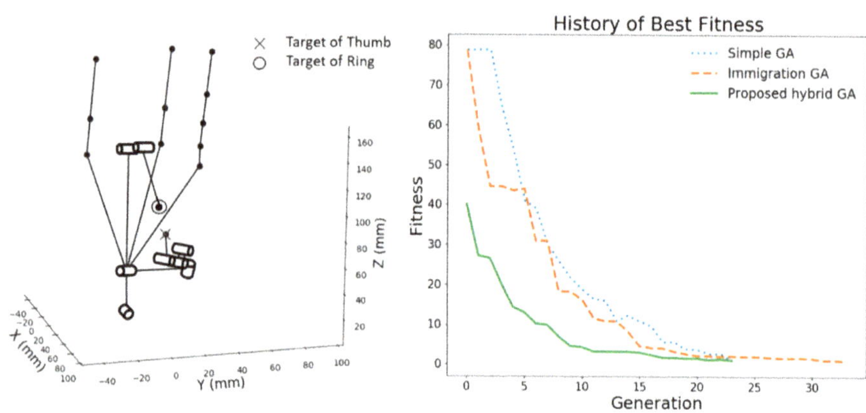

(**b**)

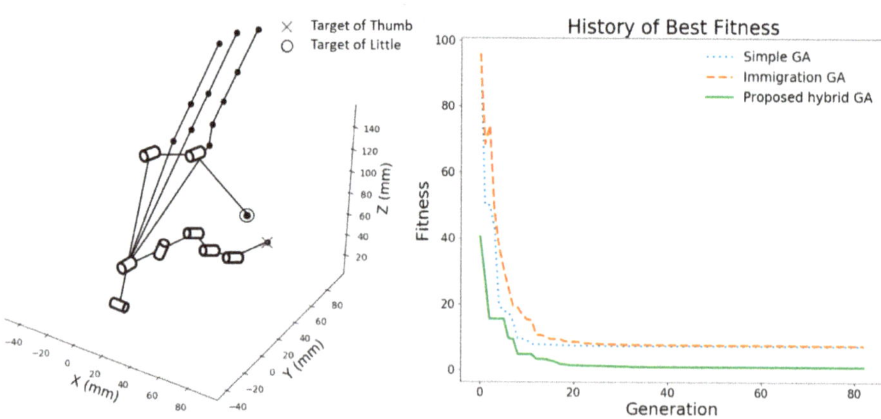

(**c**)

Figure 8. Schematic illustration showing the solution of the IK problem with (**a**) thumb and middle finger, (**b**) thumb and ring finger, and (**c**) thumb and little finger, as well as the corresponding fitness performance.

6. Conclusions

In this study, a workspace-analysis-based immigration genetic algorithm was presented to solve the inverse kinematics of a multi-fingered hand. In the offline stage, the kinematic model of fingers is built to perform forward kinematics, from which the workspace database of the fingers can be established using the Monte Carlo method. In the online stage, an estimation function is used to check the feasibility of the target point whenever an IK request is assigned. If the target point is within the workspace, the estimation function then provides an optimal set of wrist angles for the IGA to compute the solution; contrarily, the estimation function terminates further calculation. This estimation function enables the searching range to be narrowed to the neighborhood of the solution. Therefore, the distribution of the population is less widespread, and the IGA can have better initial condition and greater convergence. Numerical simulations were conducted to verify the performance of the proposed hybrid GA. For the unreachable targets, the estimation function could efficiently abort the entire computation. For the reachable targets, the proposed algorithm took fewer iterations to obtain accurate results than the simple GA and the IGA. As a whole, this study presented an algorithm to solve the multi-fingered inverse kinematics, which enables the tips of the index finger and the thumb to move to their target points concurrently; likewise, the same approach can be applied to the other fingers. The results can also be applied to the IK problem of a dual-arm robot that has degree of freedoms in the waist. The proposed method is believed to have the potential to be employed in many industrial robots.

Author Contributions: Conceptualization, methodology, software, formal analysis, visualization, and writing—original draft preparation, C.-T.L.; conceptualization, writing—review and editing, supervision, project administration, and funding acquisition, J.-Y.C. All authors have read and agreed to the published version of the manuscript.

Funding: This research was funded by the Ministry of Science and Technology of Taiwan, grant number MOST 109-2218-E-007-024.

Institutional Review Board Statement: Not applicable.

Informed Consent Statement: Not applicable.

Data Availability Statement: Not applicable.

Conflicts of Interest: The authors declare no conflict of interest.

References

1. Jiang, L.; Sun, B.; Gu, H. A Power Series Based Inverse-Kinematics Solution of Humanoid Robot Finger with Coupled Joints. *Int. J. Hum. Robot.* **2018**, *15*, 1850016. [CrossRef]
2. Manocha, D.; Canny, J.F. Efficient inverse kinematics for general 6R manipulators. *IEEE Trans. Robot. Autom.* **1994**, *10*, 648–657. [CrossRef]
3. Lee, C.G. Robot arm kinematics, dynamics, and control. *Computer* **1982**, *15*, 62–80. [CrossRef]
4. Kelemen, M.; Virgala, I.; Lipták, T.; Miková, L'.; Filakovský, F.; Bulej, V. A novel approach for a inverse kinematics solution of a redundant manipulator. *Appl. Sci.* **2018**, *8*, 2229. [CrossRef]
5. Bensalah, C.; Gonzalez-Quijano, J.; Hendrich, N.; Abderrahim, M. Anthropomorphic robotics hand inverse kinematics using estimated svd in an extended sdls approach. In Proceedings of the 2013 16th International Conference on Advanced Robotics (ICAR), Montevideo, Uruguay, 25–29 November 2013; pp. 1–7.
6. Tchon, K. Optimal Extended Jacobian Inverse Kinematics Algorithms for Robotic Manipulators. *IEEE Trans. Robot.* **2008**, *24*, 1440–1445. [CrossRef]
7. Buss, S.R. Introduction to inverse kinematics with jacobian transpose, pseudoinverse and damped least squares methods. *IEEE J. Robot. Autom.* **2004**, *17*, 16.
8. Hartenberg, R.S.; Denavit, J. A Kinematic Notation for Lower Pair Mechanisms Based on Matrices. *J. Appl. Mech.* **1955**, *77*, 215–221.
9. Ren, H.; Ben-Tzvi, P. Learning inverse kinematics and dynamics of a robotic manipulator using generative adversarial networks. *Robot. Auton. Syst.* **2020**, *124*, 103386. [CrossRef]
10. Kim, B.-H. An adaptive neural network learning-based solution for the inverse kinematics of humanoid fingers. *Int. J. Adv. Robot. Syst.* **2014**, *11*, 3. [CrossRef]

11. Al-Gallaf, E. Multi-fingered robot hand optimal task force distribution: Neural inverse kinematics approach. *Robot. Auton. Syst.* **2006**, *54*, 34–51. [CrossRef]
12. Mayorga, R.V.; Sanongboon, P. Inverse kinematics and geometrically bounded singularities prevention of redundant manipulators: An Artificial Neural Network approach. *Robot. Auton. Syst.* **2005**, *53*, 164–176. [CrossRef]
13. El-Sherbiny, A.; Elhosseini, M.A.; Haikal, A.Y. A comparative study of soft computing methods to solve inverse kinematics problem. *Ain Shams Eng. J.* **2018**, *9*, 2535–2548. [CrossRef]
14. Dereli, S.; Köker, R. A meta-heuristic proposal for inverse kinematics solution of 7-DOF serial robotic manipulator: Quantum behaved particle swarm algorithm. *Artif. Intell. Rev.* **2020**, *53*, 949–964. [CrossRef]
15. Starke, S.; Hendrich, N.; Magg, S.; Zhang, J. An efficient hybridization of genetic algorithms and particle swarm optimization for inverse kinematics. In Proceedings of the 2016 IEEE International Conference on Robotics and Biomimetics (ROBIO), Qingdao, China, 3–7 December 2016; pp. 1782–1789.
16. Durmuş, B.; Temurtaş, H.; Gün, A. An inverse kinematics solution using particle swarm optimization. In Proceedings of the International Advanced Technologies Symposium, Elazığ, Turkey, 16–18 May 2011; pp. 193–197.
17. Momani, S.; Abo-Hammour, Z.S.; Alsmadi, O.M. Solution of inverse kinematics problem using genetic algorithms. *Appl. Math. Inf. Sci.* **2016**, *10*, 225. [CrossRef]
18. Ahmad, M.; Kumar, N.; Kumari, R. A Hybrid Genetic Algorithm Approach to Solve Inverse Kinematics of A Mechanical Manipulator. *Int. J. Sci. Technol. Res.* **2019**, *8*, 1777–1782.
19. KöKer, R. A genetic algorithm approach to a neural-network-based inverse kinematics solution of robotic manipulators based on error minimization. *Inf. Sci.* **2013**, *222*, 528–543. [CrossRef]
20. Tabandeh, S.; Melek, W.W.; Clark, C.M. An adaptive niching genetic algorithm approach for generating multiple solutions of serial manipulator inverse kinematics with applications to modular robots. *Robotica* **2010**, *28*, 493–507. [CrossRef]
21. Zhou, Z.; Guo, H.; Wang, Y.; Zhu, Z.; Wu, J.; Liu, X. Inverse kinematics solution for robotic manipulator based on extreme learning machine and sequential mutation genetic algorithm. *Int. J. Adv. Robot. Syst.* **2018**, 15. [CrossRef]
22. Shengzheng, K.; Wu, H.; Yao, L.; Li, D. Coordinated workspace analysis and trajectory planning of redundant dual-arm robot. In Proceedings of the 2016 13th International Conference on Ubiquitous Robots and Ambient Intelligence (URAI), Xi'an, China, 19–22 August 2016; pp. 178–183.
23. Tajani, C.; Abdoun, O.; Lahjouji, A.I. Genetic algorithm adopting immigration operator to solve the asymmetric traveling salesman problem. *Int. J. Pure Appl. Math.* **2017**, *115*, 801–812. [CrossRef]

applied sciences

MDPI

Article

Design of a Planar Cable-Driven Parallel Robot for Non-Contact Tasks

Valentina Mattioni †, Edoardo Ida' *,† and Marco Carricato

Department of Industrial Engineering, University of Bologna, Viale Risorgimento 2, 40136 Bologna, Italy; valentina.mattioni@unibo.it (V.M.); marco.carricato@unibo.it (M.C.)
* Correspondence: edoardo.ida2@unibo.it
† These authors contributed equally to this work.

Abstract: Cable-driven parallel robots offer significant advantages in terms of workspace dimensions and payload capability. Their mechanical structure and transmission system consist of light and extendable cables that can withstand high tensile loads. Cables are wound and unwound by a set of motorized winches, so that the robot workspace dimensions mainly depend on the amount of cable that each drum can store. For this reason, these manipulators are attractive for many industrial tasks to be performed on a large scale, such as handling, pick-and-place, and manufacturing, without a substantial increase in costs and mechanical complexity with respect to a small-scale application. This paper presents the design of a planar overconstrained cable-driven parallel robot for quasi-static non-contact operations on planar vertical surfaces, such as laser engraving, inspection and thermal treatment. The overall mechanical structure of the robot is shown, by focusing on the actuation and guidance systems. A novel concept of the cable guidance system is outlined, which allows for a simple kinematic model to control the manipulator. As an application example, a laser diode is mounted onto the end-effector of a prototype to perform laser engraving on a paper sheet. Observations on the experiments are reported and discussed.

Keywords: cable-driven parallel robots; overconstrained robots; design; non-contact operations

Citation: Mattioni, V.; Ida', E.; Carricato, M. Design of a Planar Cable-Driven Parallel Robot for Non-Contact Tasks. *Appl. Sci.* **2021**, *11*, 9491. https://doi.org/10.3390/app11209491

Academic Editor: Alessandro Gasparetto

Received: 19 September 2021
Accepted: 23 September 2021
Published: 13 October 2021

Publisher's Note: MDPI stays neutral with regard to jurisdictional claims in published maps and institutional affiliations.

1. Introduction

Cable-driven parallel robots (*CDPRs* in short) combine the successful features of parallel manipulators with the benefits of cable transmissions. The payload is divided among light extendable cables, resulting in an energy-efficient system that can achieve high acceleration of the end-effector (*EE*) over a remarkably large workspace. From a structural point of view, a *CDPR* is formed by a set of actuation units, usually fixed to a frame, and a mobile platform, working as *EE* [1]. The cables, driven by the actuation units, are guided inside the robot workspace (*WS*) using a guidance system, and then connected to the mobile platform. The variation of cable lengths is responsible for the *EE* displacement throughout the robot *WS*. These features result in easily reconfigurable systems, where the *WS* can be modified by relocating the actuation and/or guidance units [2].

Nevertheless, the use of *CDPRs* in industrial environments is still limited, mainly due to the drawbacks of employing flexible cables. Indeed, cables impose unilateral constraints that can only exert tensile forces and, consequently, the *EE* cannot withstand any arbitrary external action. In addition, the viscoelastic nature of cables can cause their elongation under prolonged load application [3]. This highly non-linear behavior complicates the control of the robot and the estimation of the actual cable lengths, and therefore the determination of the platform pose through direct kinematics. To enhance the robot controllability, *CDPRs* can be overconstrained by employing a number of cables higher than the degrees of freedom (*DoFs*) of the *EE*. This allows cables to pull one against the other and to keep the overall system controllable over a wide range of externally applied loads. In other cases, to increase accessibility and reduce hardware complexity, suspended architectures with less cables than *EE DoFs* can be used, though they are more difficult to control [4,5].

Thanks to their simple architecture, high reconfigurability, easy deployability, and large workspace, *CDPRs* have been proposed in many fields of application: entertainment [6,7], logistics [8], construction [9–11], maintenance [12,13], to name a few. In general, it is possible to design a general-purpose machine that can be dedicated to different tasks by changing the *EE* tool [14,15] or, in alternative, a *CDPR* for a specific purpose can be developed. In the second case, the robot geometry, the number and configuration of cables, and, consequently, additional design constraints are established by the final task. This is the case of the deployable *CDPR* in [16]: the suspended architecture and the *EE* self-deployability ensure the high orientational capability required for the laser scanning of low-accessibility environments. As an additional example, the storage-retrieval *CDPR* in [8] is equipped with eight cables, despite the planar nature of the robot task, in order to provide a suitable machine stiffness and prevent platform tilt.

Usually, if the task requires a specific motion pattern of the *EE*, it is possible to act in two different ways to limit the *EE* range of motion. First, an external means, which physically constrains the *EE*, can be employed. This happens with the *CDPR* Pickable [17], whose *EE* is constrained to slide on a planar surface thanks to an air-bearing system. Likewise, in [13], the *EE* of the robot moves in direct contact with the building façade to perform window cleaning operations. The contact of the *EE* with the environment requires specific contact models [18], and suitable force-control strategies [19]. For the Pickable robot, undesired phenomena, such as friction, are significantly reduced by using air bearings, whereas, for the cleaning robot, friction is part of the process itself. A different approach for constraining the *EE* motion consists in adopting special cable configurations, such as the parallelogram arrangements inspired by delta robots [20–22]. This expedient allows obtaining a translational motion of the *EE* without any external constraint, provided that all cables are taut. This is useful in the case of large-scale industrial tasks, such as pick-and-place and warehousing [23], or operations that need deployable systems, such as rescue and contour crafting [24,25].

The parallelogram arrangement strategy can also be applied in non-contact tasks where the *EE* must move on a prescribed vertical plane without interacting with the surrounding environment. This is the case of welding, laser engraving, thermal treatment, and decoration or inspection of building façades, to name a few. This paper focuses on the design of a *CDPR* for this kind of applications. Specifically, an overconstrained planar machine is presented. This prototype employs a parallelogram cable arrangement to constrain the *EE* to move on a vertical plane. Four cables, closed in a parallelogram loop, are used to control the mobile platform's three degrees of freedom (*DoFs*). The cable guidance system comprises a pair of swivel pulleys whose swivel axes are orthogonal to the work surface: the intersections of these swivel axes with the work surface effectively define the cable proximal and distal points. Each cable loop is driven by one winch, whose drum has two helicoidal starts. Each winch includes a rotary encoder for cable length control. In addition, the pulley swivel axes are equipped with rotary encoders to achieve redundant kinematic measures, and cable tensions are measured with shear beam load cells integrated into the proximal anchor systems. Like other planar *CDPR* that share the idea of parallelogram mechanisms [23,26], this system allows the platform to move on a defined plane, excluding low-amplitude oscillatory motions induced by external disturbances [27,28]. On the other hand, unlike the others, the proposed system shows three advantages: the use of a single cable loop, instead of two separate cables, ensures that the parallel cable segments are in tension when controlled by a single actuator, thus reducing the number of actuators to be used; the cable anchor system includes a measurement unit to continuously monitor cable tensions and cable angles with respect to the frame; the mounting arrangement of the swivel pulleys allows to estimate the cable lengths as distances between two precisely defined points, which are the intersections between the swivel axes and the work plane, without introducing geometrical approximations.

Consequently, a simple geometric model can be adopted to control the *CDPR* without inherent geometrical errors.

The paper is organized as follows. The main aspects to be taken into consideration for the mechanical design of a *CDPR* are described in Section 2. Then, the design of the planar overconstrained *CDPR* prototype is presented in Section 3, focusing on the actuation and guidance system, and the influence of the kinematic model. In Section 4, as an application example, an experiment of the robot performing laser engraving is reported and discussed. Lastly, some conclusions are drawn.

2. Design Strategies

In most cases, cables are driven by actuators that are placed on a fixed base. Sometimes, if deployability is a requirement, they can be attached to mobile vehicles [24] or placed on the mobile platform [29]. A guidance system is used to convey cables from the actuators to the mobile platform. Many authors analyzed the design process of *CDPRs* and identified several methods for optimizing their structure and dimensions [1,30]. For a given criterion (i.e., workspace dimension, accuracy, stiffness, orientation capability), optimization procedures help determine the geometry of the *CDPR*, including the number of cables, position of cable anchor points, and platform and frame shapes [2,15,31]. Once the architecture is known, the hardware design requires developing three distinct systems: the actuation unit, the guidance system, and the mobile platform. These systems can be equipped with suitable sensors, to monitor the robot performance and obtain feedback on relevant control variables. The two most notable examples are cable lengths, which are usually estimated by using angular sensors mounted on the actuation unit motors, and cable tensions, whose measurement involves integrating force sensors in the actuation unit, guidance system, or *EE* [1].

This section shows how specific design techniques can help keep the robot model and control as simple as possible without introducing geometrical approximations. Usually, a simplified kinematic model, called standard in the following, is adopted. This model considers cables inextensible and massless and, thus, treated as line segments between two points, one on the mobile platform and the other on the frame (distal and proximal anchor points, respectively). The former is usually assumed to be fixed with respect to the moving platform, whereas the latter is fixed with respect to the robot frame. These assumptions neglect not only all effects caused by the elastic nature of cables, which may be negligible in some small-to-medium-scale applications, but also the kinematics of the guidance system. Geometrical-model errors can be limited by accounting for the pulley geometry [4,32], but the model and control complexity is usually increased. As an alternative, if the standard kinematic model is to be used without inherent geometrical errors, special design techniques of the actuation units and guidance systems can be used, as discussed in the following.

2.1. Actuation Unit

The most appropriate type of actuation is chosen depending on the *EE* motion pattern and *WS* size, defined by the task [1,33]. A rotary actuation system generally consists of a servo-actuated winch where the cable is coiled onto a cylindrical drum. This solution is easy to implement but can result in low robot position accuracy since it introduces several errors when estimating cable lengths [34]. Lengths are computed by measuring the rotation of the motor and, thus, they depend on the drum design and cable arrangement on it. Linear actuation systems reduce cable-length estimation errors, but they usually limit the size of the achievable *WS* [35], or increase cable wear [34].

A solution could be to adopt a rotary actuation system designed to be as close as possible to the model intended for its control. The main drawback of the rotary actuation, if a simple lifting winch is considered, is the fact that the cable exit direction from the drum varies, even unpredictably, as the motor rotates: this results in a non-linear transmission ratio between the motor angle and the cable length, which is usually undesirable. A constant transmission ratio can be obtained (i) by avoiding the cable overlapping onto the drum surface, (ii) by suitably grooving the drum to accommodate the cable, which is also desirable for reducing cable wear [36], and (iii) by constraining the cable to exit the drum in a fixed, known, direction. There are several solutions in the literature to achieve such desired design requirements (Figure 1).

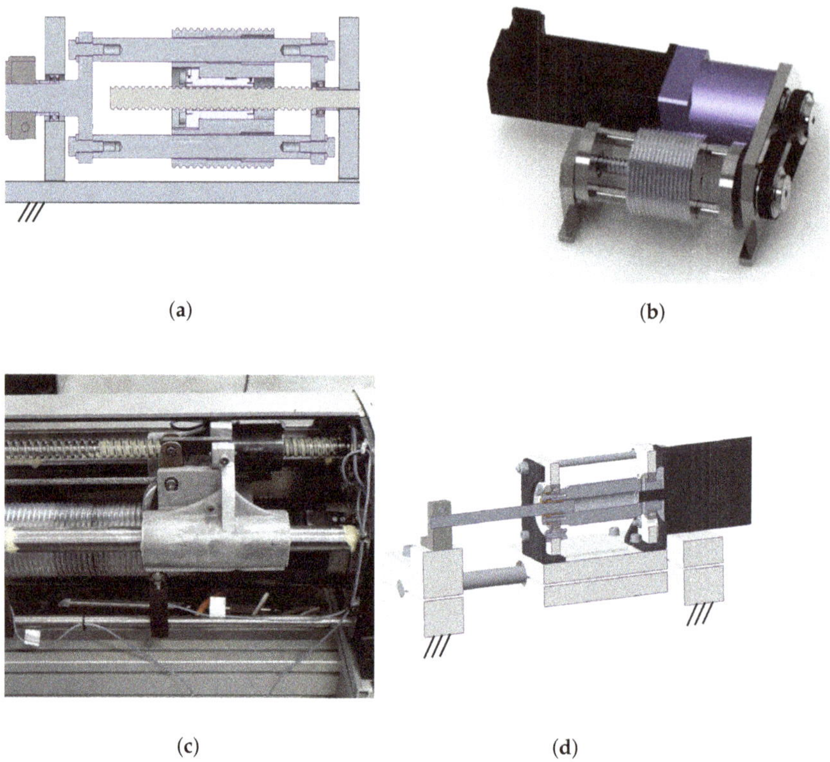

Figure 1. Winch models. (**a**) Cross section of a rototranslating drum. (**b**) Rendering of a winch with rototranslating drum kinematically equivalent to the ones in [37,38]. (**c**) Close-up view of a winch of the Ipanema *CDPR* family [14]. (**d**) Winch with translational motor-drum.

It is possible to rototranslate the drum [16,37,38] so that the cable exit point from the drum, and consequently its direction, is kept fixed with respect to the winch frame, while the cable is coiled and uncoiled. This is usually achieved by employing screw/nut joints to convert the rotational motion of the motor into rotational and translational motion of the drum, and allowing the drum to translate by mounting it on a passive prismatic joint. In [38] the drum houses a nut and two pairs of cylindrical bearings. Two rods, which are coupled with the motor shaft, parallel to the drum axis, but mounted with a radial offset with respect to the drum, pass through the bearings, thus transmitting the rotational motion, but allowing the drum to translate. The translation is caused by the drum nut being joined with a screw, which is fixed on the winch frame: a cross section of the drum system is shown in Figure 1a. The main advantages of this solution are compactness and mechanical simplicity, but this design suffers from one main drawback, namely the rods used for transmitting the motor rotations pass through the drum, thus the latter diameter needs to be large. An example of this design is shown in Figure 1b, where a timing belt is used to transmit motion from the motor to the drum, as in [37].

As an alternative, in [14] an auxiliary cable guiding device equipped with a pulley (*spooling helper*) continuously follows the variable cable exit point on the rotating drum (Figure 1c) by translating parallel to the drum axis, to ensure that the direction of the cable connecting the drum and the spooling helper is constant. The spooling helper is mounted on a support equipped with a nut and a pair of bearings, and thus slides with respect to fixed rods, driven by a screw, which is ultimately connected to the drum with a synchronous belt. Given its footprint, this solution has the primary benefit of allowing very long cables to be stored onto the drum. Still, an additional mechanical transmission, such as a timing belt, is always required to transmit drum motion to the screw.

A different solution allows the translation of the entire motor-drum system on a linear guide [39], similar to the design shown in Figure 1d. This solution has the advantage of being particularly simple and cheap, since the drum is directly connected to the motor, and the drum supports are connected to a carriage. Still, the motor mass, which needs to be translated, may limit the system's dynamic performance.

2.2. Guidance System

One of the main advantages of *CDPRs* is the possibility to locate the actuation units almost anywhere. This is possible thanks to the cable guidance system that conveys the cable, from the actuator, through the prescribed anchor points on the frame (proximal) to the anchor point on the platform (distal). For the standard model to be respected, without introducing geometrical-model errors, the cable anchor points need to be geometrically determined. An extensive review of possible construction solutions is reported in [1], where a distinction is made between the proximal and distal anchor systems.

The mechanical devices that compose the proximal guidance system should allow for large deflection angles and at the same time determine the anchor point position. Swivel (or panning) pulleys and eyelets are primarily used for this purpose. Eyelets allow for the approximate definition of a point and significant cable orientation, but cause high cable wear [40]. However, they are employed in many *CDPRs* thanks to their simplicity, and the possibly low geometrical error introduced, if the *CDPR* workspace is large [16,38,41]. Figure 2 shows the use of eyelets as proximal anchor points on the fixed frame of the *CDPR* prototype presented in [16]. Conversely, swivel pulleys can reduce cable wear and still allow for orientation in a broad direction range, since, unlike classical fixed pulleys (Figure 3a), they can rotate about an axis tangent to the pulley groove, the swivel axis (Figure 3b). This kind of solution is adopted in [39,42], and an example of exit point layout is shown in Figure 4. In general, the presence of pulleys in the guidance system makes the system reliable, but complicates the geometrical model of the robot. This issue can be faced in two ways: by adopting a more complex model, which includes pulley kinematics for *CDPR* control [4,32,43], or by ignoring the actual geometry of the system, and accepting several modeling errors. Otherwise, particular layout and orientations of the pulley system must be adopted.

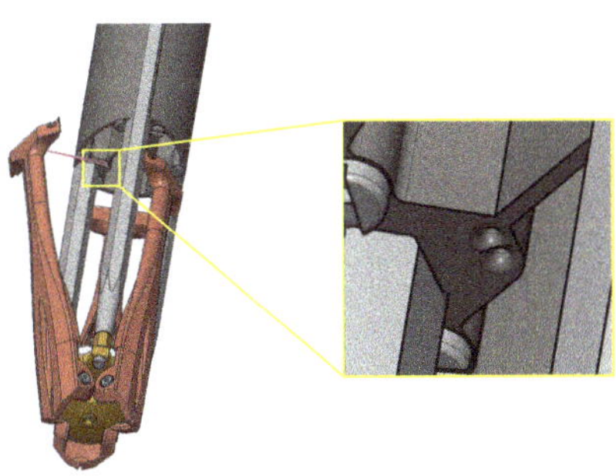

Figure 2. Model of the eyelet exit point of the *CDPR* laser scanner in [16].

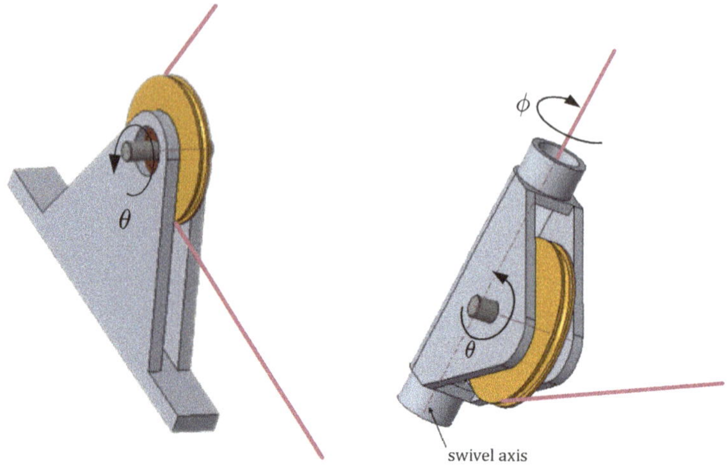

Figure 3. 3D model of pulley transmissions. (**a**) Fixed pulley. (**b**) Swivel pulley.

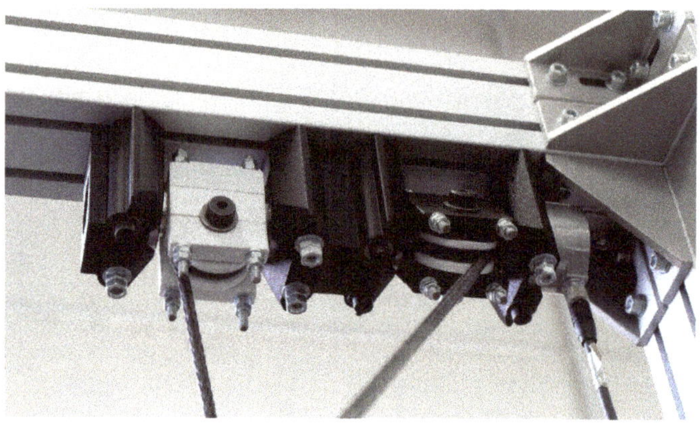

Figure 4. Application of swivel pulleys for cable proximal anchor point.

On the platform, at the distal anchor points, cables can be knotted to a fixture (i.e., an eyelet or a hook) or anchored to a universal joint: both these options allow to define a fixed point with respect to the platform. The knot solution is preferred for its simplicity (Figure 5) [16,44], but it is neither accurate nor reliable for long-term use. On the other hand, the universal joint (Figure 6) is more precise but complex and more expensive. Pulleys can also be employed if the cable needs to be deflected and not fixed at the distal anchor point. In [45], a pulley system is used at the distal anchor point on the platform to compensate for the effect of the proximal swiveling pulley geometrical model.

(a) (b)

Figure 5. Cable knotted at the anchor point on the platform of the *CDPR* laser scanner [16]. (**a**) Platform overview. (**b**) Knot detail.

(a) (b)

Figure 6. Prototypal universal joint at the anchor point on the platform of a *CDPR*. (**a**) Platform overview. (**b**) Universal joint detail.

3. Mechanical Design

The prototype presented in this section is devoted to non-contact tasks, such as laser engraving, welding, thermal treatment, decoration, and inspection of vertical surfaces, such as building façades. Thus, the main design requirements are the following: (i) it should present a planar vertical workspace, which should be easily scaled upon the redefinition of the installation locations of the guidance systems, (ii) the tool mounted on the platform should be interchangeable.

The 3D model of the designed prototype is shown in Figure 7. The machine frame (1) is formed by aluminum profiles that define the robot installation (and *WS*) limits. An actuation (2) and a proximal guidance unit (3) are placed at each of the four corners of the frame to coil and uncoil four cables and convey them to the distal guidance unit (4) attached to the mobile platform (5). The platform has $n = 3$ *DoFs* in the plane, thus, a 4-cable architecture results in an overconstrained manipulator. The constraint redundancy allows firstly to obtain a wider *WS* and then improve the *EE* controllability. Redundant actuated cables make them pull one against the other to keep the overall system under tension. Reconfigurable anchor points also characterize the prototype: cables can be arranged in a standard (Figure 8a) or crossed (Figure 8b) layout by simply un-mounting and re-mounting the distal guidance unit. As shown in Figure 8, each swivel pulley (1) on the platform, which forms the distal anchor point, can be translated into slots (2) on the support (3); the equivalent operation can be done on the proximal anchor systems on the frame. The crossed layout aims to improve the platform's orientation capability [46], if required by the task, and to withstand external torques induced by gravity (if the tool is not mounted in a barycentric position) or external disturbances.

The actuation units and guidance systems are the same for each cable, and they are described in the following subsections.

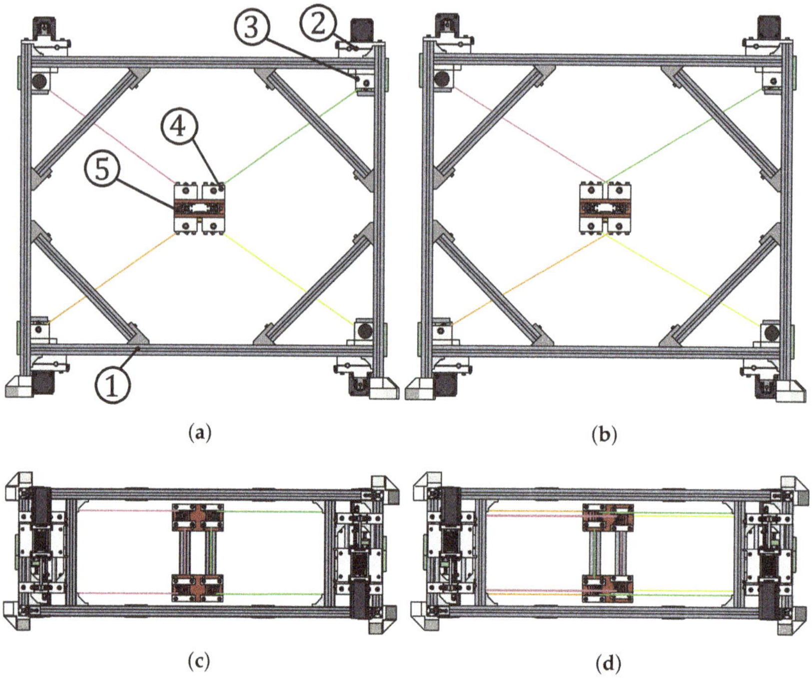

Figure 7. Planar 4-cable *CDPR* prototype with two cable arrangements: (1) frame, (2) actuation unit, (3) proximal guidance unit, (4) distal guidance unit, (5) platform. (**a**) Standard layout, front view. (**b**) Crossed layout, front view. (**c**) Standard layout, top view (**d**) Crossed layout, top view.

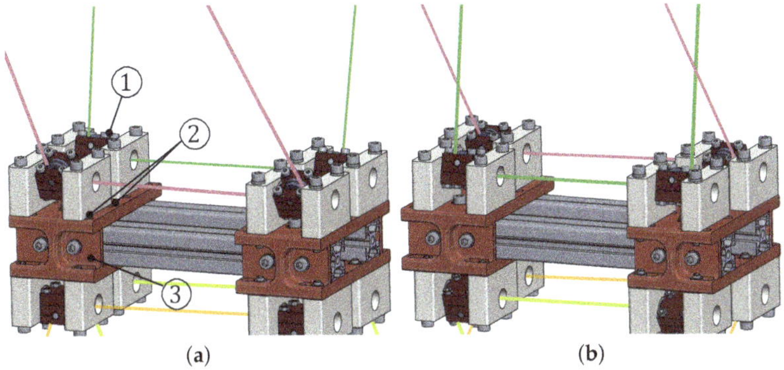

Figure 8. Reconfigurable distal anchor points on the mobile platform: (1) swivel pulley, (2) slots for the translation of the swivel pulley, (3) platform support. (**a**) Standard layout. (**b**) Crossed layout.

3.1. Actuation Unit

The cable actuation unit is a servo-actuated winch where both the motor and the drum translate along a linear guide during the motor rotation (Figure 9). As described in Section 2.1, this design choice limits the overall machine cost, at the price of poorer actuator dynamic performances: given the quasi-static requirement for the robot operation, this design choice was deemed the best. In addition, to further limit the robot cost, most components were produced by Fused Deposition Modeling (FDM) technology using PETG as base material. The choice of a rotary actuation, instead of a linear one, is supported by the project requirements reported in Section 3. In particular a rotary actuator allows for easy scalability in case of wide-workspace applications and field deployability.

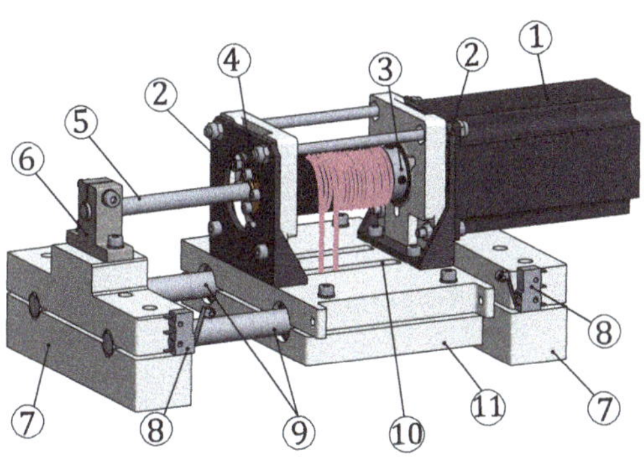

Figure 9. 3D model of the winch: (1) servo-stepper motor, (2) brackets, (3) aluminum drum, (4) brass nut, (5) screw shaft, (6) support element, (7) base plates, (8) end-stop switches, (9) circular rods, (10) cable guidance slot, (11) carriage.

A double-helicoidal-groove aluminum drum (3) is rigidly connected to the shaft of a Wantai (57HBM20-1000, 2.1 Nm holding torque) Nema 23 servo-stepper motor (1) equipped with an incremental encoder ($\approx 0.1^\circ$ resolution). Drum dimensions were chosen in order to coil 2.2 m of a 0.51 mm diameter dyneema cable (DAIWA j-braid X8, 550 N breaking tension), and limit (for the given motor) the realizable maximum tension to 160 N (cable tension safety factor of $\approx$3.5). Thus, drum diameter was selected as 27 mm, and groove pitch as 2 mm. A brass nut (4) is rigidly centered to the drum and coupled to a trapezoidal screw shaft (5), which is clamped at one end to the actuation unit frame, in a support element (6). While the motor rotates, the drum-motor system, fixed to a carriage (11) using two brackets (2), moves on a linear guide (9) made by two circular rods and linear ball bearings. The drum and the shaft have the same pitch, so the cable portions exiting the drum always have a fixed direction and position with respect to the winch frame. Two end-stop switches (8) are mounted at the physical limits of the drum-motor system, to provide a safety-stop signal to the motor controller. The winch frame is ultimately fixed to the overall robot frame with bolted connections via two plates (7).

This winch houses one of the four cables. One end of the cable is fixed to the drum employing a screw. Then, the cable is coiled onto the first start of the drum and conveyed to the guidance system through a thin slot (10) on the carriage (11). After being guided to the distal anchor point on the platform (as detailed in Section 3.2), the cable returns to the drum through the slot (10) and is then wound onto the other start of the drum and secured with another screw.

3.2. Guidance System

The cable guidance system is designed to (i) guide each of the four cables on a parallelogram-like routing (Figure 10) to ensure the motion of the *EE* in a defined plane; (ii) geometrically define the cable exit points, to apply a standard kinematic model for robot control, (iii) integrate a load sensor for cable tension measurement.

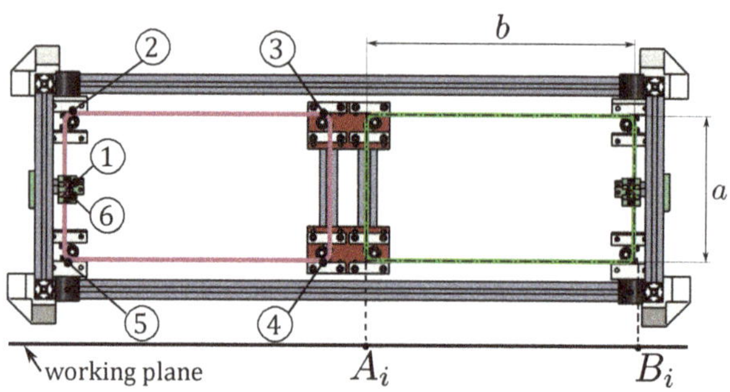

Figure 10. Layout of the guidance system of the robot in standard cable arrangement (top view): (1), (6) fixed pulleys, and (2), (3), (4), (5) swivel pulleys; *a* and *b* sides of the parallelogram defined by the cable loop.

The cable, which leaves the drum groove from a fixed point in space, in a fixed direction (see Section 3.1), is then guided on a fixed pulley (1) attached to a tension measurement unit, which will be analyzed later on. After that, to obtain a parallelogram shape, the cable passes through four identical swivel pulleys (2,3,4,5) and then, through another fixed pulley (6) (also included in the tension measurement unit) before engaging the other groove on the drum. The proposed cable-loop arrangement realized with one cable avoids possible cable slackness during *EE* motion, which may happen if the parallelogram is made of two separate cables driven by the same winch. Two of the swivel pulleys (2,5) are fixed to the base and form the proximal guidance unit, while the other two (3,4) are attached to the platform and form the distal guidance unit. The distance *a* between the two pulleys composing each group is equal. This defines the first two sides of the parallelogram, which are invariant, being distances between points of rigid bodies. The other two sides *b* are formed by cable sections that are equal and parallel when the cable is taut by design.

3.2.1. Sensor Integration

The *CDPR* guidance system is often equipped with position [47] and force sensors [48] to monitor the robot performance in real-time, if required by the control strategy. In this prototype, one of the swivel pulleys of the proximal group is equipped with an incremental encoder. The encoder is coaxial with the swivel pulley, thus it measures the angle between the cable and the frame: this feature can be used for platform pose estimation through direct kinematics [49]. In addition, the fixed pulleys of the proximal guidance group are equipped with a load sensor, a shear beam load cell (Phidget CZL635 micro load-cell, load range 20 kg). Figure 11 shows the force measurement unit: the load cell (1) is directly connected to the pulley support (2) and structurally constraints the fixed pulleys (3) to the machine frame (4) via an auxiliary bracket (5). The system is designed for the load direction of the pulley to be always orthogonal to the shear beam load cell, to ensure the optimal working conditions of the sensor. The load cell, ideally, measures the double of the cable tension value.

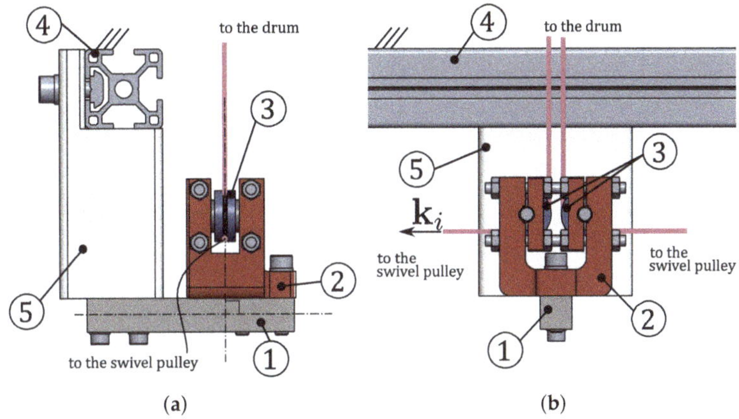

Figure 11. Force measurement unit: (1) load cell, (2) pulley support, (3) fixed pulleys, (4) machine frame, (5) auxiliary bracket. (**a**) Front view. (**b**) Side view.

3.2.2. Kinematic Model

To define the cable exit points, each of the four swivel pulleys forming the guidance system is oriented so that its swivel axis, identified by the unit vector $\mathbf{k}_i$ (Figures 11b and 12b) is orthogonal to the work plane, that is a vertical plane. The two pulleys, forming the proximal or distal guidance units, are coaxial, and the projection of their swivel axes onto the vertical plane identifies the cable exit points (B_i, for $i = 1, \ldots, 4$ for the proximal point in Figure 12a,b). Thus, the kinematic *standard* model can be used without causing geometrical modeling errors. Angle ϕ defines the rotation of the pulley about the swivel axis. $\mathbf{t}_i$ is the unit vector of the i-th cable.

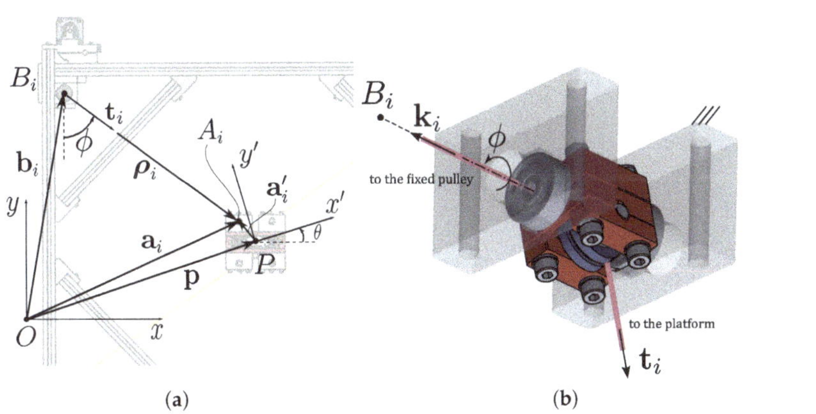

Figure 12. Standard model definition. (**a**) Geometry of the i-th constraint of the planar 4-cable *CDPR*. (**b**) Detail of a proximal swivel pulley.

Considering one of the four cable transmissions (Figure 12), Oxy and $Px'y'$ are defined as inertial and mobile frames. The mobile frame is attached to the platform center of mass P, described as $\mathbf{p}$ with respect to the inertial frame. Cables, assumed massless and inextensible, are considered as the line segments between the distal and proximal anchor points, defined by the pulley swivel axes (A_i and B_i respectively, for $i = 1, \ldots, 4$). Points B_i are described by vectors $\mathbf{b}_i$ in the inertial frame, and points A_i are described by $\mathbf{a}'_i$ and $\mathbf{a}'$ in the mobile and inertial frame, respectively. The rotation matrix $\mathbf{R}(\theta)$ represents the orientation of the moving platform in the vertical plane. The platform pose is defined as $[\mathbf{p}^T, \theta]^T$, and i-th cable vector is:

$$\boldsymbol{\rho}_i = \mathbf{a}_i - \mathbf{b}_i = \mathbf{p} + \mathbf{R}(\theta)\mathbf{a}'_i - \mathbf{b}_i, \quad \mathbf{R}(\theta) \triangleq \begin{bmatrix} cos(\theta) & -sin(\theta) \\ sin(\theta) & cos(\theta) \end{bmatrix} \tag{1}$$

For a given EE pose, the i-th cable length l_i and unit vector $\mathbf{t}_i$ can be computed as:

$$l_i = \|\boldsymbol{\rho}_i\|, \quad \mathbf{t}_i = \frac{\boldsymbol{\rho}_i}{l_i} \tag{2}$$

4. Experimental Demonstration: Laser Engraving

A prototype of a 3-DoF 4-cable *CDPR* with a *crossed layout* was built at IRMA L@B of the University of Bologna (Figure 13a). It employs 4 of the actuation systems described in Section 3.1, connected to the corners of an aluminum frame, and 4 guidance units, whose distal swivel pulleys are bolted to the robot platform. The *CDPR* has rectangular base (0.875 m $\times$ 0.700 m) and mobile platform (0.080 m $\times$ 0.100 m). The inertial frame Oxy is located in the center of the base and the moving frame $Px'y'$ is located at the center of the *EE*, coinciding with its center of mass. Its wrench-feasible translational workspace [50] is represented in Figure 13b with a regular discrete grid of 100 $\times$ 100 points, considering the platform mass equal to $m = 2.5$ kg, and $\tau_{min} = 10$ N and $\tau_{max} = 80$ N as cable tension limits.

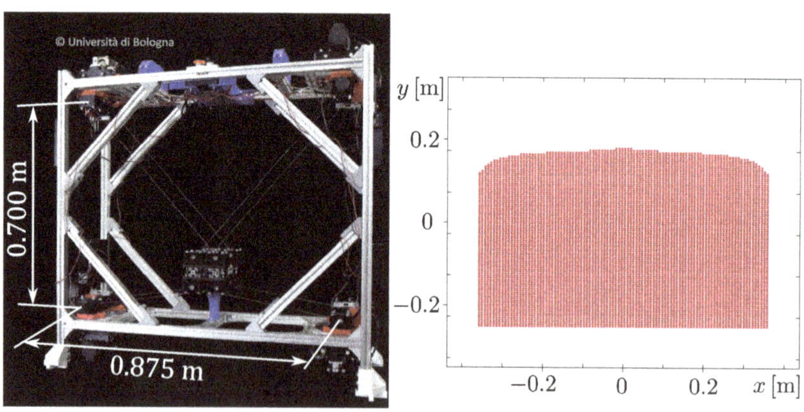

(**a**) (**b**)

Figure 13. Laser-engraving prototype at IRMA L@B. (**a**) Picture of the prototype. (**b**) Wrench-feasible translational workspace of the prototype.

The *CDPR* is controlled by a Real-Time Linux PC, which runs its control algorithm at 1 kHz rate, and feeds motor commands through Ethercat protocol [51] to low-level servo drives. The low-level controller is also developed in-house and run on a ST Nucleo-H743ZI development board at 10 kHz: in addition to controlling motor angles, this allows to directly feed a cable-tension command to the drive. This latter feature simplifies the adoption of the hybrid-input control strategy for the *CDPR* described in [50]: once an

EE-pose set-point is assigned in the Cartesian space, 3 cables are length controlled (i.e., motor angles are commanded to the drives), and one cable is tension controlled (i.e., cable tension is commanded to the drive, and the error with respect to the readings of the load cell is compensated).

A 2.5 W laser diode is mounted on the robot *EE* so that its focal axis is perpendicular to the working plane (Figure 14), thus allowing the engraving of a paper sheet (for additional details, see Supplementary Video Material). The name of our laboratory, IRMA L@B, was stylized in a way that every letter was a union of a finite number of linear segments (Figure 15a): accordingly, a quantitative comparison between the digital model and the engraved result could be performed, by comparing each segment length. A series of consecutive points $P_1, P_2, \ldots, P_n$, with $n = 15$, were extracted from the digital model (Figure 15b), and the robot reference point commanded to follow the path between them with a trapezoidal velocity profile in the task space, while keeping the orientation identically zero. If the trajectory is expressed as:

$$\mathbf{p}(t) = \mathbf{p}_i + (\mathbf{p}_{i+1} - \mathbf{p}_i)u(t), \qquad u(0) = 0,\, u(T) = 1, \qquad i = 1, \ldots, n-1 \tag{3}$$

the motion law $u(t)$ [52] takes the form:

$$u(t) = \begin{cases} \frac{1-\alpha}{2\alpha}(vt)^2, & \text{for } 0 \le t < \alpha T \\ -\frac{\alpha}{2(1-\alpha)} + vt, & \text{for } \alpha T \le t < (1-\alpha)T, \\ -\frac{2\alpha^2-2\alpha+1}{2\alpha(1-\alpha)} + \frac{1}{\alpha}vt - \frac{1-\alpha}{2\alpha}(vt)^2, & \text{for } (1-\alpha)T \le t \le T \end{cases} \qquad T = \frac{1}{v(1-\alpha)} \tag{4}$$

Since the set-points P_i of Equation (3) are set by the task, the only parameters to be determined are v and α of Equation (4). They represent, respectively, the motion law maximum speed, and the ratio between the accelerating phase duration and the total time; they are set considering the requirements of the technological operation to be performed. Since the diode laser can only be on or off, the accelerating phase duration should be as limited as possible, to avoid burning the paper when the *EE* is moving too slow; on the other hand, motors have limited power, and too quick accelerations may result in their stall: by trial and error, $\alpha = 0.1$ was established as a good compromise. The motion law maximum speed was instead calculated so as the *EE* maximum translational speed was optimal for the engraving operation, namely $||\dot{\mathbf{p}}||_{max} = 0.02$ m/s for the 2.5 W laser diode at hand; thus, by differentiating Equations (3) and (4) we get:

$$||\dot{\mathbf{p}}||_{max} = 0.02 = ||\mathbf{p}_{i+1} - \mathbf{p}_i||v \tag{5}$$

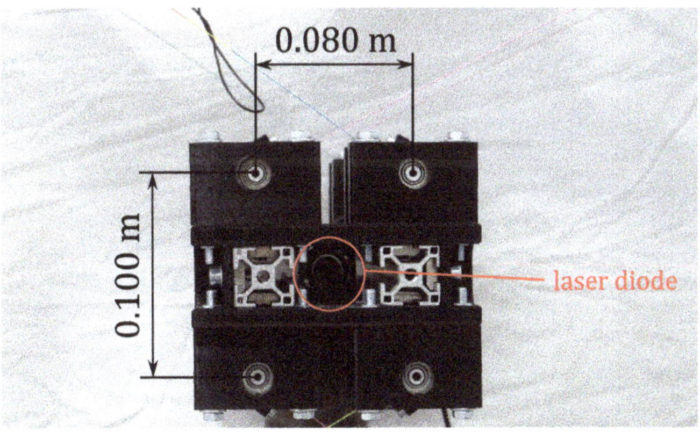

Figure 14. Robot end-effector equipped with a laser diode.

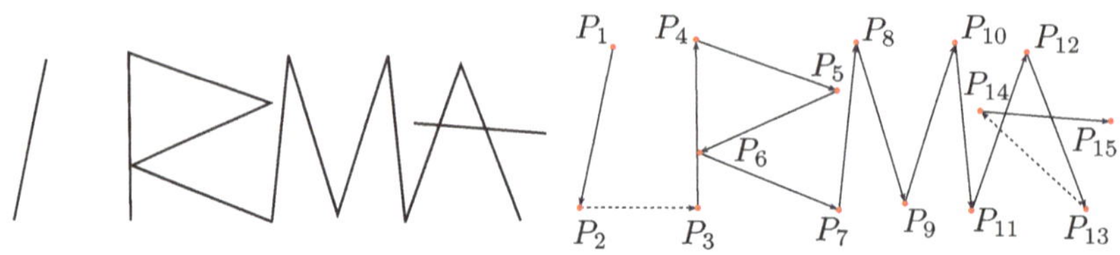

(**a**) (**b**)

Figure 15. Digital image for engraving task. (**a**) Stylized version of IRMA. (**b**) End-effector path: solid line for engraving process, dotted line for movement; P_i, with $i = 1, \ldots, 15$, consecutive points extracted from the digital image.

The engraving operation of the stylized *IRMA L@B* was then performed, and the result is shown in Figure 16. First of all, it can be noticed that rectilinearity of the segments is qualitatively good (see Figure 17a). Small out-of-plane vibrations were excited by the laser diode cooling fan when the laser was switched on and off at the beginning and the end of each segment. This is nearly unnoticeable even scaling up the engraved paper (see Figure 17b), and it was ultimately expected: a vibration-damping device, such as the one proposed in [27], can be added to prevent such small-magnitude effects.

Figure 16. Result of the laser engraving operation.

(**a**) (**b**)

Figure 17. Qualitative evaluation of the laser engraving operation. (**a**) Rectilinearity of a segment. (**b**) Platform out-of-plane oscillations induce sub-millimetric errors.

From a quantitative point of view, the ideal length $l_i^\star$ of each segment was compared to the engraved one l_i, measured on the paper with a digital caliper (maximum measuring error of 0.3 mm at full scale, 180 mm). If the error is defined as $\epsilon_{l_i} = l_i^\star - l_i$ and the percentage error is $\epsilon_{l_i,\%} = 100(l_i^\star - l_i)/l_i^\star$, the root-mean-square error was 2 mm, and the root-mean-square percentage error was 2.4% (complete results can be found in Table 1). Considering the size of the robot workspace and its prototypal nature (most of the components are 3D printed in plastic material), errors of this magnitude are reasonably attributed to an imperfect calibration, mounting errors of the guidance system on the aluminum frame, and elastic components deflection. All in all, the machine can perform the planar non-contact task with relatively good accuracy, which can be increased with a more advanced mechanical design, followed by a precise calibration.

Table 1. Comparison between ideal and engraved segment lengths.

i	1	2	3	4	5	6	7	8	9	10	11	12	13	14
$l_i^\star$ [mm]	90.8	66.4	92.7	84.2	85.1	84.6	92.7	92.7	92.7	92.7	92.7	92.7	80.9	75.2
l_i [mm]	89.4	64.4	91.2	87.4	85.5	88.3	91.2	94.6	91.3	91.6	90.5	92.6	78.3	76.9
ϵ_{l_i} [mm]	−1.4	−2.0	−1.5	3.2	0.4	3.7	−1.5	1.9	−1.4	−1.1	−2.2	−0.1	−2.6	1.7
$\epsilon_{l_i,\%}$ [%]	−1.6	−3.1	−1.6	3.7	0.4	4.4	−1.6	2.1	−1.5	−1.2	−2.3	−0.1	−3.3	2.3

5. Conclusions

This paper presented an overconstrained *CDPR* for non-contact tasks on planar vertical surfaces. The actuation and guidance systems were described and the applicability of the standard model was outlined by showing the employed design expedients. The *CDPR* prototype was built at IRMA laboratory at the University of Bologna, and used to perform laser engraving on a paper sheet. Engraving results are qualitatively good, considering the prototypal nature of the robot, and quantitatively acceptable, since they can be improved by better manufacture of mechanical components and calibration.

In the future, the proposed actuation and guidance units design will be evolved in order to achieve on-site deployability: the actuation (or guidance) unit will be mounted on a vertical façade, and the system geometry will be self-calibrated on-site after each mounting operation.

6. Patents

The preliminary design of the prototype presented in this paper is object of the patent WO2021144685.

Supplementary Materials: The following are available at https://www.mdpi.com/2076-3417/11/20/9491/s1, Video S1: This video shows the engraving process described in the experimental section of the paper.

Author Contributions: Conceptualization, V.M. and E.I.; methodology, V.M. and E.I.; software, V.M. and E.I.; validation, V.M. and E.I.; writing—original draft preparation, V.M.; writing—review and editing, V.M. and E.I.; supervision, M.C.; project administration, E.I. and M.C. All authors have read and agreed to the published version of the manuscript.

Funding: This research received no external funding.

Institutional Review Board Statement: Not applicable.

Informed Consent Statement: Not applicable.

Data Availability Statement: Data is contained within the article.

Acknowledgments: The authors would like to acknowledge Federico Zaccaria for his help in designing the prototype components.

Conflicts of Interest: The authors declare no conflict of interest.

References

1. Pott, A. *Cable-Driven Parallel Robots: Theory and Application*; Springer: Berlin/Heidelberg, Germany, 2018; Volume 120.
2. Gagliardini, L.; Caro, S.; Gouttefarde, M.; Wenger, P.; Girin, A. A Reconfigurable Cable-Driven Parallel Robot for Sandblasting and Painting of Large Structures. In *Cable-Driven Parallel Robots*; Pott, A., Bruckmann, T., Eds.; Springer International Publishing: Cham, Switzerland, 2015; pp. 275–291.

3. Piao, J.; Jin, X.; Choi, E.; Park, J.O.; Kim, C.S.; Jung, J. A Polymer Cable Creep Modeling for a Cable-Driven Parallel Robot in a Heavy Payload Application. In *Cable-Driven Parallel Robots*; Gosselin, C., Cardou, P., Bruckmann, T., Pott, A., Eds.; Springer International Publishing: Cham, Switzerland, 2018; pp. 62–72.
4. Idà, E.; Bruckmann, T.; Carricato, M. Rest-to-Rest Trajectory Planning for Underactuated Cable-Driven Parallel Robots. *IEEE Trans. Robot.* **2019**, *35*, 1338–1351. [CrossRef]
5. Idà, E.; Briot, S.; Carricato, M. Robust Trajectory Planning of Under-Actuated Cable-Driven Parallel Robot with 3 Cables. In *Advances in Robot Kinematics 2020*; Lenarčič, J., Siciliano, B., Eds.; Springer International Publishing: Cham, Switzerland, 2021; pp. 65–72.
6. Brown, G.W. Suspension System for Supporting and Conveying Equipment, such as a Camera. U.S. Patent 4,710,819, 1 December 1987.
7. Abdolshah, S.; Zanotto, D.; Rosati, G.; Agrawal, S. Performance evaluation of a new design of cable-suspended camera system. In Proceedings of the 2017 IEEE International Conference on Robotics and Automation (ICRA), Singapore, 29 May–3 June 2017; pp. 3728–3733.
8. Bruckmann, T.; Lalo, W.; Nguyen, K.; Salah, B. Development of a Storage Retrieval Machine for High Racks Using a Wire Robot. In Proceedings of the International Design Engineering Technical Conferences and Computers and Information in Engineering Conference, Chicago, IL, USA, 12–15 August 2012; Volume 4, pp. 771–780.
9. Pott, A.; Meyer, C.; Verl, A. Large-scale assembly of solar power plants with parallel cable robots. In Proceedings of the ISR 2010 (41st International Symposium on Robotics) and ROBOTIK 2010 (6th German Conference on Robotics), Munich, Germany, 7–9 June 2010; pp. 1–6.
10. Izard, J.B.; Dubor, A.; Hervé, P.E.; Cabay, E.; Culla, D.; Rodriguez, M.; Barrado, M. Large-scale 3D printing with cable-driven parallel robots. *Construct. Robot.* **2017**, *1*, 69–76. [CrossRef]
11. Bruckmann, T.; Boumann, R. Simulation and optimization of automated masonry construction using cable robots. *Adv. Eng. Inf.* **2021**, *50*, 101388. [CrossRef]
12. Izard, J.B.; Gouttefarde, M.; Baradat, C.; Culla, D.; Sallé, D. Integration of a Parallel Cable-Driven Robot on an Existing Building Façade. In *Cable-Driven Parallel Robots*; Bruckmann, T., Pott, A., Eds.; Springer: Berlin/Heidelberg, Germany, 2013; pp. 149–164.
13. De Zeeuw, D.; Spanjer, S.A.J.; Zeggelaar, K. Cleaning System for a Façade of a Building Structure. International Patent WO 2016/085330, 2 June 2016.
14. Pott, A.; Mütherich, H.; Kraus, W.; Schmidt, V.; Miermeister, P.; Verl, A. IPAnema: A family of Cable-Driven Parallel Robots for Industrial Applications. In *Cable-Driven Parallel Robots*; Bruckmann, T., Pott, A., Eds.; Springer: Berlin/Heidelberg, Germany, 2013; pp. 119–134.
15. Gouttefarde, M.; Collard, J.F.; Riehl, N.; Baradat, C. Geometry Selection of a Redundantly Actuated Cable-Suspended Parallel Robot. *IEEE Trans. Robot.* **2015**, *31*, 501–510. [CrossRef]
16. Idà, E.; Marian, D.; Carricato, M. A Deployable Cable-Driven Parallel Robot with Large Rotational Capabilities for Laser-Scanning Applications. *IEEE Robot. Automat. Lett.* **2020**, *5*, 4140–4147. [CrossRef]
17. Yang, H.; Izard, J.B.; Baradat, C.; Krut, S.; Pierrot, F.; Gouttefarde-tanich, M.; Company, O. Planar Motion Device. U.S. Patent 10,449,668, 22 October 2019.
18. Corral, E.; Moreno, R.G.; García, M.J.G.; Castejón, C. Nonlinear phenomena of contact in multibody systems dynamics: A review. *Nonlinear Dyn.* **2021**, *104*, 1269–1295. [CrossRef]
19. Reichert, C.; Bruckmann, T. Unified Contact Force Control Approach for Cable-driven Parallel Robots using an Impedance/Admittance Control Strategy. In Proceedings of the 14th IFToMM World Congress, Taipei, Taiwan, 25–30 October 2015; pp. 645–654.
20. Vu, D.S.; Barnett, E.; Zaccarin, A.M.; Gosselin, C. On the Design of a Three-DOF Cable-Suspended Parallel Robot Based on a Parallelogram Arrangement of the Cables. In *Cable-Driven Parallel Robots*; Gosselin, C., Cardou, P., Bruckmann, T., Pott, A., Eds.; Springer International Publishing: Cham, Switzerland, 2018; pp. 319–330.
21. Mottola, G.; Gosselin, C.; Carricato, M. Effect of Actuation Errors on a Purely-Translational Spatial Cable-Driven Parallel Robot. In Proceedings of the 2019 IEEE 9th Annual International Conference on CYBER Technology in Automation, Control, and Intelligent Systems (CYBER), Suzhou, China, 29 July–2 August 2019; pp. 701–707.
22. Khodadadi, N.; Hosseini, M.I.; Khalilpour, S.A.; Taghirad, H.D.; Cardou, P. Multi Objective Optimization of a Cable-Driven Robot with Parallelogram Links. In *Cable-Driven Parallel Robots*; Gouttefarde, M., Bruckmann, T., Pott, A., Eds.; Springer International Publishing: Cham, Switzerland, 2021; pp. 170–181.
23. Torres-Mendez, S.; Khajepour, A. Analysis of a High Stiffness Warehousing Cable-Based Robot. In Proceedings of the International Design Engineering Technical Conferences and Computers and Information in Engineering Conference, Buffalo, NY, USA, 17–20 August 2014; Volume 5A.
24. Bosscher, P.; Williams, R.L.; Tummino, M. A Concept for Rapidly-Deployable Cable Robot Search and Rescue Systems. In Proceedings of the ASME 2005 International Design Engineering Technical Conferences and Computers and Information in Engineering Conference, Long Beach, CA, USA, 24–28 September 2005; pp. 589–598.
25. Bosscher, P.; Williams, R.L.; Bryson, L.S.; Castro-Lacouture, D. Cable-Suspended Robotic Contour Crafting System. In Proceedings of the International Design Engineering Technical Conferences and Computers and Information in Engineering Conference, Philadelphia, PA, USA, 10–13 September 2006; pp. 777–785.

26. Wahle, S. Storage System, in Particular Rack Storage. Deutsches DE Patent 10 2010 015 530, 6 September 2021.
27. De Rijk, R.; Rushton, M.; Khajepour, A. Out-of-Plane Vibration Control of a Planar Cable-Driven Parallel Robot. *IEEE/ASME Trans. Mechatron.* **2018**, *23*, 1684–1692. [CrossRef]
28. Idà, E.; Briot, S.; Carricato, M. Natural Oscillations of Underactuated Cable-Driven Parallel Robots. *IEEE Access* **2021**, *9*, 71660–71672. [CrossRef]
29. Wang, D.; Ahn, J.; Jung, J.; Seon, J.A.; Park, J.O.; Ko, S.Y.; Park, S. Winch-integrated mobile end-effector for a cable-driven parallel robot with auto-installation. *Int. J. Control Automat. Syst.* **2017**, *15*, 2355–2363. [CrossRef]
30. Bruckmann, T.; Mikelsons, L.; Brandt, T.; Hiller, M.; Schramm, D. Wire Robots Part I: Kinematics, Analysis & Design. In *Parallel Manipulators*; Ryu, J.H., Ed.; IntechOpen: Rijeka, Croatia, 2008; Chapter 6.
31. Reichert, C.; Bruckmann, T. Optimization of the Geometry of a Cable-Driven Storage and Retrieval System. In *Robotics and Mechatronics*; Yang, R., Takeda, Y., Zhang, C., Fang, G., Eds.; Springer International Publishing: Cham, Switzerland, 2019; pp. 225–237.
32. Pott, A. Influence of Pulley Kinematics on Cable-Driven Parallel Robots. In *Latest Advances in Robot Kinematics*; Lenarcic, J., Husty, M., Eds.; Springer: Dordrecht, The Netherlands, 2012; pp. 197–204.
33. Merlet, J.P. Comparison of Actuation Schemes for Wire-Driven Parallel Robots. In *New Trends in Mechanism and Machine Science*; Viadero, F., Ceccarelli, M., Eds.; Springer: Dordrecht, The Netherlands, 2013; pp. 245–254.
34. Merlet, J.P. MARIONET, A Family of Modular Wire-Driven Parallel Robots. In *Advances in Robot Kinematics: Motion in Man and Machine*; Lenarcic, J., Stanisic, M.M., Eds.; Springer: Dordrecht, The Netherlands, 2010; pp. 53–61.
35. Surdilovic, D.; Bernhardt, R. STRING-MAN: A new wire robot for gait rehabilitation. In Proceedings of the IEEE International Conference on Robotics and Automation, ICRA'04, New Orleans, LA, USA, 26 April–1 May 2004; Volume 2, pp. 2031–2036.
36. Feyrer, K. *Wire Ropes: Tension, Endurance, Reliability*; Springer Science & Business Media: Berlin/Heidelberg, Germany 2015.
37. Pham, C.B.; Yang, G.; Yeo, S.H. Dynamic analysis of cable-driven parallel mechanisms. In Proceedings of the 2005 IEEE/ASME International Conference on Advanced Intelligent Mechatronics, Monterey, CA, USA, 24–28 July 2005; pp. 612–617.
38. Izard, J.B.; Gouttefarde, M.; Michelin, M.; Tempier, O.; Baradat, C. A Reconfigurable Robot for Cable-Driven Parallel Robotic Research and Industrial Scenario Proofing. In *Cable-Driven Parallel Robots*; Bruckmann, T., Pott, A., Eds.; Springer: Berlin/Heidelberg, Germany, 2013; pp. 135–148.
39. Rognant, M.; Courteille, E. Improvement of Cable Tension Observability Through a New Cable Driving Unit Design. In *Cable-Driven Parallel Robots*; Gosselin, C., Cardou, P., Bruckmann, T., Pott, A., Eds.; Springer International Publishing: Cham, Switzerland, 2018; pp. 280–291.
40. von Zitzewitz, J.; Fehlberg, L.; Bruckmann, T.; Vallery, H. Use of Passively Guided Deflection Units and Energy-Storing Elements to Increase the Application Range of Wire Robots. In *Cable-Driven Parallel Robots*; Bruckmann, T., Pott, A., Eds.; Springer: Berlin/Heidelberg, Germany, 2013; pp. 167–184.
41. Fang, S.; Franitza, D.; Torlo, M.; Bekes, F.; Hiller, M. Motion control of a tendon-based parallel manipulator using optimal tension distribution. *IEEE/ASME Trans. Mechatron.* **2004**, *9*, 561–568. [CrossRef]
42. Ottaviano, E.; Ceccarelli, M.; De Ciantis, M. A 4–4 cable-based parallel manipulator for an application in hospital environment. In Proceedings of the 2007 Mediterranean Conference on Control Automation, Athens, Greece, 27–29 June 2007; pp. 1–6.
43. Kraus, W.; Kessler, M.; Pott, A. Pulley friction compensation for winch-integrated cable force measurement and verification on a cable-driven parallel robot. In Proceedings of the 2015 IEEE International Conference on Robotics and Automation (ICRA), Seattle, WA, USA, 26–30 May 2015; pp. 1627–1632.
44. Hiller, M.; Fang, S.; Mielczarek, S.; Verhoeven, R.; Franitza, D. Design, analysis and realization of tendon-based parallel manipulators. *Mech. Mach. Theory* **2005**, *40*, 429–445. [CrossRef]
45. Gonzalez-Rodriguez, A.; Castillo-Garcia, F.J.; Ottaviano, E.; Rea, P.; Gonzalez-Rodriguez, A.G. On the effects of the design of cable-Driven robots on kinematics and dynamics models accuracy. *Mechatronics* **2017**, *43*, 18–27. [CrossRef]
46. Gouttefarde, M.; Krut, S.; Company, O.; Pierrot, F.; Ramdani, N. On the Design of Fully Constrained Parallel Cable-Driven Robots. In *Advances in Robot Kinematics: Analysis and Design*; Lenarčič, J., Wenger, P., Eds.; Springer: Dordrecht, The Netherlands, 2008; pp. 71–78.
47. Idá, E.; Merlet, J.P.; Carricato, M. Automatic Self-Calibration of Suspended Under-Actuated Cable-Driven Parallel Robot using Incremental Measurements. In *Cable-Driven Parallel Robots*; Pott, A., Bruckmann, T., Eds.; Springer International Publishing: Cham, Switzerland, 2019; pp. 333–344.
48. Yang Ho, W.; Kraus, W.; Mangold, A.; Pott, A. Haptic Interaction with a Cable-Driven Parallel Robot Using Admittance Control. In *Cable-Driven Parallel Robots*; Pott, A., Bruckmann, T., Eds.; Springer International Publishing: Cham, Switzerland, 2015; pp. 201–212.
49. Merlet, J.P. Direct Kinematics of CDPR with Extra Cable Orientation Sensors: The 2 and 3 Cables Case with Perfect Measurement and Ideal or Elastic Cables. In *Cable-Driven Parallel Robots*; Gosselin, C., Cardou, P., Bruckmann, T., Pott, A., Eds.; Springer International Publishing: Cham, Switzerland, 2018; pp. 180–191.
50. Mattioni, V.; Idà, E.; Carricato, M. Force-Distribution Sensitivity to Cable-Tension Errors: A Preliminary Investigation. In *Cable-Driven Parallel Robots*; Gouttefarde, M., Bruckmann, T., Pott, A., Eds.; Springer International Publishing: Cham, Switzerland, 2021; pp. 129–141.

51. Boschetti, G.; Minto, R.; Trevisani, A. Experimental Investigation of a Cable Robot Recovery Strategy. *Robotics* **2021**, *10*, 35. [CrossRef]
52. Biagiotti, L.; Melchiorri, C. *Trajectory Planning for Automatic Machines and Robots*; Springer Science & Business Media: Berlin/Heidelberg, Germany, 2008.

applied sciences

Article

Design and Implementation of Inverse Kinematics and Motion Monitoring System for 6DoF Platform

Ming-Yen Wei

Aeronautical System Research Division Simulation System Section, National Chung-Shan Institute of Science and Technology, Taichung 407, Taiwan; b515410@ncsist.org.tw; Tel.: +886-4-2702-3051

Abstract: Six-axis motion platforms have a low contraction height and a high degree of freedom. First of all, the designed six-axis crank arm platform, including the motor, reducer, crank arm, link, platform support arm, and upper and lower platforms, can be designed for different bearing requirements. Secondly, it uses a coordinate transform and kinematics theory to derive each motor rotor angle. A set of platform data acquisition (DAQ) monitoring modules was established, and the LabVIEW programming language was used to write measurement software. The monitoring items include displacement, speed, and acceleration, which can be displayed on the screen and recorded by an industrial computer in real time and dynamically. Then, an RS-485 or RS-232 communication transmission interface was used to provide the control system with the related movement information. Finally, an industrial computer combined with a motion control card was used as a control kernel to realize the control algorithms, internet module function, I/O write and read signals, firmware integration, and human–machine interface message. The experimental results validate the appropriateness of the proposed method.

Keywords: 6DoF motion platform; inverse kinematics; monitoring system; crank arm mechanisms

Citation: Wei, M.-Y. Design and Implementation of Inverse Kinematics and Motion Monitoring System for 6DoF Platform. *Appl. Sci.* **2021**, *11*, 9330. https://doi.org/10.3390/app11199330

Academic Editors: Alessandro Gasparetto, Stefano Seriani and Lorenzo Scalera

Received: 27 August 2021
Accepted: 4 October 2021
Published: 8 October 2021

1. Introduction

Simulators continue to progress toward high performance, high fidelity, and complex, realistic scenes [1]. A motion platform with six degrees of freedom (DoF) simulates the dynamic motion of several vehicles to be perceived by passengers. In addition to visual effects, sound effects, and force, adding motion to professional training simulators can make the simulations more realistic as well as improving the effectiveness of training and exempting trainees from having to readapt to the motion of real vehicles. Regarding entertainment systems, adding motion can increase excitement, enhance interaction, and increase the value of the systems. Motion platforms with six DoF are commonly driven by one of the following devices: (1) A linear actuator, which is composed of motors, belts, and lead screws. This structure includes six electric cylinders, three pneumatic cylinders, upper and lower platforms, and a platform controller. (2) A 6DoF crank arm platform, which is created by combining a crank arm mechanism—composed of a servo motor, a gear box, and a crank arm—with a reducer, a rocker arm, a platform support arm, upper and lower platforms, and a control system. Though they possess the advantages of simple control, high DoF, high efficiency, and high loading capacity (ton) [2], linear actuators have high production costs, and the height of the platform to be positioned in the center is overly high. With approximately half the development cost of linear actuators, 3DoF and 4DoF crank arm platforms have comparable high-fidelity motion, which reduces manufacturing costs, and can be applied in professional training simulators (e.g., of ships and vehicles) and entertainment simulators, which may serve as a point device at amusement parks [3,4].

In 1965, Stewart from the United Kingdom proposed a 6DoF, six-axis motion platform structure [5], which was first used as a flight simulator. The potential advantages and application prospects of parallel platforms received considerable attention and inspired numerous studies. In 1980, Fichter and Mcdowell proposed a parallel mechanism derived from Stewart's platform and discussed the inverse kinematics of said mechanism [6], which represented the earliest discussion of inverse kinematics. In 1985, Mohamed and Duffy published a study on the relationship between the end effector speed of a six-axis motion platform and the drive input speed [7]. In 1992, Kumer clarified the relationship between the speed and acceleration of a six-axis platform and the drive. In 1993, Liu proposed a comprehensive analysis of forward kinematics and inverse kinematics of a six-axis motion platform [8]. Ji et al. analyzed the inverse kinematics of a 3-RRPS 6DoF Stewart platform [9]. By that time, most kinematics problems related to the general type of 6DoF Stewart platform had been resolved. Studies after 2010 have generally addressed means of controlling parallel architectures, such as artificial neural network control [10], and predictive control [11–13]. Studies after 2010 have also explored numerical solutions. For example, the spatial position and the center of the upper platform can be obtained by applying forward kinematics [14]. In addition, [15] proposed a real-time algorithm for solving the six motors' angular update command problem caused by multithreading. The algorithm is advantageous in that it increases the computing speed and reduces the running time, which enables the motors' angular update commands of the six-axis platform to provide more timely updates to the six servo drives. The current study applied the dynamic proprioception of real-life situations to achieve a motor sense that is identical or close to that created by real vehicles. Theoretically, the motion cueing effect most closely resembles reality when the motion state of the platform is exactly that of a real vehicle [16,17]. The simulation of motion especially targeted the speed and acceleration perceptible by the human sensory system to achieve a realistic motor sense.

The majority of simulators have been designed based on the 6DoF structure of the Stewart platform [5–17], which uses linear actuators to achieve a 6DoF overall movement. However, the need to use six linear actuators is expensive. Therefore, this work proposes a six-axis crank arm mechanism for motion systems. There is no formal literature about the design and implementation of an actual crank arm mechanism platform. As in [18], this work's design was based on DIY methods to create a simple model. Limiting the range of movement through the error correction method achieves an effect similar to an integrator. In order to improve the movement stroke of the platform, this study investigated the crank arm motion platform with a monitoring system, in order to propose an inverse kinematics design for virtual cylinders. In addition, a motion monitoring system is proposed for affordable measurement of the platform. For affordability, low-cost sensors are utilized. The software offers functions such as sampling settings, impulse response display, real-time display of measured values, and log file reading and sending. To the best of our knowledge, the ideas we mention have not been presented in previously published papers [1–18]. As a result, this paper presents some new ideas on the implementation of motion control systems, including an adjustable speed control system and a middle-distance stroke control system. Related measuring instruments, such as a touch panel, an industrial computer, professional wiring harnesses, sensors, jigs, and auxiliary appliances, are placed in a self-designed waterproof storage box to facilitate portability and operation at different venues. Validated through experiments, the proposed algorithm can control different attitudes, human–machine interface operations, network packet reading, and command receiving of the platform and can achieve 6DoF control and monitoring.

2. Inverse Kinematics Design

For six-axis motion platforms with a parallel structure, when the upper platform changes directions, the crank arm mechanism can bear the evenly divided load. That is, the highly rigid structure has a high load-bearing capacity. For six-axis motion platforms with a serial design, a structure resembling a cantilever beam is formed when the connecting rod is fully extended, and each joint connected in series accumulates errors (which reduces accuracy), possesses a low bearing capacity, and easily wobbles. Accordingly, a parallel structure has greater rigidity and stability and can achieve high-precision positioning in a complex working environment. Kinematics is divided into forward and inverse kinematics. Forward kinematics refers to the rotation angles of the servo motors of a six-axis platform, and numerical methods are used to determine the position, speed, and acceleration corresponding to the platform [19]. Inverse kinematics involves the use of platform position, speed, and acceleration to calculate the rotation angle of servo motors. This is the method adopted in the current study [20]. Figure 1 presents the physical model of the six-axis crank arm motion platform, which comprises six crank arm mechanisms, an upper platform (moving platform), a lower platform (fixed platform), and six support arms. The coordinates of the upper platform, lower platform, and rotation axis of the drive shaft are (xpi, ypi, zpi), (xbi, ybi, zbi), and (xb_Mi, yb_Mi, zb_Mi), respectively, where i = 1, 2–6. Traditional inverse kinematics methods and the proposed inverse kinematics method are explained in the section below.

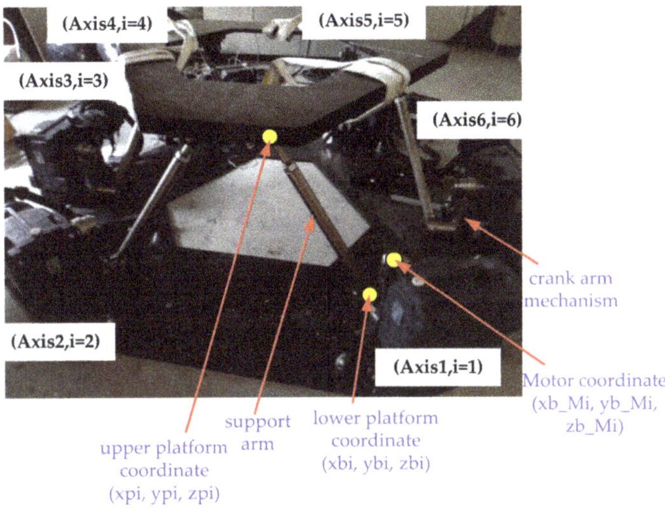

Figure 1. Photograph of the motion platform.

2.1. Traditional Inverse Kinematics

Traditional inverse kinematics changes the endpoint coordinate of the crank arm through the rotation angle of the arm [18], which requires adding together the components of the crank mapped to the platform, components of the connecting rod mapped to the platform, and the coordinates of the rotation axis of the drive shaft and then subtracting the sum from the expected upper platform coordinate to calculate its distance from the rod. Because the length of the support arm is fixed, the endpoint coordinate of the crank arm was adjusted by rotating the crank arm until the distance between the lower and upper platform coordinates aligned with the actual length of the support arm (Figure 1).

As Figure 2 illustrates, when the upper platform coordinate (xp1, yp1, zp1) of the first axis and inverse kinematics were used to calculate the expected rotating position vector (xp1_1, yp1_1, zp1_1), the coordinate corresponding to the lower platform (xb1_1, yb1_1, zb1_1) was moved from (xb1, yb1, zb1) according to the rotation angle of the crank arm. The virtual support arm ($L1$) was calculated using (xp1_1, yp1_1, zp1_1) and (xb1, yb1, zb1). If $L1$ did not equal the actual support arm (L), the two had different lengths. The M1 motor rotation command is obtained according to the change in the error value as

$$\theta_1(k+1) = \theta_1(k) + \gamma(L1(k) - L) \tag{1}$$

where k is the number of sampling points, θ_1 is the angle of the M1 motor, and γ is the weighting factor of the iteration. Then, the angle of the M1 motor had to be adjusted until $L1$ and L were of the same length. Therefore, the upper platform coordinate (xp1, yp1, zp1) was slowly moved from the virtual support arm ($L1$) to L' by controlling the M1 motor. The length of L' was calculated using Equation (1).

$$L' = \sqrt{(x_{p1\_1} - x_{b1})^2 + (y_{p1\_1} - y_{b1})^2 + (z_{p1\_1} - z_{b1})^2} \quad (2)$$

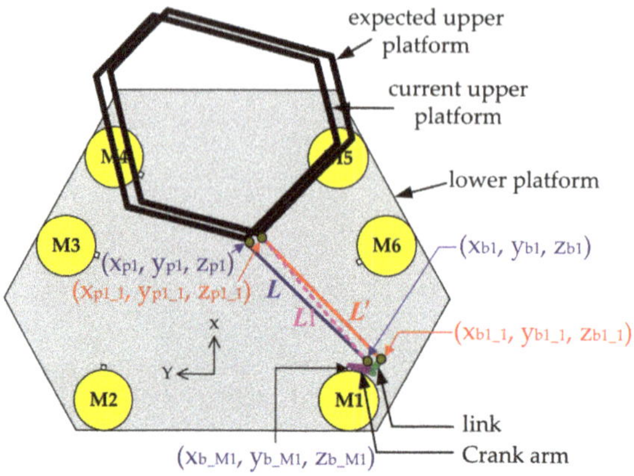

Figure 2. The schematic diagram of axis1 control.

The software ran Equation (2) until $L = L'$, which indicated that the coordinate of the lower platform (xb1_1, yb1_1, zb1_1) was correct.

2.2. Proposed Inverse Kinematics

Because traditional inverse kinematics methods fail in the position control implementation, the upper platform can only be moved approximately to adjust the length of the virtual support arm. To improve the movements of the endpoint coordinates of the motion platform crank arm, an actuating method similar to that of a linear actuator was adopted to obtain the relevant coordinates. The position of the linear actuator was set as the lower fixed point coordinate, and the upper platform coordinate could be changed by adjusting the elongation of six linear actuators. In practice, the elongation required by each electric cylinder can be calculated by subtracting the distance between the upper and lower platform coordinates with elongation from that without elongation. Therefore, this study proposes changing the motion platform of the rotary crank arm to the output end of the crank arm motor as the lower fixed point of the linear actuator, thereby determining di, which is the distance between the upper platform coordinate and the motor coordinate (xb_Mi, yb_Mi, zb_Mi). The vector of di is expressed as follows [21]:

$$d_i = q_i - b_i = T + R \cdot p_i - b_i \quad (3)$$

where T is the position vector of the upper platform coordinate relative to the lower platform, R denotes the homogeneous transformation matrix, p_i = [xpi ypi zpi] represents the position vector of the upper platform, b_i = [xb_Mi yb_Mi zb_Mi] is the position vector of the motor rotation axis, and d_i denotes the equivalent rod length.

Thus, by taking into account the relationship between the lengths of the crank, connecting rod, and support arm, the positions of the motor end and upper platform were bi and pi, respectively, s was the sum of the connecting link length (p) and support arm length (l), and the contact coordinate (A_i) between s and the crank length (a) was projected to the xy plane. In Figure 3, ϕ_i denotes the included angle between the x direction and the crank length. The known conditions, namely, s, a, $B_i = [x_{b\_Mi}\ y_{b\_Mi}\ z_{b\_Mi}]$, pi = $[x_{pi}\ y_{pi}\ z_{pi}]$, and ϕ_i, were used to calculate the angle required for each axis to rotate (θ_i), thereby controlling the different virtual cylinder lengths (di), controlling the elongation of the virtual cylinder, and determining the platform attitude required by the 6DoF platform.

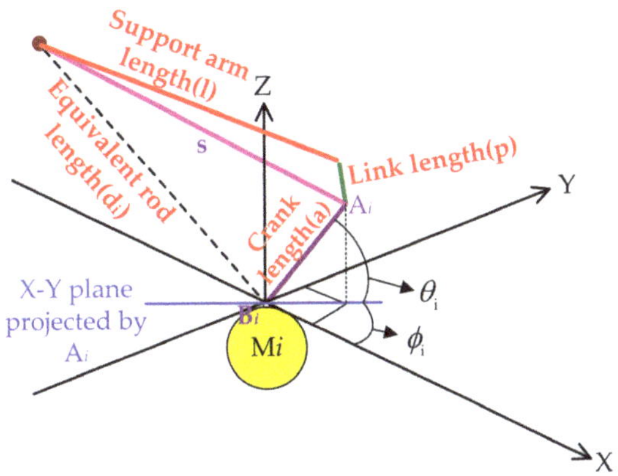

Figure 3. The relationship of axis i (i = 1–6) and upper platform coordinate.

Because the six sets of crank arm mechanisms were arranged opposite to each other, they were divided into odd and even numbers. The first, third, and fifth axes were controlled at 0–180 degrees, while the second, fourth, and sixth axes were controlled at 180–360 degrees. The odd axes of Ai are expressed as

$$x_{ai} = a \cdot \cos(\theta_i)\cos(\phi_i) + x_{b\_Mi} \tag{4}$$

$$y_{ai} = a \cdot \cos(\theta_i)\cos(\phi_i) + y_{b\_Mi} \tag{5}$$

$$z_{ai} = a \cdot \sin(\theta_i) + z_{b\_Mi} \tag{6}$$

where i = 1, 3, and 5. The even axes of Ai are expressed as

$$x_{ai} = a \cdot \cos(180 - \theta_i)\cos(180 + \phi_i) + x_{b\_Mi} \tag{7}$$

$$y_{ai} = a \cdot \cos(180 - \theta_i)\cos(180 + \phi_i) + y_{b\_Mi} \tag{8}$$

$$z_{ai} = a \cdot \sin(180 - \theta_i) + z_{b\_Mi} \tag{9}$$

where i = 2, 4, and 6. As shown in Figure 3, Pythagorean theorem is applied to find s, di, and a, which are expressed as follows:

$$s^2 = \left(x_{pi} - x_{ai}\right)^2 + \left(y_{pi} - y_{ai}\right)^2 + \left(z_{pi} - z_{ai}\right)^2 \tag{10}$$

$$d_i^2 = \left(x_{pi} - x_{b\_Mi}\right)^2 + \left(y_{pi} - y_{b\_Mi}\right)^2 + \left(z_{pi} - z_{b\_Mi}\right)^2 \tag{11}$$

$$a^2 = \left(x_{ai} - x_{b\_Mi}\right)^2 + \left(y_{ai} - y_{b\_Mi}\right)^2 + \left(z_{ai} - z_{b\_Mi}\right)^2 \tag{12}$$

Equations (11) and (12) are substituted into Equation (10) to derive

$$\begin{aligned} s^2 = d_i^2 + a^2 + 2\left(x_{b\_Mi}^2 + y_{b\_Mi}^2 + z_{b\_Mi}^2\right) - 2\left(x_{pi}x_{b\_Mi} + y_{pi}y_{b\_Mi} + z_{pi}z_{b\_Mi}\right) + 2x_{ai}\left(x_{pi} - x_{b\_Mi}\right) \\ + 2y_{ai}\left(y_{pi} - y_{b\_Mi}\right) + 2z_{ai}\left(z_{pi} - z_{b\_Mi}\right) \end{aligned} \tag{13}$$

Equations (4)−(9) are substituted into (13) to derive

$$d_i^2 + a^2 - s^2 + 2a \cdot \sin(\theta_i)\left(z_{pi} - z_{b\_Mi}\right) - 2a \cdot \cos(\theta_i) \cdot \left\{\cos(\phi_i)\left(x_{pi} - x_{b\_Mi}\right) + \sin(\phi_i)\left(y_{pi} - y_{b\_Mi}\right)\right\} = 0 \tag{14}$$

To facilitate the calculation, the equation is rewritten as follows:

$$H = E \cdot \sin(\theta_i) + K \cdot \cos(\theta_i) \quad (15)$$

where

$$H = d_i^2 + a^2 - s^2 \quad (16)$$

$$E = -2a \cdot (z_{pi} - z_{b\_Mi}) \quad (17)$$

$$K = 2a \cdot \cos(\theta_i)\{\cos(\phi_i)(x_{pi} - x_{b\_Mi}) + \sin(\phi_i)(y_{pi} - y_{b\_Mi})\} \quad (18)$$

Equation (15) belongs to the trigonometric identity, and θ_i can be written as

$$\theta_i = \sin^{-1}\left(H/\sqrt{(E^2 + K^2)}\right) - \tan^{-1}(K/E) \quad (19)$$

The aforementioned equation demonstrates that the actual rotation angles (19) of the six motors can be directly derived using the known conditions—s, a, bi = [x_{b_Mi} y_{b_Mi} z_{b_Mi}], pi = [x_{pi} y_{pi} z_{pi}], ϕ_i, and the distance (di) from the upper platform coordinate to the motor coordinate.

3. PID Controller

In the synchronous d-q frame, the voltage equations can be expressed as

$$v_d = r_s i_d + L_d \frac{d}{dt} i_d - \omega_{re} L_q i_q \quad (20)$$

and

$$v_q = r_s i_q + L_q \frac{d}{dt} i_q + \omega_{re} L_d i_d + \omega_{re} \lambda_m \quad (21)$$

where v_d is the d axis voltage, v_q is the q axis voltage, r_s is the stator resistance, i_d is the d axis current, i_q is the q axis current, L_d is the d axis inductance, L_q is the q axis inductance, d/dt is the differential operator, ω_{re} is the electrical motor speed, and λ_m is the flux linkage of the permanent magnet mounted on the shaft rotor. Through realizing $i_d = 0$ by applying field-oriented control, Equations (20) and (21) can be rewritten as

$$v_d = -\omega_{re} L_q i_q \quad (22)$$

and

$$\begin{aligned} v_q &= r_s i_q + L_q \tfrac{d}{dt} i_q + \omega_{re} \lambda_m \\ &= u_q + \omega_{re} \lambda_m \end{aligned} \quad (23)$$

The electro-magnetic torque is

$$T_e = \frac{3}{2}\frac{P}{2} \lambda_m i_q = K_T i_q \quad (24)$$

where T_e is the electro-magnetic torque of the motor, P is the number of poles of the motor, and K_T is the torque constant. The mechanical speed and position of the motor are expressed as

$$\frac{d}{dt}\omega_{rm} = \frac{1}{J_t}(T_e - T_L - B_t \omega_{rm}) \quad (25)$$

and

$$\frac{d}{dt}\theta_{rm} = \omega_{rm} \quad (26)$$

where ω_{rm} is the mechanical speed, θ_{rm} is the mechanical position, J_t is the total inertia of the motor and gearbox, T_L is the load torque, and B_t is the total viscous coefficient of the motor and gearbox. The electrical position and speed, which can be expressed as θ_{re} and ω_{re}, are shown as follows:

$$\theta_{re} = \frac{P}{2}\theta_{rm} \tag{27}$$

and

$$\omega_{re} = \frac{P}{2}\omega_{rm} \tag{28}$$

The PID control law is

$$u = K_P e(t) + K_I \int e(t)dt + K_D \frac{d}{dt}e(t) \tag{29}$$

where $e(t)$, K_P, K_I, and K_D are the control error, proportional, and integral coefficients, respectively. In Figure 4, G_θ, G_ω, and G_C of the servo driver are the PID controllers, namely, the position loop, speed loop, and current loop controllers, respectively. The control system was designed and applied to each of the six actuators independently.

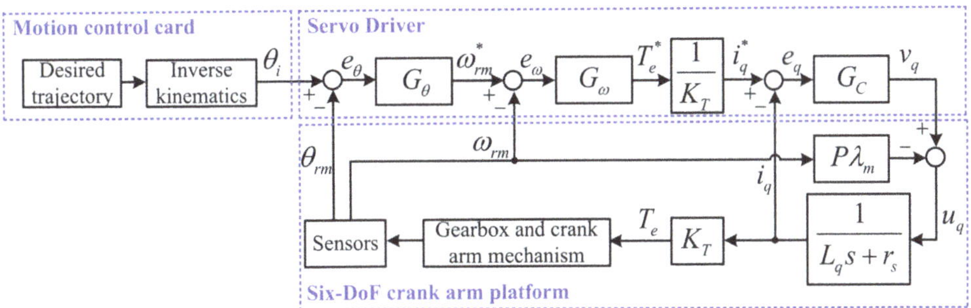

Figure 4. Block diagram of the position control system.

4. Monitoring System Design

The six-axis platform monitoring module was developed exclusively for six-axis platform testing. The industrial computer monitoring program was designed under the LabVIEW environment and the interactive property is good. The function of the system reached the expected design requirements, and the cost is low. Combined with the modular design concept, it is integrated in the waterproof box shown in Figure 5. The waterproof box contains a high-speed DAQ, an industrial computer, a touch screen, sensors, professional wiring harnesses, and jigs. The special jig allowed for convenient and rapid installation. Measurement software was written to measure and record various parameters of the six-axis motion platform at high speed.

Figure 5. Photograph of the monitoring system.

4.1. Gyroscope and Angular Velocity Meter Jig

Three angular velocity sensors were fixed on the cube to measure angular velocity in the x, y, and z directions. The attitude gyroscope and the cube were fixed on a magnetic bottom plate, the magnetic attraction was approximately 47 kg. The adherence distance was adjusted using three screws to avoid the strong impact of direct magnetic adherence from damaging the sensors. Two powerful magnets at the bottom of the jig were adhered to the six-axis platform to be tested. To avoid the impact of powerful magnetic adherence, three knobs were designed. Before adherence, the knobs were turned clockwise to the end such that the three screw threads formed three points of support. After aligning with the x direction of the platform, the jig was adhered to the platform, and the three knobs were turned counterclockwise for the bottom plate to fit closely to the platform, the jig was thus firmly adhered to the platform. The jig could be easily dissembled by turning the three knobs clockwise to the end, as shown in Figure 6.

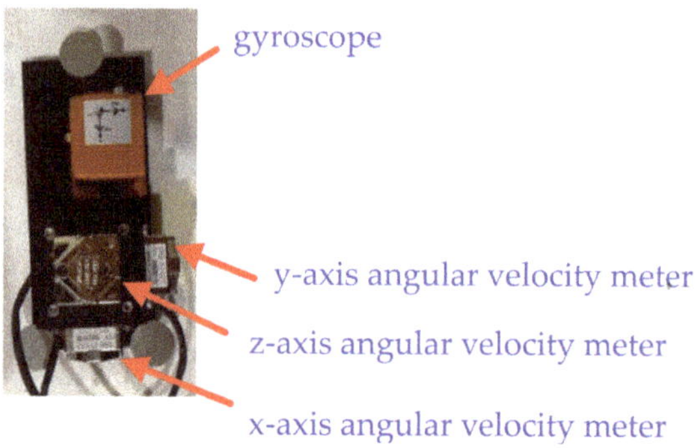

Figure 6. Gyroscope and angular velocity meters.

4.2. Laser Displacement and Angular Meters

Three three-section brackets (Figure 7a) were used to adjust the height of the x, y, and z axes of the measurement platform, while three measurement reference plates (Figure 7b) were magnetically adhered to the x, y, and z axes of the six-axis platform. The positions of the brackets or reference plates were adjusted for a red laser dot fired out of the laser displacement meter to hit near the center of the reference plate. Because the six-axis platform had a hexagonal structure design when viewed from above, the x axis was adjusted for the measurement reference plate to magnetically adhere to the x direction, and the x direction displacement was immediately read when the laser hit the measurement reference plate. The y axis was not perpendicular to any plane. Therefore, the reference plate was installed on the side of the six-axis platform, and y direction displacement was obtained through angle correction. The correction method is shown in Figure 7c.

4.3. Measurement Software

Figure 8a presents the measurement software setting form, which allows for archive path and measurement speed settings. The five sensor-related settings are as follows: angular velocity sensor and laser displacement meter receiving signals from the voltage input channels (the software enables or disables the function as well as adjusting the channel name, scale, offset, and unit), three-axis accelerometer receiving signals from the IEPE channel (the software enables or disables the function as well as adjusting the channel name, sensor sensitivity, and unit), and attitude gyroscope receiving signals from the RS232 serial port (the software sets the communication port). Figure 8 presents the real-time curve and trend curve for the sensor signals received by the high-speed DAQ. The left side of Figure 8b is the channel display. The operator can manually select the number of measurement channels for continuous and simultaneous measurement, display, and archiving.

(**a**)

(**b**)

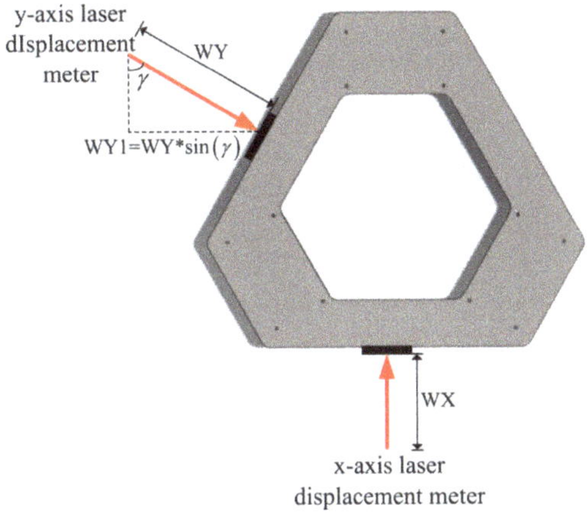

(**c**)

Figure 7. Laser displacement meter: (**a**) three-section bracket, (**b**) measurement reference plate, and (**c**) schematic diagram of x axis and y axis measurement.

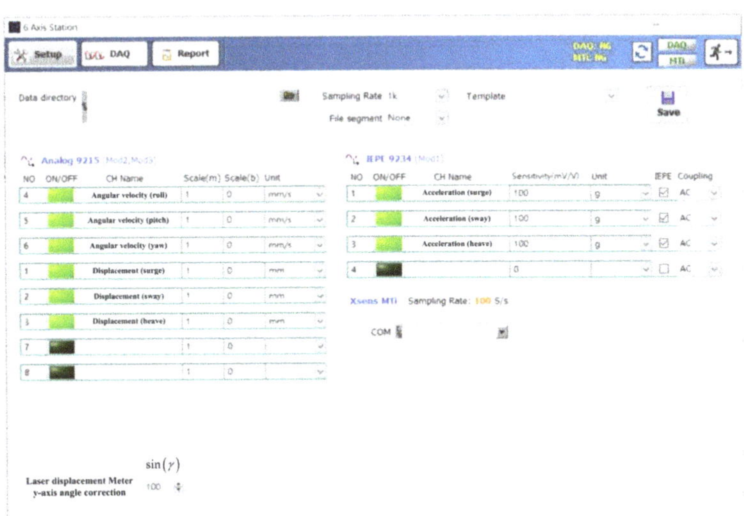

(**a**)

Figure 8. *Cont.*

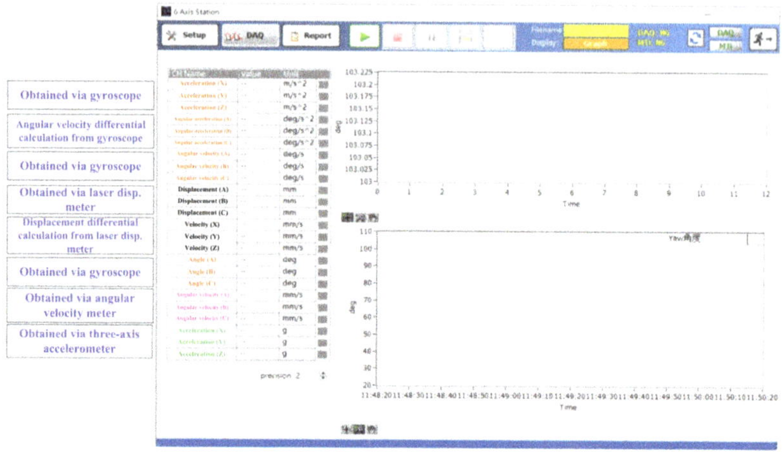

(b)

Figure 8. Operation interface: (a) software setting and (b) channels display.

5. Implementation

The implementation block diagram of the platform is displayed in Figure 9. The mechanical structure of the platform was designed to carry a simulator cab of 1500 kg or heavier (Figure 10a). The platform mechanism was divided into upper and lower platforms. The six support arms of the platform achieved the 6DoF stroke required by the simulator cab through crank arm rotation. Fisheye bearing joints were used to connect the upper end of the support arm to the upper platform, while the lower end of the support arm was coupled with the crank arm to drive the upper platform. The prototype of the crank mechanism is shown in Figure 10b. The dimensions are link length = 105 mm, support arm length = 835 mm, and crank length = 250 mm. The parameters of the motor are shown in Table 1. The gearbox ratio is 32.41. The servo motor was connected with a gearbox to magnify the torque output. A motion control card PMDK, produced by ICP Das Co., Ltd, was used with a 5-V PCI bus. The PMDK had six pulse input and output channels, six analog input and output channels, and various digital input and output channels. The PMDK was equipped with a two-wire FRnet IO module, which can be used for remote input and output control.

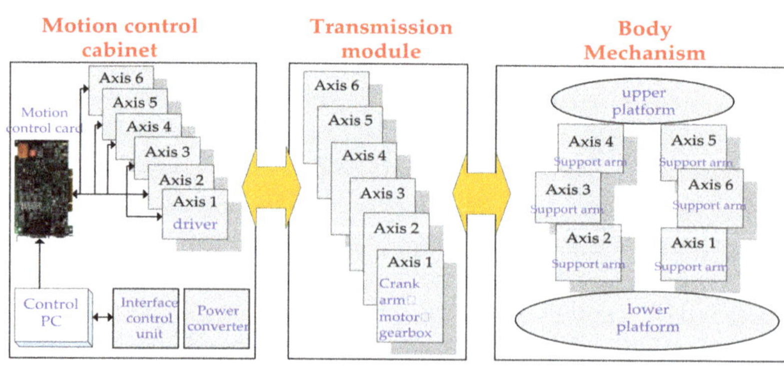

Figure 9. The block diagram of the 6DoF platform.

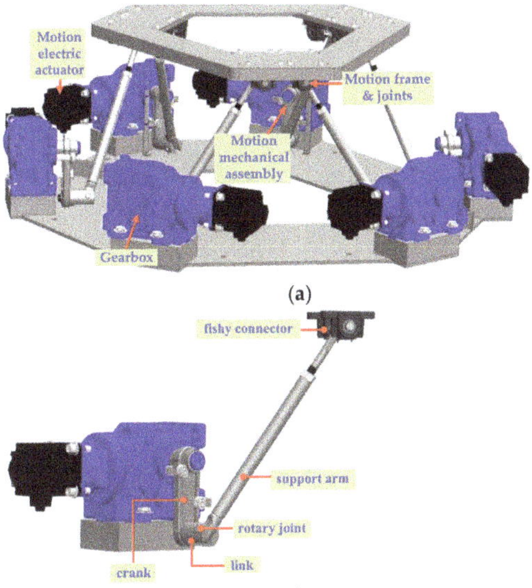

Figure 10. 3D design drawing of the 6DoF platform: (**a**) 6DoF platform, and (**b**) crank arm mechanism.

Table 1. The parameters of the motor.

Pole Number	8 Pole
Power dissipation	5.0 kW
Rated voltage	220 V
Rated load current	25.9 A
Rotor inertia	0.00488 Kg-m^2
Rated speed	2000 r/min
Maximum speed	3000 r/min
Rated torque	23.9 N.m
Maximum torque	71.6 N.m

The monitoring module is shown in Figure 11a,b. The operator opened the waterproof storage box, screwed one end of the wiring harness to the input and output connector and the other end to sensor a, and secured the measurement reference plates and three-section brackets in place (Figure 11a). Figure 11b displays the positions of the angular velocity meter and the gyroscope.

The main program conducts the initial setup, thread enabling, and mode login of the platform. Figure 12a contains a flowchart of the implementation. Under the normal mode, the login mode can be selected using the human–machine interface. When the operator discovers an abnormal alarm or when maintenance personnel need to inspect the electric cylinders, they may select the maintenance mode via the interface. Data can be read and written via Internet transmission, and a thread of network modules is produced (Figure 12b). Operators can issue 6DoF commands to the platform and transmit the 6DoF commands to the platform computer by using Internet cable transmission.

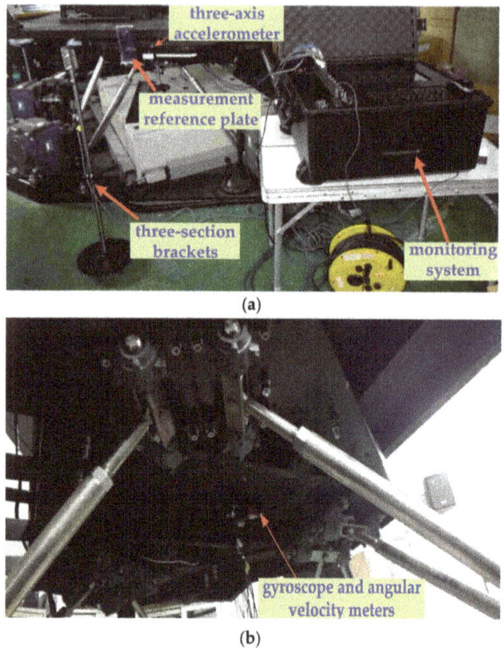

Figure 11. Photograph of the implementation system: (**a**) monitoring module and (**b**) upper platform.

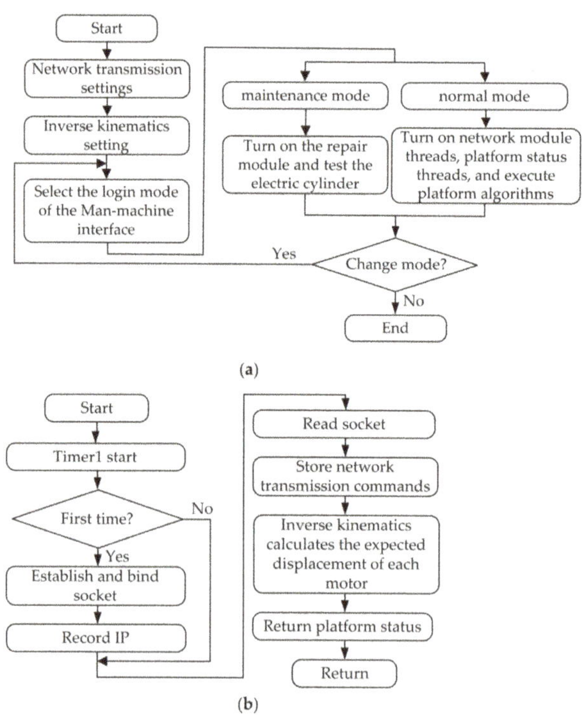

Figure 12. Flowchart of the implementation: (**a**) main program and (**b**) network module.

6. Experimental Validation

Figure 13 contains photographs of the four strokes under the control of 6DoF. This section validates the accuracy and feasibility of the proposed method by using experimental results. Traditional inverse kinematics and the proposed inverse kinematics were written into the software—only the inverse kinematics software was modified, the hardware was not—to compare the motion of the six-axis crank arm platform when the two algorithms were applied. Then, 6DoF commands were issued to the platform through the human–machine interface, and the movable range of each DoF was recorded.

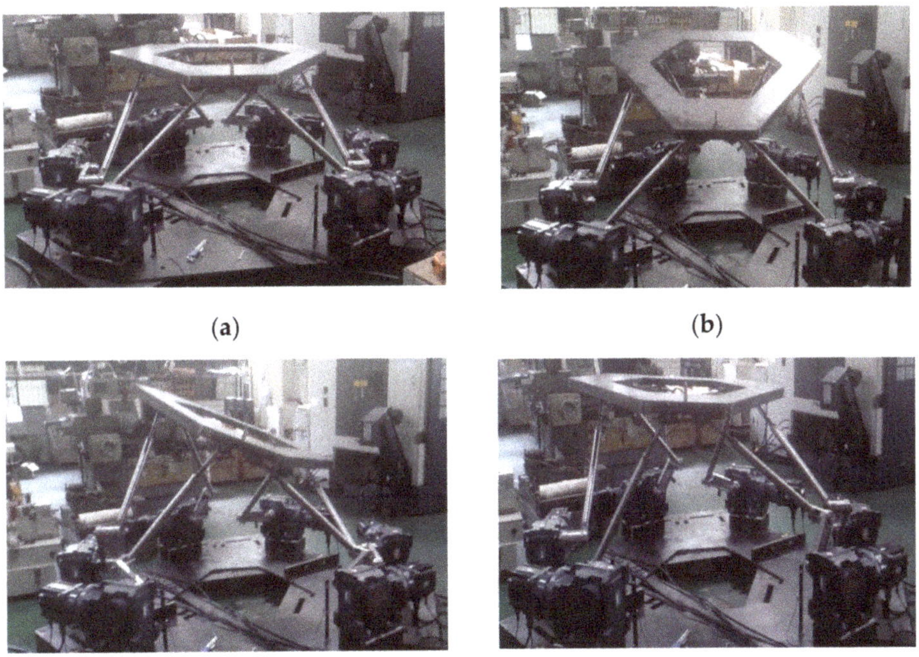

(a) (b) (c) (d)

Figure 13. The control of 6DoF: (**a**) heave, (**b**) pitch, (**c**) roll, and (**d**) surge.

Figure 14a–c are records of the relevant measurement results. The controllable angles of the two algorithms were compared, as indicated in Figure 14a. Because the traditional method used the trial-and-error method to control the lower platform coordinate, when the calculated rod length approximated the actual rod length, the movement was completed. The traditional method allowed the actuator to be operated only within the linear region, that is, a position of 60 to 180 degrees. The proposed method opted to directly map the crank arm platform to the Stewart platform, enabling the actuator operating region to reach 50 to 230 degrees. A comparison of the algorithms is presented in Figure 14b, where the subscripts "old" and "new" denote the traditional and proposed methods, respectively. As in the figure, the movable range of the proposed method was larger than that of the traditional method—a front–back distance 60 mm longer, a left–right distance 80 mm longer, and a vertical distance 160 mm longer, which effectively improved the stroke of the platform operation space. The algorithms' controllable angles of rotation—known as roll, pitch, and yaw—were compared, as in Figure 14c. The angles of rotation of the proposed algorithm were wider than those of the traditional algorithm—8 degrees wider in roll, 4 degrees wider in pitch, and 4 degrees wider in heave. An enlarged operating angle implies an increased overall platform space. Figure 12a,b are the 2 DoF $\pm$ 300 mm displacement response diagrams, where the motion control command begins at 8 s with a period of 4 s. To observe the relationship between surge and sway—a 90-degree difference—more clearly, they were plotted on the horizontal and vertical axes, respectively, forming a perfect circle (Figure 15b). Figure 16a–c present the displacement, velocity, and acceleration curves of surge conducting 150 mm of forward and reverse rotation, and the velocity was obtained using differential software. Platform operation involved complex mechanical parts as well as the simulator cab. Accordingly, a few spikes were observed in the differential velocity due to discontinuous displacement. Figure 17a,b are the response diagrams of the three translational DoF (surge, sway, and heave) at ±300 mm displacement with an angular difference of 120 degrees between any two. Regarding heave, the laser light of the displacement meter hit the rear of the upper platform. Due to the unfavorable mechanism smoothness, ripples were observed in the extracted signals. A 3D circle was formed when the three translational DoF were spatially plotted. Figure 18a–d are the response diagrams of the three other DoF at ±26 degrees with an angular difference of 120 degrees between any two. In Figure 18b, RS232 is used to read the angular velocity of the gyroscope, which has an amplitude of ±12 degrees/s. The roll, pitch, and yaw were spatially plotted, forming a 3D circle. Figure 19 shows the x, y, and z axis accelerations of the test platform, and the impulse responses were captured by the accelerometer. The platform performed x axis acceleration and deceleration alternately at 5–15 s, and the acceleration was approximately 0.5 g, whereas the platform performed y and z axis acceleration and deceleration alternately at 22–23 s and 43–55 s, respectively, and the acceleration was also approximately 0.5 g.

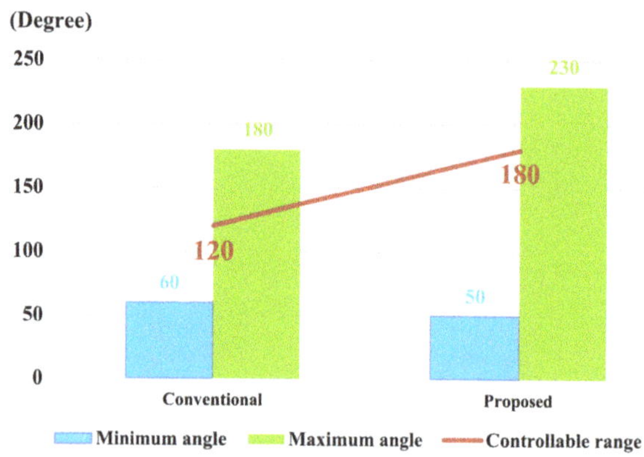

(a)

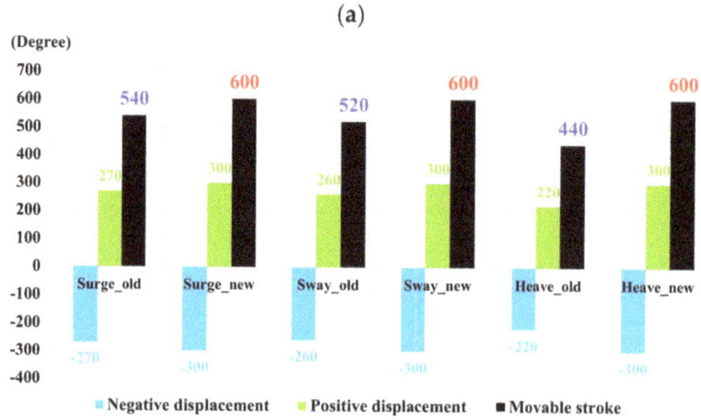

(b)

Figure 14. *Cont.*

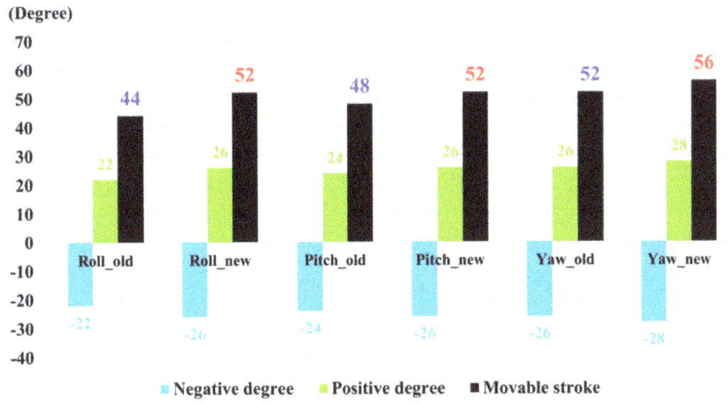

(c)

Figure 14. Comparison of the proposed method and the conventional method: (**a**) control range, (**b**) surge, sway, and heave, and (**c**) roll, pitch, and yaw.

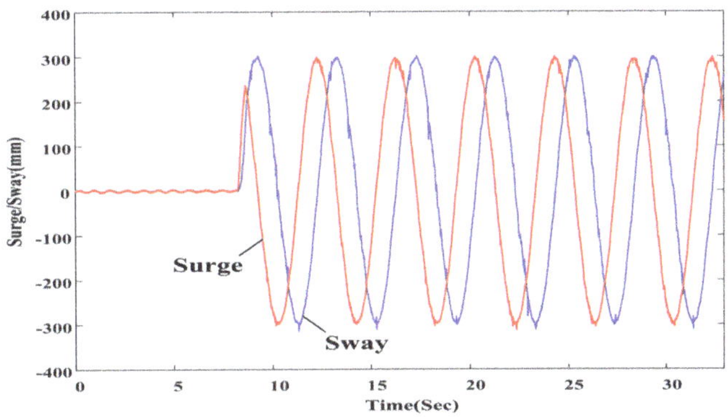

(**a**)

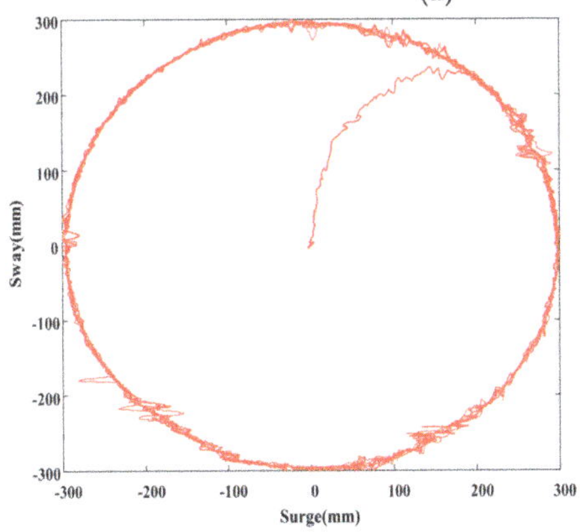

(**b**)

Figure 15. Measured displacement responses: (**a**) responses and (**b**) trajectories.

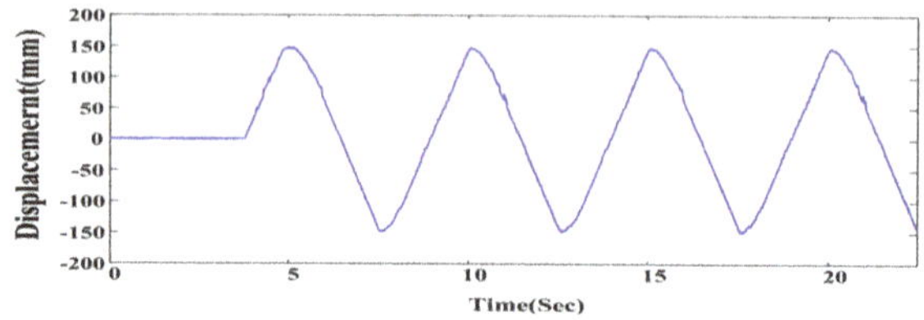

(**a**)

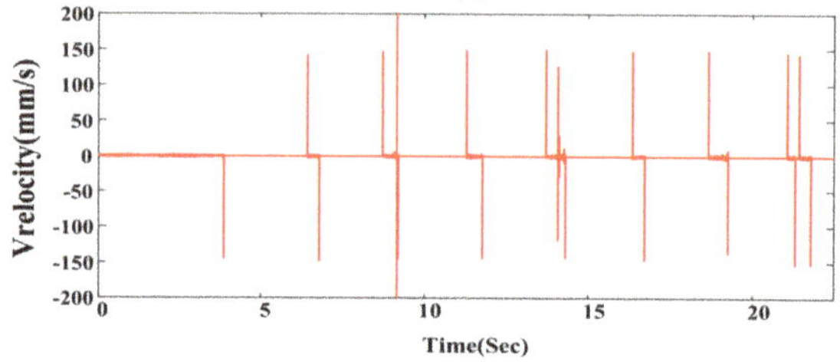

(**b**)

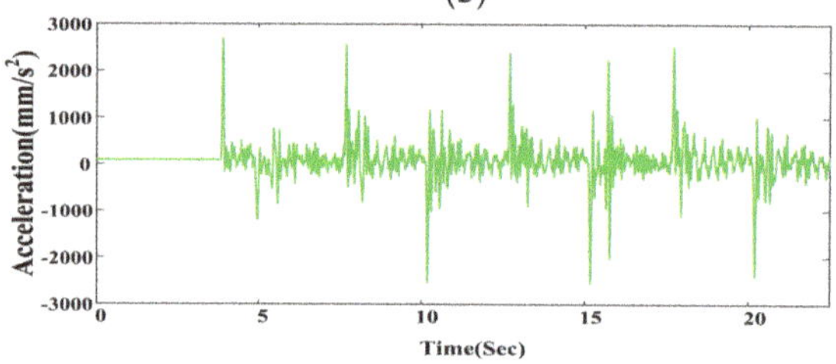

(**c**)

Figure 16. Measured triangular wave displacement responses: (**a**) displacement, (**b**) velocity, and (**c**) acceleration.

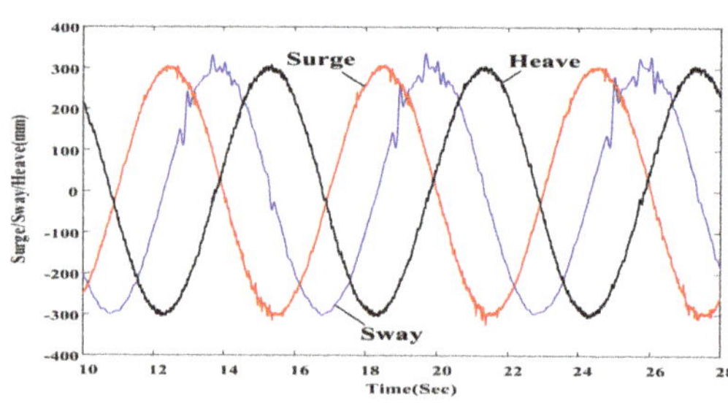

(**a**)

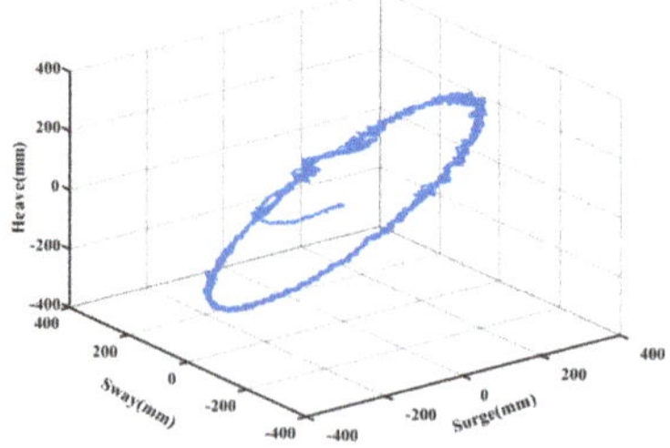

(**b**)

Figure 17. Measured responses at different time-varying commands: (**a**) 2D view and (**b**) 3D view.

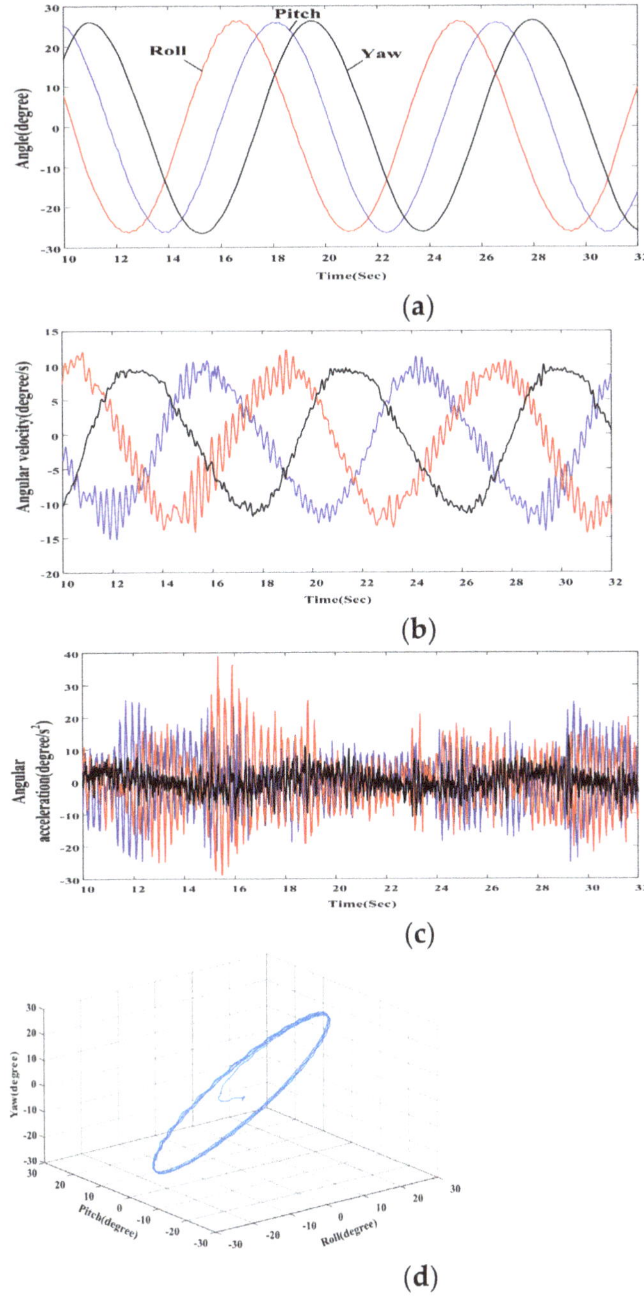

Figure 18. Measured sine wave degree responses: (**a**) angle, (**b**) angular velocity, (**c**) angular acceleration, and (**d**) 3D view.

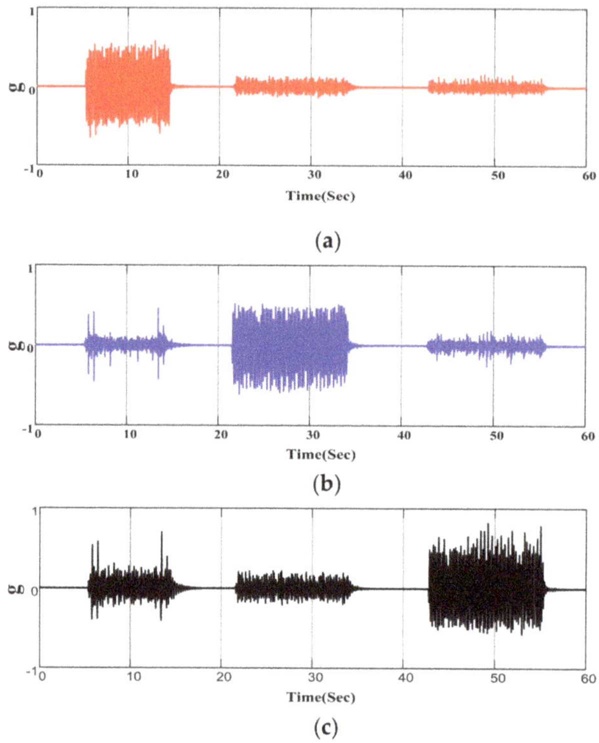

Figure 19. Measured acceleration responses: (**a**) x axis, (**b**) y axis, and (**c**) z axis.

7. Conclusions

This study discussed the inverse kinematics of a six-axis motion platform, and the development of a monitoring system. The main contribution of this study is the design of a 6DoF crank arm mechanism to replace the expensive 6DoF linear actuator Stewart platform, 3DoF crank arm platform, and 4DoF crank arm platform, and the result of obtaining a comparable movement experience to the Stewart system. The algorithm was applied to the motion control of a 6DoF platform to control the elongation of the virtual cylinder and to determine the platform attitude required by the 6DoF platform. Unlike the conventional method, which relies on approximation to obtain the required angle, this study proposed a more accurate, real-time, and highly dependent processing method. In addition, this study used LabVIEW to create a set of testing equipment that integrates different sensors and jigs to read the angle, angular velocity, displacement, and linear acceleration of the platform in motion. Subsequently, the differential equation and sampling rate were used to calculate the angular acceleration and linear velocity. The testing equipment data were transmitted to the platform computer via RS-485 or RS-232 digital communication, allowing the platform to control the state of motion at all times. Finally, the experimental results prove the proposed algorithm capable of improving the operation of the platform stroke and thus applicable to simulators with higher specification requirements. Moreover, the implementation of the platform was based on the PID controller. Because of its simple structure, strong robustness, and good stability, the PID controller is still widely used in various fields of industrial control. However, the modern control theory of the platform is not the main topic of this study that will be considered to evaluate the design in future work.

Funding: This research received no external funding.

Conflicts of Interest: The author has no conflict of interest to declare.

References

1. Urlahannan, A.; Jennings, P.; Oliveira, L.; Birrell, S. Designing an adaptive interface: Using eye tracking to classify how information usage changes over time in partially automated vehicles. *IEEE Trans. Access* **2020**, *8*, 16865–16875. [CrossRef]
2. He, Z.; Lian, B.; Li, Q.; Zhang, Y.; Song, Y.; Yang, Y.; Sun, T. An error identification and compensation method of a 6-DoF parallel kinematic machine. *IEEE Trans. Access* **2020**, *8*, 119038–119047. [CrossRef]
3. Nabi, H.N.; Ullah, S.; Munir, A. Kinematics analysis of three-degree-of freedom parallel manipulator with crank arm actuator. In Proceedings of the 2014 11th International Bhurban Conference on Applied Sciences & Technology (IBCAST), Islamabad, Pakistan, 14–18 January 2014; pp. 182–188.
4. Pan, C.T.; Sun, P.Y.; Li, H.J.; Hsieh, C.H.; Hoe, Z.Y.; Shiue, Y.L. Development of multi-axis crank linkage motion system for synchronized flight simulation with VR immersion. *Appl. Sci.* **2021**, *11*, 3596. [CrossRef]
5. Stewart, D. A platform with six degrees of freedom. *Proc. Inst. Mech. Eng.* **1965**, *180*, 371–386. [CrossRef]
6. Fichtert, E.F. A Stewart platform-based manipulator: General theory and practical construction. *Int. J. Robot. Res.* **1986**, *5*, 157–182. [CrossRef]
7. Mohammed, M.G.; Duffy, J. A direct determination of the instantaneous kinematics of fully parallel manipulators. *ASME J. Mech. Transm. Autom. Des.* **1986**, *107*, 226–229. [CrossRef]
8. Liu, K.; Fitzgerald, J.M.; Lewis, F.L. Kinematic analysis of a Stewart platform manipulator. *IEEE Trans. Ind. Electron.* **1993**, *40*, 282–293. [CrossRef]
9. Ji, C.Y.; Chenand, T.C.; Lee, Y.L. Investigation of kinematic analysis and applications for a 3-RRPS parallel manipulator. *J. Chin. Soc. Mech. Eng.* **2007**, *28*, 623–632.
10. Mohammed, A.M.; Li, S. Dynamic neural networks for kinematic redundancy resolution of parallel Stewart platforms. *IEEE Trans. Cybern.* **2016**, *46*, 1538–1550. [CrossRef] [PubMed]
11. Lara-Molina, F.A.; Rosario, J.M.; Dumur, D. Architecture of predictive control for a Stewart platform manipulator. In Proceedings of the 2010 8th World Congress on Intelligent Control and Automation, Jinan, China, 7–9 July 2010; pp. 6584–6589.
12. Bruschetta, M.; Maran, F.; Beghi, A. A Nonlinear, MPC-based motion cueing algorithm for a high-performance, nine-DoF dynamic simulator platform. *IEEE Trans. Control Syst. Technol.* **2017**, *25*, 686–694. [CrossRef]
13. Asadi, H.; Mohammadi, A.; Mohamed, S.; Qazani, M.R.C.; Lim, C.P.; Khosravi, A.; Nahavandi, S. A model predictive control-based motion cueing algorithm using an optimized nonlinear scaling for driving simulators. In Proceedings of the 2019 IEEE International Conference on Systems, Man and Cybernetics (SMC), Bari, Italy, 6–9 October 2019; pp. 1245–1250.
14. Tang, Z.; Hu, M.; Pei, Z.; Liu, L.; Zhang, J. A new numerical method for Stewart platform forward kinematics. In Proceedings of the 2016 35th Chinese Control Conference (CCC), Chengdu, China, 27–29 July 2016; pp. 6311–6316.
15. Lou, J.H.; Tseng, S.P. Developing a real-time serial servo motion control system for electric Stewart platform. In Proceedings of the 2014 International Conference on Advanced Robotics and Intelligent Systems (ARIS), Taipei, Taiwan, 6–8 June 2014; pp. 66–71.
16. Qazani, M.R.C.; Asadi, H.; Khoo, S.; Nahavandi, S. A linear time-varying model predictive control-based motion cueing algorithm for hexapod simulation-based motion platform. *IEEE Trans. Syst. Man Cybern. Syst.* **2019**, *51*, 6096–6110. [CrossRef]
17. Asadi, H.; Lim, C.P.; Mohamed, S.; Nahavandi, D.; Nahavandi, S. Increasing motion fidelity in driving simulators using a fuzzy-based washout filter. *IEEE Trans. Intell. Veh.* **2019**, *4*, 298–308. [CrossRef]
18. Available online: https://docplayer.net/34514506-How-to-build-an-arduino-powered-6dof-motion-platform.html (accessed on 26 August 2021).
19. Urlahannan, A.; Jennings, P.; Oliveira, L.; Birrell, S. A efficient numerical method for forward kinematics of parallel robots. *IEEE Trans. Access* **2019**, *7*, 128758–128766.
20. Jia, G.; Pan, G.; Gao, Q.; Zhang, Y. Research on position inverse solution of electric-driven Stewart platform based on Simulink. *J. Eng.* **2019**, *2019*, 379–383. [CrossRef]
21. Sciavicco, L.; Siciliano, B. *Modelling and Control of Robot Manipulators*; Springer: London, UK, 2000.

applied sciences

MDPI

Article

Synchronization Sliding Mode Control of Closed-Kinematic Chain Robot Manipulators with Time-Delay Estimation

Tu Thi Cam Duong [1], Charles C. Nguyen [1] and Thien Duc Tran [2,*]

1 Department of Electrical Engineering and Computer Science, School of Engineering, The Catholic University of America, 620 Michigan NE Ave., Washington, DC 20064, USA; duongt@cua.edu (T.T.C.D.); nguyen@cua.edu (C.C.N.)
2 Automatic Control Department, Ho Chi Minh City University of Technology and Education, Ho Chi Minh City 700000, Vietnam
* Correspondence: thientd@hcmute.edu.vn

Abstract: Control of closed-kinematic chain manipulators (CKCM) with uncertain dynamics is a tremendous challenge due to the synchronization among actual joints and end-effectors, limited workspace, and nonexistent closed-form solutions of forward kinematics. This paper proposes a synchronization control scheme based on the concept of sliding mode control (SMC) developed for CKCMs called nonsingular fast terminal sliding mode control (NFTSMC) in conjunction with the time-delay estimation (TDE) method to address the above issues. First, the cross-coupling error is derived by combining position errors and synchronization errors to achieve the synchronization goal and then used to form a sliding mode surface of the NFTSMC. After that, a control law is developed based on the sliding mode surface to ensure faster asymptotic convergence of the errors of both position and synchronization of the CKCMs in a finite and minimal time. Then, the TDE control scheme with no prior knowledge of manipulator dynamics is employed to estimate the unknown dynamics and disturbances and thereby reject the effects of chattering caused by the NFTSMC. Lyapunov stability theorem is employed to show that the overall system controlled by the proposed control scheme achieves asymptotic convergence of errors and system stability. The performance of the proposed control is assessed by computer simulation on a 2 degrees-of-freedom (DOF) planar CKCM manipulator and simulation results are presented and discussed.

Keywords: closed-kinematic chain manipulator (CKCM); sliding mode control (SMC); time-delay estimation (TDE); nonsingular fast terminal sliding mode control (NFTSMC); synchronization control; model-free control

Citation: Duong, T.T.C.; Nguyen, C.C.; Tran, T.D. Synchronization Sliding Mode Control of Closed-Kinematic Chain Robot Manipulators with Time-Delay Estimation. *Appl. Sci.* **2022**, *12*, 5527. https://doi.org/10.3390/app12115527

Academic Editors: Alessandro Gasparetto, Stefano Seriani and Lorenzo Scalera

Received: 4 May 2022
Accepted: 27 May 2022
Published: 29 May 2022

1. Introduction

Closed-kinematic chain manipulators (CKCMs) for which its motion is achieved in all degrees-of-freedom (DOF) by the combined motion of their active joints can provide higher positioning accuracy and greater payload handling capability than the conventional open-kinematic chain manipulators (OKCM) composed of serial linkages or rigid bodies [1–4]. Despite the above advantages, CKCMs possess several drawbacks such as synchronization among actual joints and end-effectors, limited workspace, and nonexistent closed-form solutions of forward kinematics. To address the above issues, the concept of synchronization control has been considered, and as a result, there has been much effort in the development and implementation of error synchronization-based control schemes for CKCMs. In a synchronization-based control scheme, all joints are synchronously driven to improve CKCM's performance.

The synchronization concept was first introduced in [5], and it was applied to perform tracking control of parallel robots [6–9]. Most existing synchronized control schemes are model-based such as the computed torque control (CTC) [10], adaptive control [11,12], and sliding mode control (SMC) [13,14]. In general, a SMC scheme consists of a driving component that forces the system's trajectory to reach a stable hyperplane (sliding surface) and a design of a sliding surface that assures the plant's desired error dynamics. The implementation of the above synchronized control schemes requires a precise dynamical model of the manipulator, for which its calculation is highly computationally intensive. Moreover, it was concluded that an accurate mathematical dynamic model of CKCMs is difficult to obtain. Consequently, the above model-based control schemes are not suitable for real-time control of CKCM manipulators, particularly those with more than 2 DOFs. To tackle the above dynamic modeling issue, the authors in [15,16] considered synchronized control schemes that have simple structures and do not require knowledge of the manipulator dynamics to implement their control laws. However, those control approaches proposed for parallel manipulators can only achieve asymptotic stability, which requires infinite time to converge to an equilibrium point. In order to assure finite-time convergence, the terminal SMC (TSMC) scheme was proposed in [17–19]. Then, the advantage control scheme of TSMC, called Nonsingular Fast Terminal Sliding Mode Control (NFTSMC), was introduced in [20,21]. This developed control scheme can handle the singularity and fast convergence of the system. Recently, the NFTSMC was combined with synchronization, and this control scheme was applied to a parallel robot manipulator in [7]. It used the cross-coupling error that combined both tracking errors among the active joints and synchronization errors of a parallel robot to fix the actuator's external disturbances and dynamic uncertainties. Thus, the tracking performance of the robot improved significantly. However, the gains of this control scheme are still selected based on conservative estimates of the dynamic manipulator model. Thus, it leads to complications in the highly complicated model in calculations.

Recently, a simple model-free controller called Time-Delay Estimation (TDE) was applied to CKCMs to solve the above issue. The TDE has been employed to control robot manipulators over the last decade because of its efficient computation capability [22–24]. It used time-delayed information to estimate unknown dynamics and disturbances in a sufficiently small time-delay. Lately, the TDE has been combined with Nonsingular Terminal Sliding Mode (NTSM) control [25] to provide highly robust and precise control schemes for robots with a fast convergence finite time. To our best knowledge, control schemes combining the TDE, the NFTSMC and synchronization have not been considered for controlling CKCMs.

Based on the above analysis, a simple model-free synchronization control system for CKCMs based on TDE and NFTSMC is proposed in this paper to pursue simplicity while preserving the robustness of CKCMs.

Comparing to the existing control schemes approach for robot manipulators, the contribution of this paper can be marked as the following significant points:

(1) Unlike the above-mentioned control schemes, the proposed control scheme TDE-based NFTSMC with synchronization is proposed for the first time.
(2) A new control scheme is proposed based on the combination of TDE-based NFTSMC and synchronization control.
(3) The proposed control scheme is to optimally synchronize the robot joints to minimize the synchronization errors with a NFTSMC-based controller while the robot dynamics and disturbances are estimated and compensated by a TDE-based subsystem.

This paper presents the computer simulation studies of the performance of the proposed control scheme using Matlab-Simulink. Comparative studies with other existing control schemes will be conducted.

The paper is organized as follows. Section 2 presents the structure of the proposed control scheme. Section 3 presents the control scheme analysis without TDE while Section 4 presents the description of subsystems and discusses their simplicity and efficiency. The stability and the stability provided by the control scheme is analyzed and discussed in Section 5. Section 6 presents and discusses results of computer simulation conducted to study the performance of the control scheme applied to control the motion of a 2 DOF

CKCM manipulator in comparison with other existing control schemes. Finally, Section 7 concludes the paper with a summary of the paper and final comments.

2. Structure of the Control Scheme

2.1. Kinematic Scheme of the CKCM

The structures of the 2 DOF CKCM manipulator and the frame assignment are shown in Figures 1 and 2, respectively. Figure 1 shows a two DOF CKCM manipulator, which is a special case of the n-DOF CKCM manipulator. It consists of an end-effector platform and a fixed upper platform interconnected by two links. All links act in a parallel manner and share the same payload.

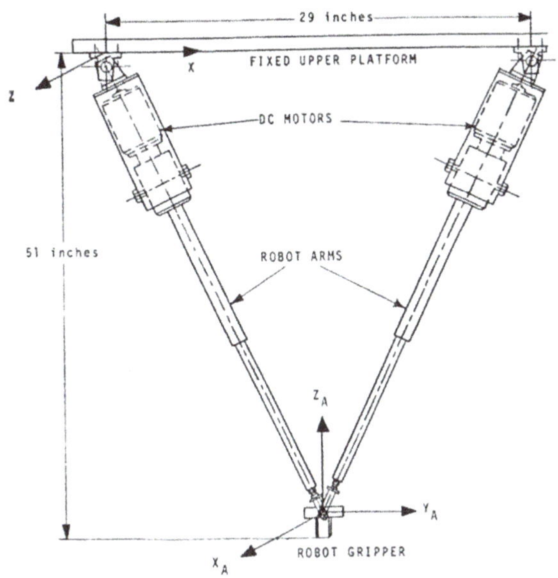

Figure 1. Architecture of the 2 DOF CKCM manipulator.

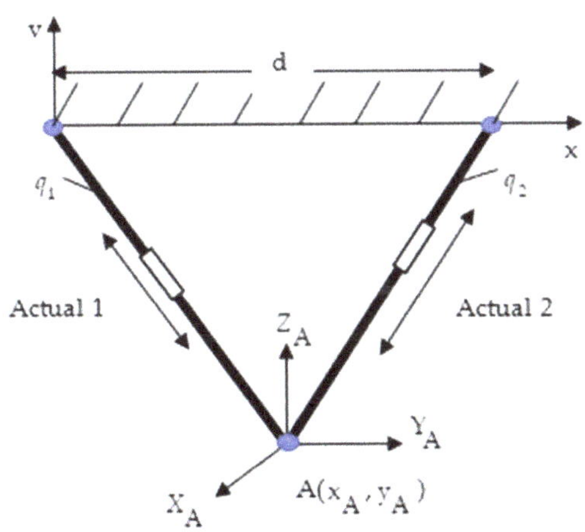

Figure 2. Frame assignment for the 2 DOF planar CKCM manipulator.

Figure 2 depicts the frame assignment for the two DOF planar CKCM manipulator with a two-dimensional coordinate system (x, y). From this figure, we obtain the following:

$$q_1^2 = x^2 + y^2 \tag{1}$$

$$q_2^2 = (d - x)^2 + y^2 \tag{2}$$

where d is the distance between the pin joints hanging the two actuators, (x, y) represents the Cartesian position of the end-effector, and q_2 and q_2 are the length of the first and second legs, respectively.

We see that (1) and (2) represent a closed-form solution for the inverse kinematics in the sense that they can be used to determine the leg lengths q_1 and q_2 that yield a desired Cartesian position (x, y).

Moreover, from (1) and (2), the Cartesian variables x and y can be obtained as follows.

$$x = \frac{q_1^2 - q_2^2 + d^2}{2d} \tag{3}$$

$$y = \frac{-\sqrt{4d^2q_1^2 - (q_1^2 - q_2^2 + d^2)}}{2d} \tag{4}$$

We see that (3) and (4) represent a closed-form solution for the forward kinematics in the sense that a Cartesian position can be determined based the actual leg lengths q_1 and q_2. We note that, due to the small number of DOFs of this manipulator, it has closed-form solutions for both its forward and inverse kinematics.

2.2. Structure of the Proposed Control Scheme

The structure of the proposed control scheme is presented in Figure 2. It mainly consists of three subsystems: the Synchronization Subsystem, the NFTSMC Subsystem. and the TDE Subsystem.

The notations used in Figure 3 are listed below:

- $\mathbf{x}_d \in \mathrm{R}^n$: the desired Cartesian configuration vector. (Note: Configuration means both position and orientation of the CKCM);
- $\mathbf{q}_d \in \mathrm{R}^n$, $\mathbf{q} \in \mathrm{R}^n$ and $\ddot{\mathbf{q}} \in \mathrm{R}^n$: the desired joint vector, actual joint vector, and actual acceleration vector, of the CKCM, respectively;
- $\mathbf{e}_c \in \mathrm{R}^n$: the synchronization error vector;
- $\mathbf{u} \in \mathrm{R}^n$: the control law vector of the NFTSMC;
- $\overline{\mathbf{M}}(\mathbf{q}) \in \mathrm{R}^{n \times n}$: constant, diagonal matrix selected by the TDE;
- $\mathbf{r} \in \mathrm{R}^n$: the output vector of the NFTSMC Subsystem;
- $\boldsymbol{\tau} \in \mathrm{R}^n$: the compensated control input vector to the CKCM;
- $\boldsymbol{\tau}_d \in \mathrm{R}^n$: the external disturbances vector;
- L: the estimate time delay of the TDE;
- $\ddot{\mathbf{q}}_{t-L} \in \mathrm{R}^n$ and $\boldsymbol{\tau}_{t-L} \in \mathrm{R}^n$: the past acceleration vector and past control input vector of the CKCM, respectively;
- $\hat{\mathbf{H}} \in \mathrm{R}^n$: the estimate of all nonlinear terms including the inertia uncertainty, Coriolis/centripetal vector, gravitational vector, friction vector, and disturbances.

The operation of the proposed control scheme applied to control the motion of an n-DOF CKCM is described as follows. The desired Cartesian vector $\mathbf{x}_d$ of the manipulator configuration (position and orientation) specified by the user or obtained by a trajectory planner is transformed to its corresponding desired joint vector $\mathbf{q}_d$ by the CKCM Inverse Kinematic Transformation. The desired joint vector $\mathbf{q}_d$ and the actual joint vector $\mathbf{q}$ (provided by the CKCM joint sensors) are supplied to the Synchronization Subsystem, which then produces the position errors $\mathbf{e}_i = \mathbf{q}_{d_i} - \mathbf{q}_i$ of every ith active joint, the synchronization error $\mathbf{e}_S$, and the cross-coupling error $\mathbf{e}_c$ between the active joints. The cross-coupling error $\mathbf{e}_c$ is then inputted to the NFTSMC subsystem that in turn based on $\mathbf{e}_c$ defines a sliding surface to achieve the above desired behavior of the errors. Next, the control law $\mathbf{u}$ composed based on the sliding surface will be a driving component forcing the system's trajectory to reach a stable hyperplane (sliding surface). This control law $\mathbf{u}$ will then serve as part of the input $\boldsymbol{\tau}$ to the CKCM and is developed to ensure asymptotic convergence of the errors of both position and synchronization of the CKCM in a finite and minimal time. The TDE subsystem uses the past control input and acceleration of the CKCM to estimate the CKCM dynamics and the disturbance torques, which are required for the implementation of the input $\boldsymbol{\tau}$ to CKCM.

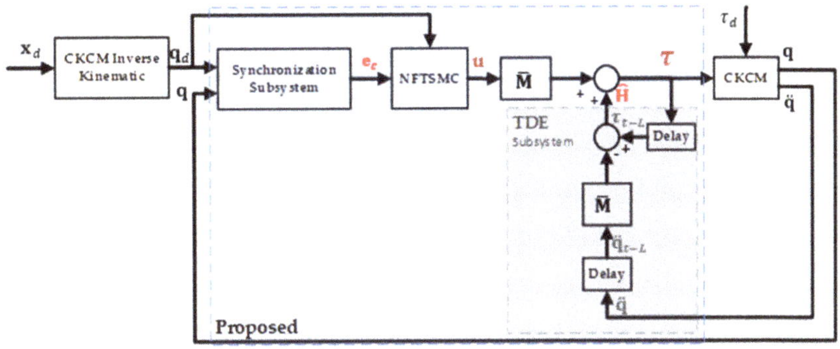

Figure 3. Structure of the Proposed Control Scheme.

3. Control Scheme Analysis without TDE

The dynamics of an n-DOF CKCM manipulator can be represented in joint-space as follows [25]:

$$\mathbf{M}(\mathbf{q})\ddot{\mathbf{q}} + \mathbf{C}(\mathbf{q},\dot{\mathbf{q}})\dot{\mathbf{q}} + \mathbf{G}(\mathbf{q}) + \mathbf{F}(\mathbf{q},\dot{\mathbf{q}}) + \boldsymbol{\tau}_d = \boldsymbol{\tau} \tag{5}$$

where $\mathbf{M}(\mathbf{q}) \in \mathrm{R}^{n\times n}$ stands for the generalized inertia matrix, $\mathbf{C}(\mathbf{q},\dot{\mathbf{q}}) \in \mathrm{R}^{n\times n}$ is the Coriolis/centripetal matrix, $\mathbf{G}(\mathbf{q}) \in \mathrm{R}^n$ is the gravitational vector, and $\mathbf{F}(\mathbf{q},\dot{\mathbf{q}}) \in \mathrm{R}^n$ is the friction forces.

Suppose system (5) can be decomposed into n decoupled systems and is presented by the following:

$$\overline{\mathbf{M}}\ddot{\mathbf{q}}+\mathbf{H}(\mathbf{q},\dot{\mathbf{q}},\ddot{\mathbf{q}}) = \boldsymbol{\tau} \tag{6}$$

where $\overline{\mathbf{M}}$ is a constant diagonal matrix, and $\mathbf{H}(\mathbf{q},\dot{\mathbf{q}},\ddot{\mathbf{q}})$ represents the necessary CKCM dynamics and the disturbance torques. Then, from (5) and (6), we obtain the following.

$$\mathbf{H}(\mathbf{q},\dot{\mathbf{q}},\ddot{\mathbf{q}})=\left[\mathbf{M}(\mathbf{q})-\overline{\mathbf{M}}\right]\ddot{\mathbf{q}}+\mathbf{C}(\mathbf{q},\dot{\mathbf{q}})\dot{\mathbf{q}}+\mathbf{G}(\mathbf{q})+\mathbf{F}(\mathbf{q},\dot{\mathbf{q}}) + \boldsymbol{\tau}_d \tag{7}$$

Now we apply an input $\boldsymbol{\tau}$ such that the following is the case.

$$\boldsymbol{\tau} = \overline{\mathbf{M}}\mathbf{u} + \mathbf{H}(\mathbf{q},\dot{\mathbf{q}},\ddot{\mathbf{q}}) \tag{8}$$

In order to implement (8), the control system must evaluate $\mathbf{H}(\mathbf{q},\dot{\mathbf{q}},\ddot{\mathbf{q}})$ which in light of (7) requires heavy computation and thereby making the proposed control scheme impractical and not suitable for real-time control applications. Consequently, some computationally efficient estimation of $\mathbf{H}(\mathbf{q},\dot{\mathbf{q}},\ddot{\mathbf{q}})$ is needed for the control scheme, which the TDE Subsystem could provide. Finally, the Synchronization Subsystem will enhance the overall performance of the control system by using the synchronization errors instead of the conventional joint errors. Detailed development of the above subsystems will be presented below.

4. Description of Subsystems

This section presents the function of the three subsystems of the control structure including the Synchronization Subsystem, the NFTSMC Subsystem, and the TDE Subsystem.

4.1. The Synchronization Subsystem

In this section, the error equations will be developed to achieve the synchronization goal. First the trajectory tracking error $\mathbf{e}_i$ of the ith active joint is defined as follows:

$$\mathbf{e}_i(t) = \mathbf{q}_{d_i}(t) - \mathbf{q}_i(t) \tag{9}$$

where $\mathbf{q}_{d_i}(t)$ and $\mathbf{q}_i(t)$ denote the desired and actual trajectories of the ith active joint, respectively.

Then, the tracking error vector $\mathbf{e}(t)$ can be written as follows.

$$\mathbf{e}(t) = [\mathbf{e}_1(t)\ \mathbf{e}_2(t) \ldots \mathbf{e}_n(t)]^T \tag{10}$$

The synchronization goal is to make the tracking errors of all active joints identical at all times, which can be achieved if the following is the case.

$$\mathbf{e}_1(t) = \mathbf{e}_2(t) = \ldots \mathbf{e}_n(t) \tag{11}$$

In order for (11) to be valid, a control scheme must be aware of all the joint errors and must control the motions of all joints, thereby resulting into possible control and communication errors and heavy real-time computational requirements. Alternatively, (11) can be satisfied by achieving its following sub-goals [8]:

$$\begin{aligned} \mathbf{e}_{s_1}(t) &= 2\mathbf{e}_1(t) - [\mathbf{e}_2(t) + \mathbf{e}_n(t)] \\ &\vdots \\ \mathbf{e}_{s_i}(t) &= 2\mathbf{e}_i(t) - [\mathbf{e}_{i+1}(t) + \mathbf{e}_{i-1}(t)] \\ &\vdots \\ \mathbf{e}_{s_n}(t) &= 2\mathbf{e}_n(t) - [\mathbf{e}_{n-1}(t) + \mathbf{e}_1(t)] \end{aligned} \tag{12}$$

where $\mathbf{e}_{s_i}(t)$ presents the synchronization errors of the ith active joint.

It is evident that if all synchronization errors in (12) are equal to zero, then the original synchronization goal stated in (11) is automatically achieved.

From (12), a synchronization error vector $\mathbf{e}_s$ can be written as follows:

$$\mathbf{e_s}(t) = \begin{bmatrix} 2 & -1 & 0 & \cdots & -1 \\ -1 & 2 & -1 & \cdots & 0 \\ \vdots & \ddots & \ddots & \ddots & \vdots \\ 0 & \cdots & -1 & 2 & -1 \\ -1 & 0 & \cdots & -1 & 2 \end{bmatrix} \begin{bmatrix} e_1(t) \\ e_2(t) \\ \vdots \\ e_{n-1}(t) \\ e_n(t) \end{bmatrix} = \mathbf{Ce}(t) \tag{13}$$

where $\mathbf{C}$ is the synchronization transformation matrix and $\mathbf{e}_s(t) = [e_{s_1}(t)e_{s_2}(t)e_{s_3}(t)\ldots e_{s_n}(t)]^T$. The cross-coupling error vector that combines both tracking errors and synchronization errors is defined as follows:

$$\mathbf{e}_c(t) = \mathbf{e}(t) + \boldsymbol{\alpha}\mathbf{e}_s(t) = (\mathbf{I} + \boldsymbol{\alpha}\mathbf{C})\mathbf{e}(t) \tag{14}$$

where $\mathbf{I}$ is the $n \times n$ identity matrix, and $\boldsymbol{\alpha}$ is an $n \times n$ diagonal positive definite matrix. Since every leading principal sub-matrix of $(\mathbf{I} + \boldsymbol{\alpha}\mathbf{C})$ has positive determinant, $(\mathbf{I} + \boldsymbol{\alpha}\mathbf{C})$ is positive definite. [26]

Remark 1. *Assuming that all the elements of matrix $\boldsymbol{\alpha}$ are very small, then if $\mathbf{e}_c(t)$ is controlled such that as $t \to \infty$, $\mathbf{e}_c(t) \to 0$ then $\mathbf{e}(t) \to 0$ and then $\mathbf{e}_s(t) \to 0$ and finally $\mathbf{e}_i(t) = \mathbf{e}_{i+1}(t)$ (synchronization goal).*

4.2. The NFTSMC and TDE Subsystems

This section presents the development of the control law of the NFTSMC Subsystem in conjunction with the TDE Subsystem.

4.2.1. Preliminaries and Notations

The preliminaries and notations can be stated as follow

$$\mathbf{x}^{[c]} = |\mathbf{x}|^c \text{sign}(x), \text{ where } c > 0$$

It can be easily verified that as $c \geq 1$, the following is the case.

$$\frac{d}{dt}\mathbf{x}^{[c]} = c|\mathbf{x}|^{c-1}\dot{\mathbf{x}}$$

The sign function is defined as follows.

$$\text{sign}(x) = \begin{cases} 1 & \text{if } x > 0 \\ 0 & \text{if } x = 0 \\ -1 & \text{if } x < 0 \end{cases}$$

The power of error vectors is defined as follows.

$$\begin{aligned}
\mathbf{e}^{[\varphi]} &:= \left(e_1^{[\varphi]}, \ldots, e_n^{[\varphi]}\right)^T \in R^n \\
\dot{\mathbf{e}}^{[\varphi]} &:= \left(\dot{e}_1^{[\varphi]}, \ldots, \dot{e}_1^{[\varphi]}\right)^T \in R^n \\
|\mathbf{e}|^{\varphi-1} &:= diag\left(|e_1|^{\varphi-1}, \ldots, |e_n|^{\varphi-1}\right) \in R^{n \times n} \\
|\dot{\mathbf{e}}|^{\varphi-1} &:= diag\left(|\dot{e}_1|^{\varphi-1}, \ldots, |\dot{e}_n|^{\varphi-1}\right) \in R^{n \times n}
\end{aligned}$$

The spectral norm $\|\mathbf{A}\|$ of a matrix $\mathbf{A} \in R^{n \times m}$ is defined as $\|\mathbf{A}\| = \sqrt{\lambda_{\max}\left\{\mathbf{A}^{\mathrm{T}}\mathbf{A}\right\}}$ where $\lambda_{\max}\left\{\mathbf{A}^{\mathrm{T}}\mathbf{A}\right\}$ is the biggest eigenvalue of $\mathbf{A}^{\mathrm{T}}\mathbf{A}$.

4.2.2. The NFTSMC and TDE Sybstems Design

First a nonsingular terminal sliding surface is defined as [20,27].

$$\mathbf{s} = \mathbf{e}_c + \mathbf{K}_1 \mathbf{e}_c^{p_1/q_1} + \mathbf{K}_2 \dot{\mathbf{e}}_c^{p_2/q_2} \tag{15}$$

where $\mathbf{K}_1$ and $\mathbf{K}_2$ are diagonal design matrices, $1 < p_1/q_1 < 2$, $1 < p_2/q_2 < 2$, and p_1, p_2, q_1, and q_2 are positive odd integers.

As discussed above, the implementation of (8) requires the computation of $\mathbf{H}(\mathbf{q},\dot{\mathbf{q}},\ddot{\mathbf{q}})$, which is in light of (7) highly computationally intensive. Thus, an estimate of $\mathbf{H}(\mathbf{q},\dot{\mathbf{q}},\ddot{\mathbf{q}})$ is needed. We assume that L is the smallest obtainable time between which $\mathbf{H}(\mathbf{q},\dot{\mathbf{q}},\ddot{\mathbf{q}})$ remains almost unchanged, such that an estimate of $\mathbf{H}(\mathbf{q},\dot{\mathbf{q}},\ddot{\mathbf{q}})$, namely $\hat{\mathbf{H}}(\mathbf{q},\dot{\mathbf{q}},\ddot{\mathbf{q}})$ is equal to $\mathbf{H}(\mathbf{q},\dot{\mathbf{q}},\ddot{\mathbf{q}})_{t-L}$ which is $\mathbf{H}(\mathbf{q},\dot{\mathbf{q}},\ddot{\mathbf{q}})$ evaluated at $(t-L)$. In other words, we obtain [22] the following:

$$\hat{\mathbf{H}}(\mathbf{q},\dot{\mathbf{q}},\ddot{\mathbf{q}}) = \mathbf{H}(\mathbf{q},\dot{\mathbf{q}},\ddot{\mathbf{q}})_{t-L} \tag{16}$$

where $\ddot{\mathbf{q}}_{t-L}$ can be computed as follows.

$$\ddot{\mathbf{q}}_{t-L} = \frac{\frac{\mathbf{q}_t - \mathbf{q}_{t-L}}{L} - \frac{\mathbf{q}_{t-L} - \mathbf{q}_{t-2L}}{L}}{L} = \frac{(\mathbf{q}_t - 2\mathbf{q}_{t-L} + \mathbf{q}_{t-2L})}{L^2} \tag{17}$$

From (6) and (16), we obtain the following.

$$\hat{\mathbf{H}}(\mathbf{q},\dot{\mathbf{q}},\ddot{\mathbf{q}}) = \boldsymbol{\tau}_{t-L} - \overline{\mathbf{M}}\ddot{\mathbf{q}}_{t-L} \tag{18}$$

Consequently, (8) can be expressed as follows.

$$\boldsymbol{\tau} = \hat{\mathbf{H}}(\mathbf{q},\dot{\mathbf{q}},\ddot{\mathbf{q}}) + \overline{\mathbf{M}}\mathbf{u} \tag{19}$$

We proceed to develop a control $\mathbf{u}$ as follows.

$$\mathbf{u} = \ddot{\mathbf{q}}_d + \frac{q_2}{p_2}[\mathbf{K}_2(\mathbf{I}+\alpha\mathbf{C})]^{-1}\left[|\dot{\mathbf{e}}_c|^{p_2/q_2-1}\right]^{-1}\left(1+\frac{p_1}{q_1}\mathbf{K}_1|\mathbf{e}_c|^{p_1/q_1-1}\right)\dot{\mathbf{e}}_c + \mathbf{K}\mathbf{s} + \mathbf{K}_{sw}\mathrm{sign}(\mathbf{s}) \tag{20}$$

where $\mathbf{K}$ and $\mathbf{K}_{sw}$ are diagonal design matrices, $\mathrm{sign}(\mathbf{s}) = (\mathrm{sign}(\mathbf{s}_1),\ldots,\mathrm{sign}(\mathbf{s}_n))^T \in \mathrm{R}^n$. $\mathbf{x}_d \in \mathrm{R}^n$.

Thus, replacing $\mathbf{u}$ in (19) by (20) and using (18), we obtain the following.

$$\boldsymbol{\tau} = \underbrace{(\boldsymbol{\tau}_{t-L} - \overline{\mathbf{M}}\ddot{\mathbf{q}}_{t-L})}_{\text{TDE}} + \underbrace{\overline{\mathbf{M}}\left[\ddot{\mathbf{q}}_d + \frac{q_2}{p_2}[\mathbf{K}_2(\mathbf{I}+\alpha\mathbf{C})]^{-1}\left[|\dot{\mathbf{e}}_c|^{p_2/q_2-1}\right]^{-1}\left(1+\frac{p_1}{q_1}\mathbf{K}_1|\mathbf{e}_c|^{p_1/q_1-1}\right)\dot{\mathbf{e}}_c + \mathbf{K}\mathbf{s} + \mathbf{K}_{sw}\mathrm{sign}(\mathbf{s})\right]}_{\text{NFTSM control}} \tag{21}$$

As indicated above in Equation (21), CKCM input $\boldsymbol{\tau}$ consists of two main components: a TDE-based input and an NFTSM input. The TDE-based component minimizes the impact of the unknown CKCM dynamics while the NFTSM component forces the cross-coupling errors $\mathbf{e}_c$ and tracking error $\mathbf{e}$ to converge to zero asymptotically. Furthermore, in light of the application of the TDE Subsystem as presented above and reflected in Figure 1, the TDE-based component can be derived with the estimate $\hat{\mathbf{H}}(\mathbf{q},\dot{\mathbf{q}},\ddot{\mathbf{q}})$ of $\mathbf{H}(\mathbf{q},\dot{\mathbf{q}},\ddot{\mathbf{q}})$ using (16), instead of having to compute $\mathbf{H}(\mathbf{q},\dot{\mathbf{q}},\ddot{\mathbf{q}})$ directly, thereby making the control scheme highly efficient.

5. Stability Analysis

This section presents the stability analysis of the control scheme using the Lyapunov Theorem. Substituting $\boldsymbol{\tau}$ in (6) by (21), using (9) and (14) and solving for $\ddot{\mathbf{e}}_c$, after rearranging some terms, we obtain the following:

$$\ddot{\mathbf{e}}_c = -\left\{\frac{q_2}{p_2}[\mathbf{K}_2]^{-1}\left[|\dot{\mathbf{e}}_c|^{p_2/q_2-1}\right]^{-1}\left(1+\frac{p_1}{q_1}\mathbf{K}_1|\mathbf{e}_c|^{p_1/q_1-1}\right)\dot{\mathbf{e}}_c + (\mathbf{I}+\alpha\mathbf{C})[\mathbf{K}\mathbf{s} + \mathbf{K}_{sw}\mathrm{sign}(\mathbf{s})]\right\} + (\mathbf{I}+\alpha\mathbf{C})\boldsymbol{\varepsilon} \tag{22}$$

where the TDE error $\boldsymbol{\varepsilon}$ is defined as follows.

$$\boldsymbol{\varepsilon} = \overline{\mathbf{M}}^{-1}[\mathbf{H}(\mathbf{q},\dot{\mathbf{q}},\ddot{\mathbf{q}}) - \hat{\mathbf{H}}(\mathbf{q},\dot{\mathbf{q}},\ddot{\mathbf{q}})] \tag{23}$$

Using (15), $\dot{\mathbf{s}}$ can be obtained as follows.

$$\dot{\mathbf{s}} = \dot{\mathbf{e}}_c + \frac{p_1}{q_1}\mathbf{K}_1|\mathbf{e}_c|^{p_1/q_1-1}\dot{\mathbf{e}}_c + \frac{p_2}{q_2}\mathbf{K}_2|\dot{\mathbf{e}}_c|^{p_2/q_2-1}\ddot{\mathbf{e}}_c \tag{24}$$

Next, we consider a candidate Lyapunov function $V = \frac{\mathbf{s}^{\mathrm{T}}\mathbf{s}}{2}$. Using (24), the derivative of V with respect to time is obtained as follows.

$$\dot{V} = \mathbf{s}^{\mathrm{T}}\dot{\mathbf{s}} = \mathbf{s}^{\mathrm{T}}\left[\dot{\mathbf{e}}_c + \frac{p_1}{q_1}\mathbf{K}_1|\mathbf{e}_c|^{p_1/q_1-1}\dot{\mathbf{e}}_c + \frac{p_2}{q_2}\mathbf{K}_2|\dot{\mathbf{e}}_c|^{p_2/q_2-1}\ddot{\mathbf{e}}_c\right] \tag{25}$$

Applying (24) in (25) provides the following.

$$\dot{V} = \mathbf{s}^{\mathrm{T}}\left\{\frac{p_2}{q_2}(\mathbf{I}+\alpha\mathbf{C})\mathbf{K}_2|\dot{\mathbf{e}}_c|^{p_1/q_1-1}[-\mathbf{K}\mathbf{s} - \mathbf{K}_{sw}\mathrm{sign}(\mathbf{s}) + \boldsymbol{\varepsilon}]\right\} \tag{26}$$

In other to achieve the asymptotic stability of $\dot{\mathbf{s}}$ about the equilibrium point $\mathbf{s} = 0$, the following conditions must be satisfied [28]:

(a) $\dot{V} < 0$ for $\mathbf{s} \neq 0$;
(b) $\lim_{|\mathbf{s}|\to\infty} V = \infty$.

Condition (b) is obviously satisfied by V. In (26), since p_2 and q_2 are positive integers and $1 < p_2/q_2 < 2$, there is $|\dot{\mathbf{e}}_{c_i}|^{p_2/q_2-1} > 0$ for $\dot{\mathbf{e}}_{c_i} \neq 0$ [29].

Thus, (26) can be presented as follows.

$$\dot{V} = \mathbf{s}^{\mathrm{T}}\left\{\frac{p_2}{q_2}(\mathbf{I} + \alpha\mathbf{C})\mathbf{K}_2|\dot{\mathbf{e}}_c|^{p_1/q_1-1}[-\mathbf{K}\mathbf{s} - \mathbf{K}_{sw}\mathrm{sign}(\mathbf{s}) + \boldsymbol{\varepsilon}]\right\} \leq \mathbf{s}^{\mathrm{T}}[-\mathbf{K}\mathbf{s} - \mathbf{K}_{sw}\mathrm{sign}(\mathbf{s}) + \boldsymbol{\varepsilon}] \tag{27}$$

The derivative of the candidate Lyapunov function (27) is negative definite if the following is the case:

$$\{\mathbf{K}_{sw}\}_{ii} > |\varepsilon_i| \tag{28}$$

where $\bullet_{ii}$ denotes ith diagonal element of $\bullet$.

Thus, if $\boldsymbol{\varepsilon}$ is bounded, the stability condition (28) ensures that the time derivative of the candidate Lyapunov function is negative and the cross-coupling error is bounded.

Using (19), (6) becomes the following.

$$\mathbf{H}(\mathbf{q},\dot{\mathbf{q}},\ddot{\mathbf{q}}) - \hat{\mathbf{H}}(\mathbf{q},\dot{\mathbf{q}},\ddot{\mathbf{q}}) = \overline{\mathbf{M}}(\mathbf{u} - \ddot{\mathbf{q}}) \tag{29}$$

Applying (29), the TDE error in (23) is given as follows.

$$\boldsymbol{\varepsilon} = \mathbf{u} - \ddot{\mathbf{q}} \tag{30}$$

From (5), the acceleration $\ddot{\mathbf{q}}$ can be determined as follows.

$$\ddot{\mathbf{q}} = \mathbf{M}^{-1}(\mathbf{q})[\boldsymbol{\tau} - \mathbf{C}(\mathbf{q},\dot{\mathbf{q}})\dot{\mathbf{q}} - \mathbf{G}(\mathbf{q}) - \mathbf{F}(\mathbf{q},\dot{\mathbf{q}}) - \boldsymbol{\tau}_d] \tag{31}$$

Substituting (31) into (30) yields the following.

$$\mathbf{M}\boldsymbol{\varepsilon} = \mathbf{M}\mathbf{u} + (\mathbf{C}(\mathbf{q},\dot{\mathbf{q}})\dot{\mathbf{q}} + \mathbf{G}(\mathbf{q}) + \mathbf{F}(\mathbf{q},\dot{\mathbf{q}}) + \boldsymbol{\tau}_d - \boldsymbol{\tau}) \tag{32}$$

Then, using (16) and (19), (32) becomes the following.

$$\mathbf{M}\boldsymbol{\varepsilon} = \mathbf{M}\mathbf{u} + (\mathbf{C}(\mathbf{q},\dot{\mathbf{q}})\dot{\mathbf{q}} + \mathbf{G}(\mathbf{q}) + \mathbf{F}(\mathbf{q},\dot{\mathbf{q}}) + \boldsymbol{\tau}_d - \overline{\mathbf{M}}\mathbf{u} - \mathbf{H}(\mathbf{q},\dot{\mathbf{q}},\ddot{\mathbf{q}})_{t-L}) \tag{33}$$

From (7), the delayed nonlinear term can be derived as follows.

$$\mathbf{H}(\mathbf{q},\dot{\mathbf{q}},\ddot{\mathbf{q}})_{t-L} = (\mathbf{M}_{t-L} - \overline{\mathbf{M}})\ddot{\mathbf{q}}_{t-L} + (\mathbf{C}(\mathbf{q},\dot{\mathbf{q}})\dot{\mathbf{q}})_{t-L} + (\mathbf{G}(\mathbf{q}))_{t-L} + (\mathbf{F}(\mathbf{q},\dot{\mathbf{q}}))_{t-L} + (\boldsymbol{\tau}_d)_{t-L} \tag{34}$$

Substituting (34) into (33) provides the following:

$$\mathbf{M}\boldsymbol{\varepsilon} = (\mathbf{M} - \overline{\mathbf{M}})\mathbf{u} - (\mathbf{M}_{t-L} - \overline{\mathbf{M}})\ddot{\mathbf{q}}_{t-L} + \boldsymbol{\Omega} \tag{35}$$

where the following is the case.

$$\boldsymbol{\Omega} = \mathbf{C}(\mathbf{q},\dot{\mathbf{q}})\dot{\mathbf{q}} + \mathbf{G}(\mathbf{q}) + \mathbf{F}(\mathbf{q},\dot{\mathbf{q}}) + \boldsymbol{\tau}_d - (\mathbf{C}(\mathbf{q},\dot{\mathbf{q}})\dot{\mathbf{q}})_{t-L} - (\mathbf{G}(\mathbf{q}))_{t-L} - (\mathbf{F}(\mathbf{q},\dot{\mathbf{q}}))_{t-L} - (\boldsymbol{\tau}_d)_{t-L} \tag{36}$$

The friction term $\mathbf{F}$ in (36) can be divided as $\mathbf{F} = \mathbf{F}_V + \mathbf{F}_C$, where $\mathbf{F}_V$ denotes that the viscous friction is continuous, and $\mathbf{F}_C$ denotes that the Coulomb friction is bounded and discontinuous at velocity reversal [25]. Next, we divide $\boldsymbol{\Omega}$ into continuous term and discontinuous term to obtain the following:

$$\boldsymbol{\Omega} \triangleq \boldsymbol{\Omega}_{\mathrm{con}} + \boldsymbol{\Omega}_{\mathrm{discon}} \tag{37}$$

where the following is the case.

$$\left.\begin{array}{l}\boldsymbol{\Omega}_{\text{con}} \triangleq \mathbf{C}(\mathbf{q},\dot{\mathbf{q}})\dot{\mathbf{q}} + \mathbf{G}(\mathbf{q}) + \mathbf{F}_V + \boldsymbol{\tau}_d - (\mathbf{C}(\mathbf{q},\dot{\mathbf{q}})\dot{\mathbf{q}})_{t-L} - (\mathbf{G}(\mathbf{q}))_{t-L} - (\mathbf{F}(\mathbf{q},\dot{\mathbf{q}}))_{t-L} - (\boldsymbol{\tau}_d)_{t-L} \\ \boldsymbol{\Omega}_{\text{discon}} \triangleq \mathbf{F}_C - (\mathbf{F}_C)_{t-L}\end{array}\right\} \tag{38}$$

If $\mathbf{C}(\mathbf{q},\dot{\mathbf{q}})\dot{\mathbf{q}} + \mathbf{G}(\mathbf{q}) + \mathbf{F}_V + \boldsymbol{\tau}_d$ is continuous and bounded, then $\boldsymbol{\Omega}_{\text{con}} = \mathbf{O}(L^2)$, where $\mathbf{O}$ is used to describe the error term in an approximation to a mathematical function [30].

In addition, the discontinuous term $\boldsymbol{\Omega}_{\text{discon}}$ is described as follows.

$$\boldsymbol{\Omega}_{\text{discon}} \le \begin{cases} \mathbf{b}, & \text{at velocity reversal} \\ 0, & \text{otherwise.} \end{cases} \tag{39}$$

Thus, $\boldsymbol{\Omega}$ is bounded by the following:

$$\boldsymbol{\Omega} \le \mathbf{b} + \mathbf{O}\left(L^2\right) \tag{40}$$

for a sufficient small L, where $\mathbf{b}$ is a constant vector. The approximation error can be made small by reducing sampling time L.

From (30), the delayed nonlinear term is given by the following.

$$\ddot{\mathbf{q}}_{t-L} = \mathbf{u}_{t-L} - \boldsymbol{\varepsilon}_{t-L} \tag{41}$$

Substituting (41) to (35) yields the following

$$\begin{aligned}\mathbf{M}\boldsymbol{\varepsilon} &= \left[(\mathbf{M}-\overline{\mathbf{M}})\mathbf{u} - (\mathbf{M}-\overline{\mathbf{M}})\ddot{\mathbf{q}}_{t-L} + (\mathbf{M}-\mathbf{M}_{t-L})\ddot{\mathbf{q}}_{t-L} + \boldsymbol{\Omega}\right] \\ &= (\mathbf{M}-\overline{\mathbf{M}})\boldsymbol{\varepsilon}_{t-L} + \left[(\mathbf{M}-\overline{\mathbf{M}})(\mathbf{u}-\mathbf{u}_{t-L}) + (\mathbf{M}-\mathbf{M}_{t-L})\ddot{\mathbf{q}}_{t-L} + \boldsymbol{\Omega}\right]\end{aligned} \tag{42}$$

Therefore, from (42), $\boldsymbol{\varepsilon}$ can be determined as follows:

$$\boldsymbol{\varepsilon} = \mathbf{E}\boldsymbol{\varepsilon}_{t-L} + \mathbf{E}\boldsymbol{\mu}_1 + \boldsymbol{\mu}_2 \tag{43}$$

where the following is the case.

$$\mathbf{E} = \mathbf{I} - \mathbf{M}^{-1}\overline{\mathbf{M}},\ \boldsymbol{\mu}_1 = \mathbf{u} - \mathbf{u}_{t-L},\ \boldsymbol{\mu}_2 = \mathbf{M}^{-1}\left[(\mathbf{M}-\mathbf{M}_{t-L})\ddot{\mathbf{q}}_{t-L} + \boldsymbol{\Omega}\right] \tag{44}$$

For a sufficiently small-time delay L, $\boldsymbol{\mu}_1$ and $\boldsymbol{\mu}_2$ are bounded.

There is a conformal mapping on the complex plane from continuous-time to discrete-time [31]. In the discrete time domain, (43) is represented as follows.

$$\boldsymbol{\varepsilon}(k) = \mathbf{E}(k)\boldsymbol{\varepsilon}(k-1) + \mathbf{E}(k)\boldsymbol{\mu}_1(k) + \boldsymbol{\mu}_2(k) \tag{45}$$

We assume $\|\mathbf{E}\| < 1$ by properly selecting $\overline{\mathbf{M}}$ [31]. Thus, the eigenvalues of $\mathbf{E}(k)$ reside inside a unit circle [32]. As a result, (45) is asymptotically bounded with bounded function $\boldsymbol{\mu}_1$ and $\boldsymbol{\mu}_2$. Therefore, $\|\mathbf{E}\| < 1$ implies the boundness of the $\boldsymbol{\varepsilon}$ in (43).

When $\boldsymbol{\varepsilon}$ is bounded, then (28) is satisfied and as a result, the candidate Lyapunov function (27) is negative definite. Thus, this assures of the boundedness of the cross-coupling error. Furthermore, when the cross-coupling error is bounded, the tracking error is bounded. Consequently, all the above errors and the TDE error $\boldsymbol{\varepsilon}$ will never grow out of bound and the system is uniformly stable.

6. Computer Simulation Study

6.1. Simulation Setup

The proposed scheme can be applied for a general n DOF CKCM manipulator. However, the implementation and application for a n DOF manipulator require massive computation effort and hardware complexity. Therefore, a 2 DOF CKCM robot manipulator that resembles a special case of the complete n DOF manipulation was designed and built for the purpose of testing of the results obtained in our projects. Hence, this article is devoted to investigate the simulation study of the obtained results on the 2 DOF manipulator.

In this section, computer simulation will be conducted to study the performance of the above NFTSMC in comparison to other existing control schemes when they are employed to control the motion of a 2-DOF CKCM.

The computer simulation study for the NFTSMC is described in the block diagram given in Figure 3 when n = 2 since this manipulator has two DOFs, and is designed in MATLAB/Simulink environment in Figure 4. When other existing control schemes are applied, then the block labeled as *Proposed* in Figure 3 is replaced by their particular control schemes. For this particular manipulator, it is noted that the length of an actuator is denoted as its joint variable. To facilitate the analysis of tracking errors in Cartesian space, the actual joint variables $\mathbf{q}$ and their joint velocities $\dot{\mathbf{q}}$ of the CKCM are converted to their corresponding Cartesian variables by using the CKCM forward kinematic transformation, which is also a closed-form solution due to the number of DOFs of this manipulator.

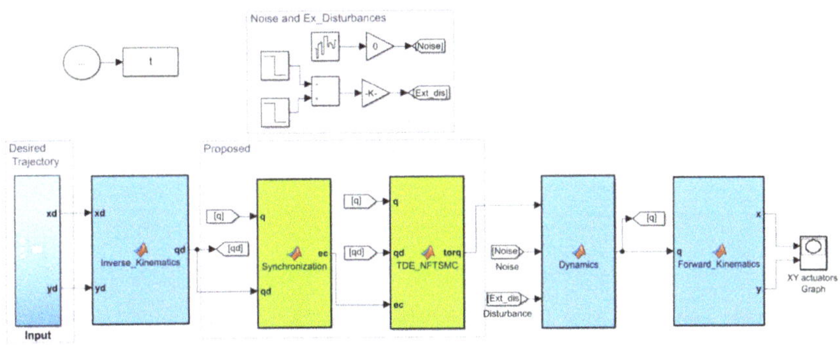

Figure 4. Control design in MatLab/Simulink environment.

MATLAB-Simulink® is used to comparatively evaluate the performance of the developed NFTSMC (Syn-TDE-NFTSMC) in comparison with four other existing control schemes including PD-based control scheme (LINEAR), TDE-based LINEAR (TDE-LINEAR), TDE-based LINEAR with synchronization errors (Syn-TDE-LINEAR), and TDE-based SMC with synchronization errors (Syn-TDE-SMC) in tracking the same motion. A brief description of the above control schemes can be found in Appendix A. After conducting numerous simulations of the above control schemes, we selected the most optimal parameters for their best tracking performance.

The parameters of the manipulator are listed in Table 1 while the control parameters of the control schemes are provided in Table 2.

Remark 2. *The parameters of the control scheme are tuned as: Tuning* $\overline{\mathbf{M}}$ *and* $\boldsymbol{\alpha}$*, diagonal matrices, by increasing the diagonal elements from small positive values, while checking the control performance by trial error. The selection of the other parameters of the proposed control scheme* p_1, p_2, q_1, q_2, $\mathbf{K}_1$, $\mathbf{K}_2$, *and* $\mathbf{K}$ *are described in* [28]. $\mathbf{K}_{sw}$ *can be selected from (28).*

The Lagrangian dynamic equations of the above manipulator is given in [33] as follows:

$$\boldsymbol{\tau} = \mathbf{M}(\mathbf{q})\ddot{\mathbf{q}}(t) + \mathbf{C}(\mathbf{q},\dot{\mathbf{q}})\dot{\mathbf{q}}(t) + \mathbf{G}(\mathbf{q}) + \mathbf{F}(\mathbf{q},\dot{\mathbf{q}}) \tag{46}$$

with

$$\boldsymbol{\tau}(t) = (\tau_1\ \tau_2)^T;\ \mathbf{q}(t) = (q_1\ q_2)^T \tag{47}$$

where τ_i denotes the joint force of the ith actuator, respectively, for i = 1,2.

Table 1. The robot parameters.

Robot Parameters	Description	Value	Unit
m	Link's total mass	4.91	kg
m_1	Link's moving part mass	0.59	kg
d	Grounds' horizontal distance	0.74	m
l_s	Link's fixed length	0.26	m
F_{V_1}	Viscous friction coefficient of the 1st link	5	N·m·s/rad
F_{V_2}	Viscous friction coefficient of the 2nd link	5	N·m·s/rad
F_{C_1}	Coulomb friction coefficient of the 1st link	5	N·m
F_{C_2}	Coulomb friction coefficient of the 2nd link	5	N·m
g	Gravitational acceleration constant	9.81	m/s^2

Table 2. The parameters of the control schemes.

Control Scheme	Control Parameters
LINEAR	$\overline{\mathbf{M}}$= diag(0.1, 0.1), $\mathbf{K}_D$= diag(200, 200), $\mathbf{K}_P$= diag(7.704 × 10^3, 7.704 × 10^3)
TDE-LINEAR	$L = 9.999 \times 10^{-4}$s, $\overline{\mathbf{M}}$ = diag(0.1, 0.1), $\mathbf{K}_D$= diag(200, 200), $\mathbf{K}_P$= diag(7.704 × 10^3, 7.704 × 10^3)
Syn-TDE-LINEAR	$L = 9.999 \times 10^{-4}$s, $\overline{\mathbf{M}}$= diag(0.1, 0.1), $\mathbf{K}_D$= diag(200, 200), $\mathbf{K}_P$= diag(7.704 × 10^3, 7.704 × 10^3), α = diag(0.5,0.5)
Syn-TDE-SMC	$L = 9.999 \times 10^{-4}$s, $\overline{\mathbf{M}}$= diag(0.1, 0.1), $\mathbf{K}_1$ = diag(0.1, 0.1), $\mathbf{K}$= diag(10, 10), $\mathbf{K}_{sw}$= diag(5, 5), α = diag(0.5, 0.5)
Syn-TDE-NFTSMC	$L = 9.999 \times 10^{-4}$s, p_1= 19, p_2= 11, q_1= 17, q_2= 9, $\overline{\mathbf{M}}$= diag(0.1, 0.1), $\mathbf{K}_1$= diag(25, 25), $\mathbf{K}_2$= diag(5, 5), $\mathbf{K}$= diag(0.1, 0.1), $\mathbf{K}_{sw}$ = diag(15, 15), α = diag(0.5, 0.5)

The inertia matrix, the Centrifugal and Coriolis forces, and the friction and the gravitational forces at two joints are given by the following:

$$\mathbf{M} = \begin{bmatrix} m_1 & 0 \\ 0 & m_1 \end{bmatrix}, \ \mathbf{C} = \begin{bmatrix} 0 & \frac{ml_s(q_2-q_1)}{3v} \\ \frac{ml_s(q_2-q_1)}{3v} & 0 \end{bmatrix}, \ \mathbf{G} = [G_1 G_2]^T \tag{48}$$

with

$$\begin{aligned} G_1 &= \left(-m_1 g\left[2v_1 q_1^2(q_1 l_s + q_2 l_s + 2q_1 q_2) - q_2 l_s v^2\right] - mgl_s\left[2q_1^2 v_1(q_1+q_2) - q_2 v^2\right]\right) \Big/ 4dq_1^2 q_2 v' \\ G_2 &= \left(-m_1 g\left[2v_1 q_2^2(q_1 l_s + q_2 l_s + 2q_1 q_2) - q_2 l_s v^2\right] - mgl_s\left[2q_2^2 v_2(q_1+q_2) - q_1 v^2\right]\right) \Big/ 4dq_2^2 q_1 v \end{aligned} \tag{49}$$

$$\mathbf{F} = \begin{bmatrix} F_{V_1}\dot{q}_1 + F_{C_1}\mathrm{sgn}(\dot{q}_1) \\ F_{V_2}\dot{q}_2 + F_{C_2}\mathrm{sgn}(\dot{q}_2) \end{bmatrix} \tag{50}$$

and the following is obtained.

$$v_1 = v_2 = q_2^2 - q_1^2 + d^2, \ v = \sqrt{4d^2 q_1^2 - v_1} \tag{51}$$

6.2. Simulation Results

The control schemes listed in Table 2 are used in the computer simulation to control the end-effector of the manipulator to track a circle specified by $x_{des}(t)$ and $y_{des}(t)$ as follows.

$$\begin{cases} x_{des}(t) = 0.3683 + 0.05\cos(\pi t + \pi/2) \\ y_{des}(t) = -0.4183 - 0.05\sin(\pi t/10 + \pi/2) \end{cases} \tag{52}$$

The results obtained from the simulation are presented in Figures 5–9 and Tables 3–5. Figure 5 shows the planar motions of the manipulator end-effector when controlled by the above control schemes while Figure 6 presents the time trajectories of the tracking errors

$\mathbf{e}(t)$ of the control schemes. Table 3 tabulates the absolute average tracking errors (AATE) of the control schemes computed by MATLAB.

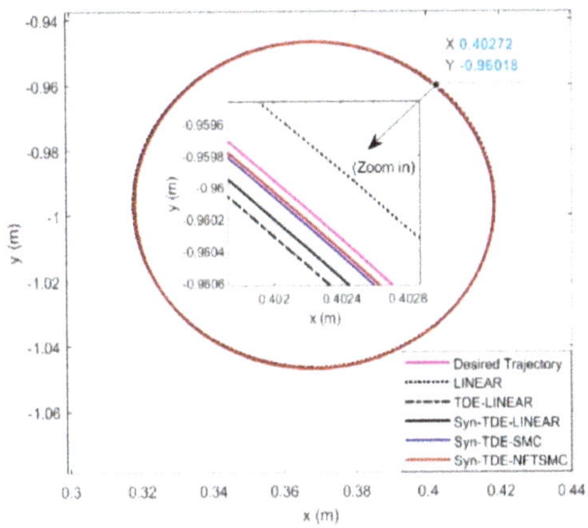

Figure 5. Trajectory tracking in X-Y plane (circular motion).

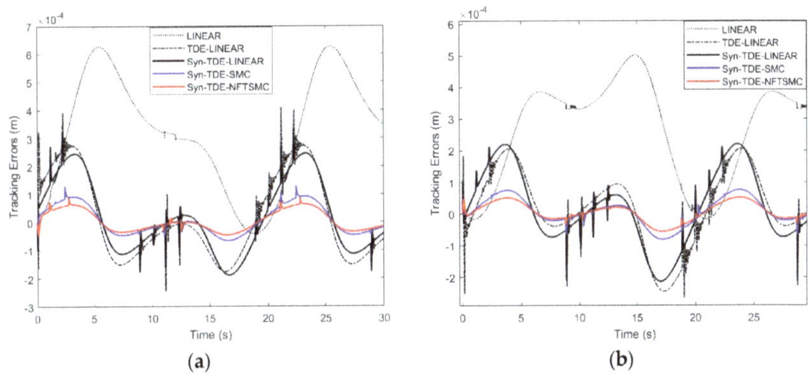

Figure 6. Time trajectories of tracking errors of the control schemes, of joint 1 (**a**) and joint 2 (**b**).

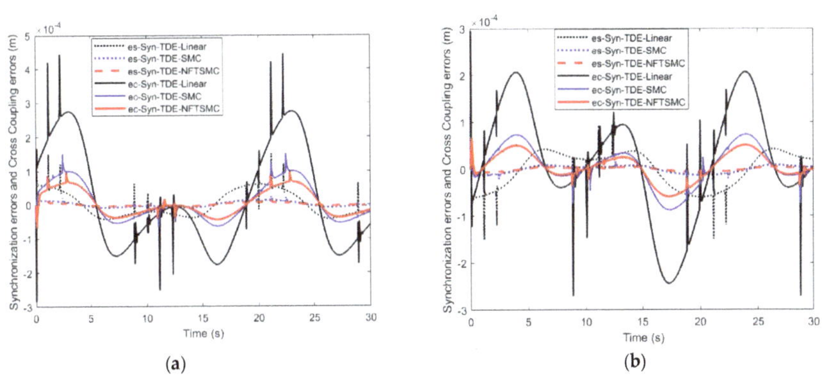

Figure 7. Time trajectories of synchronization errors and cross coupling errors of the control schemes, of joint 1 (**a**) and joint 2 (**b**).

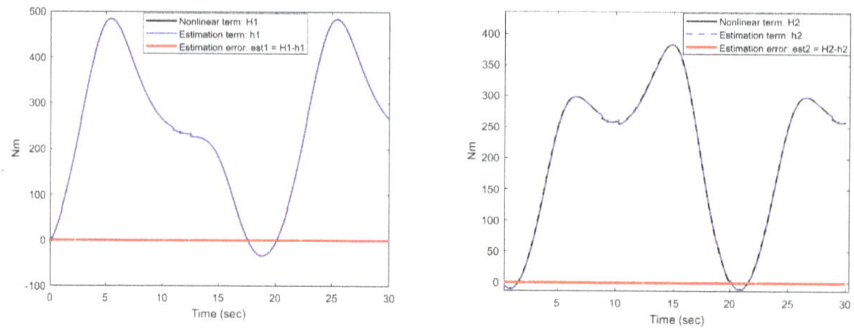

(a) (b)

Figure 8. Time trajectories of estimation errors of Syn-TDE-NFTSMC of joint 1 (**a**) and joint 2 (**b**).

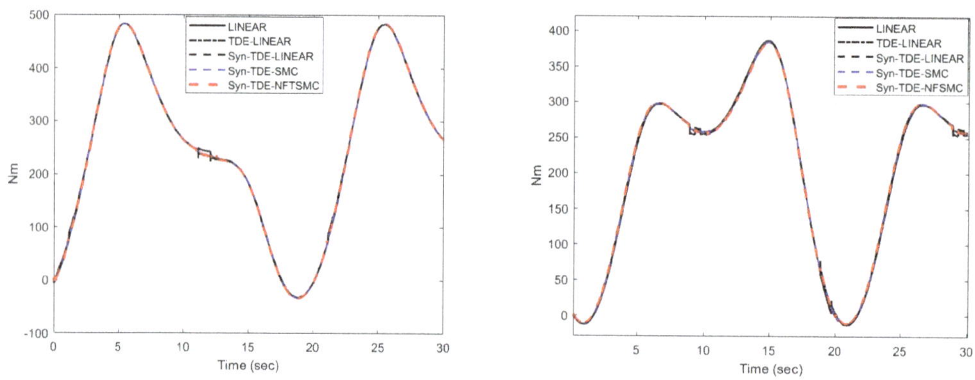

(a) (b)

Figure 9. The control input signals of the control schemes of joint 1 (**a**) and joint 2 (**b**).

Table 3. The absolute average tracking errors (AATE) of the control schemes, computed by MATLAB.

Tracking Errors	LINEAR	TDE-LINEAR	Syn-TDE-LINEAR	Syn-TDE-SMC	Syn-TDE-NFTSMC
e_1 (mm)	0.32	0.0197	0.0197	7.17×10^{-3}	4.07×10^{-3}
e_2 (mm)	0.26	0.0192	0.0192	6.95×10^{-3}	5.06×10^{-3}

Table 4. The absolute average synchronization errors (AASE) and cross-coupling errors (AACE) of the control schemes, computed by MATLAB.

Tracking Errors	Syn-TDE-LINEAR	Syn-TDE-SMC	Syn-TDE-NFTSMC
e_{s_1} (mm)	0.0206	3.82×10^{-3}	2.16×10^{-3}
e_{s_2} (mm)	0.0206	3.82×10^{-3}	2.16×10^{-3}
e_{c_1} (mm)	0.028	1.03×10^{-2}	9.07×10^{-3}
e_{c_2} (mm)	0.016	5.33×10^{-3}	4.99×10^{-3}

Table 5. The absolute average estimation errors (AAEE) of the Syn-TDE-NFTSMC computed by MATLAB.

AAEE	Syn-TDE-NFTSMC
e_{est_1} (Nm)	0.0177
e_{est_2} (Nm)	0.0172

Figure 7 presents the time trajectories of the synchronization errors ($\mathbf{e}_s(t)$) and cross-coupling ($\mathbf{e}_c(t)$) of the control schemes while Figure 8 presents the estimation errors ($\mathbf{e}_{est}(t)$) of the Syn-TDE-NFTSMC. Table 4 tabulates the absolute average synchronization errors (AASE) and cross-coupling errors (AACE) of the control schemes while Table 5 tabulates the absolute average estimation errors (AAEE) of the Syn-TDE-NFTSMC, computed by MATLAB.

The control inputs of both joints show no chattering, as shown in Figure 9a,b.

From Figures 5–7, it is seen that Syn-TDE-LINEAR provided better performance and faster error convergence than both LINEAR and TDE-LINEAR. It is meaningful that the involvement of TDE and the synchronization errors improved the performance of the control schemes. Syn-TDE-SMC showed a better tracking path than the Syn-TDE-LINEAR. Finally, we see that our proposed control scheme, Syn-TDE-NFTSMC, was on track with the fastest desired trajectory with the slightest deviation (from the desired path in Figure 5) and had the fastest error convergence compared to other existing control schemes.

From Table 3, based on the computed AATEs of the control scheme, it is clear that inclusion of TDE and the synchronization errors improved the performance of control schemes as for example the AATE of the TDE-LINEAR (0.0197 mm) is smaller than that of the LINEAR (0.32 mm). Other AATEs in the table validate the above observation. From the results presented in Table 3, we see that our proposed control scheme, namely Syn-TDE-NFTSMC, has the best tracking performance as compared to other existing control schemes due to its smallest AATEs for both joint variables.

From Table 4, based on the computed AASEs and AACEs of the control scheme, it is clear that inclusion of Syn-TDE-NFTSMC improved the performance of control schemes as for example the AASEs of the Syn-TDE-NFTSMC (2.16×10^{-3} mm) is smaller than that of the Syn-TDE-SMC (3.82×10^{-3} mm). Other AASEs and AACEs in the table validate the above observation. From the results presented in Table 4, we see that our proposed control scheme has the best tracking performance as compared to other existing control schemes due to its smallest AASEs and AACEs for both joint variables.

Figure 8a,b plotted the nonlinear term $\mathbf{H}_i = \mathbf{H}(\mathbf{q},\dot{\mathbf{q}},\ddot{\mathbf{q}})$, the estimation term $\mathbf{h}_i = \boldsymbol{\tau}_{t-L} - \overline{\mathbf{M}}\ddot{\mathbf{q}}_{t-L}$, and the estimation error $\mathbf{e}_{est_i} = \mathbf{H}_i - \mathbf{h}_i = \mathbf{H}(\mathbf{q},\dot{\mathbf{q}},\ddot{\mathbf{q}}) - \left(\boldsymbol{\tau}_{t-L} - \overline{\mathbf{M}}\ddot{\mathbf{q}}_{t-L}\right)$, respectively, of the ith active joint. It can be seen that the estimation error remains close to zero. Furthermore, from the results presented in Table 5, we see that our proposed control scheme, namely Syn-TDE-NFTSMC, has the AAEEs close to zero as 0.0177 Nm and 0.0172 Nm for both joint variables. This implies that the TDE cancels the uncertainty, and the chattering phenomenon is reduced while maintaining the tracking accuracy.

Therefore, it can be concluded that the proposed control scheme shows high-accuracy tracking performance with the model-free control performance in comparison with the other control schemes.

7. Conclusions

In this paper, we proposed a new NFTSMC scheme in which TDE was applied to efficiently compute the dynamics of a robot manipulator and disturbances required for control scheme. In addition, the synchronization errors were used instead of the conventional joint errors. A new NFTSMC law was proposed and the Lyapunov Theorem was employed to prove that the proposed control scheme is uniformly stable. The conducted computer simulation showed that the proposed control scheme provided the best tracking performance compared with other existing control schemes including LINEAR, TDE-based LINEAR, TDE-based LINEAR with synchronization errors, and TDE-based SMC with synchronization errors when tracking the same motion for a 2-DOF-CKCM.

Comparing with the existing approach, the proposed control scheme has several significant improvements:

(1) The proposed control scheme optimally synchronized the robot joints to minimize the synchronization errors with an NFTSMC-based controller.

(2) Since the proposed control scheme does not require the computation of the manipulator dynamics thanks to TDE, it is computationally efficent and is, therefore, suitable for real-time control applications.

Future work from this paper could include computer simulation study on higher DOF manipulators and experimental studies of the proposed control scheme on real manipulators.

Author Contributions: Conceptualization, T.T.C.D., C.C.N. and T.D.T.; methodology, T.T.C.D. and C.C.N.; software, T.T.C.D. and C.C.N.; validation, T.T.C.D. and C.C.N.; results analysis, T.T.C.D. and T.D.T.; investigation, C.C.N., T.T.C.D. and T.D.T.; data curation, T.T.C.D. and T.D.T.; writing—original draft preparation, C.C.N., T.T.C.D. and T.D.T.; writing—review and editing, C.C.N. and T.T.C.D.; visualization, C.C.N. and T.T.C.D.; supervision, C.C.N. and T.T.C.D. All authors have read and agreed to the published version of the manuscript.

Funding: This research received no external funding.

Institutional Review Board Statement: Not applicable.

Informed Consent Statement: Not applicable.

Data Availability Statement: Not applicable.

Conflicts of Interest: All authors announce that they have no conflict of interest in relation to the publication of this article.

Appendix A. Other Control Schemes Used in Computer Simulation

This appendix contains relevant equations of control schemes to which the proposed control scheme is compared to in our computer simulation study.

LINEAR

The LINEAR and TDE-LINEAR Control Scheme were suggested in Reference [22] given by the following.

$$\tau = \overline{\mathbf{M}}(\ddot{\mathbf{q}}_d + \mathbf{K}_D\dot{\mathbf{e}} + \mathbf{K}_P\mathbf{e}) \tag{A1}$$

where $\mathbf{K}_D$ and $\mathbf{K}_P$ are constant matrices.

TDE-based LINEAR (TDE-LINEAR)

$$\tau = \underbrace{\tau_{t-L} - \overline{\mathbf{M}}\ddot{\mathbf{q}}_{t-L}}_{\text{TDE}} + \underbrace{\overline{\mathbf{M}}(\ddot{\mathbf{q}}_d + \mathbf{K}_D\dot{\mathbf{e}} + \mathbf{K}_P\mathbf{e})}_{\text{LINEAR control}} \tag{A2}$$

TDE-based LINEAR with synchronization errors (Syn-TDE-LINEAR)

$$\tau = \underbrace{\tau_{t-L} - \overline{\mathbf{M}}\ddot{\mathbf{q}}_{t-L}}_{\text{TDE}} + \underbrace{\overline{\mathbf{M}}(\ddot{\mathbf{q}}_d + \mathbf{K}_D\dot{\mathbf{e}}_c + \mathbf{K}_P\mathbf{e}_c)}_{\text{LINEAR control}} \tag{A3}$$

TDE-based SMC with synchronization errors (Syn-TDE-SMC)

The control scheme was suggested in Reference [13] and is given by the following:

$$\tau = \underbrace{\tau_{t-L} - \overline{\mathbf{M}}\ddot{\mathbf{q}}_{t-L}}_{\text{TDE}} + \underbrace{\overline{\mathbf{M}}\left(\ddot{\mathbf{q}}_d + \mathbf{K}\dot{\mathbf{e}}_c + \mathbf{K}_{sw}\text{sign}(\mathbf{s}) + \mathbf{K}_1\mathbf{s}\right)}_{\text{SMC control}} \tag{A4}$$

where the sliding surface expressed by $\mathbf{s} = \dot{\mathbf{e}}_c + \mathbf{K}\mathbf{e}_c$.

References

1. Patel, Y.D.; George, P.M. Parallel Manipulators Applications—A Survey. *Mod. Mech. Eng.* **2012**, *2*, 57–64. [CrossRef]
2. Nguyen, C.C.; Pooran, F.J. Kinematic analysis and workspace determination of a 6 dof CKCM robot end-effector. *J. Mech. Work. Technol.* **1989**, *20*, 283. [CrossRef]
3. Nguyen, C.C.; Pooran, F.J. Dynamic analysis of a 6 DOF CKCM robot end-effector for dual-arm telerobot systems. *Robot. Auton. Syst.* **1989**, *5*, 377. [CrossRef]

4. Nguyen, C.C.; Sam, S.A.; Zhen, L.Z.; Campbell, C.E., Jr. Adaptive control of a stewart platform-based manipulator. *J. Robot. Syst.* **1993**, *10*, 657. [CrossRef]
5. Koren, Y. Cross-Coupled Biaxial Computer Control for Manufacturing Systems. *J. Dyn. Syst. Meas. Control* **1980**, *102*, 265. [CrossRef]
6. Doan, Q.V.; Le, T.D.; Vo, A.T. Synchronization Full-Order Terminal Sliding Mode Control for an Uncertain 3-DOF Planar Parallel Robotic Manipulator. *Appl. Sci.* **2019**, *9*, 1756. [CrossRef]
7. Zhao, D.; Li, S.; Gao, F. Finite time position synchronised control for parallel manipulators using fast terminal sliding mode. *Int. J. Syst. Sci.* **2009**, *40*, 829. [CrossRef]
8. Nguyen, T.T. Intelligent Control of Closed-Kinematic Chain Robot Manipulators. Ph.D. Thesis, The Catholic University of America, Washington, DC, USA, 2020.
9. Su, Y.; Sun, D.; Ren, L.; Mills, J.K. Integration of saturated PI synchronous control and PD feedback for control of parallel manipulators. *IEEE Trans. Robot. A Publ. IEEE Robot. Autom. Soc.* **2006**, *22*, 202. [CrossRef]
10. Sang, C.; Cong, S. Nonlinear computed torque control for a high-speed planar parallel manipulator. *Mechatronics* **2009**, *19*, 987. [CrossRef]
11. Chung, S.J.; Slotine, J.J.E. Cooperative robot control and synchronization of Lagrangian systems. In Proceedings of the 46th IEEE Conference on Decision and Control, New Orleans, LA, USA, 12–14 December 2007.
12. Nuno, E.; Basanez, L.; Ortega., R. Adaptive control for the synchronization of multiple robot manipulators with coupling time-delays. In Proceedings of the IEEE/RSJ International Conference on Intelligent Robots and Systems, Taipei, Taiwan, 18–22 October 2010. [CrossRef]
13. Duong, T.; Tran, T.D.; Nguyen, T.T.; Duong, V.N. Synchronization Sliding Mode Control with Time-Delay Estimation for a 2-DOF Closed-Kinematic Chain Robot Manipulator. In Proceedings of the International Conference on System Science and Engineering (ICSSE), Ho Chi Minh City, Vietnam, 26–28 August 2021. [CrossRef]
14. Bouaziz, F.A.; Bouteraa, Y.; Medhaffar, H. Distributed sliding mode control of cooperative robotic manipulators. In Proceedings of the 10th International Multi-Conferences on Systems, Signals & Devices 2013 (SSD13), Hammamet, Tunisia, 18–21 March 2013. [CrossRef]
15. Zhao, D.; Li, S.; Gao, F. Fully adaptive feedforward feedback synchronized tracking control for Stewart Platform systems. *Int. J. Control Autom. Syst.* **2008**, *6*, 689–701.
16. Su, Y.X.; Sun, D.; Ren, L.; Wang, X.; Mills, J.K. Nonlinear PD Synchronized Control for Parallel Manipulators. In Proceedings of the 2005 IEEE International Conference on Robotics and Automation, Barcelona, Spain, 18–22 April 2005. [CrossRef]
17. Zhao, D.; Li, S.; Gao, F. A new terminal sliding mode control for robotic manipulators. *Int. J. Control* **2009**, *82*, 1804. [CrossRef]
18. Zhihong, M.; Paplinski, A.P.; Wu, H.R. A robust MIMO terminal sliding mode control scheme for rigid robotic manipulators. *IEEE Trans. Autom. Control* **1994**, *39*, 2464. [CrossRef]
19. Tang, Y. Terminal sliding mode control for rigid robots. *Automatica* **1998**, *34*, 51. [CrossRef]
20. Yang, L.; Yang, J. Nonsingular fast terminal sliding-mode control for nonlinear dynamical systems. *Int. J. Robust Nonlinear Control* **2011**, *21*, 1865. [CrossRef]
21. Van, M.; Ge, S.S.; Ren, H. Finite Time Fault Tolerant Control for Robot Manipulators Using Time Delay Estimation and Continuous Nonsingular Fast Terminal Sliding Mode Control. *IEEE Trans. Cybern.* **2017**, *47*, 1681. [CrossRef]
22. Hsia, T.C.S.; Lasky, T.A.; Guo, Z. Robust independent joint controller design for industrial robot manipulators. *IEEE Trans. Ind. Electron. A Publ. IEEE Ind. Electron. Soc.* **1991**, *38*, 21. [CrossRef]
23. Baek, J.; Cho, S.; Han, S. Practical Time-Delay Control with Adaptive Gains for Trajectory Tracking of Robot Manipulators. *IEEE Trans. Ind. Electron. A Publ. IEEE Ind. Electron. Soc.* **2018**, *65*, 5682. [CrossRef]
24. Bae, H.-J.; Jin, M.; Suh, J.; Lee, J.Y.; Chang, P.-H.; Ahn, D.-s. Control of Robot Manipulators Using Time-Delay Estimation and Fuzzy Logic Systems. *J. Electr. Eng. Technol.* **2017**, *12*, 1271. [CrossRef]
25. Maolin, J.; Jinoh, L.; Pyung, H.C.; Chintae, C. Practical Nonsingular Terminal Sliding-Mode Control of Robot Manipulators for High-Accuracy Tracking Control. *IEEE Trans. Ind. Electron. A Publ. IEEE Ind. Electron. Soc.* **2009**, *56*, 3593. [CrossRef]
26. Kelly, R.; Davila, V.S.; Loría, A. *Control of Robot Manipulators in Joint Space*, 1st ed.; Springer: London, UK, 2005; pp. XXVI, 426. [CrossRef]
27. Xu, Z.; Huang, W.; Li, Z.; Hu, L.; Lu, P. Nonlinear Nonsingular Fast Terminal Sliding Mode Control Using Deep Deterministic Policy Gradient. *Appl. Sci.* **2021**, *11*, 4685. [CrossRef]
28. Feng, Y.; Yu, X.; Man, Z. Non-singular terminal sliding mode control of rigid manipulators. *Automatica* **2002**, *38*, 2159. [CrossRef]
29. Shtessel, Y.; Edwards, C.; Fridman, L.; Levant, A. *Sliding Mode Control and Observation*; Birkhäuser: Basel, Switzerland, 2014.
30. Su, W.-C.; Drakunov, S.V.; Ozguner, U. An O(T/sup 2/) boundary layer in sliding mode for sampled-data systems. *IEEE Trans. Autom. Control* **2000**, *45*, 482. [CrossRef]
31. Spong, M.; Vidyasagar, M. Robust linear compensator design for nonlinear robotic control. *IEEE Trans. Robot. Autom. A Publ. IEEE Robot. Autom. Soc.* **1987**, *3*, 345. [CrossRef]
32. Brunton, S.L.; Kutz, J.N. *Data-Driven Science and Engineering*; Cambridge University Press: Cambridge, UK, 2019. [CrossRef]
33. Pooran, J.F. Dynamics and Control of Robot Manipulators with Closed-Kinematic Chain Mechanism. Ph.D. Thesis, The Catholic University of America, Washington, DC, USA, 1989.

applied sciences

MDPI

Article

Estimating Natural Frequencies of Cartesian 3D Printer Based on Kinematic Scheme

Ekaterina Kopets [1], Artur Karimov [1], Lorenzo Scalera [2] and Denis Butusov [1,*]

[1] Youth Research Institute, Saint Petersburg Electrotechnical University "LETI", 197376 Saint Petersburg, Russia; eekopets@etu.ru (E.K.); aikarimov@etu.ru (A.K.)
[2] Polytechnic Department of Engineering and Architecture, University of Udine, 33100 Udine, Italy; lorenzo.scalera@uniud.it
* Correspondence: dnbutusov@etu.ru

Abstract: Nowadays, 3D printers based on Cartesian kinematics are becoming extremely popular due to their reliability and inexpensiveness. In the early stages of the 3D printer design, once it is chosen to use the Cartesian kinematics, it is always necessary to select relative positions of axes and linear drives (prismatic joints), which would be optimal for the particular specification. Within the class of Cartesian mechanics, many designs are possible. Using the Euler–Lagrange formalism, this paper introduces a method for estimating the natural frequencies of Cartesian 3D printers based on the kinematic scheme. Comparison with the finite element method and experimental validation of the proposed method are given. The method can help to develop preliminary designs of Cartesian 3D printers and is especially useful for emerging 3D-printing technologies.

Keywords: 3D printer; Cartesian kinematics; vibration analysis; additive manufacturing; mechanical design

Citation: Kopets, E.; Karimov, A.; Scalera, L.; Butusov, D. Estimating Natural Frequencies of Cartesian 3D Printer Based on Kinematic Scheme. *Appl. Sci.* **2022**, *12*, 4514. https://doi.org/10.3390/app12094514

Academic Editors: Abílio M.P. De Jesus and Giangiacomo Minak

Received: 23 February 2022
Accepted: 25 April 2022
Published: 29 April 2022

1. Introduction

Robotic systems are becoming more and more popular in various applications since they perform fast and accurate operations while decreasing production costs and reducing tedious and potentially hazardous tasks. One rapidly developing field of application for robots is additive manufacturing, often referred to as 3D printing [1,2]. Having emerged as a tool for rapid prototyping, nowadays 3D printing is also widely used for various industrial applications [3,4].

Industrial robots are highly diverse in many parameters, and several classifications of them can be proposed [5]. Usually, researchers identify six significant kinematics of industrial robots [6]: articulated, SCARA, Cartesian, Parallel (or Delta), Cylindrical, Spherical.

Cartesian mechanical design became a predominant configuration in 3D printing [7,8]. The main advantage of Cartesian robots for additive manufacturing is the lowest cost compared to the other types of robots with the same accuracy and repeatability. The second popular design for 3D printers is the Delta robot, almost as inexpensive as Cartesian but more demanding on production quality and calibration accuracy. This issue sufficiently limits its popularity [9]; nevertheless, special implementation of this type of robot may yield unique advantages, e.g., energy efficiency [10]. Articulated robots gained the most popularity in such applications like welding, assembly, handling, and inspecting [11] and are used only in specific 3D printing applications [12]. SCARA robots, as well as cylindrical robots, are relatively rare in 3D printing but have some prospects in this field [13,14].

As can be expected for the rapidly developing area, the level of scientific comprehension of additive manufacturing sufficiently lags behind the practice. For example, a study of publications on the Design for additive manufacturing (DfAM) found that the main content of these works is guidelines, rules and certain practical aspects, papers are mainly concentrated in a small number of clusters and the level of international collaboration in them is weak [15]. There is a similar situation with publications on the theory of 3D printer design. This paper tries to introduce one more important theoretical aspect of additive manufacturing into scientific discourse.

The printing rate of the 3D printer is one of the most critical consumer characteristics, directly affecting its user experience and economic returns. A recent study by J. Go et al. proposed that three main factors limit the building rate in FDM/FFF technology: the rate of filament feed, the rate of filament heating, and velocity of the printhead positioning [16]. The first and second factors have been addressed in several recent studies. For example, a novel rotatable extruder with a slot-shaped nozzle was designed to increase the filament feed rate [17], and the heating rate was sufficiently increased by introducing a laser heating system [18]. In the article [19], the authors propose to use fuzzy adaptive control for balanced thermal distribution during extrusion, in order to achieve the highest quality rapid prototyping. However, increasing the speed of printhead positioning is probably the most challenging problem in mass-market printers. Despite the technical opportunity of using rigid and heavy CNC machine frames providing better printing quality at high speeds, the design of 3D printers always implies a trade-off between mechanical stiffness and the printer cost, which should be as minimal as possible for economic reasons. This is especially crucial in FDM/FFF 3D printers since this technology is the most inexpensive and the most demanded in the 3D printer mass market. Therefore, many various designs are present on the market. Unfortunately, almost each of them suffers from certain flaws limiting the building rate sufficiently below the theoretical maximum determined by the properties of the printing material.

The vibration characteristics of 3D printers and components directly affect the accuracy of the work, as reported in several studies, e.g., [20–22]. The main problem in FDM/FFF 3D printing originating from mechanical imperfectness is ghosting (ringing): the emergence of a repetitive pattern on the printed detail surface caused by mechanical vibration of the printer frame. This matter limits the maximal acceleration and speed available to a particular 3D printer [23,24]. Another defect also caused by mechanical imperfectness is corner swell, an outgrowth of the deposed material on a corner [24,25]. An example of these defects is given in Figure 1. They are negligible within a specific range of positioning settings, but printing errors become significant and practically unacceptable when they exceed particular values.

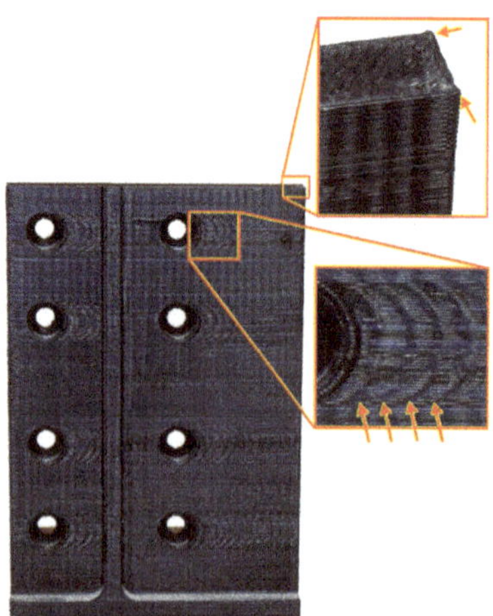

Figure 1. Example of corner swell (upper box) and ghosting (lower box). Arrows mark typical variants of these defects.

The magnitude of the resulting vibrations directly depends on the rigidity of the structure, which is also affected by many other factors [26]. Nevertheless, reducing the vibration amplitude of the system and increasing the natural frequency of vibrations can improve the quality of 3D printing.

Several recent studies are dedicated to detecting and decreasing these effects [24,27,28]. The most effortless way to do that is to increase the natural frequency of the printer head vibration due to proper mechanical design, leading to decreasing mode shape displacements [29]. A more elaborate solution is using active vibration control, which is a common practice in CNC milling machines [30,31]. For 3D printing, some solutions based on feed-forward control have been reported recently applied to vibration-prone 3D printers [32,33]. In addition, open-source Klipper firmware allows using one of the feed-forward resonance compensation algorithms for all compatible 3D printers [34].

However, how does an engineer understand which design is the most vibration-tolerant at the stage of preliminary design, where neither a complete CAD/CAE model exists nor any experimental specimens are available? This paper proposes using natural frequencies as the feature for distinguishing vibration-tolerant designs from vibration-prone ones. This feature depends on the relative positions and mobility of 3D printer parts. We propose a method for natural frequencies estimation from the 3D printer kinematic scheme, study in detail its bounds of applicability and give several illustrative examples including decision making on the 3D printer construction.

The paper is organized as follows. Section 2 describes the approach to calculating the natural frequency for a given Cartesian design. Section 3 presents examples of natural frequency analysis for Anycubic i3 Mega (Shenzhen Anycubic Technology Co., Ltd., Shenzhen, China) The natural frequencies have been obtained theoretically using the proposed approach and experimentally, confirming its applicability. Section 4 gives brief conclusions.

2. Vibration Analysis of 3D Printers

This section introduces kinematic schemes for 3D printers. We show that a simple approach based on a flexible joint model is applicable in the case of composite structures with considerably stiff beams, such as constructions of the aluminum extruded profile, often used in custom 3D printing buildings.

2.1. Problems of Modal Analysis for Constructions with Bolted Joints

Knowing its dimensions and material, the vibration characteristics of any monolithic component can be calculated with one of the well-established approaches. In the case of beams, the basic tool for their normal mode estimation is the elastic beam theory, including the Euler–Bernoulli, Rayleigh, Timoshenko, and other models [35]. However, stiffness calculation for the composite structures with this approach is a nontrivial problem [36]. Usually, estimation of vibration frequency distinguishes several types of boundary constraints for beam couplings: hinged, fixed, sliding, and free. They imply an absolute coupling stiffness. In practice, components are usually coupled with different bracing (bolts, angles, screws, etc.). In such systems, joints, backlashes, surface friction, and nonlinear deformations play a huge role, making the problem of normal mode estimation extremely complex. Only very recent studies deal with efficient calculations of natural frequencies of bolted structures, given their nonlinear character. For example, the work [37] proposes a machine learning approach to nonlinear modal analysis. The paper [38] describes an artificial neural network for predicting parameters of normal contact stiffness, penetration limit, and contact. Recent work [39] proposed an efficient algorithm for estimating damping in bolted joints providing good accuracy compared to simulations in commercial finite element software. Nevertheless, this solution is still not implemented in commercially available engineering packages.

In several papers, the determination of natural frequencies for bolted structures using FEM analysis shows unsatisfactory results, with errors ranging from 9 to 46.9% in determining natural frequencies [40]. A technique of using equivalent material can be applied to improve the accuracy[41]. Equivalent lower-dimensional models can also be efficient for simulating bolted constructions [42]. In civil engineering, to avoid detailed analysis of each element joint, a concept of effective stiffness is introduced reflecting the overall stiffness of a joint [43].

In our previous study, we did not find low-frequency vibrations for the 3D printer Anycubic i3 Mega using finite-elements analysis [44] since we did not simulate joints, which played a leading role in the occurrence of this low-frequency mode.

Therefore, we will consider an approach based on the equivalent model of the entire construction. Since we consider the preliminary design stage, we will omit any implementation details and focus on the effective stiffness of the joint. We will use the term "stiffness" for simplicity, meaning exactly "effective stiffness". Nevertheless, the proposed approach is valid only if the beams can be treated as absolutely stiff. Therefore, we must ensure that at least their first normal modes are far above the normal modes of the entire setup, i.e., the condition must be satisfied

$$w_b >> w, \tag{1}$$

where w_b is the beam first normal mode, w is the setup lower frequency. Elastic beam theory provides a well-established tool for verifying this assumption via estimating normal beam modes. The following section briefly recalls the main concepts of the Euler–Bernoulli beam theory and its application to our problem.

2.2. Euler–Bernoulli First Normal Modes

The natural frequency of the beam using the Euler Bernoulli theory is calculated by the formula [45]:

$$\omega = a^2\sqrt{\frac{EI}{\rho AL}}, \tag{2}$$

where E is elastic modulus, I is the second moment of the beam's cross-section area, ρ is beam material density, A is the cross-sectional area, and a is a wavenumber of Euler–Bernoulli mode.

Let us calculate the natural frequency of a 20 × 20 mm aluminum profile fixed at one end. The properties of the beam are:

$$L = 0.38\ \text{m}, A = 1.6 \times 10^{-4}\ \text{m}^2, I = 0.7 \times 10^{-8}\ \text{m}^4,$$

$$\rho = 2700\ \text{kg m}^{-3}, E = 7.1 \times 10^{10}\ \text{N m}^{-1}.$$

Wavenumber a for a beam fixed at one end and corresponding to the first mode equals 1.875 [45].

$$f = \frac{\omega}{2\pi} = \frac{1.875^2}{2\pi}\sqrt{\frac{7.1 \times 10^{10} \times 0.7 \times 10^{-8}}{2700 \times 1.6 \times 10^{-4} \times 0.38}} = 135.86\ \text{Hz}.$$

In our previous study, we found the experimental natural frequency of the aluminum profile beam to test the theoretical results (see Figure 2). The beam was attached to a fixed base using hidden aluminum profile corner brackets.

We investigated two options for fastening the beam to a fixed base: using one or two hidden brackets. The first mode frequency for the case with one hidden bracket was 22.1 Hz, and for the case with two hidden brackets, it was 45.2 Hz. As a result, the beam's natural frequency value calculated with the Euler–Bernoulli approach was significantly higher than the values obtained experimentally (the theoretical value is 135.86 Hz).

It means that almost all energy received by the beam from the impact hammer was turned into vibration caused by the joint flexibility rather than the beam elasticity. Furthermore, various beam mounting options affect the obtained values of natural frequencies, which verifies this assumption.

The flexible mounting option is not considered in the Euler–Bernoulli theory; however, it can be extended to such a case. Meanwhile, theoretical calculation confirms that the condition (1) is satisfied. Experimental results show that vibration on Euler–Bernoulli's first normal mode of the beam is almost indistinguishable compared to the vibration caused by a finite joint stiffness k. So we can conclude that the condition (1) allows considering the aluminum profile beams used in the 3D printer construction as a rigid body.

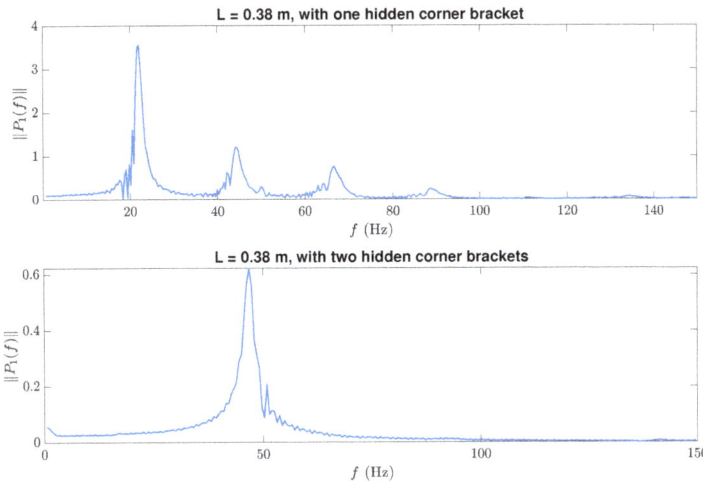

Figure 2. Experimental frequencies of the beam. (**Top**): one hidden bracket for fastening to a base is used, several peaks emerge because of the joint stiffness non-linearity [46]. (**Bottom**): two hidden brackets are used, the vibration is close to harmonic.

2.3. How Cartesian 3D Printer Kinematic Scheme Affects Its Dynamics

Cartesian 3D printers consist of two independent parts: the frame with the printhead mounted on it and the printing surface (the printbed). These parts move relative to each other. The system requires the mobility of 3D printer parts along three orthogonal axes to be functional.

Using the standard graphical notation of kinematic schemes [47], we can depict a kinematic structure of any Cartesian design as a very simple object with three translational degrees of freedom provided by three orthogonal groups of prismatic joints (redundant systems will not be considered). Examples of such schemes are given in Figure 3, left column. Figure 3a refers to the simplest structure used in very cheap 3D printers, Figure 3b refers to popular gantry kinematics with one closed kinematic chain, Figure 3c presents a variant of more stiff kinematics, containing three closed kinematic chains. These kinematic schemes figure out only how the structure is intended to move.

It is not obvious from each kinematic scheme in the left column, which one is better in terms of vibration tolerance unless we consider an additional factor: a finite stiffness of joints between beams, linear guides and other elements of construction. As the simplest approximation, these joints can be considered spherical joints with internal stiffeners. Such a point of view is reflected in Figure 3d–f. From these extended schemes, the advantage of closed kinematic chains becomes obvious: the more spherical joints are involved in possible angular movement the higher the overall stiffness is.

For better illustration, we introduce a simple formalism describing differences in relative axis mobility and axes structure for 3D printers and present all possible kinematic designs within this formalism in Appendix A.

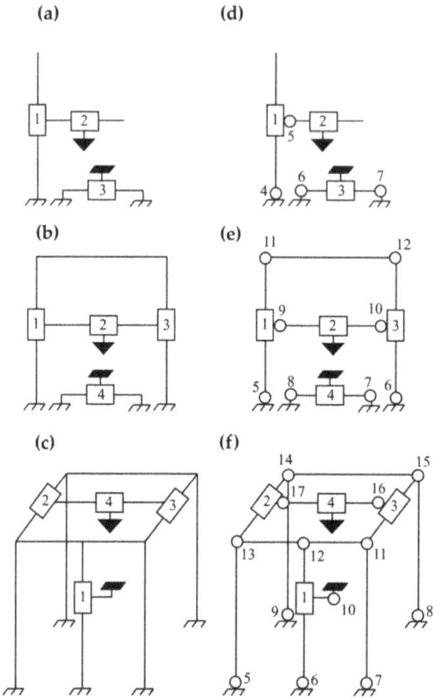

Figure 3. Examples of kinematic schemes for Cartesian 3D printers. Rectangles denote prismatic joints, circles denote spherical joints, a black triangle denotes the printhead, and a black parallelepiped denotes the printbed. Panels (**a**–**c**) refer to functional kinematic schemes, while the panels (**d**–**f**) additionally show frame and guide joints that cannot be considered absolutely stiff.

Now, let us consider how the kinematics of the 3D printer affects its vibration characteristics and therefore printing quality. Denote the vector of angular rotations in spherical joints as $\boldsymbol{\theta} = (\theta_1, \theta_2, \ldots, \theta_n)^\top$, and a vector of linear translations in prismatic joints in local coordinate systems associated with each direction of prismatic joints as $\mathbf{q} = (x_L, y_L, z_L)^\top$, where N is the number of spherical joints and $n \le N$ is the number of rotational degrees of freedom, and the number of translational degrees of freedom is 3. Denote the global coordinates of the 3D printer working point as $\mathbf{x} = (x, y, z)^\top$. The overall dynamics of the 3D printer correspond to a linearized ODE:

$$\begin{cases} J(\mathbf{x})\ddot{\boldsymbol{\theta}} + C_\theta \dot{\boldsymbol{\theta}} + K_\theta \boldsymbol{\theta} = \mathbf{f}_\theta(\mathbf{x}, \dot{\mathbf{x}}, \ddot{\mathbf{x}}, \mathbf{q}, \dot{\mathbf{q}}, \ddot{\mathbf{q}}), \\ \mathbf{x} = L(\mathbf{q}, \boldsymbol{\theta}), \\ M\ddot{\mathbf{x}} + C_x \dot{\mathbf{x}} + K_x \mathbf{x} = \mathbf{f}_x(\mathbf{x}, \dot{\mathbf{x}}, \ddot{\mathbf{x}}, \mathbf{q}, \dot{\mathbf{q}}, \ddot{\mathbf{q}}), \\ \mathbf{x} = L(\mathbf{q}, \boldsymbol{\theta}), \\ \mathbf{e} = \mathbf{q} - \mathbf{x}, \end{cases} \tag{3}$$

where $J(\mathbf{x})$ is a diagonal matrix of moments of inertia, M is a diagonal matrix of masses, C_θ, C_x are diagonal matrices of damping coefficients, K_θ, K_x are matrices of linearized stiffness, $\mathbf{f}_\theta$ is a vector of disturbance torques, $\mathbf{q}$ is a vector of relative translations in local coordinate systems attached to the prismatic joints, $\mathbf{f}_x$ is a vector of disturbance forces, $\mathbf{x}$ is an absolute position of the 3D printer working point, $\mathbf{e}$ is the positioning error. The kinematic scheme determines the number of degrees of freedom and the function $J(\mathbf{x})$.

Let us show more exactly how dynamical error is related to mechanical movement. Consider a case where stiffness matrices are diagonal, so each equation of the system (3) is independent. Then, one rotational degree of freedom of a linearized system is described as follows. Let the system be affected by the torque $\mathbf{f}_\theta = Ae^{j\Omega t}$, where Ω is the frequency of vibrations induced by the 3D printer parts' translational movements. In case of small angles the function L is linear, so we imply $e = L\theta$, $K_\theta = k$, $C_\theta = 0$. The Equation (3) then reads:

$$\begin{cases} J\ddot{\theta} + k\theta = Ae^{j\Omega t} \\ \mathbf{e} = L\theta. \end{cases} \tag{4}$$

Substituting a known solution $\theta(t) = \theta_1 e^{j\omega t} + \theta_2 e^{j\Omega t}$ into the Equation (4), where $\omega = \sqrt{k/J}$ is a natural frequency of the printer vibration, we obtain:

$$\theta_2 = \frac{A}{J(\omega^2 - \Omega^2)}. \tag{5}$$

The error amplitude maximal value is $e^* = L(|\theta_1| + |\theta_2|)$. The low-frequency vibrations caused by massive 3D printer parts movements satisfy the condition $\omega^2 > \Omega^2$. The Equation (5) explains the profit from the natural frequency enhancement: increasing the natural frequency lowers down the amplitude of the dynamical error.

The way how real Cartesian systems are implemented suggests using more complicated structures than the ones shown in Figure 3. A typical Cartesian system can be illustrated by the gantry-type kinematics often used in architectural 3D printers or CNC milling machines. Its functional kinematic scheme is given in Figure 4a following literature on CNC machine design [48,49]. Meanwhile, a more detailed kinematic scheme given in Figure 4b shows some issues with practical implementations. In addition to actuated prismatic joints, each axis is supplied with passive linear translation mechanisms for several reasons: to decrease the load on actuated joints, to prevent skew and jamming in actuated joints, and to improve the overall stiffness. The latter is important in the context of the current paper. A study for the CNC machine [49] confirms the importance of passive joint stiffness.

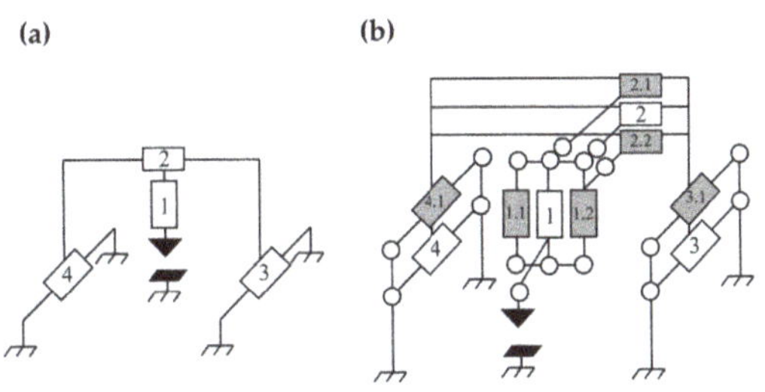

Figure 4. Practical implementation of a Cartesian system. Panel (**a**) shows a functional kinematic scheme of a Cartesian gantry CNC machine and panel (**b**) gives a detailed kinematic scheme of its implementation with actuated prismatic joints (shown with white rectangles) and passive rail guides (shown with grey rectangles). Circles denote passive spherical joints with internal stiffeners, a black triangle denotes the printhead, and a black parallelepiped denotes the printbed

The engineering practice of 3D printer designing implies using multiple additional passive joints as well. Several types of passive linear guides are common including round rods with ball bearing carriages, profiled rails with ball bearing carriages and aluminum profiles with wheeled carriages. Their properties such as backlashes, precision and price are highly variable and should be considered individually for each manufacturer and each design, and their application is also highly variable. For example, in inexpensive gantry 3D printers a brintbed with round rod guides is used in (Prusa Research, Prague, Czech Republic) and Anycubic i3 Mega (Shenzhen Anycubic Technology Co., Ltd., Shenzhen, China) 3D printers [50,51], while Sovol SVO3 (Sovol 3D, Shenzhen, China) has a wheeled carriage on an aluminum profile [52]. None of these designs should be preferred without an additional investigation.

2.4. Flexible Joint Model

The approach described in the previous section based on the flexible joint model is applicable to any construction made of rigid metal beams, in which natural frequencies are much higher than the natural frequencies emerging due to joint imperfectness. Unless other is claimed, we assume the system is strictly linear and symmetric and stiffness matrices in (3) are diagonal, so vibrations, which would occur in at least two orthogonal planes (denote them $\theta_x(t)$ and $\theta_y(t)$) are independent and similar in case of similar initial conditions. Therefore, we first consider only a planar motion in one selected plane (denote the angle $\theta(t)$) and only one corresponding frequency, and then extend this approach to more complex cases.

Vibration model of the beam. A joint model based on a linear flexible joint is proposed to find the natural frequency of vibrations, a. When the length of the beam is relatively small, it can be assumed that the structural profile is rigid. Then, according to Newton's second law, the motion of the beam is expressed in the following linear differential equation:

$$J\ddot{\theta} + k\theta = 0, \tag{6}$$

where $J = \frac{1}{3}M_T L^2$ is the beam moment of inertia, M_T is the total mass of the beam, L is the beam length, and k is the torsional stiffness of the joint, θ is the angular displacement (see Figure 5).

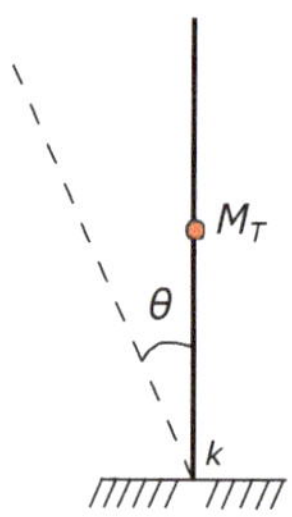

Figure 5. Model of the beam.

The analytical solution of Equation (6) has the form:

$$\theta(t) = \theta(0)e^{j\omega t}. \tag{7}$$

Substituting (7) into (6), we obtain:

$$-J\omega^2 + k = 0,$$

hence, the vibration frequency of the extruded profile beam is expressed as:

$$\omega = \sqrt{\frac{3k}{M_T L^2}} = \frac{1}{L}\sqrt{\frac{3k}{M_T}}. \tag{8}$$

Equation (6) can be used to determine the stiffness of the joint by empirically determining the first mode frequency of the system and expressing the stiffness as a ratio:

$$k = \frac{1}{3}\omega^2 L^2 M_T. \tag{9}$$

In the case of linear stiffness, (9) provides a simple way to determine k from experimentally estimated ω. In the case of nonlinear stiffness, the identification procedure should involve the waveform shape and would be more complicated [46].

Vibration model of the gantry structure. To find the natural vibration frequency of the gantry structure shown in Figure 6. Let us determine two independent mechanical stiffness of the joints. The stiffness of the lower joint is denoted by k_1, and the stiffness of the upper joint is denoted by k_2. The mass of the vertical beam is indicated M_1, and the mass of the horizontal beam is represented by M_2.

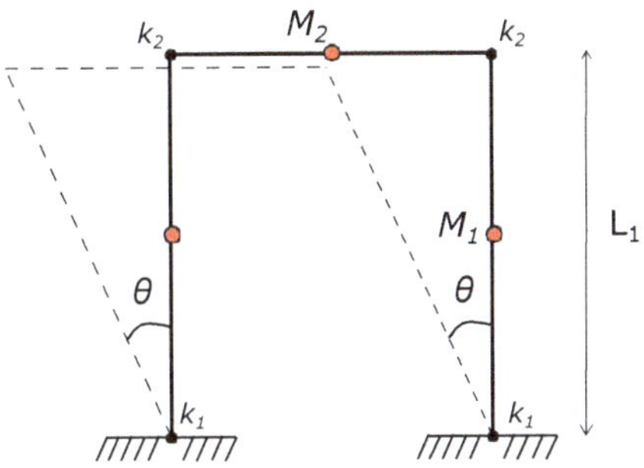

Figure 6. Model of the portal.

The moment of inertia of the beam is equal to $\frac{L_1}{2}$:

$$J_1 = \frac{M_1 {L_1}^2}{3}$$

The kinetic energy of motion of the gantry structure in this case is equal to:

$$K = \frac{2J_1\dot{\theta}^2}{2} + \frac{M_2(\dot{\theta}L_1)^2}{2} = \frac{M_1 L_1^2}{3}\dot{\theta}^2 + \frac{M_2 L_1^2}{2}\dot{\theta}^2,$$

the potential energy:

$$P = \left(\frac{2k_1\theta^2}{2} + \frac{2k_2\theta^2}{2}\right) = (k_1 + k_2)\theta^2.$$

The Lagrange function, where $\mathcal{L}$ is the Langrangian, is equal to the difference of kinetic and potential energy:

$$\mathcal{L} = K - P = \left(\frac{M_1}{3} + \frac{M_2}{2}\right)(L_1\dot{\theta})^2 - (k_1 + k_2)\theta^2,$$

the Euler–Lagrange equation is written as:

$$\frac{d}{dt}\frac{\partial\mathcal{L}}{\partial\dot{\theta}} - \frac{\partial\mathcal{L}}{\partial\theta} = 0$$

from where

$$\left(\frac{2M_1}{3} + M_2\right)\frac{d}{dt}({L_1}^2\dot{\theta}) + 2(k_1 + k_2)\theta = 0,$$

and

$$\left(\frac{2M_1}{3} + M_2\right){L_1}^2\ddot{\theta} + 2(k_1 + k_2)\theta = 0.$$

Using (7), we find the equation of the frequency of natural vibrations of the gantry structure:

$$\omega = \frac{1}{L_1}\sqrt{\frac{2(k_1+k_2)}{\frac{2M_1}{3}+M_2}}. \tag{10}$$

It should be noted that the length of the horizontal beam does not affect the frequencies of the gantry structure.

Vibration model of a parallelepiped. For simplicity, in the structure shown in Figure 7, the stiffness of the joints can be assumed and equal to k_1 and k_2. The length of all vertical beams is equal to L_1.

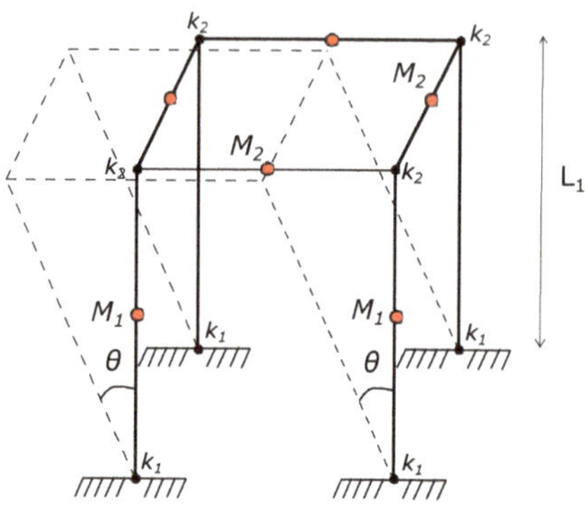

Figure 7. Model of the parallelepiped.

The moments of inertia of the vertical and horizontal beams are equal, respectively, to:

$$J_1 = \frac{1}{3}M_1{L_1}^2, J_2 = M_2{L_1}^2.$$

The kinetic energy of motion of the parallelepiped structure is equal to:

$$K = \frac{4J_1\dot{\theta}^2}{2} + \frac{4J_2\dot{\theta}^2}{2},$$

potential energy:

$$P = \left(\frac{4(k_1+k_2)\theta^2}{2}\right) = 2(k_1+k_2)\theta^2.$$

The Lagrange equation after transformations will be:

$$\frac{4}{3}M_1L_1^2\ddot{\theta} + 4M_2L_1^2\ddot{\theta} + 4(k_1+k_2)\theta = 0.$$

The frequency of the parallelepipedal structure, in this case, is equal to:

$$\omega = \sqrt{\frac{k_1+k_2}{\frac{1}{3}M_1{L_1}^2 + M_2{L_1}^2}} = \frac{1}{L_1}\sqrt{\frac{k_1+k_2}{\frac{1}{3}M_1+M_2}}. \tag{11}$$

Taking into account several degrees of freedom. Usually, 3D printers have two, three or more degrees of freedom, including oscillations in directions of X and Y axes, oscillations of the printbed independently from the frame, moreover, some printer designs provide horizontal and vertical elements of the frame to oscillate independently as well, like the open kinematic chain design shown in Figure A3b. If we use linearized models and also assume that the stiffness matrices in the Equation (3) are diagonal, treatment of multiple degrees of freedom is fairly simple: each rotational degree of freedom is considered as if the other degrees of freedom are fixed. For example, let us calculate the gantry structure vibration in another plane, orthogonal to the plane considered before. The kinetic energy of motion of the gantry structure in this case is exactly the same

$$K = \frac{M_1 L_1^2}{3}\dot{\theta}^2 + \frac{M_2 L_1^2}{2}\dot{\theta}^2,$$

but the potential energy, due to the fact that upper joints are not involved in the oscillations, is

$$P = k_1\theta^2.$$

This results in the frequency:

$$\omega = \frac{1}{L_1}\sqrt{\frac{2k_1}{\frac{2M_1}{3} + M_2}}. \tag{12}$$

To simplify this approach, we present an Algorithm 1 for calculating Cartesian structures. For example, for the gantry and parallelepiped structures considered before longitudinal beams are vertical, transverse beams are horizontal, so Formulas (10)–(12) could be easily derived without Lagrangians.

Algorithm 1: The algorithm for calculating Cartesian structure natural frequency.

Input: $\{m_i\}, \{L_i\}, \{k_i\}$.
Output: ω.

1. Select the plane in which the vibration of the structure will be investigated. Select an appropriate degree of freedom.
2. Decompose the construction into primitives:
   (a) n longitudinal beams of length L_i with mass m_{Li}, length L_{Li} and moment of inertia: $J_{Li} = \frac{m_{Li}{L_{Li}}^2}{3}$.
   (b) m transverse beams and lumped masses with the distance from the base to the center of mass L_{Ti}, mass m_{Ti} and moment of inertia $J_{Ti} = m_{Ti}{L_{Ti}}^2$.
3. Calculate the stiffeners $k_j, j = 1 \ldots N$ (N is the number of joints).
4. Estimate the frequency using the formula:

$$\omega = \sqrt{\frac{\sum_{j=1}^{N} k_j}{\sum_{i=1}^{n} J_{Li} + \sum_{i=1}^{m} J_{Ti}}}$$

3. Analysis of Sample Designs

The current section describes the application of the proposed method for estimating natural frequencies of 3D printers to some practically valuable examples.

3.1. Experimental Analysis of a Gantry 3D Printer

It is possible to find the first lower mode for an existing 3D printer using the proposed approach. A gantry design Anycubic i3 Mega was chosen as an example. A distinctive feature of this 3D printer is the presence of a relatively heavy printhead mounted on two horizontal cylindrical guides working as a whole due to the rigid coupling of their ends (see Figure 8). The length of the vertical beams is labeled L_1, and the height at which the printhead is located is labeled L_2. This design has three types of joints with different stiffness.

The kinetic energy of motion of the gantry 3D printer is equal to:

$$K = \frac{M_1 L_1^2}{3}\dot{\theta}^2 + \frac{M_2 L_1^2}{2}\dot{\theta}^2 + \frac{M_H L_2^2}{2}\dot{\theta}^2,$$

potential energy of motion:

$$P = \left(\frac{2k_1\theta^2}{2} + \frac{2k_2\theta^2}{2} + \frac{2k_3\theta^2}{2}\right) = (k_1 + k_2 + k_3)\theta^2.$$

The Lagrange equation after transformations will look like this:

$$\left(\frac{2M_1{L_1}^2}{3} + M_2{L_1}^2 + M_H{L_2}^2\right)\ddot{\theta} + 2(k_1 + k_2 + k_3)\theta = 0.$$

The natural frequency of oscillation of the 3D gantry printer can be expressed using the formula:

$$\omega = \sqrt{\frac{2(k_1 + k_2 + k_3)}{\frac{2M_1{L_1}^2}{3} + M_2{L_1}^2 + M_H{L_2}^2}}. \quad (13)$$

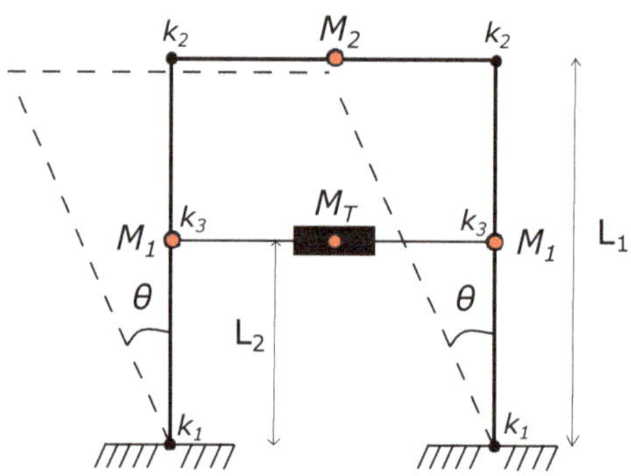

Figure 8. Anycubic i3 Mega kinematic scheme.

This formula can be also obtained using the Algorithm 1. Using the Formula (13), we can find the relationship between the printhead position on the Z-axis and the natural frequency of oscillations. A testbench for vibration data acquisition was assembled to measure the vibrations of the head, bed, and frame of the 3D printer. The testbench includes the following components: NI PXI 1042 controller with the PXI-4461 module for collecting vibroacoustic signals, piezoelectric accelerometer IMV VP-4200, accelerometer power amplifier, and custom mount for accelerometer printed on a 3D printer, which allows placing the sensor parallel to any of three axes X, Y, Z (see Figure 9).

A series of experimental data were recorded with a sampling rate of 1000 Hz. The program for recording data into a file was implemented in the LabVIEW environment. The data received from the accelerometer was processed in the MATLAB environment using the built-in FFT function.

We investigated the dependence of the natural frequency of oscillation on the height of the printhead. Since the printhead and related components involved in Z-axis movement (guides, belts, mounts) have a total mass of 1.7 kg, which is greater than the mass of the vertical beams (0.7 kg), the location of the printhead on the Z-axis affects the 3D printer natural frequency, much.

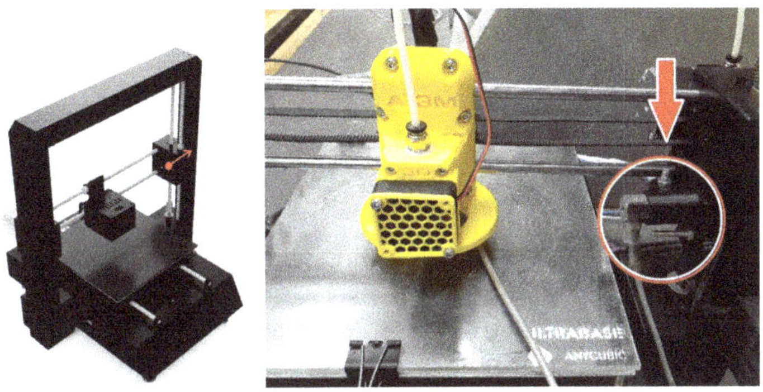

(a) (b)

Figure 9. (**a**) Model of the 3D printer Anycubic i3 Mega. A circle marks a place of the accelerometer mounting, an arrow marks the direction of axis sensitivity (**b**) Example of the accelerometer mounting.

In the experiment, we used the g-code to set a short-term extruder movement at each of the positions along the Z axis with a step of 10 mm, and the maximum height at which the printhead was positioned was 200 mm. Printer movement gave an impulse to the printer frame impact with the spectrum as close to that experienced by a 3D printer during printing as possible. The accelerometer was mounted on the X axis of the 3D printer in the right corner, see Figure 9. This position allows determining the frequency of frame vibrations affecting the printhead directly during operation.

The total stiffness of the 3D printer at zero position was found using Equation (13) to validate the theoretical approach based on the flexible joint model. Further, the theoretical natural frequencies of vibrations were found. Figure 10 shows the dependence of the experimental and theoretical natural frequencies on the position of the printhead along the Z axis.

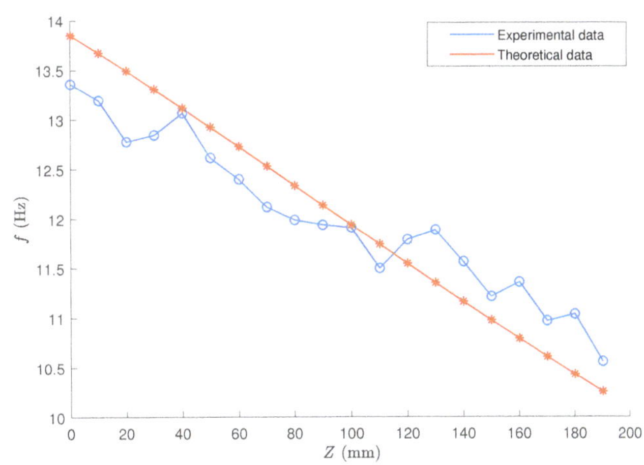

Figure 10. The first mode of the Anycubic i3 Mega 3D printer.

Figure 10 shows that the proposed approach successfully predicts the first mode of the Anycubic i3 Mega. The relative error of the theoretical frequency was 2–6 percent (see Table 1).

Table 1. The first mode of the Anycubic i3 Mega 3D printer.

Z, mm	Theoretical, Hz	Experimental, Hz	Relative Error,%
0	13.85	13.36	3.54
10	13.67	13.19	3.52
20	13.49	12.78	5.31
30	13.31	12.84	3.48
40	13.12	13.07	0.38
50	12.92	12.62	2.38
60	12.73	12.40	2.58
70	12.53	12.12	3.26
80	12.33	11.99	2.79
90	12.14	11.94	1.62
100	11.94	11.91	0.23
120	11.74	11.50	2.07
130	11.55	11.79	2.08
140	11.36	11.89	4.69
150	11.16	11.57	3.59
160	10.98	11.22	2.18
170	10.79	11.36	5.27
180	10.61	10.97	3.41
190	10.43	11.04	5.82
200	10.26	10.56	2.97

3.2. Preliminary Analysis of a New Design

This subsection shows how natural frequency can be calculated from the classification scheme for an arbitrary 3D printer.

Suppose, natural frequencies of cubic printers depicted in Figure A2a,b are predicted. From Algorithm 1 it follows that the natural frequency equals to

$$f = \frac{1}{2\pi}\sqrt{\frac{\sum_i k_i}{\sum_i J_i}}, \tag{14}$$

where k_i is i-th joint stiffness, and J_i is i-th moment of inertia.

The formula for the frequency of open Z-axis design is:

$$f_1 = \frac{1}{2\pi}\sqrt{\frac{4k_P + 4k_P + 2k_L + 2k_L}{\frac{4M_v L_1^2}{3} + M_B L_2^2 + (M_H + 5M_h + 4M_M + 3M_R)L_1^2 + \frac{2M_G L_1^2}{3}}}, \tag{15}$$

where M_G is the guide mass and M_P is the profile mass. Let the mass of a horizontal beam equal $M_h = \rho_P L_h$, where L_h is the length of the horizontal beam, and $M_v = \rho_P L_1$ is the mass of a vertical beam.

From this and Equation (15) we can obtain:

$$f_1 = \frac{1}{2\pi}\sqrt{\frac{8k_P + 4k_L}{\frac{4\rho_P L_1^3}{3} + M_B L_2^2 + (M_H + 4M_M + 3\rho_R L_h)L_1^2 + 5\rho_P L_h L_1^2 + \frac{2\rho_G L_1^3}{3}}}. \tag{16}$$

Substituting values from Table 2 yields:

$$f_1 = 32.0 \text{ Hz}.$$

The frequency of the kinematics with closed Z-axes is calculated using the following formula (supposing that one motor is used for actuating both linear drives of the closed Z-axis on a printbed):

$$f_2 = \frac{1}{2\pi}\sqrt{\frac{8k_P + 8k_L}{\frac{4\rho_P L_1^3}{3} + M_B L_2^2 + (M_H + 4M_M + 3\rho_R L_h)L_1^2 + 5\rho_P L_h L_1^2 + \frac{4\rho_G L_1^3}{3}}}. \tag{17}$$

Substituting values from Table 2, another value is obtained:

$$f_2 = 34.9 \text{ Hz}.$$

The resulting difference is relatively small, so we can conclude that using the open Z-axis is reasonable for the design from the point of view of natural frequency in XY plane, and a closed Z will not yield much more stiffness.

Table 2. Parameters of the evaluated design.

Parameter	Value	Units
Profile angle stiffness k_P	3700	N·m/rad
Linear guide mount angle stiffness k_L	2000	N·m/rad
Profile density ρ_P	0.85	kg/m
Linear rail guide density ρ_R	1.45	kg/m
Linear 8 mm guide density ρ_G	0.4	kg/m
Stepper motor mass M_M	0.26	kg
Printhead mass M_H	0.7	kg
Printbed mass M_B	0.4	kg
Vertical beam length L_1	0.4	m
Z position of the printbed L_2	0.3	m
Horizontal beam length L_h	0.4	m

3.3. *Comparison of Common Designs*

To compare the seven most common 3D printer designs, their CAD/CAE models have been developed. A more detailed description of these designs is given in Appendix A.

First, we used the finite-element method (FEM) to calculate the natural frequencies of the printers assuming that all joints are absolutely stiff. Then, we used the proposed method based on the kinematic schemes of the same structures and calculated the approximate lowest frequencies of these printers using Algorithm 1 with the parameters given in Table 2.

Figure 11 shows small images of the printers in one panel with their ordinal numbers, and Figure 12 shows the corresponding lowest natural frequencies found with two methods: the FEM and the proposed one.

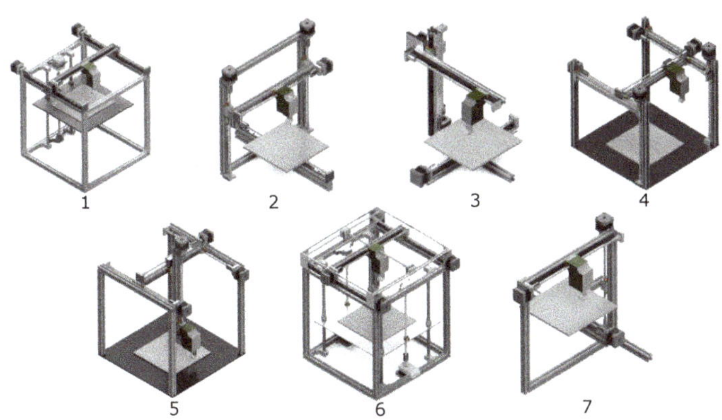

Figure 11. Common printer designs used in the investigation.

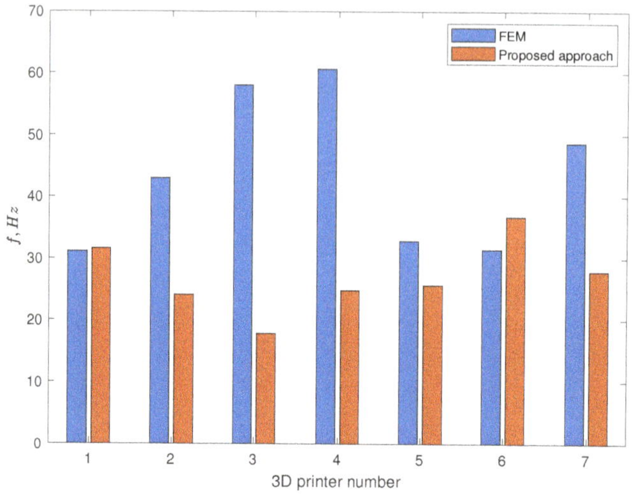

Figure 12. Lowest normal modes of 3D printer structures found by the FEM and by the proposed approach based on a flexible joint model.

The experiment found that for the cuboid structures 1 and 6 the results of our method and FEM approximately coincide (see Table 3). This shows that profile elasticity has an almost similar impact on the vibration as the flexibility of the joints. For the other designs, the FEM overestimates the vibration frequency in comparison with the proposed method. This can be explained by the fact that these constructions are more light-weighted, and in the case of absolutely stiff joints the overall stiffness is high. In practice, flexibility in joints make such structures more vibration-prone, and this is confirmed by engineering practice: only a few cheapest 3D printers have an open kinematic chain structure 3, and structure 2 is also becoming less common in recent years.

Table 3. Modes 1–3 of the printers obtained by the FEM and the lowest mode by the proposed method.

№	Mode 1, Hz	Mode 2, Hz	Mode 3, Hz	Proposed Method, Hz
1	31.12	31.38	53.99	31.59
2	42.97	45.50	95.37	24.14
3	58.04	67.02	103.5	17.81
4	60.62	84.89	117.4	24.83
5	32.83	40.81	77.71	25.67
6	31.45	36.12	60.58	36.72
7	48.73	50.38	96.97	27.93

Of course, these results should be extrapolated on real designs with care. However, they clearly show the limitation of the FEM when joint flexibility is not taken into account. This is especially important when novel additive manufacturing machines are developed with heavy printheads, such as painting robots [53].

3.4. Example of Cost Calculation

As an example of calculating the cost of an FDM 3D printer, a gantry printer was chosen, shown in Figure A3a, with a frame made of 20 × 20 mm aluminum profiles. The length of horizontal and vertical aluminum profiles is 400 mm. The length of lead screws was chosen as 350 mm. Table 4 shows the cost (minimum and maximum) of the main components of a 3D printer. The difference in the cost of the components is conditioned by their different quality, brands and other features. Among the components, there are obligatory ones used in any configuration of a 3D printer and not depending on the kinematic scheme: the power supply unit (PSU), the printbed, the printhead and the controller board.

Table 4. The cost of a 3D printer in Figure A3a.

Component	Amount	Cost of 1 Piece, $	Total Cost, $	Cost Ratio
Aluminum profile	2400 mm	3–5 (1000 mm)	7.2–12	6–3%
Lead screw	5 pc.	2–5	10–25	8–6%
Rail guide system	4 pc.	7–25	28–100	22%
Stepper motor	4 pc.	5–15	20–60	15–13%
PSU	1 pc.	5–20	5–20	4%
Print bed	1 pc.	15–50	15–50	12–11%
Print Head	1 pc.	15–80	15–80	12–18%
Controller Board	1 pc.	30–100	30–100	23–22%
All components excluding additional spare parts			130.2–447	

The economic effect of creating a 3D printer with the specific kinematic scheme increases when cheap obligatory components are used, because in this case a rail guide system, motors, etc. will affect the total cost of the printer much more significantly. In the case of using top-level obligatory components, a more stiff kinematic scheme will not result in relatively much higher expenses, so it should be preferred. The exception here is the rail guide system, which is a rather costly component, but it can be replaced with cheaper cylindrical guides in many cases.

4. Conclusions

The paper proposes a simple method for calculating 3D printer natural frequencies based on the kinematic scheme. It is applicable in the case of high stiffness of elongated elements such as profiles, guides, and beams. This method allows replacing elaborate finite-element analysis of the detailed model on a preliminary design stage. The accuracy of this approach is relatively high and feasible for engineering purposes; the provided examples confirm its practical applicability. Several most common designs have been compared via their lowest natural frequency, and the variant with the greater value of vibration frequency is preferred using this technique. Additionally, we give an example of calculating the FDM printer cost. We conclude that using expensive obligatory parts such as the power unit, the printhead and the controller, may reduce the relative cost of mechanical components and therefore make using a more stiff kinematic scheme more reasonable. An example of decision-making concerning the type of 3D printer is given.

We also propose a classification scheme for the Cartesian mechanical systems used for additive manufacturing which considers variants of already existing structures and examples not yet applied to 3D printing. It can be used by developers of various additive machines, such as desktop 3D printers, architectural 3D printers, and others, to simplify the engineering process in a preliminary design stage. The algorithm for natural frequency estimation can be used for computer search of proper designs using optimization. A table of stiffness of various joint components obtained from the experiment can be developed which would help engineers to find optimal design solutions.

To summarize, the paper makes the following contributions. First, the paper provides theoretical and practical confirmation of the applicability of the flexible joint model and the necessity to take bolted joint flexibility into account during finite element analysis, which is a non-trivial task in most moderns software packages such as Fusion 360. Second, the paper gives a simplified and handy algorithm for natural frequency estimation of the 3D printer which does not require the application of the Euler–Lagrange formula directly and is suitable for both manual and machine usage. Third, a classification scheme for Cartesian 3D printers is proposed using two criteria: the type of kinematic chain associated with each axis (open/closed) and the mobility of each axis' elements (actuated/fixed) connected to the print head and the print bed.

Further work will be dedicated to testing some non-existing designs and considering other decision-making criteria, such as the system tolerance to the inaccuracy of the 3D printer components, in addition to the natural frequency criterion.

Author Contributions: Conceptualization, A.K. and D.B.; data curation, D.B.; formal analysis, A.K. and L.S.; investigation, E.K.; methodology, A.K.; project administration, A.K. and D.B.; resources, A.K. and L.S.; software, E.K. and A.K.; supervision, A.K. and D.B.; validation, L.S. and D.B.; visualization, E.K.; writing—original draft, E.K. and A.K.; writing—review and editing, L.S. and D.B. All authors have read and agreed to the published version of the manuscript.

Funding: This research received no external funding.

Institutional Review Board Statement: Not applicable.

Informed Consent Statement: Not applicable.

Data Availability Statement: Not applicable.

Acknowledgments: The authors are thankful to anonymous Referees for their insightful reviews and useful recommendations.

Conflicts of Interest: The authors declare no conflict of interest.

Abbreviations

The following abbreviations are used in this manuscript:

CNC	Computer Numerical Control
PXI	PCI eXtension for Instrumentation

Appendix A. Classification of Cartesian Designs: Principles and Examples

The usefulness of a specific classification scheme for practice depends on whether it is accurate enough to systematize all the existing knowledge and is capable of accommodating new subjects [54]. Basics of kinematic scheme topology classification and methods for mobility investigation are reported in papers [55–57]. For 3D printing, a detailed classification of possible designs may be helpful for engineers at the early design stage to choose the most feasible kinematics, for sellers to organize their catalogs, for purchasers to identify their requirements more efficiently, for researchers to place their novel designs among existing ones.

Cartesian 3D printers consist of two independent parts: the frame with the printhead mounted on it and the printbed.To distinguish between the coordinate axes of the printhead and the printbed, we denote them as X_H, Y_H, Z_H for the printhead and X_B, Y_B, Z_B for the printbed. These parts are separated in space by the printed part and move relative to each other. The system requires the mobility of 3D printer parts along three orthogonal axes to be functional. For example, if the coordinate system as a whole is stationarily attached to the printbed, then the printhead must move along the X_H, Y_H and Z_H axes. Hereafter, for brevity, we will use the term "axis" not only for the direction in space but also to denote a set of mechanical elements (beams, guides) designed to move the printer's working body along a given coordinate axis.

Kinematics of modern Cartesian printers implies variation of mobility and immobility of the axes. In our analysis, the actuated axis is designated (A) and the fixed axis is designated (F).

Table A1 shows all possible combinations of the printhead and the printbed axis mobility. Each row in the table represents a kinematics variant with three actuated axes. Because creating each layer of the printed part takes place in the XY plane, the Ox and Oy axes of the printhead and the printbed are interchangeable. Thus, the third row is synonymous with the fourth row, the fifth row is interchangeable with the seventh row, and only one version of each pair will be considered further.

Also, two mounting options for the actuated axes were considered:

- Open axes (A_O)—actuated axes that have one attachment point to a fixed base or elements of other axes. This corresponds to the open kinematic chain.
- Closed axes (A_C)—actuated axes that have at least two attachment points to a fixed base or elements of other axes. This corresponds to a closed kinematic chain.

The closed axis provides more rigidity to the construction but needs a more complicated design with additional guides, motors, etc.

Table A1. Variability in axis mobility of the Cartesian 3D printer. Cells for the printhead are filled gray. Rows 4 and 7 are not investigated further.

№	Actuated Axis			Fixed Axis		
1	X_H	Y_H	Z_H	X_B	Y_B	Z_B
2	X_H	Y_H	Z_B	X_B	Y_B	Z_H
3	X_H	Y_B	Z_H	X_B	Y_H	Z_B
4	X_B	Y_H	Z_H	X_H	Y_B	Z_B
5	X_H	Y_B	Z_B	X_B	Y_H	Z_H
6	X_B	Y_B	Z_H	X_H	Y_H	Z_B
7	X_B	Y_H	Z_B	X_B	Y_B	Z_B
8	X_B	Y_B	Z_B	X_H	Y_H	Z_H

Appendix A.1. Classification of Cartesian Designs

In this subsection, we will overview all possible combinations of fixed and actuated axes and make a complete classification of Cartesian 3D printers. An example of the visual representation of each kinematics is presented in Appendix A in Figures A5–A10. It should be noted that the kinematic schemes can be depicted differently depending on where the entire structure is mounted (e.g., standing on the table or attached to the wall). All illustrated constructions shown in Appendix A are fixed on the floor. Tables A2–A7 show kinematic schemes with different variations. Since the X and Y axes are interchangeable, designs numbered 2 and 8 (marked yellow (*)) as well as 6 and 7 (marked red (**)) in some tables are the same, and one solution from each pair can be omitted. In Tables A4 and A5, there are no interchangeable designs because the axes X_B and Y_H (or X_H and Y_B) belong to different 3D printer components.

Table A2. Fixed axes X_B, Y_B, Z_B and actuated axes X_H, Y_H, Z_H.

№	X_B	Y_B	Z_B	X_H	Y_H	Z_H
1	F	F	F	A_O	A_O	A_O
2 *	.	.	.	A_O	A_C	A_C
3	.	.	.	A_C	A_C	A_O
4	.	.	.	A_C	A_C	A_C
5	.	.	.	A_O	A_O	A_C
6 **	.	.	.	A_O	A_C	A_O
7 **	.	.	.	A_C	A_O	A_O
8 *	F	F	F	A_C	A_O	A_C

Table A3. Fixed axes X_B, Y_B, Z_H and actuated axes X_H, Y_H, Z_B.

№	X_B	Y_B	Z_B	X_H	Y_H	Z_H
1	F	F	A_O	A_O	A_O	F
2 *	.	.	A_C	A_O	A_C	.
3	.	.	A_O	A_C	A_C	.
4	.	.	A_C	A_C	A_C	.
5	.	.	A_C	A_O	A_O	.
6 **	.	.	A_O	A_O	A_C	.
7 **	.	.	A_O	A_C	A_O	.
8 *	F	F	A_C	A_C	A_O	F

Table A4. Fixed axes X_B, Y_H, Z_B and actuated axes X_H, Y_B, Z_H.

№	X_B	Y_B	Z_B	X_H	Y_H	Z_H
1	F	A_O	F	A_O	F	A_O
2	.	A_C	.	A_O	.	A_C
3	.	A_C	.	A_C	.	A_O
4	.	A_C	.	A_C	.	A_C
5	.	A_O	.	A_O	.	A_C
6	.	A_C	.	A_O	.	A_O
7	.	A_O	.	A_C	.	A_O
8	F	A_O	F	A_C	F	A_C

Table A5. Fixed axes X_B, Y_H, Z_H and actuated axes X_H, Y_B, Z_B.

№	X_B	Y_B	Z_B	X_H	Y_H	Z_H
1	F	A_O	A_O	A_O	F	F
2	.	A_C	A_C	A_O	.	.
3	.	A_C	A_O	A_C	.	.
4	.	A_C	A_C	A_C)	.	.
5	.	A_O	A_C	A_O	.	.
6	.	A_C	A_O	A_O	.	.
7	.	A_O	A_O	A_C	.	.
8	F	A_O	A_C	A_C	F	F

Table A6. Fixed axes X_H, Y_H, Z_B and actuated axes X_B, Y_B, Z_H.

№	X_B	Y_B	Z_B	X_H	Y_H	Z_H
1	A_O	A_O	F	F	F	A_O
2 *	A_O	A_C	.	.	.	A_C
3	A_C	A_C	.	.	.	A_O
4	A_C	A_C	.	.	.	A_C
5	A_O	A_O	.	.	.	A_C
6 **	A_O	A_C	.	.	.	A_O
7 **	A_C	A_O	.	.	.	A_O
8 *	A_C	A_O	F	F	F	A_C

Table A7. Fixed axes X_H, Y_H, Z_H and actuated axes X_B, Y_B, Z_B.

№	X_B	Y_B	Z_B	X_H	Y_H	Z_H
1	A_O	A_O	A_O	F	F	F
2 *	A_O	A_C	A_C	.	.	.
3	A_C	A_C	A_O	.	.	.
4	A_C	A_C	A_C	.	.	.
5	A_O	A_O	A_C	.	.	.
6 **	A_O	A_C	A_O	.	.	.
7 **	A_C	A_O	A_O	.	.	.
8 *	A_C	A_O	A_C	F	F	F

Appendix A.2. Survey of Existing Mechanical Designs

A review of existing 3D printer solutions showed that many of the variants presented in Table A2 of the kinematics are not used in practice, except for numbers three and four. Examples of a additive systems with closed X_H, Y_H and open Z_H axes (see Figure A1a) are the machine painting systems [53,58]. The use of an open Z_H here is due to the small movement requirement on the Z coordinate. Examples of 3D printers with closed axes X_H, Y_H, and open Z_H (see Figure A1a) primarily are building 3D printers, e.g., from companies Specavia ATM [59], Winsun [60], COBOD [61], etc. Examples of 3D printers with closed axes X_H, Y_H, and Z_H are VORON 2.4 [62] and a custom printer with Core-XYZ kinamtics [63].

Creating 3D printers with bed fixation is sometimes not a convenient technical solution but rather a forced one. The choice of this configuration for architectural printers can be explained by the fact that we can not move the platform on which the building stands.

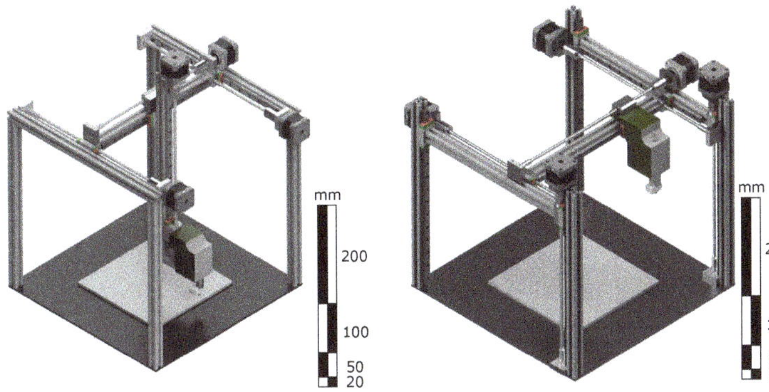

(a) (b)

Figure A1. 3D model of 3D printers with fixed axes X_B, Y_B, Z_B and actuated axes X_H, Y_H, Z_H (**a**) with closed X_H, Y_H and open Z_H axes (**b**) with closed axes X_H, Y_H, Z_H.

Table A3 illustrates two variants used in practice. Both variants are the most popular among all designs. An example of a 3D printer with closed X_H, Y_H and open Z_B axes (see Figure A2a) is Anycubic 4max Pro [51] or Ultimaker 3. An example of a 3D printer with closed X_H, Y_H, Z_B axes (see Figure A2b) is the Total Z Anyform 3D printer [64].These 3D printers differ only in the closed Z axis of the table, which affects the stiffness of the table. As a rule, 3D printers with a closed Z_B axis are often significantly more expensive, which is due to using more precise ball screw drives instead of a trapezoidal screw and is not fully determined by the printer kinematics.

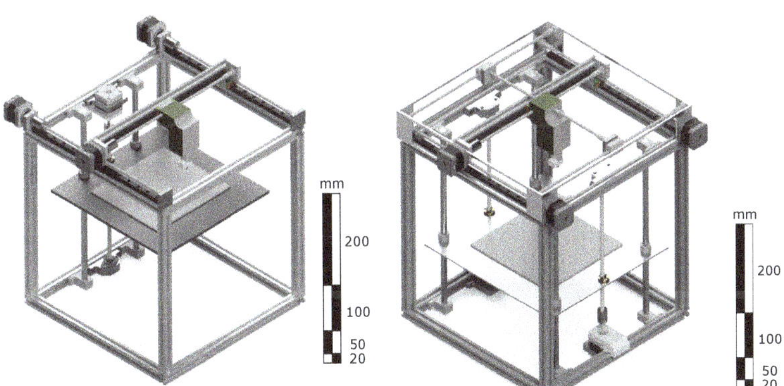

(a) (b)

Figure A2. 3D model of 3D printers with fixed axes X_B, Y_B, Z_H and actuated axes X_H, Y_H, Z_B (**a**) with closed X_H, Y_H and open Z_B axes (**b**) with closed X_H, Y_H, Z_B axes.

Table A4 shows two variants used in practice. The first variant is the one with closed axes X_H, Y_B and open Z_H. This variant is trendy in the 3D-printers market due to its cheapness and simplicity of construction. Examples of a this type of 3D printers are

Wanhao Duplicator i3 [65] and Anycubic i3 Mega [51] (see Figure A3a). The second variant with open X_H, Z_H axes and open Y_B (see Figure A3b) is a simplified version of the previous model, and one of the cheapest variations of Cartesian 3D printers. An example of such a 3D printer is the Wanhao Duplicator i3 Mini 3D printer [66].

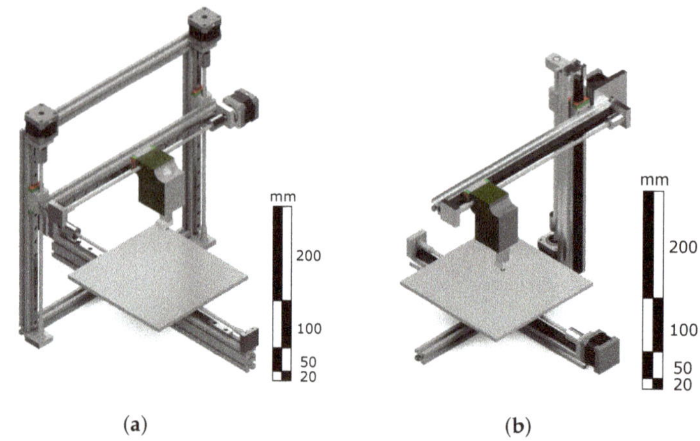

Figure A3. 3D model of 3D printers with fixed axes X_B, Y_H, Z_B and actuated axes X_H, Y_B, Z_H (**a**) with closed axes X_H, Y_B and open Z_H (**b**) with open X_H, Z_H axes and open Y_B.

Examining the designs from Table A5, we found only one variant of the existing 3D printer with two actuated printbed axes. It is Felix 3.0 [67], a 3D printer with open Y_B, Z_B and closed X_H (see Figure A4). The manufacturer claims that the 3D printer can develop a fairly high printing speed thanks to this design compared to its counterparts.

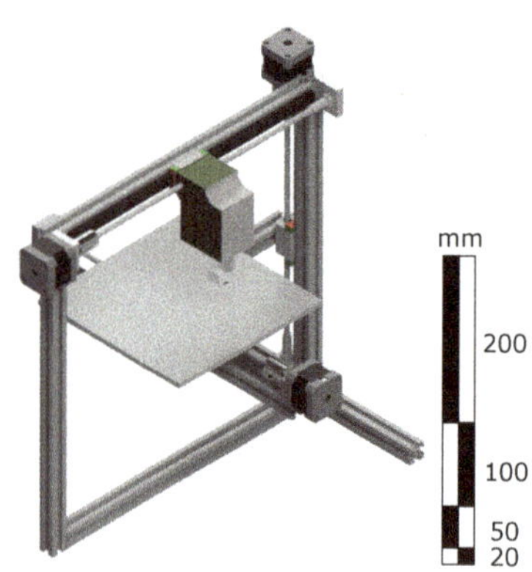

Figure A4. 3D model of 3D printer with fixed axes X_B, Y_H, Z_H and actuated axes X_H, Y_B, Z_B with closed axis X_H, and open Y_B, Z_B.

Three-dimensional printers with the designs shown in Tables A6 and A7 were not found. Probably, such designs are not quite suitable for 3D printing. Moving the printing surface along all three axes can affect print quality since the printbed with the printed detail is much heavier than the printhead. However, it is possible that designs of this type are or will be used for other additive technologies, where complete fixation of the printhead is required due to its size and mass. A recent study published the design of a prototype 3D printer with a fully fixed extruder. However, this printer is not Cartesian and is not commercially available [68].

Now, make a complete classification scheme of the existing Cartesian 3D printers. It is given in Table A8. The classification scheme shows each case's fixed and actuated axes and the closed and open axes for each actuated axis.

Table A8. Full classification of existing Cartesian 3D printers.

№	X_B	Y_B	Z_B	X_H	Y_H	Z_H
1	F	F	A_O	A_C	A_O	F
2	F	A_C	F	A_C	F	A_O
3	F	A_C	F	A_O	F	A_O
4	F	F	F	A_C	A_C	A_O
5	F	F	F	A_C	A_C	A_C
6	F	F	A_C	A_C	A_C	F
7	F	A_O	A_O	A_C	F	F

Only seven of the 40 possible kinematics designs (8 interchangeable) were identified among existing designs. Table A8 shows that the actuated axes are often closed for greater structural stiffness. Most of the actuated axes relate to the printhead.

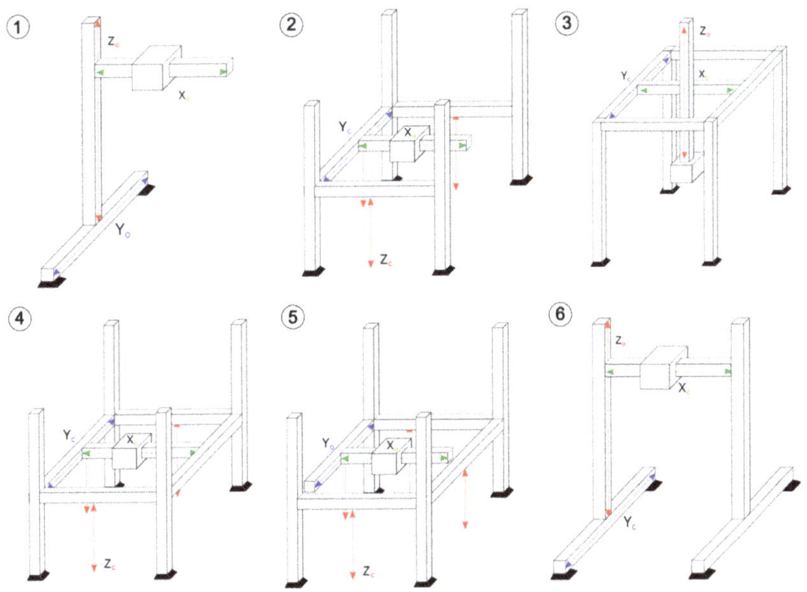

Figure A5. Fixed axes X_B, Y_B, Z_B and actuated axes X_H, Y_H, Z_H. Existing designs are 3 and 4.

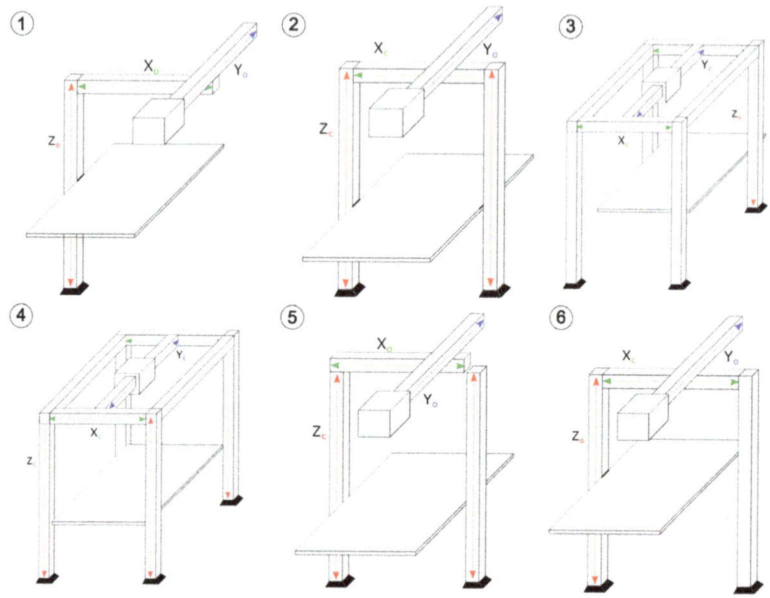

Figure A6. Fixed axes X_B, Y_B, Z_H and actuated axes X_H, Y_H, Z_B. Existing designs are 3 and 4.

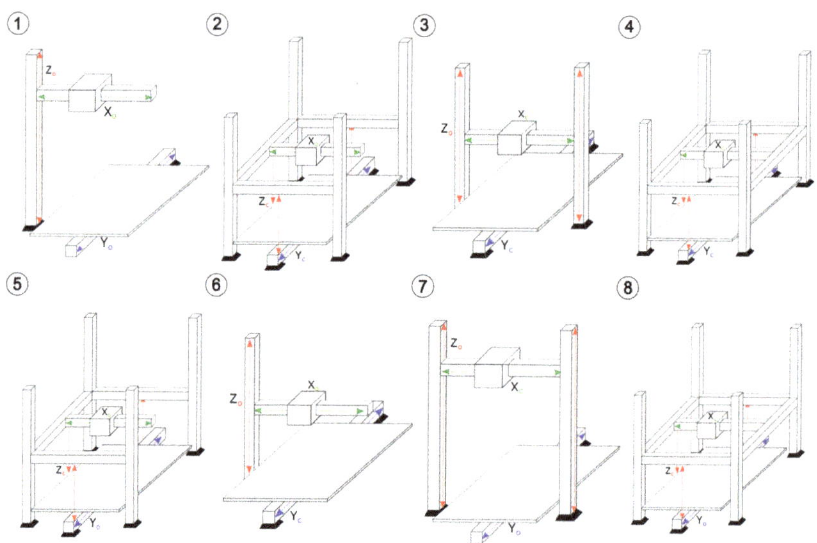

Figure A7. Fixed axes X_B, Y_H, Z_B and actuated axes X_H, Y_B, Z_H. Existing designs are 1 and 3.

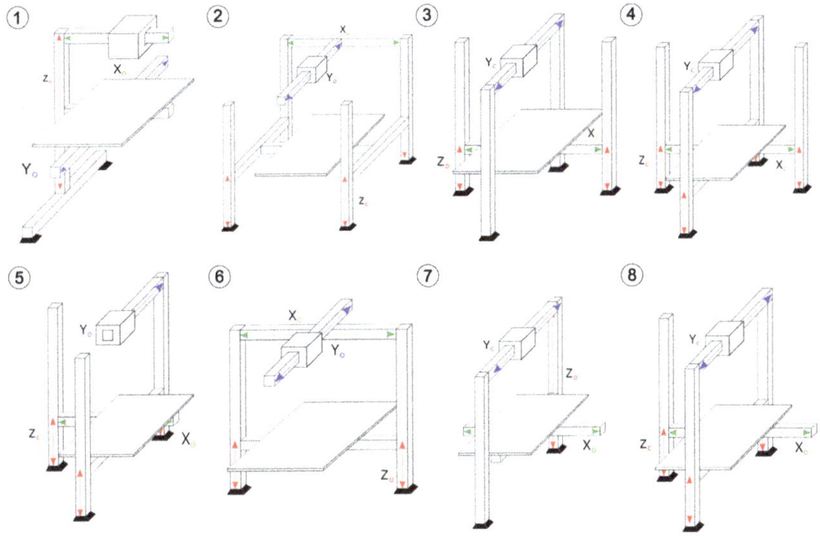

Figure A8. Fixed axes X_B, Y_H, Z_H and actuated axes X_H, Y_B, Z_B. Existing design is 7.

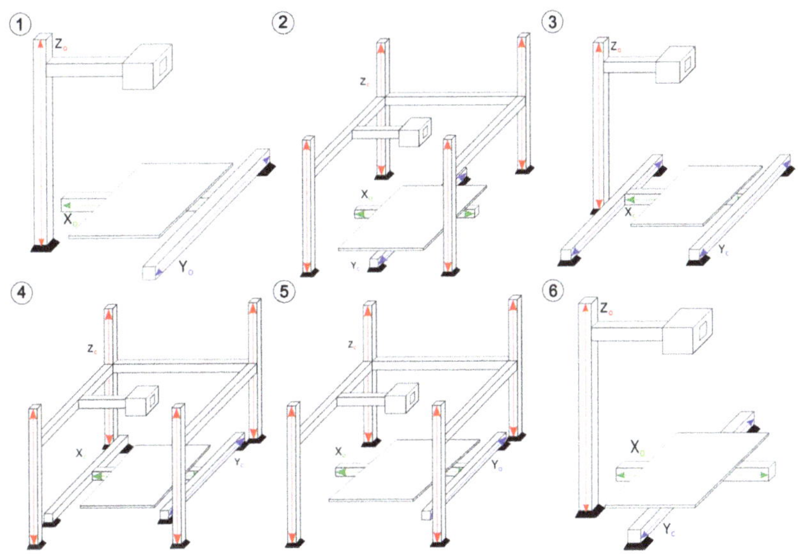

Figure A9. Fixed axes X_H, Y_H, Z_B and actuated axes X_B, Y_B, Z_H.

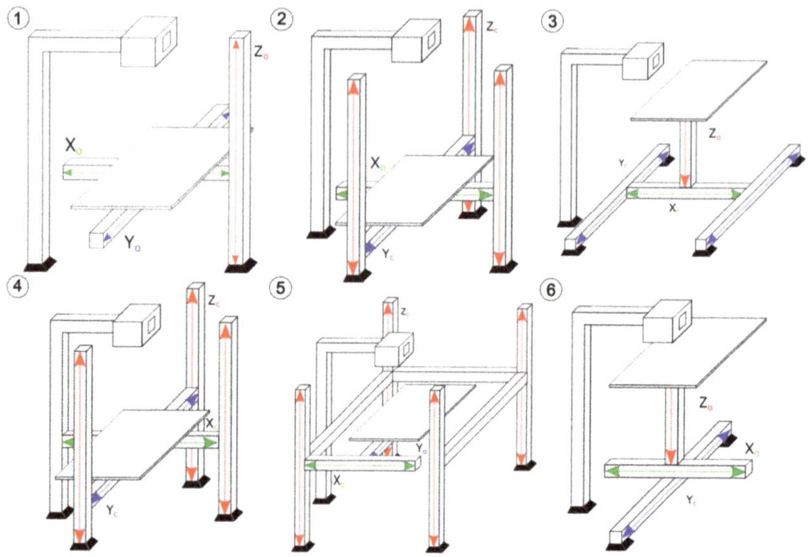

Figure A10. Fixed axes X_H, Y_H, Z_H and actuated axes X_B, Y_B, Z_B.

Cartesian 3D printers						
Head			Bed			
X	*Y*	*Z*	*X*	*Y*	*Z*	
C	C	O				1. Specavia ATM, COBOD, Winsun
C	C	C				2. Voron 2.4, Core-XYZ
C	C				O	3. Anycubic 4max Pro, Wanhao GR2, Ultimaker 3
C	C				C	4. Total Z Anyform, Paise 3D Pro2
C		O		C		5. Wanhao Duplicator i3, Anycubic i3 Mega
O		O		C		6. Wanhao Duplicator i3 Mini, Winbo SH-105
C				O	O	7. Felix 3.0

Figure A11. Classification scheme for existing Cartesian 3D printer designs, yellow C stands for closed axis, red O stands for open axis.

References

1. Qattawi, A.; Alrawi, B.; Guzman, A. Experimental optimization of fused deposition modelling processing parameters: A design-for-manufacturing approach. *Procedia Manuf.* **2017**, *10*, 791–803.
2. Sharma, K.; Kumar, K.; Singh, K.R.; Rawat, M. Optimization of FDM 3D printing process parameters using Taguchi technique. *IOP Conf. Ser. Mater. Sci. Eng.* **2021**, *1168*, 012022. [CrossRef]
3. Chand, R.; Sharma, V.S.; Trehan, R.; Gupta, M.K. A physical investigation of dimensional and mechanical characteristics of 3D printed nut and bolt for industrial applications. *Rapid Prototyp. J.* **2022**. [CrossRef]
4. Khorasani, M.; Ghasemi, A.; Rolfe, B.; Gibson, I. Additive manufacturing a powerful tool for the aerospace industry. *Rapid Prototyp. J.* **2021**, *28*, 87–100. [CrossRef]

5. Dobra, A. General classification of robots. Size criteria. In Proceedings of the 2014 23rd International Conference on Robotics in Alpe-Adria-Danube Region (RAAD), Smolenice, Slovakia, 3–5 September 2014; pp. 1–6.
6. Wilson, M. *Implementation of Robot Systems: An Introduction to Robotics, Automation, and Successful Systems Integration in Manufacturing*; Butterworth-Heinemann: Oxford, UK, 2014.
7. Zhang, G.Q.; Li, X.; Boca, R.; Newkirk, J.; Zhang, B.; Fuhlbrigge, T.A.; Feng, H.K.; Hunt, N.J. Use of industrial robots in additive manufacturing-a survey and feasibility study. In Proceedings of the ISR/Robotik 2014—41st International Symposium on Robotics, Munich, Germany, 2–3 June 2014; pp. 1–6.
8. Types Of FDM 3D Printers: Cartesian, Corexy, and More. Available online: https://all3dp.com/2/cartesian-3d-printer-delta-scara-belt-corexy-polar/ (accessed on 3 September 2021).
9. Hsieh, C.T. Investigation of delta robot 3d printer for a good quality of printing. In *Applied Mechanics and Materials*; Trans Tech Publications Ltd.: Stafa-Zuerich, Switzerland, 2017; Volume 870, pp. 164–169.
10. Carabin, G.; Scalera, L.; Wongratanaphisan, T.; Vidoni, R. An energy-efficient approach for 3D printing with a Linear Delta Robot equipped with optimal springs. *Robot.-Comput.-Integr. Manuf.* **2021**, *67*, 102045. [CrossRef]
11. Vojić, S. Applications of collaborative industrial robots. *Mach. Technol. Mater.* **2020**, *14*, 96–99.
12. Mohammad, A.E.K.; Hong, J.; Wang, D. Design of a force-controlled end-effector with low-inertia effect for robotic polishing using macro-mini robot approach. *Robot.-Comput.-Integr. Manuf.* **2018**, *49*, 54–65. [CrossRef]
13. McClinton, E.; Wylie, B.; Moore, C.A. Creating a High and Homogenous Resolution Workspace for SCARA Based 3D Printers. *IOP Conf. Ser. Mater. Sci. Eng.* **2019**, *689*, 012021. [CrossRef]
14. Alamsyah, F.A. The Kinematics Analysis of Robotic Arm manipulators Cylindrical Robot RPP Type for FFF 3D Print using Scilab. *IOP Conf. Ser. Mater. Sci. Eng.* **2019**, *494*, 012100. [CrossRef]
15. Obi, M.U.; Pradel, P.; Sinclair, M.; Bibb, R. A bibliometric analysis of research in design for additive manufacturing. *Rapid Prototyp. J.* **2022**. [CrossRef]
16. Go, J.; Schiffres, S.N.; Stevens, A.G.; Hart, A.J. Rate limits of additive manufacturing by fused filament fabrication and guidelines for high-throughput system design. *Addit. Manuf.* **2017**, *16*, 1–11. [CrossRef]
17. Löffler, R.; Koch, M. Innovative extruder concept for fast and efficient additive manufacturing. *IFAC-PapersOnLine* **2019**, *52*, 242–247. [CrossRef]
18. Go, J.; Hart, A.J. Fast desktop-scale extrusion additive manufacturing. *Addit. Manuf.* **2017**, *18*, 276–284. [CrossRef]
19. Xu, J.; Liu, K.; Liu, Z.; Zhang, F.; Zhang, S.; Tan, J. Electrothermal response optimization of nozzle structure for multi-material rapid prototyping based on fuzzy adaptive control. *Rapid Prototyp. J.* **2022**. [CrossRef]
20. Pilch, Z.; Domin, J.; Szłapa, A. The impact of vibration of the 3D printer table on the quality of print. In Proceedings of the 2015 Selected Problems of Electrical Engineering and Electronics (WZEE), Kielce, Poland, 17–19 September 2015; pp. 1–6.
21. Ribeiro, G.d.S.; Silveira, Z. Structural and Modal Analysis of a Desktop 3-D Printer Considering an Interchangeable Mini Screw Extrusion Head. Available online: https://www.researchgate.net/publication/289484273_Structural_and_modal_analysis_of_a_desktop_3-D_printer_considering_an_interchangeable_mini_screw_extrusion_head (accessed on 1 April 2022).
22. Verma, G. *Autodesk Fusion 360 Black Book*; BPB Publications: New Delhi, India, 2018.
23. Okwudire, C.E.; Huggi, S.; Supe, S.; Huang, C.; Zeng, B. Low-level control of 3D printers from the cloud: A step toward 3D printer control as a service. *Inventions* **2018**, *3*, 56. [CrossRef]
24. Levinskas, L. Research of Dimensional Accuracy and Surface Quality Dependency on Printing Speed of FDM 3D Printed Parts. Master's Thesis, Kauno Technologijos Universitetas, Kaunas, Lithuania, 2019.
25. Cohen, Z. Speed Limits: The Architectural Design Possibilities of the 3D Printed Corner. In Proceedings of the 107th Annual ACSA Conference Proceedings, Pittsburgh, PA, USA, 28–30 March 2019; ACSA Press: Washington, DC, USA, 2019.
26. Srivastava, A.; Gautam, C.; Bhan, N.; Dayal, R. Vibration Analysis of Cantilever shaped 3D printers. *IOP Conf. Ser. Mater. Sci. Eng.* **2019**, *594*, 012020. [CrossRef]
27. Cole-Pinnix, J. Optimal Control of Process Parameters in Additive Manufacturing (3D Printing) Using Six Sigma (DMAIC) Methodology. Ph.D. Thesis, North Carolina Agricultural and Technical State University, Greensboro, NC, USA, 2021.
28. Galati, M.; Minetola, P.; Marchiandi, G.; Atzeni, E.; Calignano, F.; Salmi, A.; Iuliano, L. A methodology for evaluating the aesthetic quality of 3D printed parts. *Procedia CIRP* **2019**, *79*, 95–100. [CrossRef]
29. Zhang, D.; Li, Z.; Qin, S.; Han, S. Optimization of vibration characteristics of fused deposition modeling color 3D printer based on modal and power spectrum method. *Appl. Sci.* **2019**, *9*, 4154. [CrossRef]
30. Ford, D.G.; Myers, A.; Haase, F.; Lockwood, S.; Longstaff, A. Active vibration control for a CNC milling machine. *Proc. Inst. Mech. Eng. Part C J. Mech. Eng. Sci.* **2014**, *228*, 230–245. [CrossRef]
31. Caixu, Y.; Haining, G.; Xianli, L.; Liang, S.Y.; Lihui, W. A review of chatter vibration research in milling. *Chin. J. Aeronaut.* **2019**, *32*, 215–242.
32. Duan, M.; Yoon, D.; Okwudire, C.E. A limited-preview filtered B-spline approach to tracking control–With application to vibration-induced error compensation of a 3D printer. *Mechatronics* **2018**, *56*, 287–296. [CrossRef]
33. Ramani, K.S.; Edoimioya, N.; Okwudire, C.E. A Robust Filtered Basis Functions Approach for Feedforward Tracking Control—With Application to a Vibration-Prone 3-D Printer. *IEEE/ASME Trans. Mechatron.* **2020**, *25*, 2556–2564. [CrossRef]
34. Klipper Documentation. Resonance Compensation. Available online: https://www.klipper3d.org/Resonance_Compensation.html (accessed on 8 October 2021).

35. Labuschagne, A.; van Rensburg, N.J.; Van der Merwe, A. Comparison of linear beam theories. *Math. Comput. Model.* **2009**, *49*, 20–30. [CrossRef]
36. Fukuoka, T.; Nomura, M.; Sugano, N. Experimental investigation and finite element analysis of the free vibration problem of bolted joint by taking account of interface stiffness. *J. Environ. Eng.* **2009**, *4*, 101–111. [CrossRef]
37. Worden, K.; Green, P. A machine learning approach to nonlinear modal analysis. *Mech. Syst. Signal Process.* **2017**, *84*, 34–53. [CrossRef]
38. Hattori, G.; Serpa, A.L. Contact stiffness estimation in ANSYS using simplified models and artificial neural networks. *Finite Elem. Anal. Des.* **2015**, *97*, 43–53. [CrossRef]
39. Zare, I.; Allen, M.S. Adapting a contact-mechanics algorithm to predict damping in bolted joints using quasi-static modal analysis. *Int. J. Mech. Sci.* **2021**, *189*, 105982. [CrossRef]
40. Uwajambo, P.; Thakur, N. *Numerical and Experimental Modal Analysis of CNC Machine*; Blekinge Institute of Technology: Karlskrona, Sweden, 2021.
41. Zhang, K.; Li, G.; Gong, J.; Wan, F. Modal Analysis of Bolted Structure Based on Equivalent Material of Joint Interface. *Materials* **2019**, *12*, 3004. [CrossRef]
42. Jamia, N.; Jalali, H.; Taghipour, J.; Friswell, M.; Khodaparast, H.H. An equivalent model of a nonlinear bolted flange joint. *Mech. Syst. Signal Process.* **2021**, *153*, 107507. [CrossRef]
43. Odrobiňák, J.; Idunk, R.; Bačinský, T. Study on stiffness of composite beam-to-column joints. *Procedia Eng.* **2014**, *91*, 268–273. [CrossRef]
44. Kopets, E.E.; Kolev, G.Y.; Vatnik, V.M.; Karimov, A.I.; Rybin, V.G. Mechanical Vibration Analysis of a Gantry 3D Printer. In Proceedings of the 2021 IEEE Conference of Russian Young Researchers in Electrical and Electronic Engineering (ElConRus), St. Petersburg, Russia, 26–29 January 2021; pp. 956–960.
45. Han, S.M.; Benaroya, H.; Wei, T. Dynamics of transversely vibrating beams using four engineering theories. *J. Sound Vib.* **1999**, *225*, 935–988. [CrossRef]
46. Karimov, A.I.; Kopets, E.; Nepomuceno, E.G.; Butusov, D. Integrate-and-Differentiate Approach to Nonlinear System Identification. *Mathematics* **2021**, *9*, 2999. [CrossRef]
47. Marghitu, D.B. *Kinematic Chains and Machine Components Design*; Gulf Professional Publishing: Oxford, UK, 2005.
48. Basova, Y.; Nutsubidze, K.; Ivanova, M.; Slipchenko, S.; Kotliar, A. Design and numerical simulation of the new design of the gripper for manipulating of the rotational parts. *Diagnostyka* **2018**, *19*, 11–18. [CrossRef]
49. Zhang, H.; Zhao, W.; Du, C.; Liu, H.; Zhang, J. Dynamic modeling and analysis for gantry-type machine tools considering the effect of axis coupling force on the slider–guide joints' stiffness. *Proc. Inst. Mech. Eng. Part B J. Eng. Manuf.* **2016**, *230*, 2036–2046. [CrossRef]
50. Original Prusa i3 MK3S+ Kit. Available online: https://www.prusa3d.com/product/original-prusa-i3-mk3s-kit-3/ (accessed on 24 April 2022).
51. Anycubic 4Max Pro 2.0. Available online: https://www.anycubic.com/collections/max-series (accessed on 3 September 2021).
52. Sovol SV03. Available online: https://sovol3d.com/products/sv03 (accessed on 24 April 2022).
53. Karimov, A.I.; Kopets, E.E.; Rybin, V.G.; Leonov, S.V.; Voroshilova, A.I.; Butusov, D.N. Advanced tone rendition technique for a painting robot. *Robot. Auton. Syst.* **2019**, *115*, 17–27. [CrossRef]
54. Batley, S. *Classification in Theory and Practice*; Chandos Publishing: Oxford, UK, 2014.
55. Wang, X.; Baron, L.; Cloutier, G. Topology of serial and parallel manipulators and topological diagrams. *Mech. Mach. Theory* **2008**, *43*, 754–770. [CrossRef]
56. Müller, A. Generic mobility of rigid body mechanisms. *Mech. Mach. Theory* **2009**, *44*, 1240–1255. [CrossRef]
57. Lu, Y.; Hu, B.; Sun, T. Analyses of velocity, acceleration, statics, and workspace of a 2 (3-SPR) serial-parallel manipulator. *Robotica* **2009**, *27*, 529–538. [CrossRef]
58. Helms, C. Van gobot. *Sci. Am.* **2018**, *318*, 9.
59. Specavia ATM. Available online: http://specavia.pro/catalog/ (accessed on 3 September 2021).
60. Winsun 3D Printer. Available online: http://www.winsun3d.com/En/ (accessed on 3 September 2021).
61. The BOD. Europe's First 3D Printed Building. Available online: https://cobod.com/the-bod/ (accessed on 3 September 2021).
62. Voron 2.4 3D Printer Assembly Manual. Available online: https://github.com/VoronDesign/Voron-2/raw/Voron2.4/Docs/2.4_Assembly_Manual.pdf (accessed on 3 September 2021).
63. Core-XYZ First Motion with Modded REPETIER Firmware!—Video Uploaded by 1kreature. Available online: https://youtu.be/Ii0lUfQ8pJI (accessed on 3 September 2021).
64. 3D-printer AnyForm L250-G3(2X). Available online: https://totalz.ru/products/equipment/universal-fdm/anyform-l250-g3-2x/ (accessed on 3 September 2021).
65. Wanhao 3D Printer. Available online: https://https://www.wanhao3dprinter.com/Index.html/ (accessed on 3 September 2021).
66. Wanhao i3 Mini. Available online: https://wanhao.store/products/wanaho-i3-mini/ (accessed on 3 September 2021).
67. Felix Pro 3 Highlights. Available online: https://www.felixprinters.com/pro-3-highlights/ (accessed on 3 September 2021).
68. Tlegenov, Y.; Hong, G.S.; Lu, W.F. Nozzle condition monitoring in 3D printing. *Robot.-Comput.-Integr. Manuf.* **2018**, *54*, 45–55. [CrossRef]

applied sciences

MDPI

Article

Design and Simulation of a Vision-Based Automatic Trout Fish-Processing Robot

Hossein Azarmdel [1,2], Seyed Saeid Mohtasebi [2,*], Ali Jafary [2], Hossein Behfar [3] and Alfredo Rosado Muñoz [1,*]

1 Department of Electronics Engineering, University of Valencia, Burjassot, 46100 Valencia, Spain; azarmdel@alumni.uv.es or h.azarmdel@ut.ac.ir
2 Department of Agricultural Machinery Engineering, Faculty of Agricultural Engineering and Technology, University of Tehran, Karaj 3158777871, Iran; jafarya@ut.ac.ir
3 Department of Biosystems Engineering, Faculty of Agriculture, University of Tabriz, Tabriz 5166616471, Iran; behfar@tabrizu.ac.ir
* Correspondence: mohtaseb@ut.ac.ir (S.S.M.); alfredo.Rosado@uv.es (A.R.M.); Tel.: +98-912-1066902 (S.S.M.); +34-963543808 (A.R.M.)

Featured Application: Due to the broad automation in food industry, the need to design and manufacture automate devices with high capabilities in the fish and aquatic processing industries is of great importance. This system is introduced to process trout fish in four steps in an integrated machine including, belly cutting, beheading, gutting, and cleaning processes bsed on the machine vision. The applied grippers stabilize the fish for any trout processing operations. In this system, trout moves inside a canal along the machine, and all the operations are conducted on the fish. Therefore, this system can be applied in the fish supplying markets and fish processing factories, which facilitate the fish cleaning process in different fish sizes.

Abstract: Today, industrial automation is being applied in a wide range of fields. The initial modeling of robots and mechanical systems together with simulation results in optimal systems. In this study, the designed system is simulated to obtain the required velocities, accelerations and torques of the actuating arms in a vision-based automatic system. Due to the slippery skin of fish and the low friction coefficient, it is not easy to design an optimal tool to handle fish. Since the fish-processing operation is undertaken step by step and provides fish stability, it is essential that the gripper enables different processing operations along the system. The proposed system performs belly-cutting, beheading, gutting, and cleaning stages for different fish sizes, based on the extracted dimensions of the vision system. In the head-cutting section, the average speed of the actuator jack was considered as 500 mm s^{-1}. Under these conditions, the maximum required force to provide this speed was 332.45 N. In the belly-cutting subsystem, the required torque for the stepper motor resulted in 1.79–2.15 N m. Finally, the maximum required torque for the gutting stepper motor was calculated as 0.69 N m in the tested processing capacities.

Keywords: trout; fish processing machine; simulation; vision based system

Citation: Azarmdel, H.; Mohtasebi, S.S.; Jafary, A.; Behfar, H.; Rosado Muñoz, A. Design and Simulation of a Vision-Based Automatic Trout Fish-Processing Robot. *Appl. Sci.* **2021**, *11*, 5602. https://doi.org/10.3390/app11125602

Academic Editor: Alessandro Gasparetto

Received: 14 April 2021
Accepted: 9 June 2021
Published: 17 June 2021

1. Introduction

Trout, with the scientific name of *Oncorhynchus mykiss*, is from the salmon genus. The recent technology in the food industry is developed in different fields, such as fully automated and intelligent systems based on vision. Since the speed of the processing function, together with the accuracy and quality of the final product, is required, recent systems are equipped with vision-based equipment, especially in the products with different sizes and shapes.

Fish sorting increases the final quality of the product but requires an initial investment and carries with it structural complexities [1]. In small and large fish centers, fish species in high demand are usually available in different sizes. In non-processed fresh fish, some prefer gutted fish; others opt for gutted and beheaded, or just in fillets. It should be noted that trout scales are finer than other species, and it can be processed without scaling.

Due to the broad automation in industry, the need to design and manufacture automated devices with high capabilities in the fish and aquatic processing industries is of great importance. Because of the importance of fish consumption to the food industry, there is a need to design and develop fish-processing systems with higher capabilities than the current models. In fish-processing factories and fish supplying markets, it is necessary to properly stabilize the fish for cleaning operations. Due to the slippery skin and the low friction coefficient, it is usually challenging to control the fish constantly. To increase the system capacity, it is necessary to process the fish with functioning arms while the fish are passing in front of the processing tools.

The initial modeling of the robots and mechanical systems together with simulation results in optimal real systems. In this condition not only are the links and joints and actual movements of the arms and actuators traced, but also the required forces and torques are determined in the most optimized condition. This needs defining the initial conditions of the friction coefficients, the movement limitations, required initial speeds and accelerations. In a study conducted by Beiranvand [2], the minimum length sliding mode control of a three degrees of freedom (DOF) Cartesian parallel robot was modeled and identified considering the virtual flexible links. In another study, kinematic optimal design of a partially compliant four-bar linkage using the elliptic integral solution was applied [3]. An analytical model of a compact flexure mechanism for translational motion together with the designing process was conducted by Hao [4].

A conventional machine vision system (MVS) includes an image-capturing unit (camera), image processing unit, and statistical analysis [5]. Machine vision has been used in the field of aquaculture for classification, sex determination, and quality assessment. This non-destructive method was also applied in fish post-processing steps like removing unwanted parts and cutting the fillet edges. The following are examples of research conducted in this area. To adapt the fish cleaning system to the individual characteristics of each fish, it is necessary to provide special tools for cutting, gutting, cleaning, and gripping fish of different sizes. Machine vision as the main detection and processing tool was applied by Atienza-Vanacloig for tuna growing stages [6]. This method was also applied to develop a fish counting algorithm while passing from the embedded system [7]. It was also applied for fish fillet quality and freshness assessment [8]. Fish freshness can be detected according to the color features of the eyes and the gills [9]. The fish length and size determination algorithm was developed by Miranda [10] and Muñoz-Benavent [11]. In other research, the size, direction, and heterogeneity of the fish were measured based on edge analysis to develop the tools for grading bass species [12].

The mechanical systems and cutting arms are designed in different fish-processing machines. In a design, the researchers developed a head-cutting fish cleaner. In their prototype, the fish was held using two metal plates to separate the fish head in the appropriate position [13]. In order to cut the head, they proposed a mechanism to keep the head in the right position while gripping the body. The designed clamp only allowed the system to cut the head. In this system, the fish was held steady and the head-cutting tool cuts the head by moving toward the fish.

In another system, a machine for fish head cutting was designed [14]. In this machine, the fish were moved using a conveyor chain system with stainless steel semicircular seats. In this model, as the fish passes over the conveyor, the holding arm from the top of the line approaches the moving fish to stabilize it and for beheading processes. Like the previous model, this machine was only able to cut the fish head. The difference between this proposal and the model designed by Buckingham lies in the processing condition of the fish (stable or moving fish) [13]. Due to the high inertia of the cutting tools, it is preferred to process the fish while moving with minimum movement of cutting tools. On the other hand, the capacity of the system is high if the fish moves in front of the cutting arms. Therefore, the system used by De Silva [14] takes priority in terms of meeting the proper dynamic conditions compared to the system designed by Buckingham [13].

Later, Lang [15] proposed a device for cutting salmon heads. In this machine, the fish were placed manually in their seat and then transported to the cutting position by a hydraulic jack. In order to provide linear motion, a rotary motor was used to convert rotational motion to linear motion. This device was also developed to catch the fish for beheading.

Ketels [16] proposed a system for correctly placing the fish in the processing line during head cutting. In this design, they used metal seats mounted on the chains. Since the sorted fish were fed to this system, the fish conveyor seats were designed in fixed sizes. Therefore, in the fish of different sizes, this mechanism cannot be used unless specific conveyor chains are considered for each of the different sizes of the fish.

In another fabricated design, conveyor belts consisting of metal seats were used [17], as applied by Ketels [16]. In addition to the clamp designed for beheading, another clamp was used to separate the fishbone. It should be noted that, although the designed device was able to perform two different operations, two different types of clamp were used for each step of beheading and deboning. In the proposed system in the present study, in addition to the ability to perform various steps of belly cutting, gutting, and beheading, it will be possible to perform all these steps in one integrated system.

Urushibara [18] designed and fabricated a device for fish beheading and gutting without cutting the belly. In this design, the fish were transported while lying on their sides to cut the head and evacuate the belly contents. In another system, a device to grip the fish and clean the belly contents was applied [19]. In this system, contrary to the model proposed by Paulson [20], the fish were transported downward between two support straps along with the device. Similarly, a device to catch the fish for removing blood clots near the spinal cord was used [21]. This tool is only for catching fish to suck the blood and blood clots, and it is a single-use clamp in the fish-processing industry. This clamp was applied in a fish filleting machine.

A flow tool for extracting belly content was designed and fabricated [22]. The designers have offered this tool for cutting fish fillets, but they have also suggested this tool for evacuating the belly content. In this system, as the fish moves towards the cutting tool, the blade contacts the belly content and removes it from the belly cavity. It was necessary to separate the fish head from the body in the previous step.

In a design proposed by Finke [23], a device with attached material to the spinal cord with two sets of rotating disks was applied. The first set of the discs guides and presses the fish, and the other set under the fish performs the cleaning operation from beneath the fish. The second set of the disks were placed opposite with angles. In this design, the fish is pulled from the tail, and the belly is downward. In order to use this tool, it is necessary to clean the belly and remove the head in the previous step. Among the described systems, only in the designs proposed by Paulson [20] and Ryan [19], a combination of mechanical and electronic sensors has been applied to detect the dimensions of the fish, which can be applied to extract the fish size and cutting points. Since in each of these systems, only a single operation of cutting or gutting is undertaken, using these sensors is sufficient for dimension determination.

In Paulson's design [20], the head of the fish remains on the fish, and only the gutting process with the details of belly-cutting, suction, and belly cleaning steps are undertaken in a single system. Also, in the system designed by Ryan [19], the head of the fish is removed in the previous stage, and subsequently, by passing through the conveyor belt, the gutting process is performed. These sensors are used while cutting and cleaning operations are performed on the belly because the exact head-cutting position requires special methods like machine vision.

Then, an automated line for slaughtering and cutting fish heads using machine vision was designed [24]. To slaughter the fish, the pneumatic jacks were equipped to narrow blades. The general structure of the device used in this study is similar to those applied by

Storbeck [25] with the difference that, in addition to all devices applied as mechanical tools, the stunning technology enabled the feeding of alive fish so that they can be processed simultaneously. It was also necessary to identify the fishtail and head sides in order to make an injury on the right side.

In the current system, an automated system is designed for trout beheading, gutting, and cleaning. As this system consists of mechanical, controlling, and machine vision sections, the whole system needs an accurate design and synchronization. The mechanical system was designed in CATIA software (Dassault Systems Corporation) and simulated in Automated Dynamic Analysis of Mechanical Systems (ADAMS) simulation software owned by MSC Software Corporation. The controlling algorithm was developed in Siemens TiaPortal15 software and tested to solve any errors. Extracted data from the machine vision system are sent to the control unit by the Ethernet TCP/IP protocol. Finally, by installing magnetic and inductive sensors, pneumatic cylinders and actuators, and user interface panel, the machine functioning was synchronized and calibrated.

Since the system is supposed to process fish of different dimensions, machine vision can provide a precise cutting point determination. With such system, not only the fish dimension is defined, but also the head cutting, belly-cutting length and gutting points are defined based on the fish fin segmentation. Since the fish dimensions are different, there is a high risk of collision between the arms and the grippers as they pass in front of each functioning station. Therefore, designing a system capable of functioning different cutting and cleaning actions on fish of different sizes in an integrated machine is the main goal of this research. The detailed goals are presented as follows:

1. Designing a trout fish-processing system capable of processing the fish in different dimensions.
2. Developing the system into a machine at which different cleaning steps can be performed in one integrated system (belly-cutting, beheading, gutting, and cleaning processes).
3. Designing a system that is adapted with online data extracted from the online image processing technique.
4. Extracting the required forces, torques, and displacements by simulating the real movements.
5. Finally, fabricating and selecting the proper functioning motors and pneumatic jacks of the system based on the applied tools (mechanical design, machine vision, control algorithms, and simulation).

2. Material and Methods

This system is composed of different units which are combined in an integrated system for trout processing. The machine vision system as the main processing and data acquisition unit is mounted in the initial part of the machine. The mechanical design of the system and final fabrication processes are described in detail in the following.

2.1. Machine Vision and Dimension Determination

This system is designed, simulated, and fabricated to process trout in an online condition. The main part of the system is the image processing setup in which the images are captured and the cutting points and dimensions are extracted using the developed algorithm. To develop the cutting point determination algorithm, an imaging case with a light-emitting diode (LED) was designed. We applied a video camera (Basler, daA1280-54uc, USB3) for the online process. To capture the images, as the fish was passing in front of the camera sensor, the signal was sent to the programmable logic controller (PLC) and the frame capturing command was sent to the MATLAB software to store the frame for fish segmentation and dimension extraction. As soon as the dimensions were extracted, they were sent to the controlling unit with the TCP/IP protocol. The machine vision components, segmentation process, and cutting point determination are presented in Figure 1.

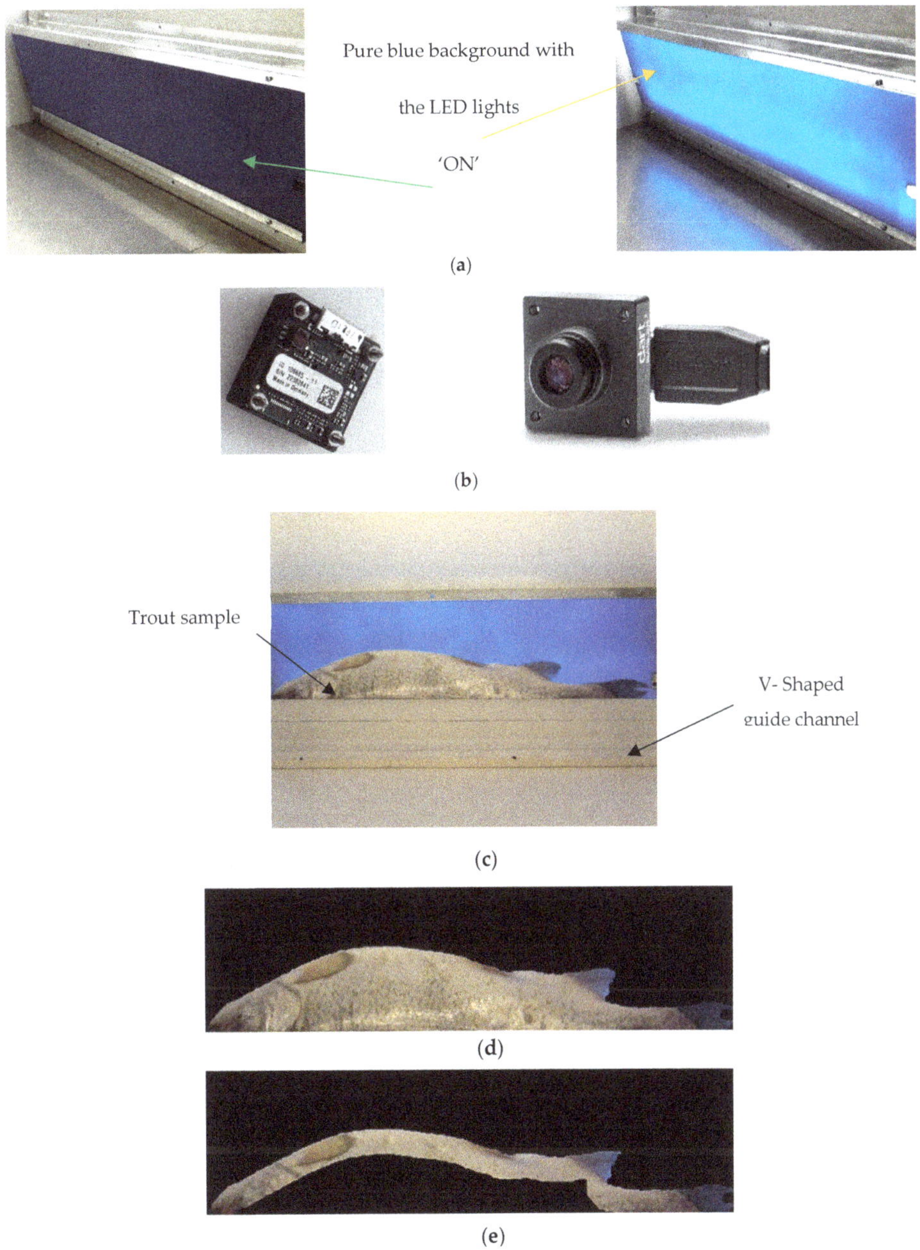

Figure 1. The image processing setup: (**a**) pure blue background, (**b**) the video camera, (**c**) inside of the imaging case with the sample, (**d**) segmented fish, and (**e**) trout fin shear to define the cutting points and functioning lengths.

The cutting points and extracted dimensions are presented in Figure 2. The head and belly-cutting regions will be considered as the operational points of the system. L_1 to L_4 are head-cutting point, belly-cutting start point, the body length without tail, and total length, respectively. The operational width of the trout is shown as W.

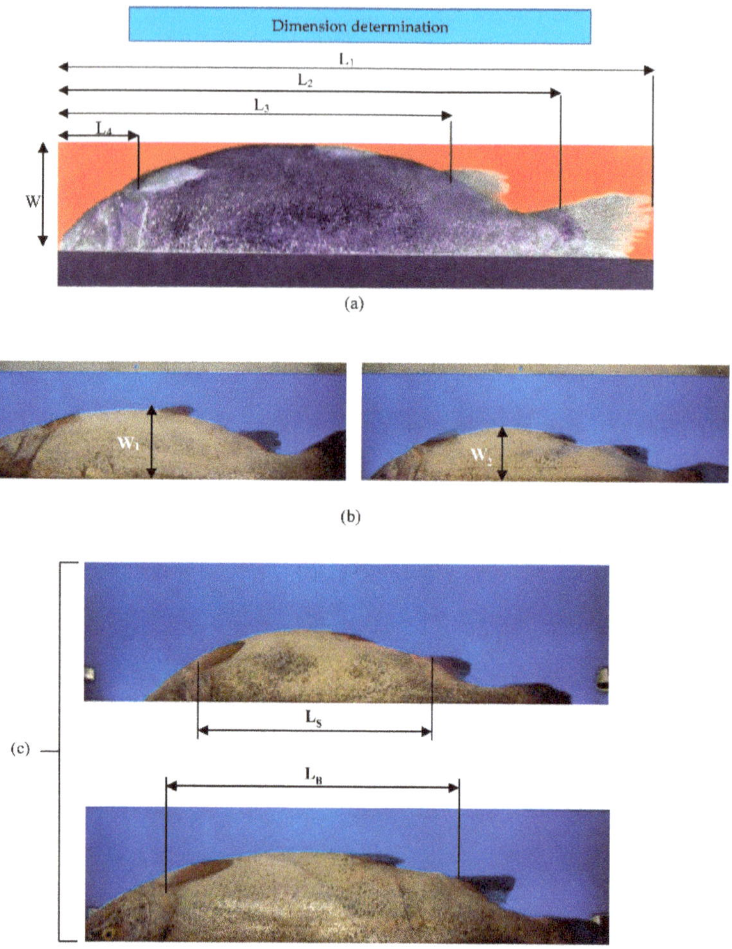

Figure 2. Trout fish dimensions: (**a**) determining trout dimensions (L_1 to L_4 are head-cutting point, belly-cutting start point, trout length without the tail, and the total length, respectively and W is the fish width), (**b**) comparing the width in different fish samples (W_1 and W_2 are different trout width), and (**c**) comparing the functioning length of actuating arms in two fish samples with different lengths (L_S and L_B are functioning length for small and big fish).

2.2. Designing the Mechanical Parts

In the first step, the initial model was designed by CATIA software. Then, all parts, such as conveyor chain and gears, were defined by entering the exact dimensions of the parts in the software. To enable all fish-processing steps in the system, we considered the dimensions of the chassis: 2.8 m in length, 0.85 m in height, and a width of 0.7 m. The total design rationale is presented in Figure 3.

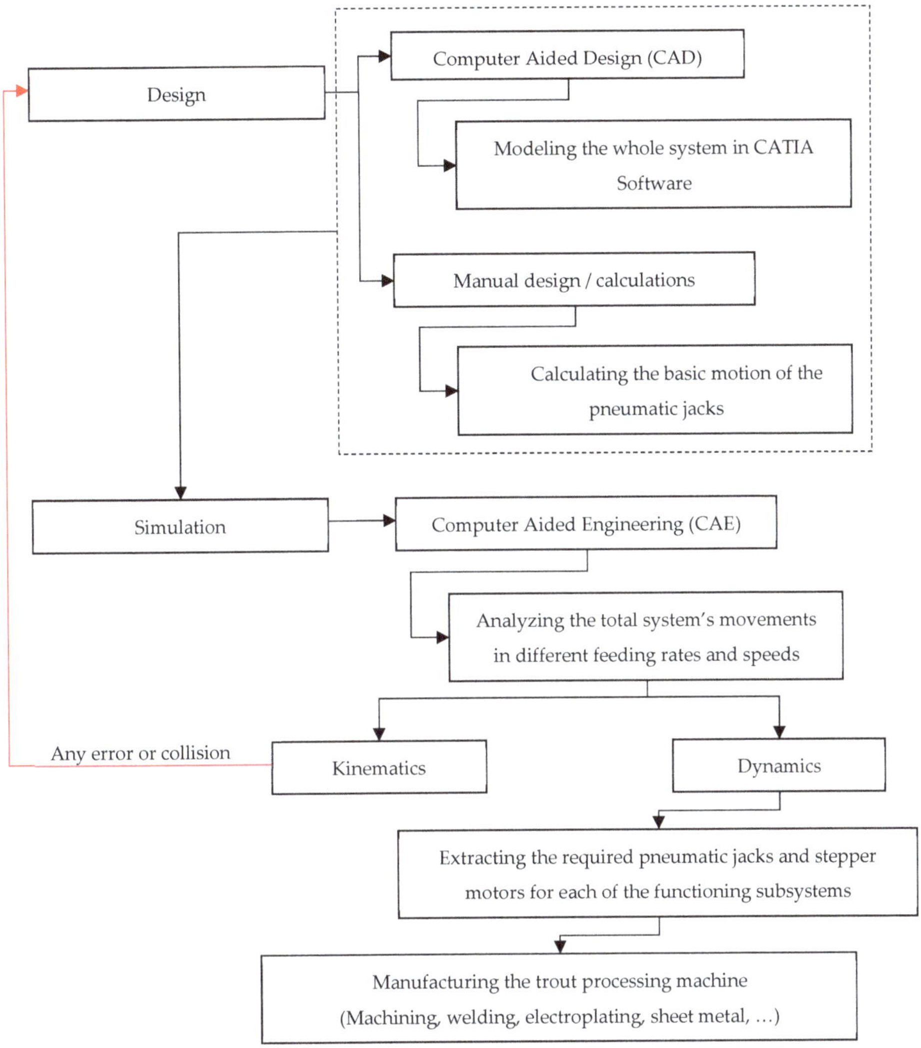

Figure 3. The design rationale of the trout-processing system with applied methods and software.

2.2.1. Fish-Carrying Subsystem

One of the most critical points in designing a system with the ability to perform different trout-processing stages is to consider a holding place in the fish body to avoid interfering with cutting and cleaning tools during the various cleaning stages. Therefore, the tail of the fish was selected as the appropriate area.

The fish carrier subsystem consists of four main parts:

1. Fish grippers;
2. Gripper guides and fish canal;
3. Driver AC motor;
4. Gripper conveyer.

In order to stain and increase the wear resistance of steel gears, hard chrome electroplating operation was used. To transmit the motor's power to the grippers, we selected a stainless steel chain with a length of 4800 mm. By dividing this length into five equal parts, a distance D_g of 960 mm was obtained. Therefore, five grippers were designed, by which five fish are processed in one turn. Figure 4 shows the sketch and schematic view of the conveyor chain and grippers in detail.

(a)

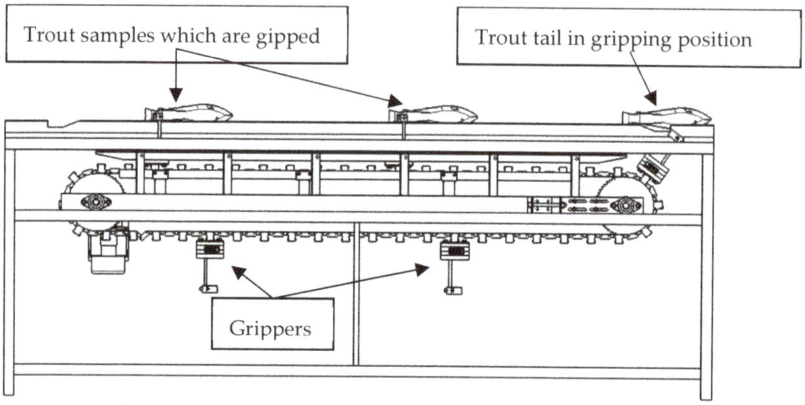

(b)

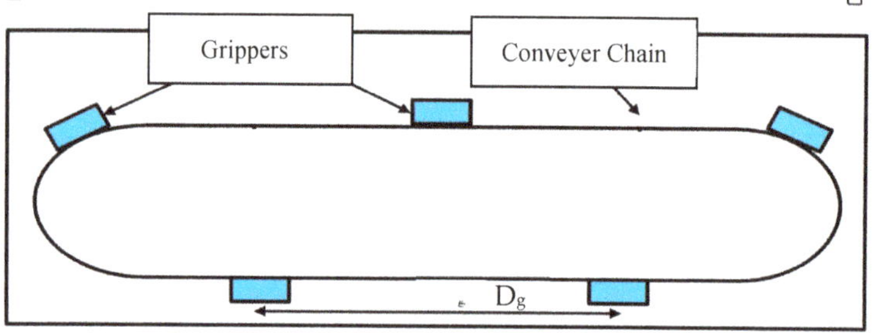

Figure 4. (**a**) System sketch, and (**b**) conveyor chain and the grippers with a distance "D_g" between the grippers.

This figure shows the gripper position and the conveyor chain with the gripper distance (Dg). If a complete cycle of the conveyor chain is performed for one minute, the number of five fish per minute ($M_{F\_}5$) will be processed by the machine. Under these conditions, the linear velocity of the carrier chain will be calculated using Equation (1):

$$v = \frac{x}{t} \tag{1}$$

In this equation, 'v' is the linear velocity (mm s^{-1}), 'x' is the length of the carrier chain (mm), and 't' is one complete path of the chain (60 s). Based on Equation (1), we considered five fish to be processed in one minute; the linear velocity will be calculated 80 mm s^{-1}. Other linear velocity values in different processing capacities are presented in Table 1.

Table 1. The linear velocities in different system capacities.

Fish Cleaning Machine Capacity (M_F)							
System capacity (fish/minute)	4	5	6	7	8	9	10
Linear velocity of the conveyer chain (mm s^{-1})	53.33	66.66	80	93.33	106.66	120	133.33

As shown in this table, by increasing the system capacity from four to 10 fish per minute, the linear velocity increased from 53.33 mm s^{-1} to 133.33 mm s^{-1}. These values are the base inputs for design and simulation of the system. The complete design of the trout-processing system is presented in Figure 5. The subsystems of the machine are described in detail below.

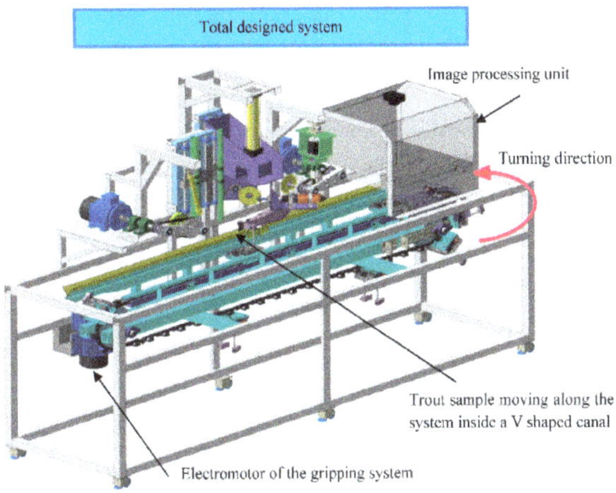

Figure 5. The complete design of the trout-processing machine.

2.2.2. Belly-Cutting Subsystem

The function of this subsystem is to make a longitudinal incision in the trout belly from the beginning of the anal fin to the pectoral fin. After extracting the required dimensions, the first step in device operation is to make a longitudinal incision in the abdomen of the fish from the anal fin to the pectoral fin based on fish size. By sending the commands from the control unit to the stepper motor, the carrier arm moves downwards, and then, the rotating blade cuts the belly. Figure 6 shows the position of the belly-cutting subsystem in the machine together with its components.

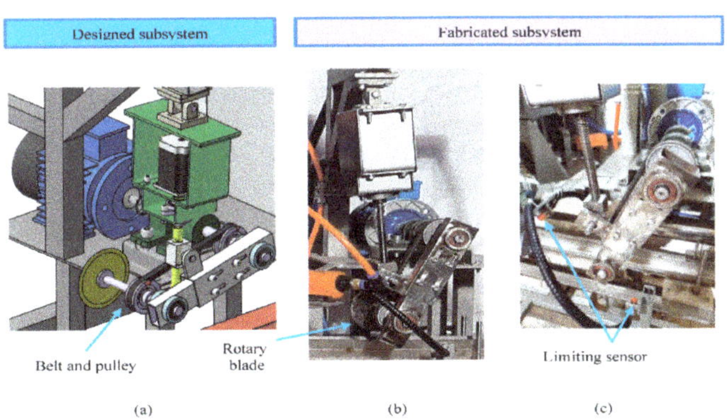

Figure 6. Belly-cutting subsystem: (**a**) designed subsystem, (**b**) fabricated belly-cutting components, and (**c**) limiting sensor.

2.2.3. Head-Cutting Subsets

Since the fish is moving along the system, it is necessary to provide the best head-cutting condition. In order to have a proper cut, a fast and precise cut must be developed.

Therefore, a pneumatic jack was used for the head cutter subset. As mentioned before, when the grippers reach the beginning of the guide rails, the fingers contact the rails and the fishtail stocks between the gripper clamps. At this point, any clamped trout is pulled along the system. Therefore, to define the system processing capacity (fish per minute); it is necessary to determine the optimal working speed. By knowing the linear velocity of the conveyor, the penetration velocity of the cutting blade will be investigated. In fact, the speed at which the cutting blade approaches the fish is directly related to the fish transferring speed. The designed head-cutting subsystem is presented in Figure 7.

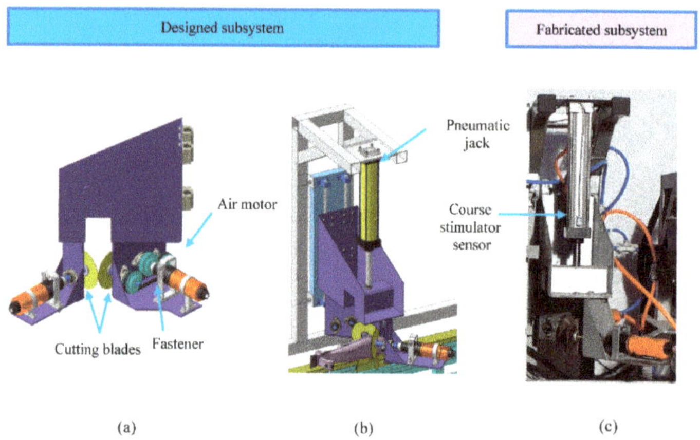

Figure 7. Head-cutting subsystem: (**a**) the position of the head-cutting subsystem in the machine, (**b**) isometric view of the subsystem, and (**c**) right view of the subsystem.

2.2.4. Gutting Subsystem

When the fish reaches the gutting position, the stepper motor is activated by receiving the start command from the control unit and rotates in clockwise and counterclockwise directions depending on fish size. In case the size of the fish is determined as small by the machine vision section, the narrower tube enters the fish belly. In fact, the stepper motor is rotated counterclockwise, and due to the engagement of the pinion with the rack gear, the set of sliders together with the suction tube move toward the fish. The gutting subsystem is presented in Figure 8.

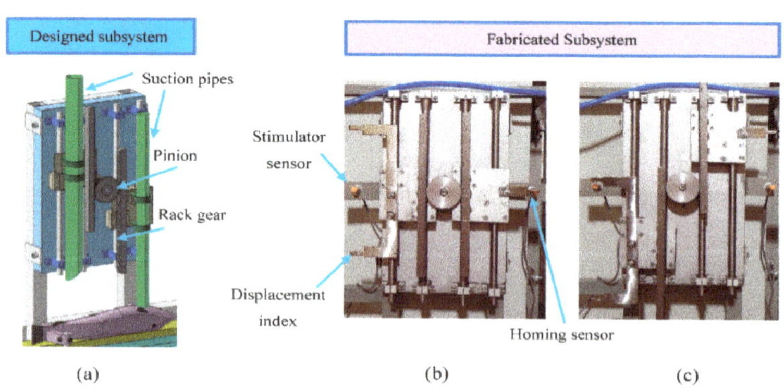

Figure 8. Gutting subsystem: (**a**) designed subsystem, (**b**) homing position with the applied sensor in the right side, and (**c**) functioning position with the limit sensors in the left side of the subsystem.

2.2.5. Cleaner Subsystem

This subset is designed to clean the inside of the fish belly and remove any material attached to the fish spine. When the fish reaches the cleaning position, the command is transferred to the pneumatic jack from the control unit and performs the cleaning operation by lowering the jack. The time duration in which the pneumatic jack is extended is directly related to the fish length. As the fish passes in front of the cleaning station, the jack retracts to the initial position. The cleaner subsystem is presented in Figure 9.

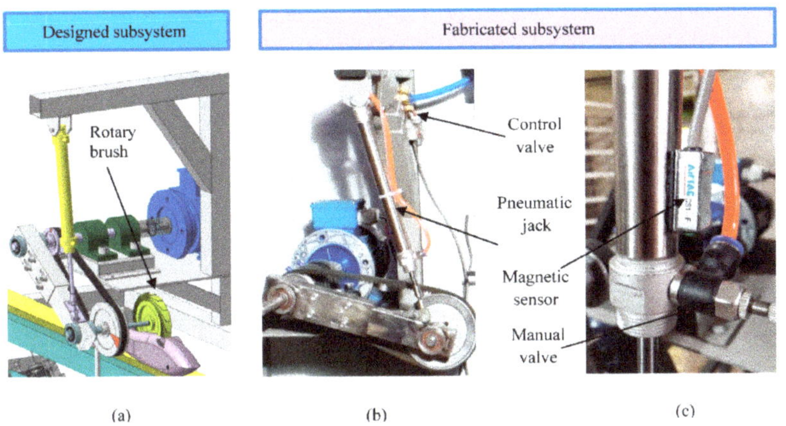

Figure 9. Fish cleaning subset: (**a**) designed subsystem, (**b**) fabricated subsystem, and (**c**) applied sensor.

The function analysis is presented in Figure 10. As shown in this figure, at the initial step, the fish is gripped by a clamp. Then the fish is carried along the machine for further processing. By passing the fish in front of the camera sensor, the signal is sent to the controller. Then, the controller sends the received data to the interface (Matlab software) via TCP_IP protocol. A frame of the camera is captured immediately and saved for further process. In the next step, the extracted data return to the controller and are reserved specifically for each trout sample. As the gripped fish is carried along the machine, the sensors in each of the stations are stimulated and receive the related information from the labeled pack of the data. Regarding the length and feeding rate, the functioning arms cut or clean the trout with subsequent delays and speeds.

2.3. *Device Simulation, Kinetic and Dynamic Analysis and Extracting the Torques of the Motors and Forces of the Pneumatic Jacks*

After designing the system in CATIA software, it was necessary to choose the motors and pneumatic jacks. Therefore, it is required to calculate the required torque and rotational speed of stepper motors and the force and speed of pneumatic jacks.

The simplest method is to design all the parts and define their weights to calculate the forces using the conventional method and numerical calculations. This method is applicable when the dynamic parameters are calculated for a single part or, in some cases, the limited number of parts within the relation. Using this method to determine accurate inertia reaction forces in a multi-body system faces a significant challenge unless the problem is simplified. It should be noted that in the simplification method, in addition to the time-consuming computational steps, the reaction forces and the dynamic and static friction conditions will not be the same as the real condition. Therefore, to have closer results to the real operating conditions and reduce the numerical calculations and increase output accuracy, the designed system in CATIA software was imported to ADAMS 2017 software and simulated for dynamic analyses. The simulation steps in the software are presented in Figure 11. The simulation process and extracting dynamic diagrams are presented in Figure 12.

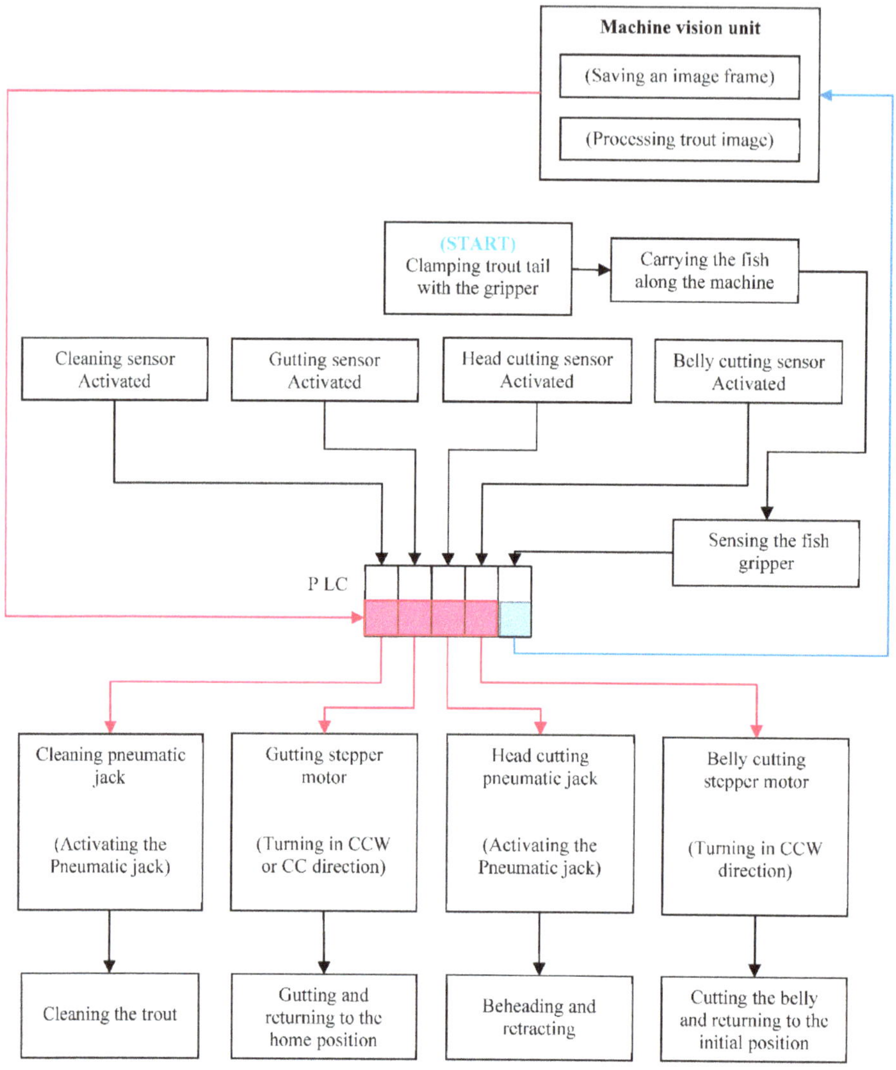

Figure 10. Function analysis of the trout fish-processing system.

After importing the designed model into simulation software, the first step is to name all parts. The related constraints were assigned for all the parts, and any possible errors were solved. At the next step, the associated motions for the moving parts were considered. For this purpose, a rotational motion was considered for the stepper motor's hinge in the belly-cutting and gutting subsystems. Likewise, for the linear moving parts, pneumatic jacks, and suction tubes, the linear constraints were considered.

In some parts of the device, to determine the constraints and types of movement, it is necessary to consider the collision between the components involved. This collision was considered between the pinion and the rack gear in the gutting subsystem. The Columbus friction model was used to determine static and dynamic friction between the parts. As the power screw and nut in the belly-cutting subsystem are both made of steel, grease lubrication will be used between these parts. The static friction coefficient was considered to be 0.16 for steel–steel contact [26]. Also, the dynamic friction coefficient was set to 0.08 [27]. For other dry parts of the device, the values of static and dynamic coefficients of friction were considered to be 0.1 and 0.2, respectively.

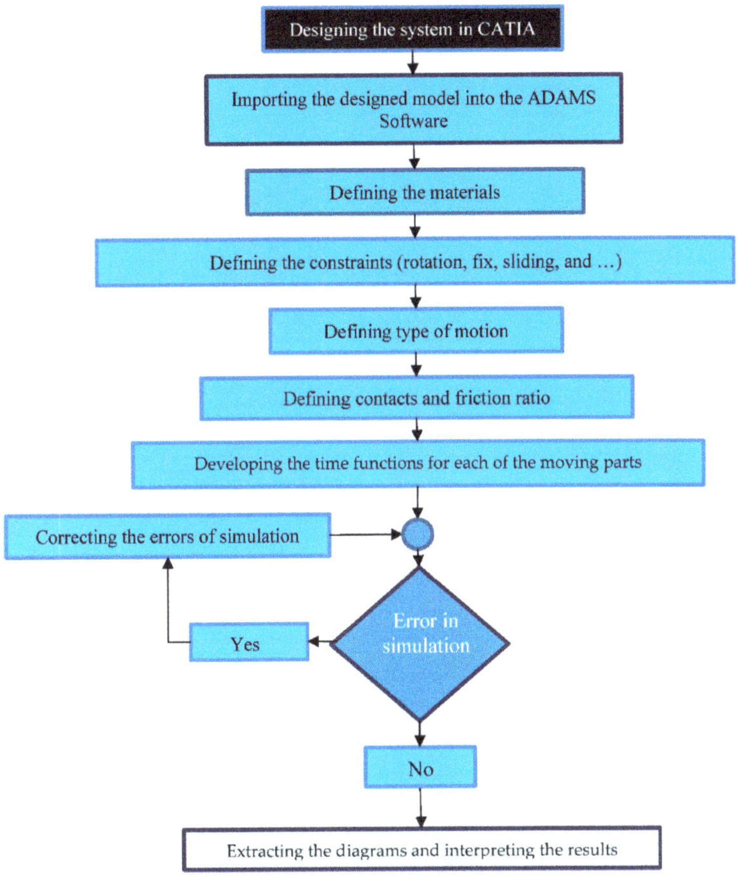

Figure 11. Simulation steps in Automated Dynamic Analysis of Mechanical Systems (ADAMS) software for dynamic analyses of the trout-processing system.

In order to determine the motion of the moving parts, the "Step" function was applied. As the fish enters the machine, trout size is extracted and the cutting points are defined using machine vision. The rotary cutter creates the longitudinal cut in the belly-cutting subsystem in the first stage. In the next step, the trout is beheaded and gutted. In this regard, it was necessary to consider the sequences of each operator in the simulation.

Therefore, four-movement steps were considered for each of the four operators of the device based on the movement time and rotational and linear speeds. Each pair of steps was considered for each operator's round trip. For example, the belly-cutting stepper motor start runs based on the written step function. As soon as the arm reaches the end of the path, the second step is applied to control the speed and acceleration. The same steps were also considered for the returning path. As the operator reaches the end of the path in the first run, the arm stops equal to the time required for cutting the trout length. This time is considered between the second and third steps. By finishing the cutting process, the arm returns to the initial position based on the defined speed to complete the fourth step. In this step, when the cutting blade arm reaches the end of the path, a stop command is issued to the stepper motor. In the simulation process, for all of the arms and moving parts, a motion function is considered so that all the contacts, collisions, and moving paths are visible. In such a condition, the timing of arms and actuators is set in an optimized condition. Therefore, by trial and error, the motions in the real system face the least errors and collisions.

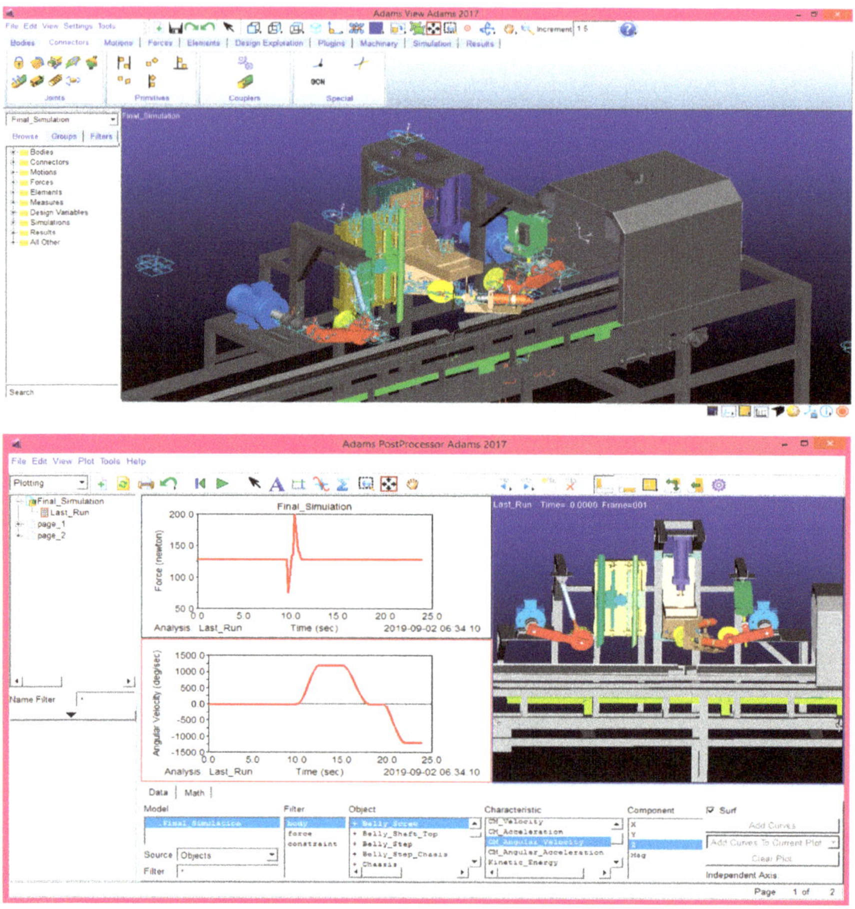

Figure 12. Simulation process and extracting the dynamic diagrams.

Since the minimum and the maximum number of fish is considered to be 4–10 fish per minute, the simulation of the device is performed for the minimum and maximum number and the average number of the fish (4, 7, and 10 fish per minute). To perform the simulation in these conditions, some preliminary considerations and calculations are required. For example, it was necessary to determine fish transfer speed in the canal in terms of fish per minute, because the operator's speed, performing all calculations, and writing step-functions will be undertaken based on fish transfer speed.

2.3.1. Calculations Related to the Stepper Motors of the Belly-Cutting and Gutting Subsystems

To determine the required torque to actuate the cutting arm and gutting tubes, the system was simulated. Considering the working conditions in the system, all the necessary parameter and considerations such as static and dynamic friction coefficients, any collisions, and all the constraints, were incorporated and motion simulation was performed.

Since the linear velocity of the fish varies in different feeding capacities, it is necessary to bring the cutting arm closer to the fish concerning linear transmission speed. It should be mentioned that the belly-cutting time at which the arm is in the cutting position also concerns the fish transfer speed. In addition to determining motor torque, rotational speed, acceleration, and energy consumption were obtained. Finally, the maximum torque required for system functioning at a capacity of 10 fish per minute was selected to determine stepper motor characteristics in both belly-cutting and gutting subsystems.

A power screw was used to move the cutting arm. Due to the self-locking property of the power screw in the non-working condition, no force will be applied on the stepper motor in idle condition. Since the gutting subset has a different structure, it was necessary to use a stepper motor with a gearbox system due to the inertia of the moving parts so that the suction tubes would stop as soon as the motor stops.

2.3.2. Calculations Related to Pneumatic Jacks of Head-Cutting and Cleaner Subsystems

Two pneumatic jacks were applied in a trout-processing system in head-cutting and cleaning subsystems. As the piston moves downwards during the extraction step, the jack is expected to prevent the subset from free falling. To know whether the pneumatic jack will empower or prevent free fall, we investigated the piston velocity in free-fall conditions from a height of 180 mm. The falling situation is presented in Figure 13. In this figure, IP, SP, and FP are the initial, start, and final positions of the head-cutting subsystem.

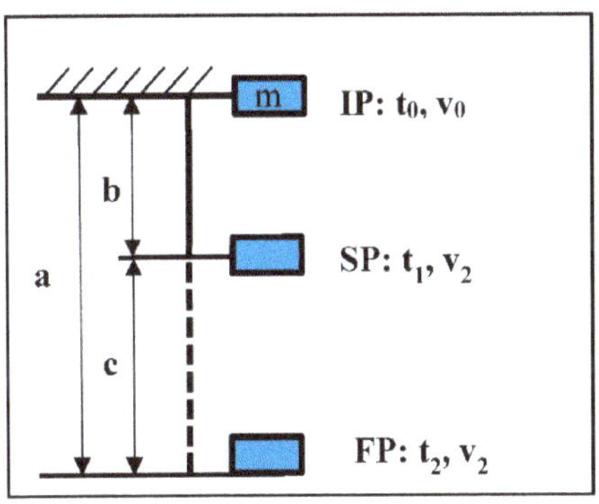

Figure 13. Free falling condition of the head-cutting subset: (**a**) distance from the home position of the head-cutting subset to the end of the cutting path, (**b**) the length from the home position to the initial contact of the blade and trout (80 mm), and (**c**) head-cutting width. In this figure, IP, SP, and FP are the initial, start, and final positions.

The following are the equations for calculating the final velocity of the head-cutting subset in a free-fall condition. According to these relationships, considering the weight of all stimulus parts was regarded as a single mass. By simplifying the falling conditions, the penetration speed in the initial contact and termination step was calculated.

$$U_0 + K_0 = U_2 + K_2 \tag{2}$$

$$m \cdot g \cdot h_0 + \frac{1}{2} \cdot m \cdot {v_0}^2 = m \cdot g \cdot h_2 + \frac{1}{2} \cdot m \cdot {v_2}^2 \tag{3}$$

$$g \cdot h_0 = \frac{1}{2} \cdot {v_2}^2 \tag{4}$$

$$9.81 \times 0.18 \times 2 = {v_2}^2 \tag{5}$$

$$v_2 = 1.88 \tag{6}$$

In these relations, 'U' and 'K' are potential and kinetic energies in terms of (J), m is body mass (kg), 'v' represents body velocity (m s^{-1}), 'h' is the falling height (m), and 'g' represents gravitational acceleration (m s^{-2}). According to the energy conservation law, the sum of kinetic and potential energy at two points at the beginning and end of the motion path must be equal (Equations (2) and (3)). After simplifying the relationship, the velocity value at the end of the path resulted in 1.88 m s^{-1}. Assuming that the head-cutting subset has a free fall, by substituting the velocity values for 'v_1' and 'v_2', the required time for complete cut will be obtained using Equations (7) and (8).

$$v_1 = g \cdot t_1 + v_0 \tag{7}$$

$$v_2 = g \cdot t_2 + v_0 \tag{8}$$

By setting the values of 1.25 m s^{-1} and 1.89 m s^{-1} for 'v_1' and v_2 in Equations (7) and (8), t_1 and t_2 were calculated as 0.13 s and 0.19 s, respectively. The blades will travel the distance from IP to FP (180 mm) in 0.19 s in the vertical direction. Also, the head-cutting blades will travel 'a' distance 'b' in 0.13 s. At this distance, the cutting blades will descend freely to reach the fish in the SP position. Considering the fish's average width of 100 mm, the cutting time from SP to FP resulted in 0.06 s. It should be noted that the penetration velocities in the SP and FP region are high values; this value is also higher than the maximum speed provided by the pneumatic jack.

Comparing the velocity values in different states, it is not allowed to consider different speed ranges for the piston. Obviously, the least amount of longitudinal displacement of the fish during the cutting process will be achieved at the highest blade penetration speed and the lowest processing capacity. Therefore, a penetration velocity of 500 mm s^{-1} was considered as the final velocity for cutting blade. Under these conditions, as this speed is considered maximum, the other feeding rates will be processed even better than the maximum feeding rate of 10 fish per minute. By comparing the velocity and time of the free fall with the maximum speed piston speed, it is expected that the piston must exert the opposite force against the free fall. This can also be seen in the output of the ADAMS software. By defining the required force, the pneumatic jack was selected for both the head-cutting and cleaning subsystems. Since the total displacement range of the head-cutting subset is 180 mm, a 250 mm course jack was selected. It should be noted that the amount of applied force was different in jack extraction and retraction courses. The amount of supplied force in extraction and retraction steps is calculated by Equations (9) and (10), respectively.

$$F_e = \frac{\pi \times D^2}{4} \cdot P \cdot g \tag{9}$$

$$F_r = \frac{\pi \times \left(D^2 - d^2\right)}{4} \cdot P \cdot g \tag{10}$$

In these equations, 'D' and 'd' are the outside and inside diameter of the piston rod in millimeters, 'g' is the gravitational acceleration (9.81 m s^{-2}), 'Fe' and 'Fr' are the force of the jack in the extraction and retraction course in Newton 'N', and P is the air pressure in terms of 'bar', respectively. Since the one-way pneumatic jack exerts more force in the extraction course, it is preferred to use the cylinder in the retraction mode until the next fish arrives, the jack can return to its original position slower than the extraction course.

Assuming the fish moves at a constant speed in the guided canal, the faster the blade penetrates the fish body; the more precise the head cut will result. Knowing that the force exerted by the jack resists the free fall of the cutting sub-section, especially at the end of the course, the designed model is entered into ADAMS software, and all the necessary constraints, their movements, and friction coefficients were entered into the model. Among the tested speeds for the jack, the maximum required force was obtained for the speed of 500 mm s^{-1}. After determining the amount of force, the next step is to select the pneumatic jack. The following items will be considered in selecting the jack:

1. Design factor;
2. Jack force;
3. Load coefficient in vertical operation mode;
4. Air pressure;
5. Speed coefficient.

Considering the design factor, the maximum force was considered 1.5 times greater than the value given by ADAMS software. Therefore, the force is considered equal to 332.45 N in further calculations. If the pneumatic jack is considered to work in a horizontal position with a pressure of 0.8 MPa, a jack with a bore diameter of 32 mm will be a proper option. If the jack is used in the vertical position, the bearing load by the jack will be reduced to 50%. This value can be reduced by up to 20% at high operating speeds. In fact, in addition to considering the design factor, the vertical load factor must also be considered. Also, in order to have an optimal performance at low air pressures, the air pressure of 0.4 MPa was considered in calculations. The pneumatic jack bore size selection graph is presented in Figure 14.

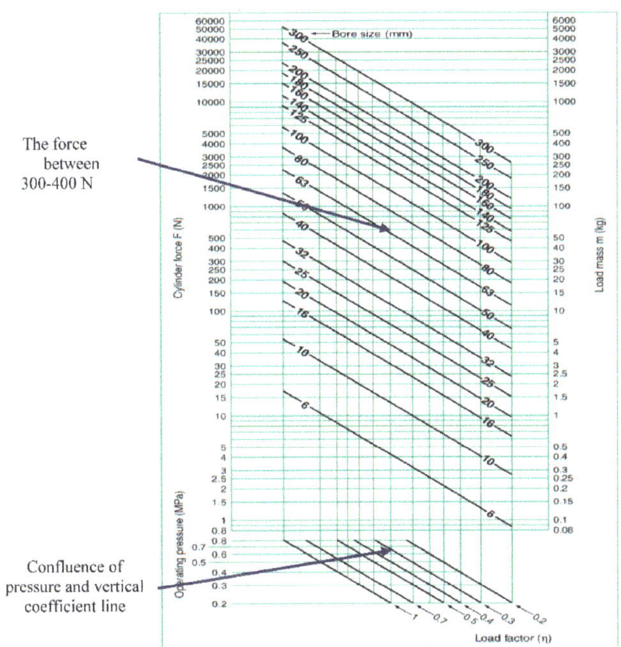

Figure 14. Pneumatic jack bore size selection graph based on the applied force, vertical operating coefficient, and operating pressure.

As shown in lower part of Figure 14, by considering the air pressure and vertical operation coefficient, a vertical line was defined. By tracing this line and contacting bore size of 32 mm, the supplied force between 100–150 N was anticipated which was less than the desired force of 332.45 N. Considering the vertical load factor as 0.5 at a working pressure of 0.4 MPa and the impact of the vertical force, an air jack with a 32 mm diameter can supply force values between 100 N and 150 N, which is less than the required value (N45/332).

As a final option, a jack with a bore size of 50 mm can supply the desired force values between 300–400 N. A similar process was performed to determine the second jack in the cleaning subsystem. Hence, at a load factor of 0.5 and air pressure of 0.4 MPa, a jack with a 20 mm diameter to supply active force between 30–40 N was selected in maximum fish-processing rate (10 fish per minute).

2.4. Fabricated System

The final trout-processing machine is presented in Figure 15a. The final system was fabricated after the motion simulation. The components of the system are presented in detail. This system also contains a controlling and machine vision section beside the mechanical design and machine vision section. The rear view and the controlling panel are presented in Figure 15b,c, respectively. In order to control the moving parts, different electronic devices are required: start/stop switches, emergency disconnect switch, sensors, controller, stepper motors, and drives. Inductive and magnetic sensors were used to control operating arms and stepper motor movements. Because the device's arms' movement in each of the working steps require unique control, the device control algorithm was written in three parts: initial setup, manual operation, and online automatic function. To develop the controlling algorithm, Siemens TIA Portal automation software and Siemens controller was used.

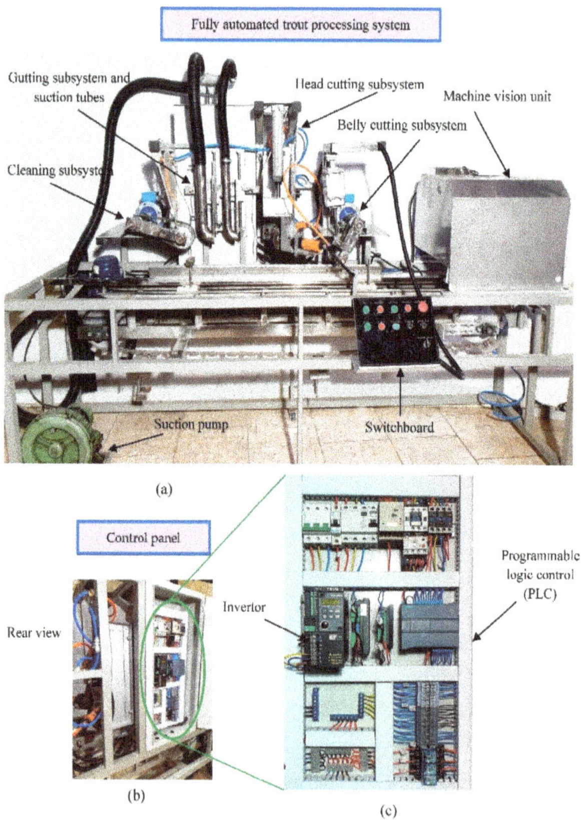

Figure 15. The fabricated trout-processing system: (**a**) the completed mechanical and electronics parts, (**b**) the rear view, and (**c**) controlling panel.

Since inductive sensors are applied in each of the five subsets (machine vision, belly-cutting, head-cutting, gutting, and cleaning subsets), in case there is no fish in the gripper, the dimensions extracted by machine vision will be calculated as "0", and the system operators will be standby state without any function (Supplementary Material).

3. Results and Discussion

Since the system is designed, simulated, and fabricated based on the simulation data, the results of the simulation are presented as follows. These calculations are conducted on the piston speed, selecting the belly-cutting stepper motor, choosing the head-cutting actuating piston, defining the gutting stepper motor's characteristics, and the required force for the cleaning subsystem. The related results are presented below.

3.1. Piston Speed Calculations

In addition to the analytical results generated by the software, some of the results were calculated in the initial step. This part of the results is related to the working capacity and processing quality of the operators during machine functioning. The results are related to the amount of fish movement in each of the capacities of the device. In Table 2, trout transfer length along the machine is presented in each feeding rate.

Table 2. Fish movement along the machine based on the feeding rate (fish per minute).

Pneumatic Cylinder		Fish Movement along the Machine Based on the Feeding Rate (Fish Per Minute)						
Speed (mm s^{-1})	t (s)	MF-4	MF-5	MF-6	MF-7	MF-8	MF-9	MF-10
100	1	53.33	66.67	80.00	93.33	106.67	120.00	133.33
200	0.50	26.67	33.33	40.00	46.67	53.33	60.00	66.67
300	0.32	17.78	22.22	26.67	31.11	35.56	40.00	44.44
400	0.25	13.33	16.76	20.00	23.33	26.67	30.00	33.33
500	0.20	10.67	13.33	16.20	18.67	21.33	24.80	26.67
600	0.17	8.89	11.11	13.33	15.56	17.78	20.00	22.22
700	0.14	7.62	9.52	11.43	13.33	15.24	17.14	19.05
800	0.13	6.67	8.33	10.00	11.67	11.85	15.00	16.67
900	0.11	5.93	7.41	8.89	10.37	14.22	13.33	14.81
Free fall	0.06	3.20	4.00	4.80	5.60	6.40	7.20	8.00

As shown in this table, by increasing the feeding rate, the fish transfer speed increases. Since it is desirable to increase the capacity of the system, the appropriate solution is to increase the penetration rate of the blades in the fish. The longitudinal displacement of fish at MF-4 and MF-10 capacity with the piston velocity with 100 mm s^{-1} was calculated as 53.33 mm and 133.33 mm, respectively. It should be noted that moving the trout as 133.33 mm while moving in front of the head-cutter blades causes separation from the gripper or creates an undesirable cut in the head.

By increasing the blade penetration rate to 900 mm s^{-1}, the displacement values in both MF-4 and MF-10 capacities were calculated as 5.93 mm and 14.81 mm, respectively. These values are suitable for cutting the head, but it should be noted that increasing the speed of the pneumatic jack increases the inertia in the actuator sections, so the rod speed such as 400 mm s^{-1} and 500 mm s^{-1} is suitable for head cutting. The amount of fish movement in the minimum and the maximum capacity of the device at the speed of 500 mm s^{-1} are 10.67 mm and 26.67 mm, respectively. The head of the fish will be cut with a slight deviation from the vertical line, in which the fillet near the neck of the fish will be separated along with the body.

Figure 16 shows fish movement at different jack speeds. As shown in this figure, the displacement values at a particular capacity, such as the F-4, decrease non-linearly by increasing jack speed, so that at piston speeds in the range between 100–200 mm s^{-1}, fish displacement decreases with a high slope while the slope decreases at speeds of 600–1000 mm s^{-1}.

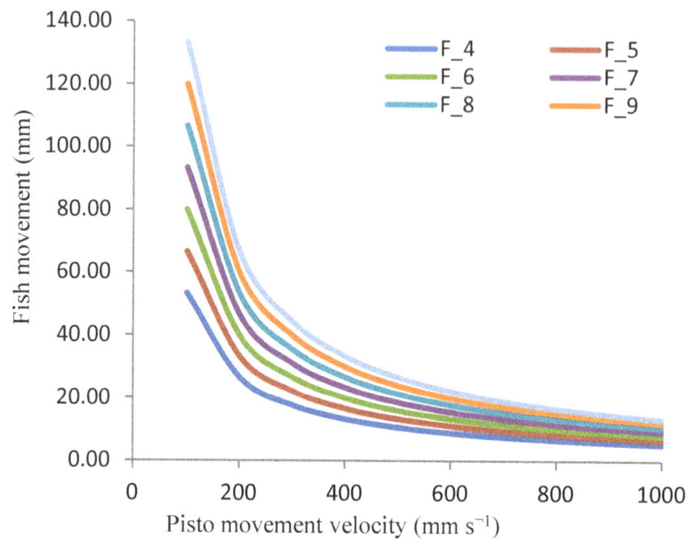

Figure 16. Fish movement at different jack speeds.

This figure also shows that, at low transfer speeds, there is a large difference between the fish movement values, whereas as jack speed increases, the difference between these values decreases. In other words, considering the balance between the jack speed and the amount of fish displacement, it is appropriate to consider higher velocities of the carrier jack, but it should be noted that at higher jack velocities, the difference between the displacement values decreases. Due to the higher inertia of the moving parts and the high vibrations in the system, electing jack speeds such as 400–500 mm s^{-1} is a suitable speed for the head-cutting blade set.

3.2. *Results of Simulation*

The results of system simulation are presented in the following. In order to reach an optimized design and observe the force, torque, and speed trends, the system was tested in three feeding rates. These capacities are the minimum, maximum, and the average number of fish per minute (4, 7, and 10 Fish/min), respectively.

3.2.1. Selecting the Stepper Motor of the Belly-Cutting Subsystem

The belly-cutting subsystem is a subset of the machine in which a longitudinal incision is made from the anal fin to the fish head using a rotating blade mounted on the main arm.

In Figures 17–19 the motion simulation results are presented for the belly-cutting subset. In these figures, the characteristics of the required stepper motor are determined. These characteristics are the required torque for cutting operations, angular velocity, angular acceleration, and energy consumption in three feeding rates. In these figures, blue dotted, green dashed, and red solid are the components in X, Y, and Z directions.

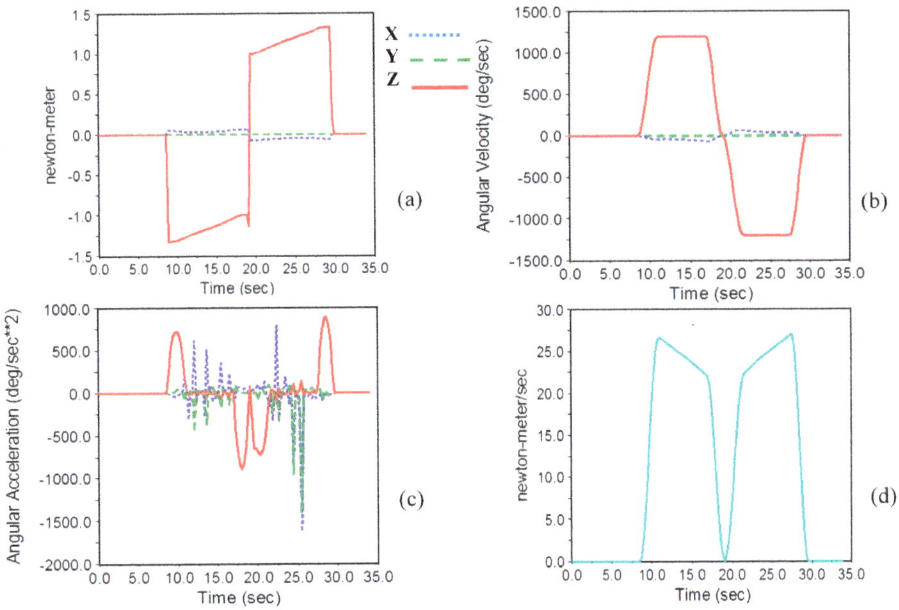

Figure 17. Results of ADAMS software simulation to determine the motor characteristics of the belly-cutting subset (4 fish per minute). In this figure, (**a**–**d**) are torque, angular velocity, angular acceleration, and required power respectively.

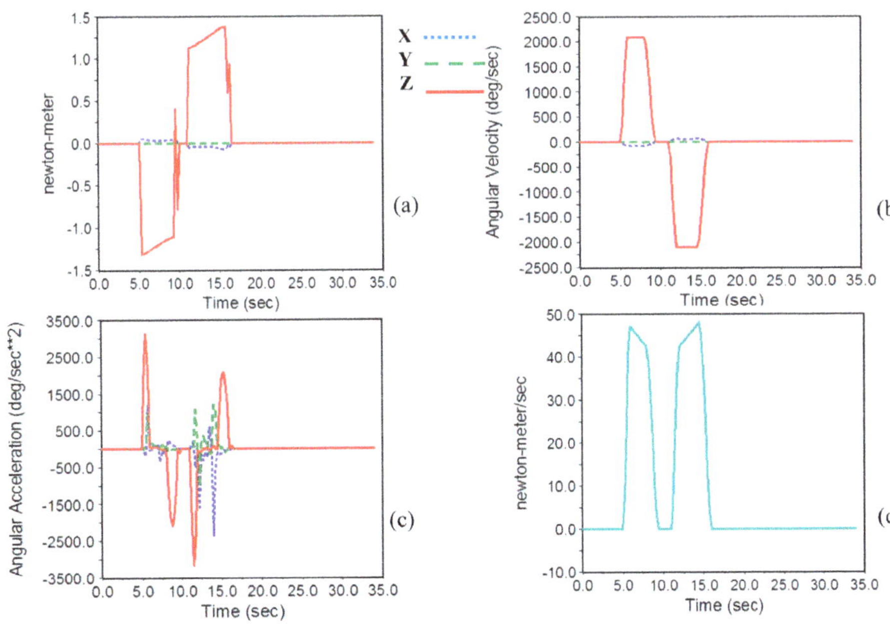

Figure 18. Results of ADAMS software simulation to determine the motor characteristics of the belly-cutting subset (seven fish per minute). In this figure, (**a**–**d**) are torque, angular velocity, angular acceleration, and required power respectively.

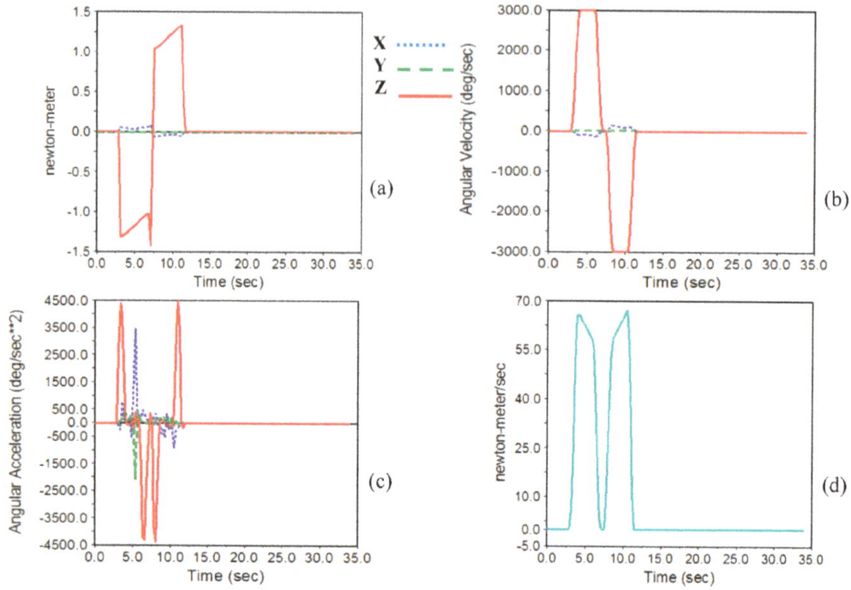

Figure 19. Results of ADAMS software simulation to determine the motor characteristics of the belly-cutting subset (10 fish per minute). In this figure, (**a**–**d**) are torque, angular velocity, angular acceleration, and required power respectively.

As shown in these figures, the angular velocity and acceleration around the *Z*-axis are maximum. Also, the required energy increases by increasing angular velocity. The maximum torque values in three capacities of 4, 7, and 10 fish per minute were 1.334 N m, 1.378 N m, and 1.431 N m, respectively. Therefore, to select the required stepper motor maximum amount of torque (10 Fish/min) was selected. By considering the design coefficient of 1.25–1.5, the required torque for the stepper motor is in the range of 1.79–2.145 N m. Therefore, a Nema 23 stepper motor (57PH20-0.2 kg m) stepper motor with a torque 1.96 N m was selected. Considering this stepper motor, a design factor of 1.37 is obtained, which is acceptable for the device operation.

In Figure 17b–d the angular velocity, acceleration, and energy consumption are presented, respectively. As shown in these figures, by increasing the fish feeding rate, the blades are about to penetrate the belly at higher speeds, requiring higher velocity and acceleration. As shown in these three figures, the maximum angular velocity, acceleration, and energy consumption are related to 10 fish per minute. Due to the sufficient time interval between the fish grippers, there is enough time to return to its initial position.

3.2.2. Selecting the Pneumatic Jack of the Head-Cutting Subsystem

The results for the simulation are shown in Figures 20–22. In order to ensure successful head cutting, it is necessary to synchronize the fish feeding rate and cutter penetration. In fact, the faster the jack penetrates and the slower the fish moves, the better the cut is done. Although the lower transfer speeds result in a proper head cut, it should be noted that it results in low system operation capacity.

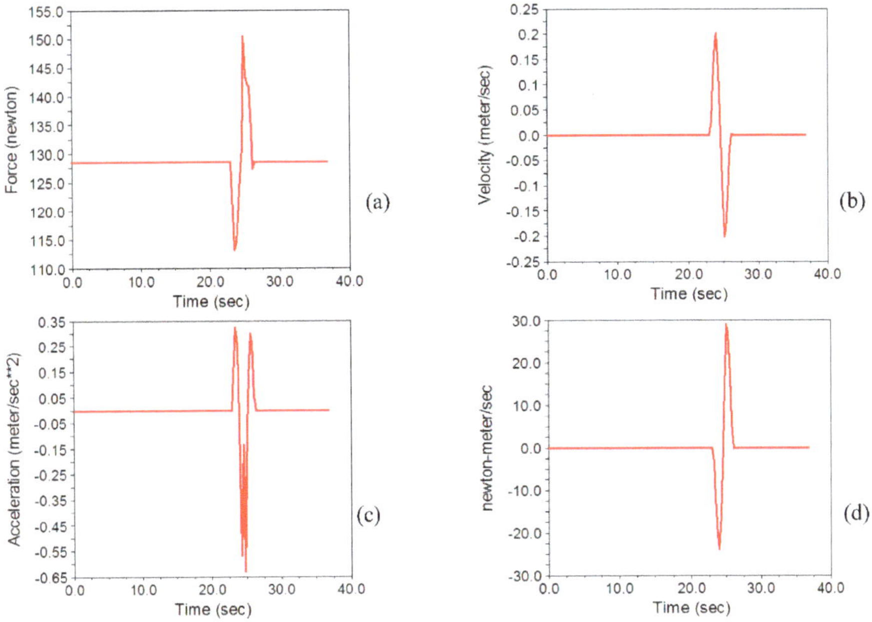

Figure 20. Results of ADAMS software to determine the characteristics of the jack of the head-cutting subset (four fish per minute). In this figure, (**a**–**d**) are force, velocity, acceleration, and required power respectively.

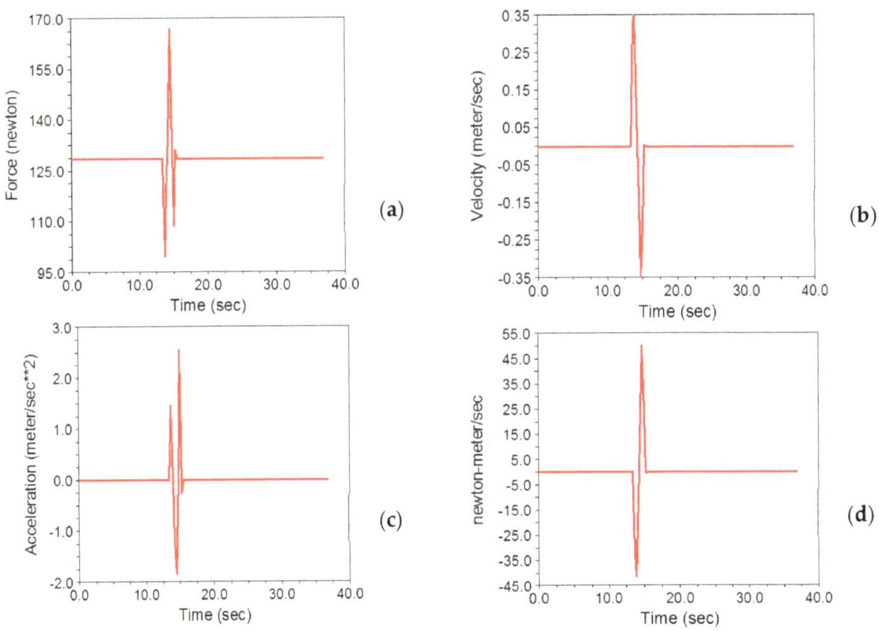

Figure 21. Results of ADAMS software simulation to determine the motor characteristics of the gutting subset (seven fish per minute). In this figure, (**a**–**d**) are force, velocity, acceleration, and required power respectively.

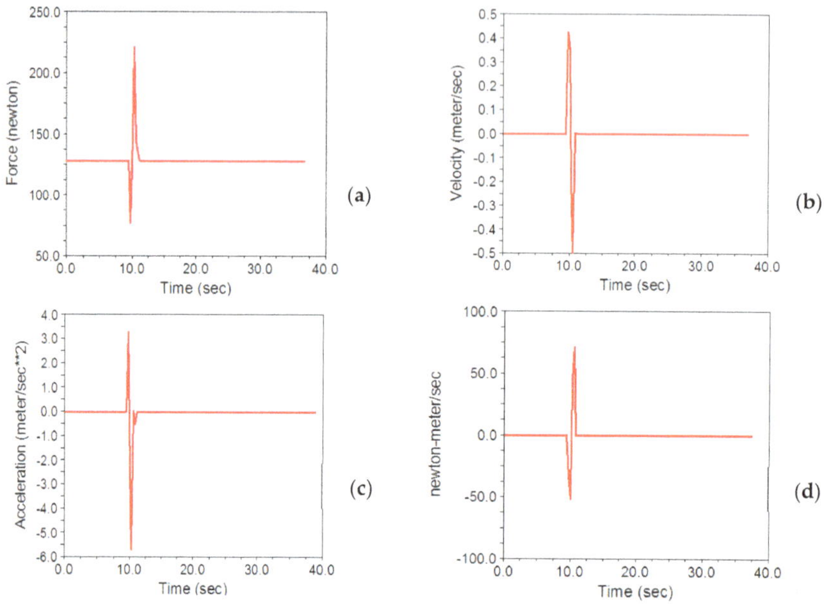

Figure 22. Results of ADAMS software to determine the characteristics of the head-cutting subset jack (10 fish per minute). In this figure, (**a**–**d**) are force, velocity, acceleration, and required power respectively.

By comparing the final velocity of the free fall for the head-cutting jack at the end of the cut path, it is obvious that the air jack prevents the moving parts from free-falling. The maximum required force in three capacities resulted in 149.83 N, 164.33 N, and 221.63 N, respectively. As shown in Figure 20, the amount of required force for the jack in the stop mode is between 127.5–130 N to compensate for the subset weight.

When the fish reaches the head-cutting position, the jack is activated and moves downward. As moving downwards, the required force decreases, and by reaching the end of the cutting path, the force increases to the maximum amount of 149.83 N. This force value resulted in a feeding rate of four fish per minute and the jack penetration speed of 200 m s^{-1}. Besides the required force and speed of the pneumatic jack, the acceleration and energy consumption in the simulation time interval in the three working capacities are shown in Figure 20c,d. Similarly, the amount of required force in 7 and 10 fish per minute resulted in 164.33 N and 212.63 N at the blade penetration speeds of 350 m s^{-1} and 500 m s^{-1}, respectively. Similar results for the two penetration velocities of 350 m s^{-1} and 500 m s^{-1} are shown in Figures 21 and 22. The maximum required force to provide a speed of 500 mm s^{-1} resulted in 2221.63 N. By multiplying this force by 1.5 as the design coefficient, the final force size on which the jack resulted in 332.45 N. Refer to the diagram presented in Figure 22.

Based on the operational position of the jack (vertical or horizontal) and air pressure, the size of the air jack was determined based on the maximum force required to move the head-cutting subset (332.332 N). Since the supplied force in extraction mode is more than the retraction, the pneumatic jack was selected based on the amount of the force that the jack can supply in retraction mode so that it can supply the required force while the jack ascends toward the initial position. On the other hand, it should be considered that the operating speed of the jack is one of the most important characteristics for determining the size of the jack. By following the line obtained from the intersection of the axis related to the vertical operating coefficient of 0.5 and the axis of air pressure, the 0.4 MPa vertical line is determined.

By contacting this line with the force axis equal to 332.45 N, a cylinder with an internal diameter of 50 mm was selected. The minimum displacement of 180 mm was considered to provide successful head cutting and avoid any contact with the moving grippers, but to ensure any contact, among two different courses of 200 mm and 250 mm, the larger size was installed in the system. Therefore, the penetration rate of 500 m s^{-1} was considered as the final penetration rate of the cutting blades. This condition will be the desired condition for head cutting in lower feeding rates. On the other hand, choosing this speed can result in an acceptable head cutting without any considerable fillet loss.

3.2.3. Selecting the Stepper Motor of the Gutting Subsystem

The motion simulation results for selecting the motor characteristics in three feeding rates are presented in Figures 23–25, respectively. In all the figures, the required torque, angular velocity, angular acceleration, and energy consumption are presented for each of the feeding rates. The diagrams present both clockwise and counterclockwise motions.

As shown in these figures, the maximum required torque to move the suction pipes in a vertical direction resulted in the highest working capacity. In this section, when the fish reaches the suction position, the stepper motor rotates in clockwise and counterclockwise directions depending on the fish dimension. As soon as the tubes open, the material inside the trout belly is evacuated. Using two discharge pipes, the required torque for lifting two pipes is less than the condition at which the stepper motor controls one suction tube.

The required torque increases by increasing the angular velocity; whereas, in the maximum angular velocity, the maximum amount of required torque in the three input velocities was 0.226 N m, 0.386 N m, and 0.465 N m, respectively. Since the maximum required torque resulted in 0.465 N m, the required torque was estimated to be 0.698 N m by considering the design coefficient of 1.5. As the pinion is connected directly to the rack gear, a gearbox mounted stepper motor was applied to move the suction pipes. Therefore, a stepper motor with a reduction ratio of 1:18 with a total maximum torque of 0.8 N m was applied.

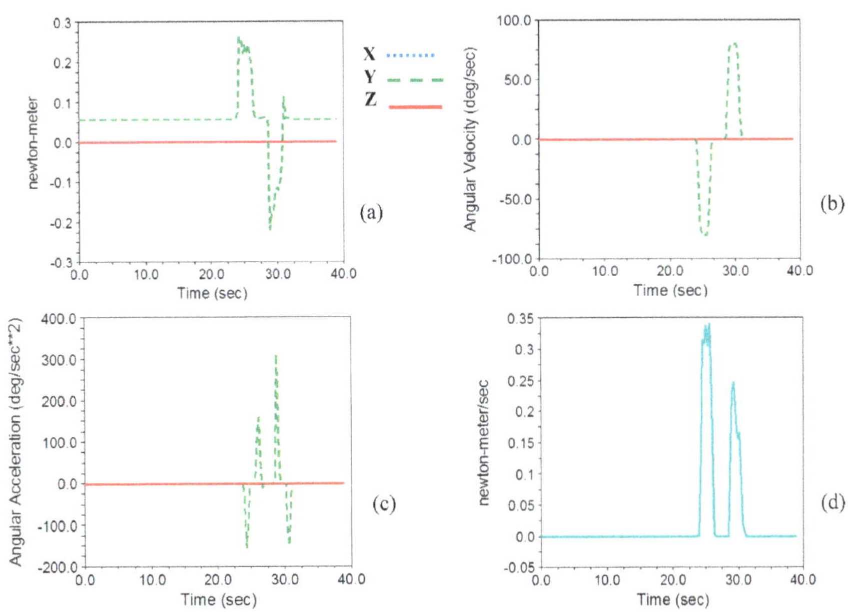

Figure 23. Results of ADAMS software simulation to determine the motor characteristics of the gutting subset (four fish per minute). In this figure, (**a**–**d**) are torque, angular velocity, angular acceleration, and required power respectively.

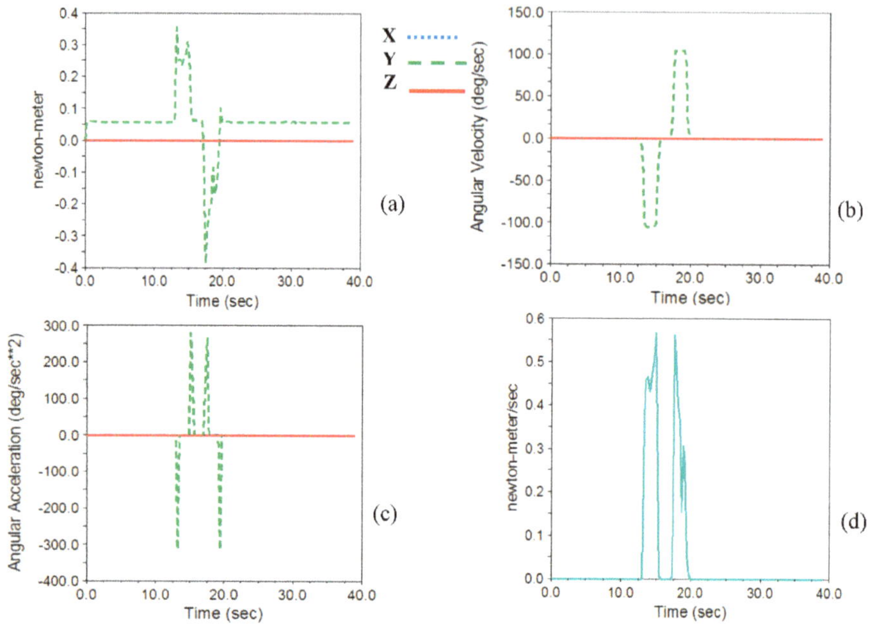

Figure 24. Results of ADAMS software simulation to determine the motor characteristics of the gutting subset (seven fish per minute). In this figure, (**a**–**d**) are torque, angular velocity, angular acceleration, and required power respectively.

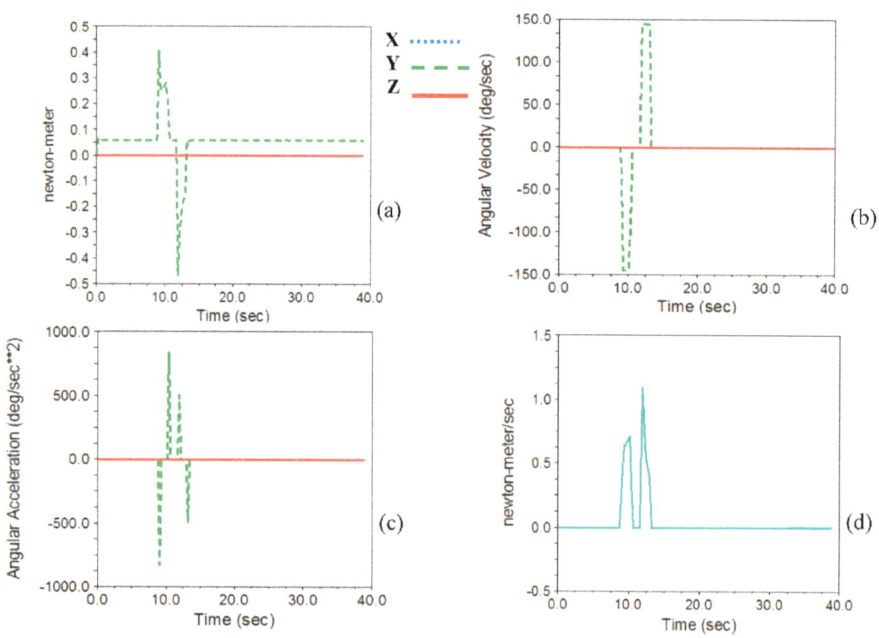

Figure 25. Results of ADAMS software simulation to determine the motor characteristics of the gutting subset (10 fish per minute). In this figure, (**a**–**d**) are torque, angular velocity, angular acceleration, and required power respectively.

3.2.4. Selecting the Pneumatic Jack of the Cleaning Subsystem

After evacuating the belly content, it is necessary to clean any sticky materials attached to the spinal cord and belly cavity. A rotary brush with a ferry-ribbed edge was driven with an AC electromotor. In order to move the belly cleaner set, it was necessary to use a jack to raise and lower the rotating brush to avoid hitting the fish grippers.

The final stage of the fish processing in this system is to clean the belly content. As long as the clamp reaches the cleaning subsystem, the cutter descends to clean the belly. As long as the fish is moving towards the end of the machine, depending on the size of the fish, the arm is in an operating position based on the fish length. As soon as the fish cleans, the arm returns to the initial position. The results for the output of ADAMS software in three capacities are presented in Figures 26–28. In these figures blue dotted, green dashed, and red solid lines are the components in X, Y, and Z directions.

In these diagrams, the force diagram in three directions of x, y, and z (diagram a), the velocity and acceleration (diagrams b and c), and the total force values (diagram d) are presented. As shown in these figures, unlike the jack used in the head-cutting subset, in addition to applying force along the z-axis, force components are also observed in both the x and y directions. The maximum required force to move the arm at the capacity of four, seven, and 10 fish per minute was 20.55 N, 21.26 N, and 25.84 N, respectively. Multiplying the maximum force by 1.5 as the design coefficient, the final force results in 38.76 N. By referring to the diagram presented in Figure 28, the size of the jack was determined based on working position and air pressure by referring to the calculated force (38.76 N). Like the pneumatic jack of the head-cutting subsystem, the retraction mode was selected as the criteria to select the jack. Since the arm operates at a considerable angle to the horizontal line, the vertical operation coefficient was 0.5.

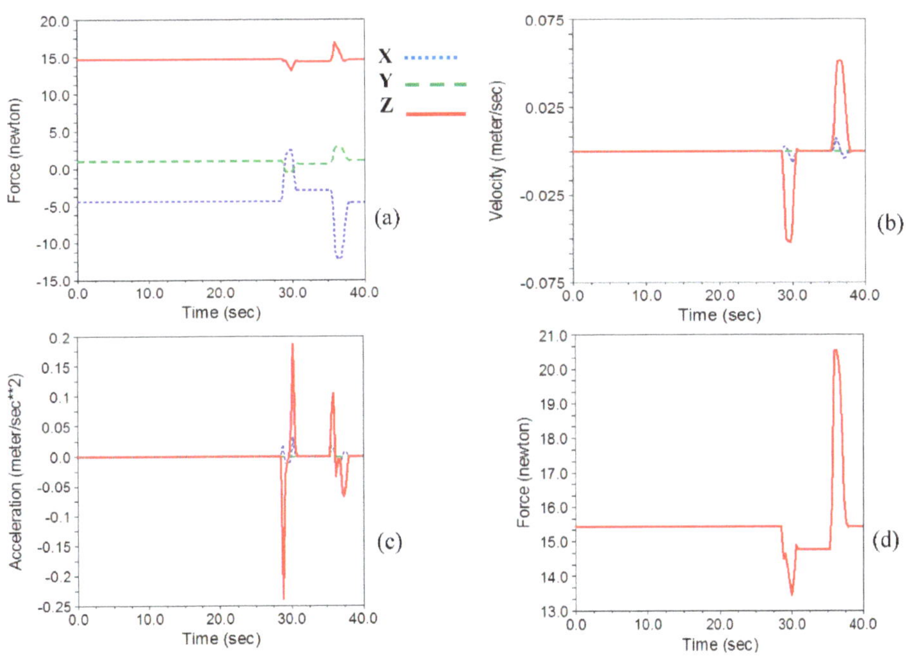

Figure 26. Results of ADAMS software simulation to determine the characteristics of the pneumatic jack of the cleaning subsystem (four fish per minute). In this figure, (**a–d**) are force, velocity, acceleration, and resultant of the force in three direction respectively.

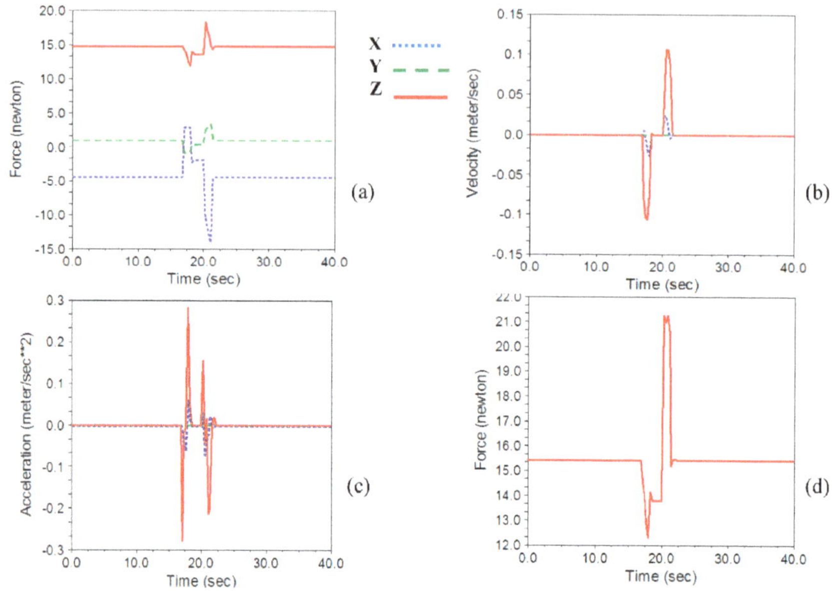

Figure 27. Results of ADAMS software simulation to determine the characteristics of the pneumatic jack of the cleaning subsystem (seven fish per minute). In this figure, (**a**–**d**) are force, velocity, acceleration, and resultant of the force in three direction respectively.

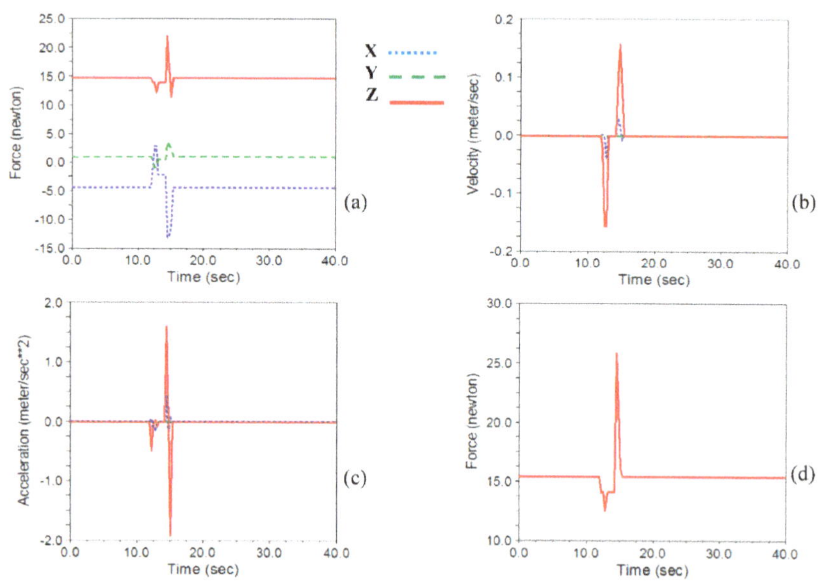

Figure 28. Results of ADAMS software simulation to determine the characteristics of the pneumatic jack of the cleaning subsystem (10 fish per minute). In this figure, (**a**–**d**) are force, velocity, acceleration, and resultant of the force in three direction respectively.

Pursuing the intersection line of vertical operation coefficient and air pressure of 4 bar, the required force value (37.86 N) falls between the possible standard bore sizes of 16 and 20 mm. Finally, a jack with 20 mm was selected. The method of selecting the jack is that by following the line obtained from the intersection of the axis related to the vertical operating coefficient of 0.5 and the air pressure of 4 bar, the upper limit bore size (20 mm) was selected. To avoid any contact of rotary brush with the moving grippers, the minimum vertical displacement length was 100 mm, so a pneumatic jack with the course of 150 mm was selected.

4. Conclusions

In this study a system capable of performing belly-cutting, beheading, gutting, and cleaning steps was designed, simulated and fabricated. As the customers choose the fish size, the image processing system is applied to measure trout size and extract the precise cutting points. In order to prevent any contact between the grippers and the system arms, the automated controlling system was designed to control the machine arms to process different cleaning operations on fish of different sizes together with providing the system security. Using this integrated system, it is possible to perform several fish-processing steps. Due to the slippery skin of the fish and the low friction coefficient, a clamping gripper was designed. By initial design and motion analysis, the most appropriate model to meet the required functions was achieved. In the next step, the designed model was simulated. This process resulted in proper and optimal force extraction in the beheading and cleaning subsystems and the required torque in the belly-cutting and gutting subsystems. By designing this system, simulating and analyzing the motion, an integrated machine with the highest efficiency of driver units was fabricated.

Supplementary Materials: The video of the functioning system is also available online at https://www.mdpi.com/article/10.3390/app11125602/s1.

Author Contributions: Conceptualization, H.A., and H.B.; methodology, H.A., H.B.; software, H.A., A.J.; validation, S.S.M., and A.R.M.; formal analysis, H.A., A.J.; investigation, A.J., and H.B.; resources, S.S.M., and A.R.M.; data curation, S.S.M., and A.R.M.; writing—original draft preparation, H.A.; writing—review and editing, S.S.M., and A.R.M.; visualization, H.A., S.S.M., and A.R.M.; supervision, S.S.M., and A.R.M.; project administration, S.S.M.; funding acquisition, H.A., and S.S.M. All authors have read and agreed to the published version of the manuscript.

Funding: This work was funded by the University of Tehran Science and Technology Park, Iran under grant number"96034". The APC cost was funded by the University of Valencia.

Institutional Review Board Statement: Ethical review and approval were waived for this study due to the use of non-alive fish coming from fish shops and usual fish processing procedures were applied.

Informed Consent Statement: Not applicable.

Conflicts of Interest: The authors declare no conflict of interest.

References

1. Booman, A.; Parin, M.; Zugarramurdi, A. Efficiency of size sorting of fish. *IJPE* **1997**, *48*, 259–2652. [CrossRef]
2. Beiranvand, A.; Kalhor, A.; Masouleh, M.T. Modeling, identification and minimum length integral sliding mode control of a 3-DOF cartesian parallel robot by considering virtual flexible links. *Mech. Mach. Theory* **2021**, *157*, 104183. [CrossRef]
3. Xu, K.; Liu, H.; Yue, W.; Xiao, J.; Ding, Y.; Wang, G. Kinematic modeling and optimal design of a partially compliant four-bar linkage using elliptic integral solution. *Mech. Mach. Theory* **2021**, *157*, 104214. [CrossRef]
4. Hao, G.; He, X.; Awtar, S. Design and analytical model of a compact flexure mechanism for translational motion. *Mech. Mach. Theory* **2019**, *142*, 103593. [CrossRef]
5. Hong, H.; Yang, X.; You, Z.; Cheng, F. Visual quality detection of aquatic products using machine vision. *Aquac. Eng.* **2014**, *63*, 62–71. [CrossRef]
6. Atienza-Vanacloig, V.; Andreu-García, G.; López-García, F.; Valiente-González, J.M.; Puig-Pons, V. Vision-based discrimination of tuna individuals in grow-out cages through a fish bending model. *Comput. Electron. Agric.* **2016**, *130*, 142–150. [CrossRef]

7. Hernández-Ontiveros, J.M.; Inzunza-González, E.; García-Guerrero, E.E.; López-Bonilla, O.R.; Infante-Prieto, S.O.; Cárdenas-Valdez, J.R.; Tlelo-Cuautle, E. Development and implementation of a fish counter by using an embedded system. *Comput. Electron. Agric.* **2018**, *145*, 53–62. [CrossRef]
8. Issac, A.; Dutta, M.K.; Sarkar, B. Computer vision based method for quality and freshness check for fish from segmented gills. *Comput. Electron. Agric.* **2017**, *139*, 10–21. [CrossRef]
9. Lalabadi, H.M.; Sadeghi, M.; Mireei, S.A. Fish freshness categorization from eyes and gills color features using multi-class artificial neural network and support vector machines. *Aquacult. Eng.* **2020**, *90*, 102076. [CrossRef]
10. Miranda, J.M.; Romero, M. A prototype to measure rainbow trout's length using image processing. *Aquacult. Eng.* **2017**, *76*, 41–49. [CrossRef]
11. Muñoz-Benavent, P.; Andreu-García, G.; Valiente-González, J.M.; Atienza-Vanacloig, V.; Puig-Pons, V.; Espinosa, V. Enhanced fish bending model for automatic tuna sizing using computer vision. *Comput. Electron. Agric.* **2018**, *150*, 52–61. [CrossRef]
12. Costa, C.; Antonucci, F.; Boglione, C.; Menesatti, P.; Vandeputte, M.; Chatain, B. Automated sorting for size, sex and skeletal anomalies of cultured seabass using external shape analysis. *Aquacult. Eng.* **2013**, *52*, 58–64. [CrossRef]
13. Buckingham, R.; Graham, A.; Arnarson, H.; Snaeland, P.; Davey, P. Robotics for de-heading fish–a case study. *Ind. Robot.* **2001**, *24*, 302–309. [CrossRef]
14. De Silva, C.; Wickramarachchi, N. An innovative machine for automated cutting of fish. *IEEE-ASME T Mech.* **1997**, *2*, 86–98. [CrossRef]
15. Lang, H.; Wang, Y.; de Silva, C.W. (Eds.) An automated industrial fish cutting machine: Control, fault diagnosis and remote monitoring. In Proceedings of the 2008 IEEE International Conference on Automation and Logistics, Qingdao, China, 1–3 September 2008.
16. Ketels, D. Apparatus for Positioning Fish for Heading. U.S. Patent 7467995, 23 December 2008.
17. Kragh, H. Fish Processing Machine with Alignment Devices and Methods for Realigning Fish during Processing. U.S. Patents US20100240289A1, 13 April 2007.
18. Urushibara, S.; Yusa, K. Method and Apparatus for Gutting Fish. U.S. Patents CA2007425A1, 1 September 1991.
19. Ryan, R.M. Fish Processing System and Method. U.S. Patent 20130316632, 6 August 2013.
20. Paulsohn, C.; Dann, A.; Rüsch, R.; Brandt, M. Tool, Device, and Method for Gutting Fish Opened at the Stomach Cavity. U.S. Patents CA2650073C, 12 November 2010.
21. Kowalski, W. Method for Removing Blood Released during Filleting from the Backbone of Fish, and Device for Removing such Blood. U.S. Patent 8956205B2, 17 February 2015.
22. Braeger, H. Method and Device for Automatically Extracting Flesh from Fish. U.S. Patent 9888697B2, 13 February 2018.
23. Finke, H.; Jurs, M.; Grabau, T. Apparatus and Method for Automatically Obtaining Flesh from Beheaded and Gutted Fish. U.S. Patent US8814637B2, 26 August 2018.
24. Bondø, M.S.; Mathiassen, J.R.; Vebenstad, P.A.; Misimi, E.; Bar, E.M.S.; Toldnes, B.; Ove østvik, S. An automated salmonid slaughter line using machine vision. *Ind. Robot.* **2011**, *38*, 399–405. [CrossRef]
25. Storbeck, F.; Daan, B. Fish species recognition using computer vision and a neural network. *Fish Res.* **2001**, *51*, 11–15. [CrossRef]
26. Godfrey, D. Friction of greases and grease components during boundary lubrication. *Asle Trans.* **1964**, *7*, 24–31. [CrossRef]
27. Gray, D.E. American institute of physics handbook. *Am. J. Phys.* **1964**, *32*, 389. [CrossRef]

applied sciences

MDPI

Article

Experimental Verification of Real-Time Flow-Rate Estimations in a Tilting-Ladle-Type Automatic Pouring Machine

Yuta Sueki [1] and Yoshiyuki Noda [2,*]

[1] Course of Mechanical Engineering, Integrated Graduate School of Medicine, Engineering and Agricultural Sciences, University of Yamanashi, Takeda 4-3-11, Kofu 400-8511, Japan; g18dts02@yamanashi.ac.jp
[2] Faculty of Engineering, Graduate Faculty of Interdisciplinary Research, University of Yamanashi, Takeda 4-3-11, Kofu 400-8511, Japan
* Correspondence: noday@yamanashi.ac.jp

Abstract: This paper discusses a real-time flow-rate estimation method for a tilting-ladle-type automatic pouring machine used in the casting industry. In most pouring machines, molten metal is poured into a mold by tilting the ladle. Precise pouring is required to improve productivity and ensure a safe pouring process. To achieve precise pouring, it is important to control the flow rate of the liquid outflow from the ladle. However, due to the high temperature of molten metal, directly measuring the flow rate to devise flow-rate feedback control is difficult. To solve this problem, specific flow-rate estimation methods have been developed. In the previous study by present authors, a simplified flow-rate estimation method was proposed, in which Kalman filters were decentralized to motor systems and the pouring process for implementing into the industrial controller of an automatic pouring machine used a complicatedly shaped ladle. The effectiveness of this flow rate estimation was verified in the experiment with the ideal condition. In the present study, the appropriateness of the real-time flow-rate estimation by decentralization of Kalman filters is verified by comparing it with two other types of existing real-time flow-rate estimations, i.e., time derivatives of the weight of the outflow liquid measured by the load cell and the liquid volume in the ladle measured by a visible camera. We especially confirmed the estimation errors of the candidate real-time flow-rate estimations in the experiments with the uncertainty of the model parameters. These flow-rate estimation methods were applied to a laboratory-type automatic pouring machine to verify their performance.

Keywords: flow-rate estimation; automatic pouring machine; extended Kalman filter

Citation: Sueki, Y.; Noda, Y. Experimental Verification of Real-Time Flow-Rate Estimations in a Tilting-Ladle-Type Automatic Pouring Machine. *Appl. Sci.* **2021**, *11*, 6701. https://doi.org/10.3390/app11156701

Academic Editors: Alessandro Gasparetto, Stefano Seriani and Lorenzo Scalera

Received: 11 June 2021
Accepted: 14 July 2021
Published: 21 July 2021

1. Introduction

Pouring processes in the casting industry can be dangerous to workers because they perform the task of handling molten metal with extremely high temperatures. Accordingly, the need for automation of the pouring process has promoted improvements in the work environment in this context [1–4]. A tilting-ladle-type automatic pouring machine, where molten metal is poured into the mold by tilting the ladle automatically, is employed as one such automated pouring process. It can be installed relatively easily as part of the pouring process because it essentially employs the same pouring method as a manual handling of this task [5]. As a control system for the tilting-ladle-type automatic pouring machine, the teaching-and-playback control approach has been practically applied [6–8]. The pouring process requires the precision pouring of molten metal into the pouring basin of the mold to improve productivity and user safety. However, doing so can be difficult because the pouring flow rate is indirectly changed by tilting the ladle [9–11]. Furthermore, the molten metal can spill from the mold due to the falling trajectory of the outflow liquid varying in accordance with the flow rate changes.

To solve these problems, control technologies have been developed to improve pouring precision. The sloshing suppression controls of the liquid in the ladle were developed for suppressing the liquid vibration while the liquid is poured into the mold [12,13]. For keeping the pouring mouth of the ladle in a fixed position, a forward-and-backward and an up-and-down motions of the ladle were controlled synchronously to the ladle tilting [12]. Moreover, the sloshing caused in the ladle's back-tilting-motion was suppressed by the input shaping approach in which the input command is shaped for generating the anti-phase sloshing in [12,13]. The sloshing caused in the pouring motion was modeled simply by a pendulum model in [14,15]. In addition, the sloshing was suppressed by combining the partial inverse dynamics approach and PID controller. The parameters of PID controller was designed using the metaheuristic search algorithm. The sloshing in the liquid container handled by a robotic manipulator was modeled by a spherical pendulum model in [16,17]. For suppressing the sloshing, the reference trajectories with the container's position and orientation were shaped by an exponential filter, the parameters of which were designed using the spherical pendulum. The sloshing caused in the ladle's forward-tilting-motion was suppressed by making the angular velocity in the ladle's tilting motion smaller [18]. The control system for the liquid height in the pouring basin was constructed using the audio vibration sensing and the deep neural network [19]. The neural network in the pouring robot system was trained with a real-world pouring dataset with multi-modal sensing data, which contains more than 3000 recordings of audio, force feedback, video, and trajectory data of the human hand that performs the pouring task. The pouring skill of human was emulated by the robot with the haptic device using the parametric hidden Markov model [20]. In this approach, the human tele-operated the robot using a haptic device, and data from the demonstrations were statistically encoded by a parametric hidden Markov model. The Gaussian mixture regression was used at the reproduction in the robotic playback motion. The angular velocity of ladle tilting was optimized using the optimization approach with a Computational Fluid Dynamics (CFD) simulator for suppressing the casting defects [21]. The flow-rate feedforward control, based on the mathematical model of the pouring process, was developed in [22]. The falling position control of the liquid outflow from the ladle was developed to achieve precise pouring of the liquid into a steady position inside the pouring basin of the mold [23]. Additionally, a control method for positioning the ladle as low as possible while pouring the liquid was proposed in [24]. In these falling position control approaches, the ladle is moved in vertical and horizontal directions. However, the liquid can still spill from the pouring basin of the mold due to the splashing inside the ladle, which is caused by the movement of the ladle [25]. Therefore, the optimal positioning of the ladle which minimizes the amount of spilled liquid from the pouring basin of the mold was developed in [25,26]. In these studies, a control system was constructed using the flow-rate feedforward control. Pouring precision can be degraded by disturbances that emerge from variable pouring conditions. To suppress the influence of such disturbances, flow-rate feedback control has to be established. However, it can be difficult to directly measure the flow rate of poured molten metal because sensors e.g., the flow meter, can be damaged by the high temperature of the molten metal. Therefore, to measure the pouring flow rate, the real-time flow-rate estimation approach using the Kalman filter (KF), specifically an extended Kalman filter (EKF), was developed in [27,28]. However, this approach can only be applied to smooth shaped ladles, e.g., a fan-shaped ladle, due to the requirements of the Jacobian matrix used in the EKF. Nonetheless, ladles with complicated shapes for which it is difficult to derive a Jacobian matrix, have been used in practical settings. Thus, flow-rate estimation with an unscented Kalman filter (UKF), which does not require a Jacobian matrix, was proposed in [29]. Moreover, the flow-rate estimation method was integrated with the flow-rate feedback control system in [30]. However, it is difficult to implement and perform a real-time computation using an industrial controller, which has low computational power. To apply a real-time flow-rate estimation to an automatic pouring machine with a complicatedly shaped ladle, and implement the flow-rate estimation method in an industrial controller, the simplified flow-rate estimation method in which the steady-state KFs (SSKFs) and the EKF are decentralized to the motor systems and the pouring process, respectively, was proposed in our previous study [31]. The decentralization of Kalman filters (DKFs) were integrated with the flow-rate feedback control system in [32]. As a remaining issue, the appropriateness of the flow rate estimation method via DKFs as a real-time pouring flow-rate measurement in the automatic pouring machine must be verified. In [31], the effectiveness of the flow rate estimation by the DKFs was verified in the experiment with the ideal condition as the pouring process model in the DKFs is identified with the experimental pouring. However, some uncertainties in the model parameters identification with practical pouring can arise. It is especially difficult to identify the tilting angle of the ladle at the start of the liquid outflow, which can be influenced by the surface tension and density of the liquid in the ladle [32]. Moreover, in order to verify the effectiveness of the flow rate estimation, the true flow rate of the liquid outflow from the ladle in the experiment should be measured. It is, however, difficult to obtain the true flow rate of the liquid outflow.

Therefore, in the current study, we verified the appropriateness of the flow-rate estimation method using DKFs by comparing it with two other existing real-time flow-rate estimations. We confirmed the noise levels and the estimation errors of the candidate real-time flow-rate estimations in the experiments with/without the error between the ideal and the actual tilting angle at the start of the liquid outflow. As evaluation of the estimation error, the estimated flow rates were compared with the simulated flow rate which represents faithfully the experimental pouring with the model of pouring process. In one of the compared methods, the flow rate was estimated by differentiating the weight of the outflow liquid measured by the load cell with respect to time [33]. In this approach, a low-pass filter was applied for reducing the noise of the measured data. In addition, the angular velocity of ladle tilting motion was controlled by the fuzzy rules referring to the filtered data. In the second estimation method, the flow rate was estimated using a visible camera [34–36]. In the approaches [35,36], the liquid height in the target container was measured by the stereo camera or RGB-D camera for estimating the pouring state. The flow line of the inflow liquid in the clear target container was measured by the visible camera [37]. Therefore, in the case of the pouring situation using the clear ladle, the visible camera is able to estimate the flow rate of the outflow liquid by measuring the liquid volume in the ladle in real time. In the casting industry, since the ladle generally consists of gypsum and metal, detecting the liquid volume in the ladle using a visible camera is difficult. In this study, we used water as the target liquid and a clear acrylic container as a ladle. As such, we were able to discern the liquid volume in the ladle from the projected liquid area, as captured by the visible camera. This flow-rate estimation method cannot, however, be applied to practical pouring processes using molten metal. Nonetheless, we applied this flow-rate estimation method for verifying the appropriateness of the flow-rate estimation method by DKFs. Most practical automatic pouring machines have a rotary encoder for measuring the tilting angle of the ladle and a load cell for measuring the weight of the outflow liquid from the ladle. DKFs in which the flow rate can be estimated indirectly using the rotary encoder and the load cell are useful for the control of automatic pouring machine in practical use.

The remainder of this paper is structured as follows. The tilting-ladle-type automatic pouring machine used in this study is introduced in Section 2. The mathematical model for the pouring motion is derived in Section 3, and compensation for the measured weight by the load cell is presented in Section 4. The flow-rate estimation method by DKFs is described in detail in Section 5. The additional two flow-rate estimation methods are discussed in Section 6. In Section 7, the appropriateness of the flow-rate estimation by DKFs is verified experimentally by comparing the process with two other flow-rate estimations. Concluding remarks are presented in the final section.

2. Automatic Pouring Machine

The tilting-ladle-type pouring machine used in this study is presented in Figure 1. In the pouring machine, the ladle could be transferred according to two dimensions (the y- and z-axes) and could also be rotated (in the Θ direction). Each direction had a direct current (DC) servomotor to drive in the velocity control mode. In the y- and z-axes, the driving force of each motor was amplified by a ball-and-screw mechanism. The transfer distance and the tilting angle of the ladle could be measured by the rotary encoders installed in the motors. These motor drivers for driving the servomotors communicated with the controller through a controller area network bus.

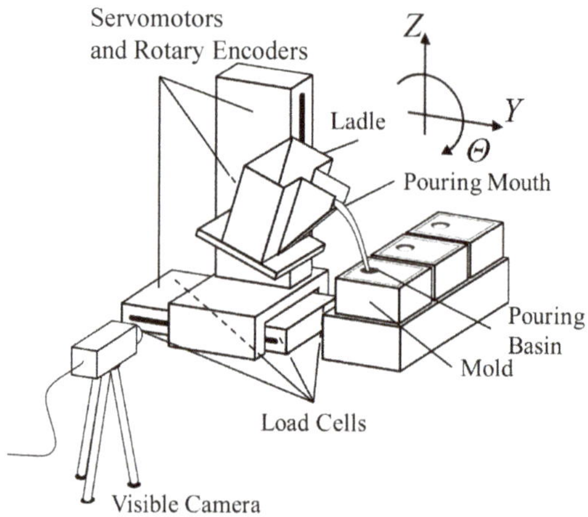

Figure 1. Automatic pouring machine.

The center of the ladle's rotation shaft was placed near the center of gravity to avoid the increase in the capacity of the motor for rotating the ladle. This mechanism has been used with the recent practical automatic pouring machines. While rotating the ladle around the center of gravity, the tip of the ladle's pouring mouth moved in a circular trajectory. By moving the tip of the pouring mouth, it became difficult to precisely pour the molten metal into the mold. Therefore, the position of the tip of the pouring mouth had to be invariably controlled while pouring; this was achieved using synchronous control of the y- and z-axes during rotational motion around the ladle's Θ-direction [38]. The weight of the outflow liquid could then be measured by the load cell equipped to the base of the pouring machine. In the load cell system, four sensing terminals were located on the four corners of the base of the pouring machine. The maximum measuring error of this load cell is 0.05 kg. The weight rate of the outflow liquid cannot be measured by the load cell. The amplifier of the load cell communicates with the controller by a serial communication method. The load cell data can be obtained with the sampling interval 0.02 s using the serial communication. The splash of the liquid in the ladle can be caused by varying the liquid shape in the pouring motion with tilting the ladle. For suppressing the splash, the input shaping approaches were proposed in [12,13]. However, it is difficult to suppress the splash in the different conditions from the parameters' setting. The splash can be suppressed by making the variation of the liquid shape smaller. We applied the splash suppression approach by limiting the angular velocity of the ladle tilting [18]. In the preliminary experiments, the amplitude of the angular velocity has been limited within 12 deg/s for suppressing the splash in the expected pouring motion using this automatic pouring machine.

In this study, the target liquid was water, and a clear acrylic ladle was used to visualize its inside; the side area of liquid in the ladle was captured by a visible camera while pouring the liquid. The camera was located 1.5 m from the ladle in the lateral direction for capturing the whole area of the ladle. In the camera system, the images with 512 × 480 pixels were captured with the frame rate 30 fps. The processing time for obtaining the side area of the liquid in the ladle from the captured image is 0.06 s.

3. Mathematical Models of Pouring Motion

The pouring motion is represented by the block diagram in Figure 2. In the motorized pouring motion, the input command was applied to the motor for tilting the ladle. The liquid was poured from the tilted ladle, and the weight of outflow liquid was measured by the load cell.

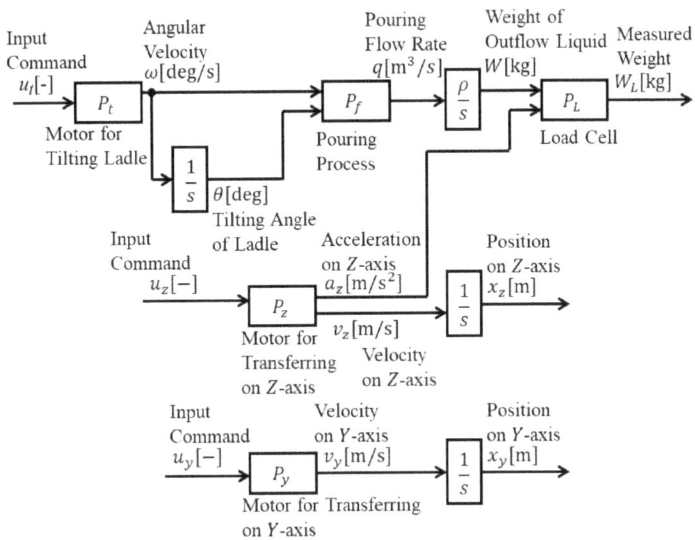

Figure 2. Block diagram of motorized pouring motion.

3.1. Motor Model for the Tilting Ladle

In Figure 2, motor model P_t for tilting the ladle is simplified as a first-order lag system and an integrator, which can be given as follows:

$$\frac{d\omega(t)}{dt} = -\frac{1}{T_{mt}}\omega(t) + \frac{K_{mt}}{T_{mt}}u_t(t), \tag{1}$$

$$\frac{d\theta(t)}{dt} = \omega(t), \tag{2}$$

where ω deg/s is the angular velocity of the tilting ladle, u_t is the input command applied to the motor, θ deg is the angle of the tilting ladle, T_{mt} s is the time constant, and K_{mt} deg/s is the gain constant. The time constant and the gain constant can be identified by a step response method. In this method, three step input commands were given as u_t = 2, 4, and 6. The time and the gain constants were obtained from the response for each step input command. We determined these constants by averaging the obtained constants. In this study, identification experiments obtained $T_{mt} = 0.022$ s and $K_{mt} = 0.980$ deg/s.

3.2. The Pouring Process Model

The pouring process model P_f in Figure 2 represents the volume balance of the topmost liquid volume in the ladle, which can be shown as the input–output relation from angular velocity ω to flow rate q m^3/s of the outflow liquid. The cross section of the pouring process at tilting angle θ is presented in Figure 3, where the volume balance of the topmost liquid volume in the ladle can be given as follows:

$$\frac{dV_r(t)}{dt} = -q(t) - \frac{\partial V_s(\theta(t))}{\partial \theta}\omega(t), \ (V_r \geq 0, \ \theta \geq \theta_s), \tag{3}$$

where V_r m^3 is the liquid volume over the pouring mouth, V_s m^3 is the liquid volume under the pouring mouth, and θ_s deg is the angle of the ladle at the start of the liquid outflow. Accordingly, volume V_r m^3 can be approximated as follows:

$$V_r(t) \approx A(\theta(t))h(t), \ (h(t) \geq 0), \tag{4}$$

where A m^2 is the upper surface of the liquid in the ladle, and h m is the height of the liquid over the pouring mouth. As presented in Figure 3, surface A is changed by tilting angle θ deg of the ladle.

Using Bernoulli's principle, flow rate q at liquid height h is given as follows:

$$q(t) = c\int_0^{h(t)} L_f(h_a)\sqrt{2gh_b}dh_b, \ (0 < c \leq 1, \ h_a = h(t) - h_b), \tag{5}$$

where L_f m is the width of the pouring mouth at height h_a m from the bottom edge of the pouring mouth (see Figure 4a), h_b m is the depth at the pouring mouth from the surface of the liquid in the ladle, g m/s^2 is the acceleration of gravity, and c is the flow-rate coefficient. Figure 4a and Equation (5) show the relation between the flow rate and the liquid height in the pouring mouth with the generalization of cross-sectional shape. In more detail, since the geometry of the pouring mouth is represented on the basis of the bottom edge of the pouring mouth, the width L_f is defined by the function of the height h_a. On the one hand, the flow velocity $\sqrt{2gh_b}$ depends on the depth h_b from the surface of the liquid. In design of the flow-rate control, the relation in Equation (5) can be implemented by the interpolation method [22] or the look-up table [39]. Accordingly, the flow rate $q(t)$ can be calculated by giving the liquid height $h(t)$ to the implementation of Equation (5) each sampling time. In the case that the cross-sectional shape of the pouring mouth is rectangle as shown in Figure 4b, the flow rate q shown in Equation (5) can be simplified as follows:

$$q(t) = \frac{2}{3}cL_f\sqrt{2gh(t)^3}, \ (0 < c \leq 1). \tag{6}$$

In this study, since the ladle with the rectangular pouring mouth was used in the experiment, Equation (6) was applied as the relation between the flow rate and the liquid height on the pouring mouth. The flow-rate coefficient c can be identified by fitting the simulation result to the experimental result of the weight of the outflow liquid measured by the load cell. In this study, we obtained $c = 0.75$.

From Equations (3), (4) and (6), the dynamics of liquid height over the pouring mouth in the pouring process were derived as follows:

$$\frac{dh(t)}{dt} = -\frac{q(h(t))}{A(\theta(t))} - \frac{1}{A(\theta(t))}\left(\frac{\partial V_s(\theta(t))}{\partial \theta(t)} + \frac{\partial A(\theta(t))}{\partial \theta(t)}h(t)\right)\omega(t), \ (h \geq 0, \ \theta \geq \theta_s). \tag{7}$$

The actual weight W kg of the outflow liquid can be represented as:

$$\frac{dW(t)}{dt} = \rho q(t), \tag{8}$$

where ρ kg/m^3 is the density of the liquid. In this study, since we used water as a target liquid, $\rho = 1.0 \times 10^3$ kg/m^3 was applied.

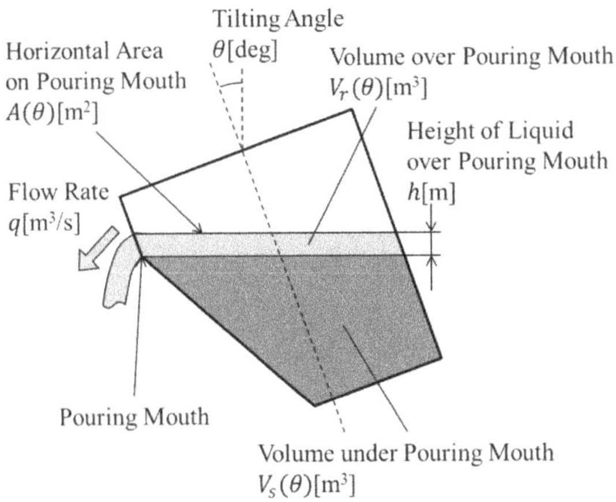

Figure 3. Cross section of the pouring process.

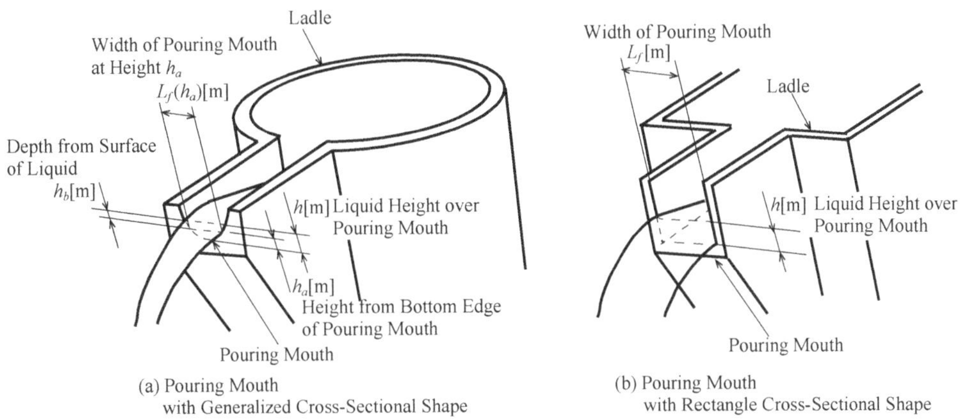

Figure 4. Parameters on pouring mouth.

3.3. Load Cell Model

The dynamics of the load cell can be simplified as a first-order lag system. Therefore, the load-cell model P_L can be given as follows:

$$\frac{dW_L(t)}{dt} = -\frac{1}{T_L}W_L(t) + \frac{1}{T_L}\left\{W(t) - \frac{M}{g}a_z(t)\right\}, \tag{9}$$

where W_L kg is the weight of the outflow liquid measured by the load cell, and T_L s is the time constant of the load cell. The time constant can be identified by fitting the simulation and experimental results of the liquid pouring. In this study, the identification experiments obtained $T_L = 0.16$ s. Furthermore, a_z m/s^2 was the acceleration for transferring the ladle on the z-axis, and M kg was the gross weight of the ladle, the liquid in the ladle, and the actuator for transferring the ladle on the z-axis. We assumed that the weight variation of the liquid in the ladle while pouring had been sufficiently smaller than the gross weight M. Accordingly, the gross weight M was given as a constant parameter that could be obtained before pouring. Thus, the gross weight in this study is given as $M = 14.93$ kg.

3.4. Motor Model for Transferring the Ladle on the y- and z-Axes

The motor models P_y and P_z are given as follows:

$$\frac{dv_i(t)}{dt} = -\frac{1}{T_{mi}}v_i(t) + \frac{K_{mi}}{T_{mi}}u_i(t), \ (i = y, z), \quad (10)$$

$$\frac{dx_i(t)}{dt} = v_i(t), \ (i = y, z), \quad (11)$$

where v_i m/s is the velocity of the ladle, and x_i[m] is the position of the ladle. T_{mi} s is the time constant, and K_{mi} m/s is the gain constant. Index i in Equations (10) and (11) refers to the direction for transferring the ladle; y and z refer to the direction on the y- and z-axes, respectively. The time constants and the gain constants can be identified by the step response method with the same procedure of the Θ-axis. In this study, $T_{mi} = 0.05$ s and $K_{mi} = 0.997$ m/s were obtained as the same parameters on each axis by the identification experiments. Acceleration a_z m/s^2 on the z-axis can be represented as follows:

$$a_z(t) = \frac{dv_z(t)}{dt}. \quad (12)$$

3.5. Synchronous Control for Transferring and Rotating the Ladle

Since the rotation shaft of the ladle is placed near the center of gravity, the tip of the pouring mouth in the ladle moved in a circular trajectory, making it difficult to precisely pour the liquid into the mold (see Figure 5a). To rotate the ladle around the tip of the pouring mouth, the ladle is transferred synchronously on the y- and z-axes while tilting it (see Figure 5b). The synchronous controller can be described as follows:

$$r_y = L_a \cos\theta_a - L_a \cos(\theta_a - \theta), \quad (13)$$

$$r_z = L_a \sin\theta_a - L_a \sin(\theta_a - \theta), \quad (14)$$

where L_a m is the length from the tip of the pouring mouth in the ladle to the rotation center of the ladle; θ_a is the angle between the line segment with length L_a and a horizontal line; and r_y m and r_z m are the reference trajectories on the y- and z-axes for rotating the ladle around the tip of the pouring mouth. These reference trajectories were applied to the position feedback controllers on each axis.

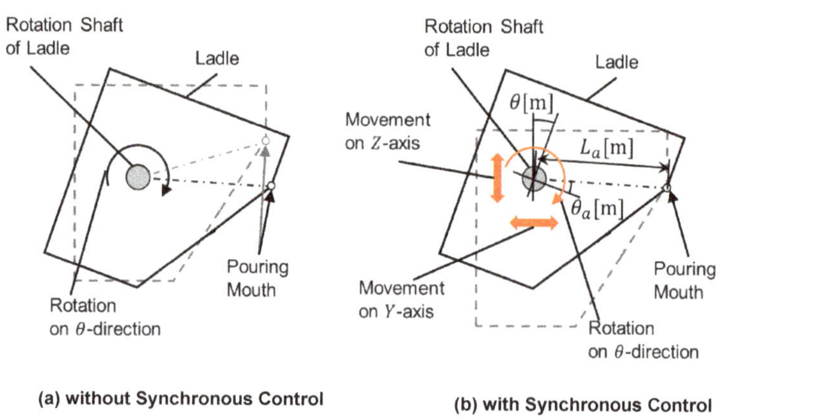

Figure 5. Motion of ladle with/without synchronous control.

4. Compensation of Measured Weight by Load Cell

In an automatic pouring machine, as presented in Figure 1, the load cell for measuring the weight of the liquid in the ladle was equipped at the bottom of the automatic pouring machine. Consequently, the weight measured by the load cell was influenced by the movement of the ladle on the z-axis (see Equation (9)) due to the synchronous control noted in the previous section.

To obtain only the weight of liquid outflow from the ladle, the influence of the movement on the z-axis was subtracted from the measured weight as follows:

$$W_{Lc}(t) = W_L(t) - W_{Lz}(t), \tag{15}$$

where W_{Lc} kg is the compensated weight, and $W_{Lz}(t)$ is the weight transformed from acceleration a_z m/s^2 of the ladle's movement on the z-axis and can be estimated as follows:

$$\frac{dW_{Lz}(t)}{dt} = -\frac{1}{T_L}W_{Lz}(t) + \frac{1}{T_L}\frac{M}{g}a_z(t). \tag{16}$$

The acceleration can be estimated by the SSKF, which is derived in the following section.

Figure 6a presents the results of the measured weight by the load cell. The sampling interval of the load cell data are 0.02 s. The black and magenta solid lines indicate the weights before and after compensation, respectively. The dashed line indicates the weight of the outflow liquid in the simulation using the pouring flow-rate model with weight compensation. The vibration of 4.5 Hz has appeared in the measured weight. It is caused by the resonant vibration of the automatic pouring machine excited by the pouring motion. In the design of flow-rate estimation, the vibration can be regarded as the process noise of the load cell. Figure 6b presents the angular velocity of the motor for tilting the ladle. Figure 6c,d present the acceleration of movement of the ladle on the z-axis and the transformed weight, respectively. As presented in Figure 6, the influence of movement on the z-axis was reduced to within the compensated weight.

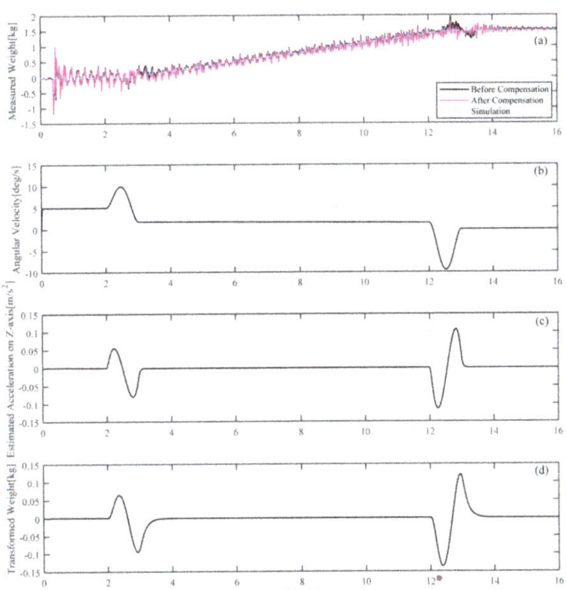

Figure 6. Experimental results of the weight compensation system. (**a**) Weight data measured by load cell; (**b**) Angular velocity of ladle tilting; (**c**) Acceleration on z-axis estimated by SSKF; (**d**) Weight data transformed from estimated acceleration on z-axis using Equation (16).

5. Flow Rate Estimation by Decentralization of Kalman Filters

5.1. The Design of Flow Rate Estimation

In the first flow-rate estimation method [28], the EKF was applied to the pouring motion using the input command for tilting the ladle to the measured weight of the outflow liquid. This estimation method is applicable only where the ladle has a smooth shape (e.g., fanned), since a complicated shape does not satisfy the twice differentiability concerning the tilting angle. The Jacobian matrix in the EKF requires this twice differentiable of the model parameters obtained from the ladle shape. To avoid the twice differentiability of the model parameters concerning the tilting angle and simply construct the flow-rate estimation system in an automatic pouring machine with complicatedly shaped ladle, the SSKFs and EKF were decentralized to the motor systems and the pouring process, respectively, in [31]. The motor system and the pouring process are sequentially connected as shown in Figure 2. The angular velocity is added as the input in the pouring process. In addition, the angle of the ladle can be detected by the rotary encoder. Therefore, in the case that the angular velocity of the ladle tilted by the motor system on the Θ-axis and the acceleration of the ladle transferred by the motor system on the z-axis can be estimated precisely by the SSKFs, and the pouring flow rate can also be estimated precisely by the EKF with only the pouring process model [31]. A block diagram of flow-rate estimation by DKFs is presented in Figure 7.

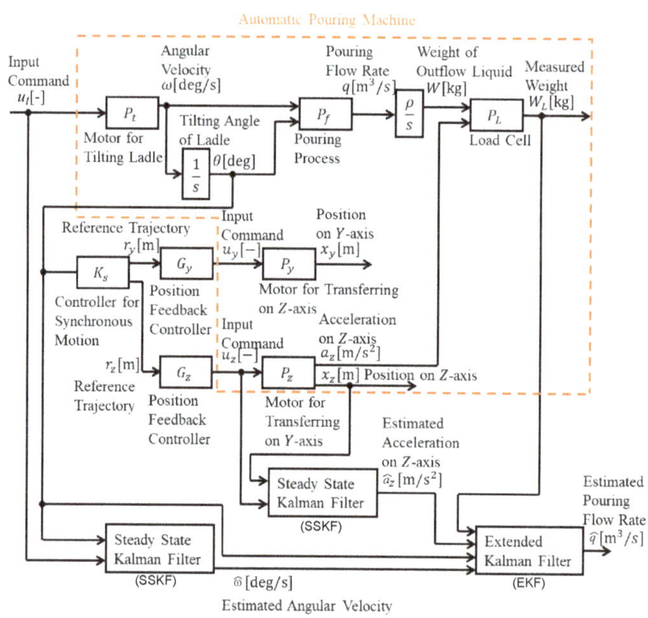

Figure 7. Block diagram of flow-rate estimation by DKFs.

In this approach, a discrete-time SSKF was applied for estimating the angular velocity of the motor for tilting the ladle. The discrete-time state equation of the motor system can be given as follows:

$$\begin{bmatrix} \omega_{n+1} \\ \theta_{n+1} \end{bmatrix} = \begin{bmatrix} 1-\frac{T_s}{T_{mt}} & 0 \\ T_s & 1 \end{bmatrix} \begin{bmatrix} \omega_n \\ \theta_n \end{bmatrix} + \begin{bmatrix} \frac{T_s K_{mt}}{T_{mt}} \\ 0 \end{bmatrix} u_{tn}, \quad (17)$$

$$y_n = \begin{bmatrix} 0 & 1 \end{bmatrix} \begin{bmatrix} \omega_n \\ \theta_n \end{bmatrix}, \quad (18)$$

where T_s s represents the sampling interval and is given as 0.020 s in this study. The angular velocity estimated by the discrete-time SSKF with the model of the motor system in Equation (18) is applied to the estimation of the pouring flow rate described later.

Similarly, a discrete-time SSKF was applied to estimate acceleration for transferring the ladle on the z-axis and is represented as follows:

$$\begin{bmatrix} v_{zn+1} \\ x_{zn+1} \end{bmatrix} = \begin{bmatrix} 1-\frac{T_s}{T_{mz}} & 0 \\ T_s & 1 \end{bmatrix} \begin{bmatrix} v_{zn} \\ x_{zn} \end{bmatrix} + \begin{bmatrix} \frac{T_s K_{mz}}{T_{mz}} \\ 0 \end{bmatrix} u_{zn}, \tag{19}$$

$$y_{zn} = \begin{bmatrix} 0 & 1 \end{bmatrix} \begin{bmatrix} v_{zn} \\ x_{zn} \end{bmatrix}. \tag{20}$$

The acceleration on the z-axis was estimated as:

$$a_{zn} = -\frac{1}{T_{mz}} v_{zn} + \frac{K_{mz}}{T_{mz}} u_{zn}. \tag{21}$$

We designed an EKF for estimating the pouring flow rate. The discrete-time state equation of the pouring process can be represented as follows:

$$x_{n+1} = f(x_n, u_n), \tag{22}$$

$$y_n = \eta(x_n), \tag{23}$$

where

$$x = \begin{bmatrix} h & W & W_L \end{bmatrix}^{\mathrm{T}}, \tag{24}$$

$$u = \begin{bmatrix} \hat{\omega} & \hat{a}_z \end{bmatrix}^{\mathrm{T}}, \tag{25}$$

$$f(x,u) = \begin{bmatrix} \left(1 - \frac{T_s}{A(\theta)} \frac{\partial A(\theta)}{\partial \theta} \hat{\omega}\right) h - \frac{T_s q(h)}{A(\theta)} - \frac{T_s}{A(\theta)} \frac{\partial V_s(\theta)}{\partial \theta} \hat{\omega} \\ W + T_s \rho q(h) \\ (1 - \frac{T_s}{T_L} W_L) + \frac{T_s}{T_L} (W - \frac{M}{g} \hat{a}_z) \end{bmatrix}, \tag{26}$$

$$\eta(x) = W_L. \tag{27}$$

The input vector u consists of the angular velocity $\hat{\omega}$ on the Θ-axis and the acceleration $\hat{a}_z$ on the z-axis estimated by the SSKFs. Then, the Jacobian matrices used for updating the Kalman gain in the EKF can be given as:

$$\frac{\partial f(x)}{\partial x} = A_f = \begin{bmatrix} \alpha_{11} & 0 & 0 \\ \alpha_{21} & \alpha_{22} & 0 \\ 0 & \alpha_{32} & \alpha_{33} \end{bmatrix}, \tag{28}$$

$$\frac{\partial \eta(x)}{\partial x} = C_f = \begin{bmatrix} 0 & 0 & 1 \end{bmatrix}, \tag{29}$$

where

$$\alpha_{11} = 1 - \frac{T_s}{A(\theta)} \left(\frac{\partial q(h)}{\partial h} + \frac{\partial A(\theta)}{\partial \theta} \hat{\omega} \right),\ \alpha_{21} = T_s \rho \frac{\partial q(h)}{\partial h},\ \alpha_{22} = 1,\ \alpha_{32} = \frac{T_s}{T_L},\ \alpha_{33} = 1 - \frac{T_s}{T_L}.$$

The derivative of the pouring flow rate q to the liquid height h can be derived from Equation (5) as follows:

$$\frac{\partial q(h)}{\partial h} = c L_f(h) \sqrt{2gh}. \tag{30}$$

Based on the Jacobian matrices in Equations (28) and (29), it was confirmed that twice differentiability of the model parameters according to the ladle shape was not required. Therefore, this estimation method could be applied to a complicatedly shaped ladle.

Following this, the pouring flow rate $\hat{q}$ m^3/s could be estimated by substituting the estimated liquid height on the pouring mouth as follows:

$$\hat{q}(\hat{h}(t)) = c\int_0^{\hat{h}(t)} L_f(h_a)\sqrt{2gh_b}dh_b. \quad (31)$$

In the EKF, the time-update equations can be represented as follows:

- Predict

$$\begin{aligned}\hat{x}_n^- &= f(\hat{x}_{n-1}),\\ P_n^- &= A_{fn-1}P_nA_{fn-1}^{\mathrm{T}} + Q_f,\end{aligned}$$

- Update

$$\begin{aligned}G_n &= P_n^-C_{fn}^{\mathrm{T}}(C_{fn}P_n^-C_{fn}^{\mathrm{T}} + R_f)^{-1},\\ \hat{x}_n &= \hat{x}_n^- + G_n\{y_n - \eta(\hat{x}_n^-)\},\\ P_n &= (I - G_nC_{fn}^{\mathrm{T}})P_n^-,\end{aligned}$$

where $\hat{x}^-$ is a priori state estimate, $\hat{x}$ is a posteriori state estimate, and Q_f and R_f represent covariance matrices of the system noise and the observation noise, respectively. Furthermore, G_n is the Kalman gain, and P^- and P are a priori error covariance and a posteriori error covariance, respectively.

5.2. Simulations

Flow-rate estimation by DKFs was performed in the simulation for the tilting-ladle-type automatic pouring machine. The ladle used in this study is presented in Figure 8.

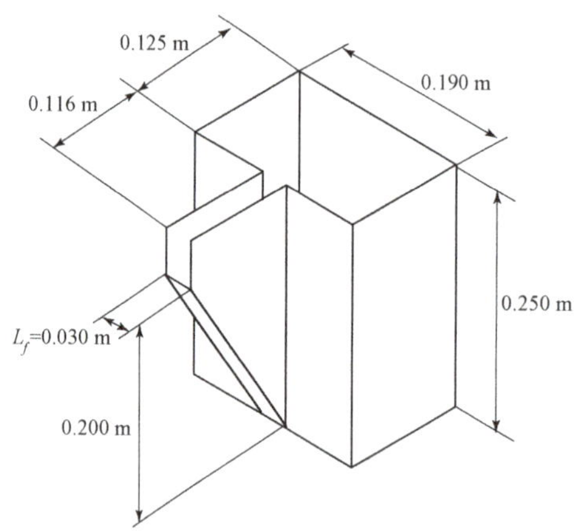

Figure 8. Geometry of ladle (inside dimension).

The model parameters of the ladle were obtained from the three-dimensional data of the ladle. In Figure 9, the horizontal area $A(\theta)$ m^2 on the pouring mouth and volume $V_s(\theta)$ m^3 under the pouring mouth (presented in Figure 3) are indicated in (a) and (b), respectively. Figure 9c,d are the derivatives of (a) and (b) for the tilting angle, respectively.

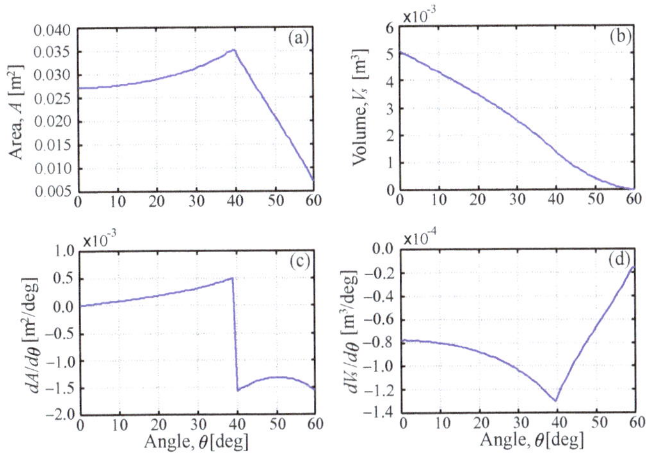

Figure 9. Model parameters of ladle. (**a**) Horizontal area on pouring mouth; (**b**) Volume under pouring mouth; (**c**) Derivative of horizontal area A with respect to angle θ; (**d**) Derivative of volume V_s with respect to angle θ.

The derivatives $\partial A(\theta)/\partial\theta$ and $\partial V_s(\theta)/\partial\theta$ can be derived as follows:

$$\frac{\partial A(\theta)}{\partial\theta} = \frac{A(\theta(k+1)) - A(\theta(k))}{\Delta\theta}, \quad \frac{\partial V_s(\theta)}{\partial\theta} = \frac{V_s(\theta(k+1)) - V_s(\theta(k))}{\Delta\theta}, \tag{32}$$

where $\Delta\theta$ is the sampling interval of the tilting angle, and k is the sampling number, which has the relation of $\theta(k) = k\Delta\theta$. In this study, we used $\Delta\theta$ = 1.0 deg. The volume $V_s(\theta)$ is decreased with increasing the tilting angle θ. The horizontal area $A(\theta)$ is increased with increasing the tilting angle until reaching the bottom of the ladle, $\theta \leq 40$ deg. In the tilting angle over 40 deg, it is decreased with increasing the tilting angle. In particular, it is decreased as the quadratic curve by increasing the area of the channel to the pouring mouth.

To estimate the angular velocity of the motor for tilting the ladle, the covariance of process noise Q_t and the covariance of observation noise R_t were assumed as follows:

$$Q_t = \mathrm{diag}(1.0 \times 10^{-3}\ \mathrm{deg}^2/\mathrm{s}^2, 1.0 \times 10^{-7}\ \mathrm{deg}^2), \tag{33}$$
$$R_t = 1.84 \times 10^{-7}\ \mathrm{deg}^2. \tag{34}$$

To estimate the acceleration of the movement of the ladle on the z-axis, the covariance of process noise Q_z and covariance of the observation noise R_z were assumed as follows:

$$Q_z = \mathrm{diag}(1.0 \times 10^{-8}\ \mathrm{m}^2, 1.0 \times 10^{-4}\ \mathrm{m}^2/\mathrm{s}^2), \tag{35}$$
$$R_z = 5.0 \times 10^{-6}\ \mathrm{m}^2. \tag{36}$$

Similarly, to estimate the state of the pouring process, the covariance of the process noise Q_f and the covariance of the observation noise R_f were assumed as follows:

$$Q_f = \mathrm{diag}(25\ \mathrm{m}^2, 1.0\ \mathrm{kg}^2, 1.0\ \mathrm{kg}^2) \times 10^{-11}, \tag{37}$$
$$R_f = 5.0 \times 10^{-4}\ \mathrm{kg}^2. \tag{38}$$

In the simulations, the feedforward flow-rate control [22] was constructed to realize the desired flow rate, The reference flow rate applied to the flow rate control is presented in Figure 10.

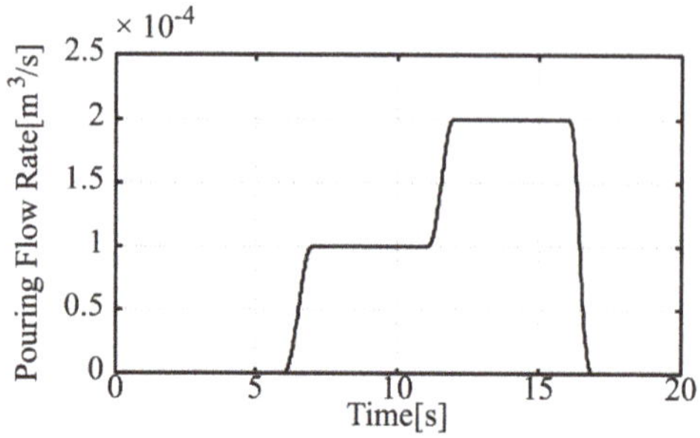

Figure 10. Reference pouring flow rate.

The simulation results obtained without any disturbances are presented in Figure 11, where (a) indicates the input command applied to the motor for tilting the ladle, and (b) and (c) indicate the angular velocity and the tilting angle of the ladle, respectively.

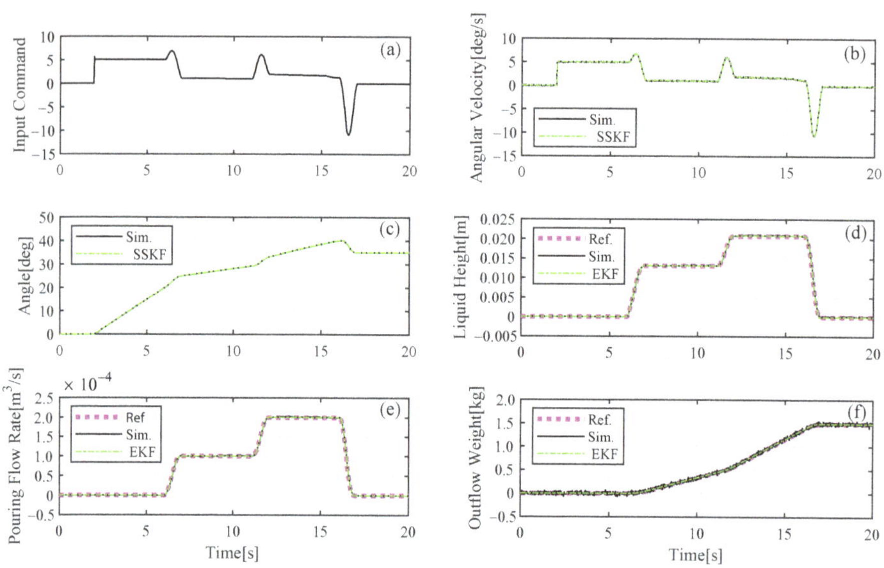

Figure 11. Simulation results of flow-rate estimation with error of tilting angle +0 deg at the start of liquid outflow. (**a**) Input command to motor for tilting ladle; (**b**) Angular velocity of tilting ladle; (**c**) Angle of tilting ladle; (**d**) Liquid height on pouirng mouth; (**e**) Flow rate of outflow liquid from ladle; (**f**) Weight of outflow liquid from ladle.

The black solid lines are the results simulated using the motor model, and the chained lines are the results estimated using the SSKF. Furthermore, (d–f) indicate the liquid height, the pouring flow rate, and the outflow weight, respectively. The dotted magenta lines are the reference values, and the black solid lines are the results simulated using the pouring process model. The chained green lines are the results estimated using the EKF. In Figure 11b, the amplitude of angular velocity of the ladle is within 12 deg/s. Therefore, the splash of the liquid in the ladle could be suppressed into the level that has no influence on the pouring motion. As presented in Figure 11e, the flow rate could be precisely estimated by the flow-rate estimation using the DKFs.

The following disturbance simulations were performed. In the practical pouring process, the volume of molten metal in the ladle was uncertain due to temperature changes in the metal. Therefore, an error between the ideal liquid volume, designed in the control system, and the actual liquid volume occurred (see Figure 12).

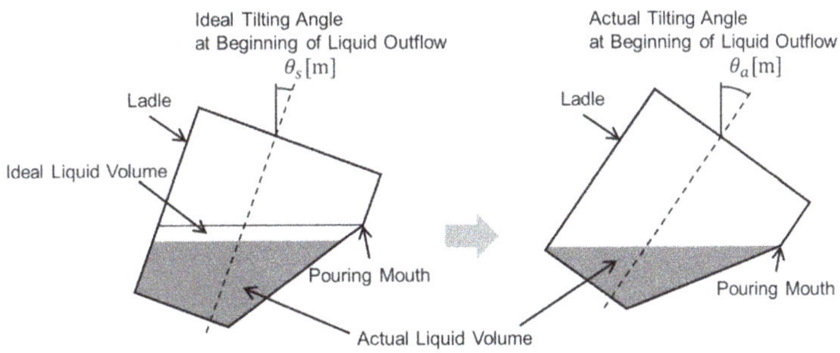

Figure 12. Tilting angle of ladle at the start of liquid outflow.

The error between the ideal and actual tilting angles at the start of the liquid outflow increased alongside an increase in the error between the ideal and actual liquid volumes. In the disturbance simulations, the ideal tilting angle θ_s at the start of the liquid outflow was given as 20 deg, and the error between the ideal and actual tilting angles at the start of the liquid outflow was +3 deg. The simulation results are presented in Figure 13, where the graphs are illustrated similar to Figure 11.

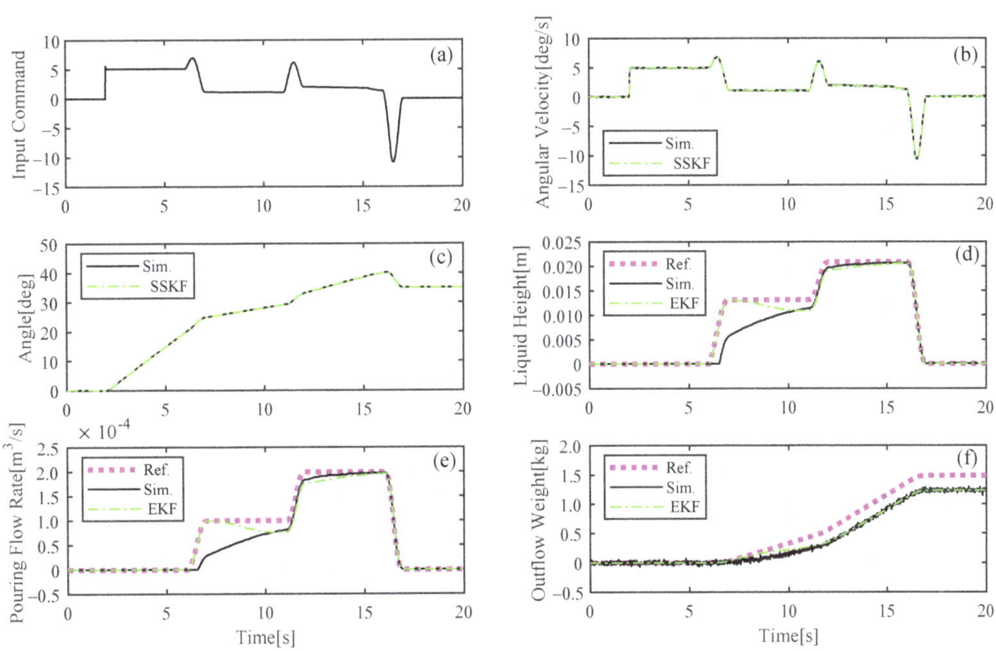

Figure 13. Simulation results of flow rate estimation with error of tilting angle +3 deg at the start of liquid outflow. (**a**) Input command to motor for tilting ladle; (**b**) Angular velocity of tilting ladle; (**c**) Angle of tilting ladle; (**d**) Liquid height on pouirng mouth; (**e**) Flow rate of outflow liquid from ladle; (**f**) Weight of outflow liquid from ladle.

As presented in Figure 13e, the error between the reference and the simulated flow rates was the result of disturbance. However, the estimated flow rate converged rapidly to the simulated flow rate. In addition, the error between the reference and the simulated outflow weights was increased by the disturbance as shown in Figure 13f. However, the estimated outflow weight tracked precisely to the simulated outflow weight. DKFs can estimate robustly the outflow weight, even if the disturbances are occurred in the pouring motion.

6. Other Flow-Rate Estimation Methods for Comparison

To verify the appropriateness of the flow-rate estimation by DKFs, we constructed two additional types of flow-rate estimation.

6.1. Flow-Rate Estimation by Differentiating Load Cell Data

Flow rate was estimated by differentiating the weight of the outflow liquid measured by the load cell. However, because the load cell data included a significant level of noise, it was processed by a low-pass filter. This flow-rate estimation method could be described as follows:

$$\frac{dW_{lf}}{dt} = -\omega_{lf}W_{lf} + \omega_{lf}W_L, \tag{39}$$

$$q_{lf} = -\frac{\omega_{lf}}{\rho}W_{lf} + \frac{\omega_{lf}}{\rho}W_L, \tag{40}$$

where W_{lf} is the weight of the outflow liquid processed by the low-pass filter, ω_{lf} is the cut-off frequency of the low-pass filter, ρ kg/m^3 is the liquid density, and q_{lf} m^3/s is the estimated flow rate. In this study, the cut-off frequency is given as ω_{lf} = 1.5 rad/s for noise reduction. In the experimental implementation, the method described in Equations (39) and (40) could be represented by the discrete time equations as follows:

$$W_{lfn+1} = (1 - \omega_{lf}T_s)W_{lfn} + \omega_{lf}T_sW_{Ln}, \tag{41}$$

$$q_{lfn} = -\frac{\omega_{lf}}{\rho}W_{lfn} + \frac{\omega_{lf}}{\rho}W_{Ln}, \tag{42}$$

where T_s s is the sampling interval. In this study, it is given as $T_s = 0.020$ s.

6.2. Flow-Rate Estimation Using a Visible Camera

To estimate the flow rate by attaching different sensor to the load cell, we conducted the flow-rate estimation using a visible camera. In this approach, the side area of liquid in the ladle was measured using a visible camera. As presented in Figure 14, the shaded parts refer to the measurement areas, and it was assumed that these areas were on the same plane for simplifying measurement.

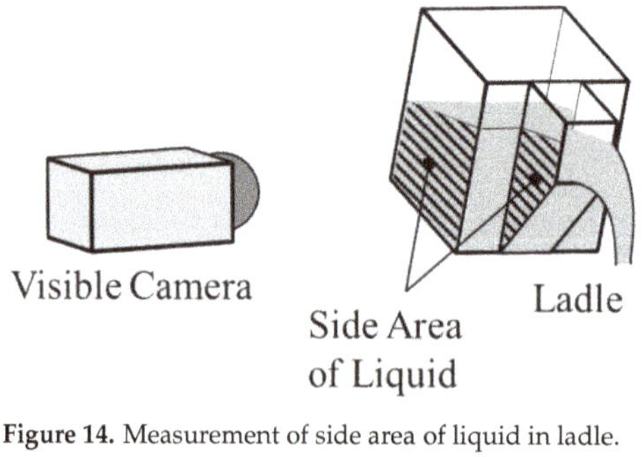

Figure 14. Measurement of side area of liquid in ladle.

Figure 15 presented the parameters of the ladle for obtaining the liquid volume in the ladle. In Figure 15, D_1 m and D_2 m denote the depth of the liquid in the ladle, and A_1 m^2 and A_2 m^2 are the side areas of the liquid in the ladle.

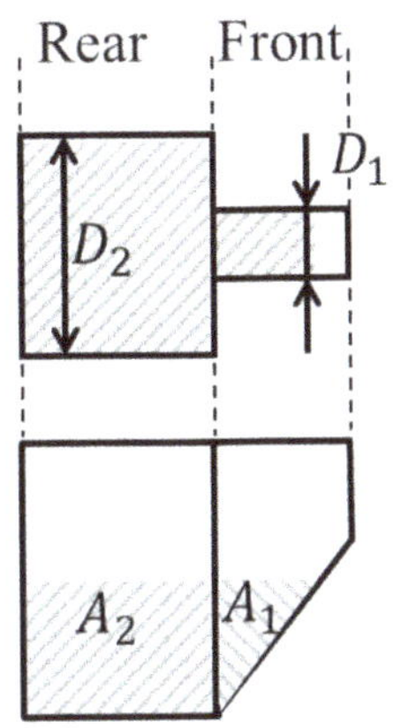

Figure 15. Parameters of ladle for obtaining liquid volume by a visible camera.

To estimate the pouring flow rate, the volume V_c m^3 of the liquid in the ladle was calculated as follows:

$$V_c(\theta) \quad = \quad V_1(\theta) + V_2(\theta), \tag{43}$$

where V_1 m^3 and V_2 m^3 are the volumes of the liquid in the front and rear parts of the ladle, respectively, as follows:

$$V_1(\theta) \quad = \quad A_1(\theta)D_1, \;\; V_2(\theta) = A_2(\theta)D_2.$$

The side areas A_1 m^2 and A_2 m^2 are varied with the tilting angle θ deg of the ladle. The depths D_1 m and D_2 m are constant without regard to the tilting angle of the ladle.

Then, the pouring flow rate q_c m^3/s could be estimated by differentiating volume V_c with respect to time and can be denoted as follows:

$$q_c \quad = \quad \frac{dV_c(\theta(t))}{dt}. \tag{44}$$

In the experimental implementation, the differential form described in Equation (44) could be represented by the backward difference as follows:

$$q_{cn} \quad = \quad \frac{V_{cn}(\theta_n) - V_{cn-1}(\theta_{n-1})}{T_c}, \tag{45}$$

where T_c s is the sampling interval of the visible camera and is given as 0.06 s in this study.

This approach is difficult to apply in the practical pouring process in the casting industry because the ladle generally consists of gypsum and metal, which means that the liquid volume in the ladle cannot be measured by the visible camera. However, our purpose in this study is to verify the appropriateness of the flow-rate estimation method by DKFs, which is able to be applied in the practical automatic pouring machine. Therefore, in this study, since we used water as the target liquid and a clear acrylic container as a ladle, we could apply this flow-rate estimation approach using the visible camera to the automatic pouring machine.

7. Experimental Verifications

The flow-rate estimation method by DKFs was applied to the laboratory-type automatic pouring machine (see Figure 16), and the appropriateness of this flow-rate estimation approach was verified via the pouring experiments described in the current section.

The conditions of the experiments were the same as for the simulations noted in Section 5.2.

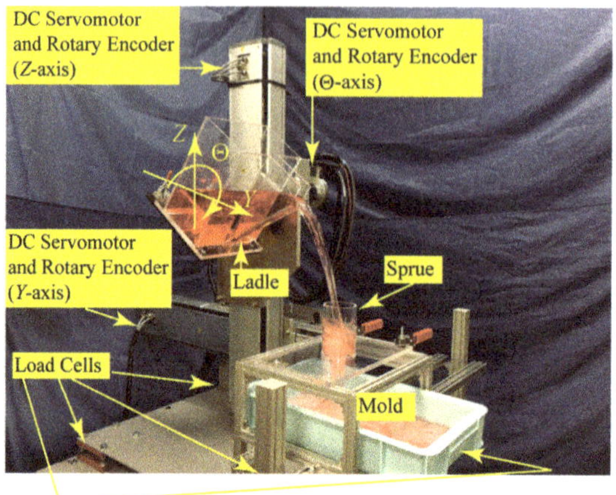

Figure 16. Laboratory-type automatic pouring machine.

The flow-rate feedforward controller was also applied to control the flow rate based on the reference pattern (see Figure 10). In the first experiment, the flow-rate estimation was performed using an ideal condition, with no error between the ideal and the actual tilting angle at the start of the liquid outflow. The experimental results are presented in Figure 17, where (a) is the input command applied to the motor for tilting the ladle; (b) and (c) indicate the ladle's angular velocity and angle, respectively; and (d) and (e) indicate the liquid height in the pouring mouth of the ladle and the pouring flow rate, respectively. In Figure 17d,e, the dotted magenta lines indicate the reference patterns, and the dashed cyan lines indicate the simulation results obtained from the pouring process model by applying the results of (b) and (c). The chained green lines show the estimated results gained by using the EKF in the flow-rate estimation via DKFs. Figure 17f presents the weight of the outflow liquid. The black solid line is the weight measured by the load cell, and the remaining lines are shown in the same manner as (d) and (e). The state variables in the pouring motion were estimated in real time, as presented in Figure 17. To verify the appropriateness of the flow-rate estimation by DKFs, we applied the two other types of flow-rate estimation methods described in Section 6. Moreover, to verify the validity of the system parameters in DKFs, we also applied DKFs with the different covariance of process noise Q_f to Equation (37). The covariance of the process noise was given as follows:

$$Q_f = \mathrm{diag}(250\ \mathrm{m}^2, 1.0\ \mathrm{kg}^2, 1.0\ \mathrm{kg}^2) \times 10^{-11}. \tag{46}$$

The process noise in the liquid height on the pouring mouth was increased as compared with Equation (37). In Figure 18a–d, the black solid lines are the simulation result obtained from the pouring process model, and these results were the same as the dashed line in Figure 17e. The solid green lines in Figure 18a–d indicate the results of flow-rate estimations using the EKF with the covariance of the process noise in Equation (37) that in Equation (46), the derivative of the load cell data, and the visible camera, respectively. As presented in Figure 18a,b, the estimated results were similar to the simulated result. The result in Figure 18c shows that the flow rate exhibited a higher level of noise compared with the result estimated by the EKF. Additionally, the response was delayed by the low-pass filter applied for noise reduction. The flow rate of the outflow liquid from the ladle does not have a negative value (i.e., $q \geq 0$ m^3/s). However, it was confirmed that the flow rate during the back-tilting motion of the ladle from 16 s to 17 s indicates the negative value as shown in Figure 18c. This response was caused by the fact that the load cell data was influenced by the movement of the ladle on the *z*-axis as shown in Figure 6.

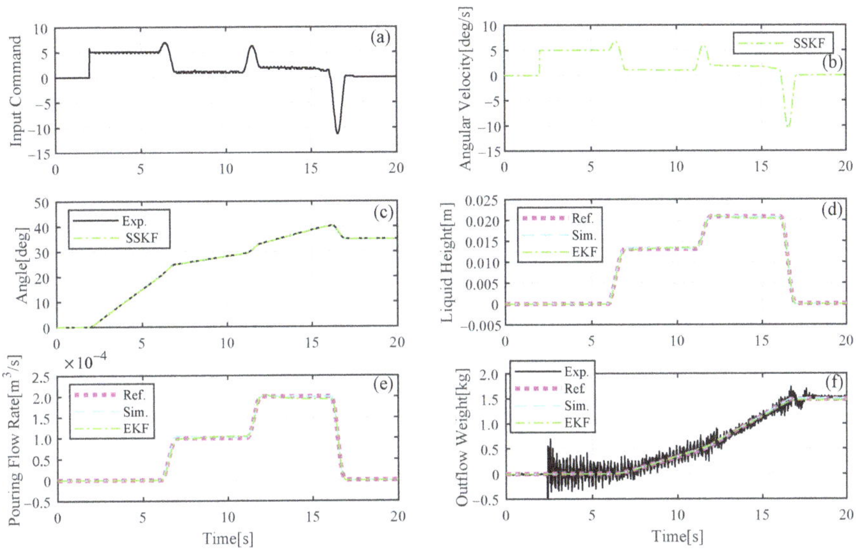

Figure 17. Experimental results using automatic pouring machine (tilting angle error at the start of liquid outflow: +0 deg). (**a**) Input command to motor for tilting ladle; (**b**) Angular velocity of tilting ladle; (**c**) Angle of tilting ladle; (**d**) Liquid height on pouring mouth; (**e**) Flow rate of outflow liquid from ladle; (**f**) Weight of outflow liquid from ladle.

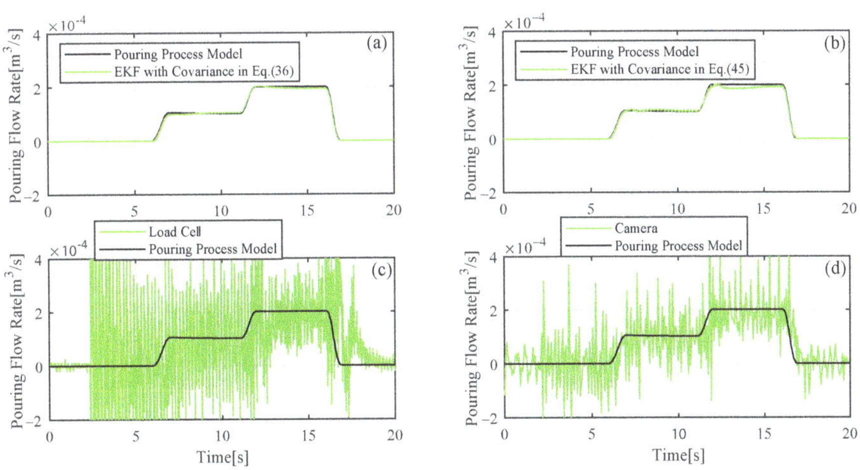

Figure 18. Comparison of flow-rate estimations in experiments with tilting angle error at the start of liquid outflow: +0 deg. (**a**) Estimated flow rate by EKF with covariance as shown in Equation (37); (**b**) Estimated flow rate by EKF with covariance as shown in Equation (46); (**c**) Estimated flow rate by differentiating weight of outflow liquid measured by load cell; (**d**) Estimated flow rate by visible camera.

The noise in the flow rate estimated by the visible camera (as presented in Figure 18d) was smaller than that estimated by differentiating the load cell data (see Figure 18c). Hence, we confirmed that the flow rate could be estimated roughly using the visible camera.

The real-time flow-rate estimation is required to process within 0.02 s for realizing the high-precision real-time flow rate control [32]. The processing times in the flow-rate estimations using DKFs and the derivative of the load cell data were within 0.02 s. The processing time in the flow-rate estimation using the visible camera was 0.06 s by taking the time for the frame rate of the camera and the image processing.

Figure 19 presents the experimental results for the pouring conditions alongside the tilting angle error at the start of the liquid outflow which was +3 deg. Figure 20 presents the estimated results of the flow rate for comparing the three types of flow-rate estimation, as discussed in previous sections.

These figures are shown in the same manner as in Figures 17 and 18. In Figure 20d, the trends of flow-rate estimation by the visible camera are similar to the simulation result. On the other hand, errors between the DKFs estimation results and the simulation result occurred at the start of pouring (see Figure 20a,b). However, these errors can be potentially reduced by updating the KF process. As shown in Figure 20a, the flow-rate after 11 s can be estimated precisely by DKFs with the covariance of the process noise in Equation (37). Moreover, the estimation result of the flow rate by DKFs with the covariance of the process noise in Equation (46) can track the simulation result faster than that in Equation (37). As presented in Figure 20, the noise in the flow rate estimated by DKFs was the smallest in the estimation methods performed in this study.

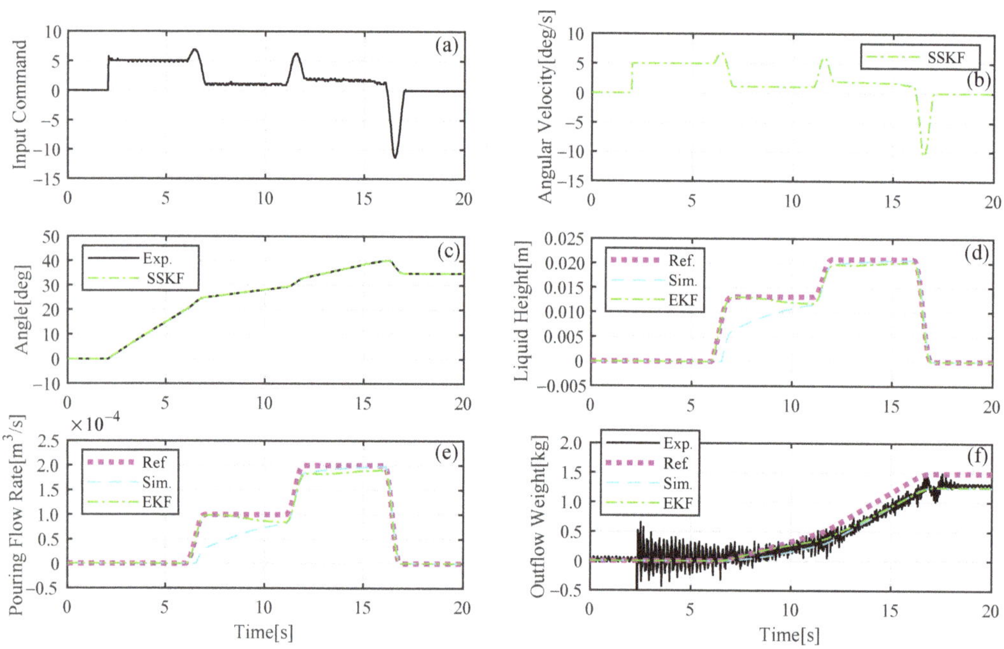

Figure 19. Experimental results using automatic pouring machine (tilting angle error at the start of liquid outflow: +3 deg). (**a**) Input command to motor for tilting ladle; (**b**) Angular velocity of tilting ladle; (**c**) Angle of tilting ladle; (**d**) Liquid height on pouring mouth; (**e**) Flow rate of outflow liquid from ladle; (**f**) Weight of outflow liquid from ladle.

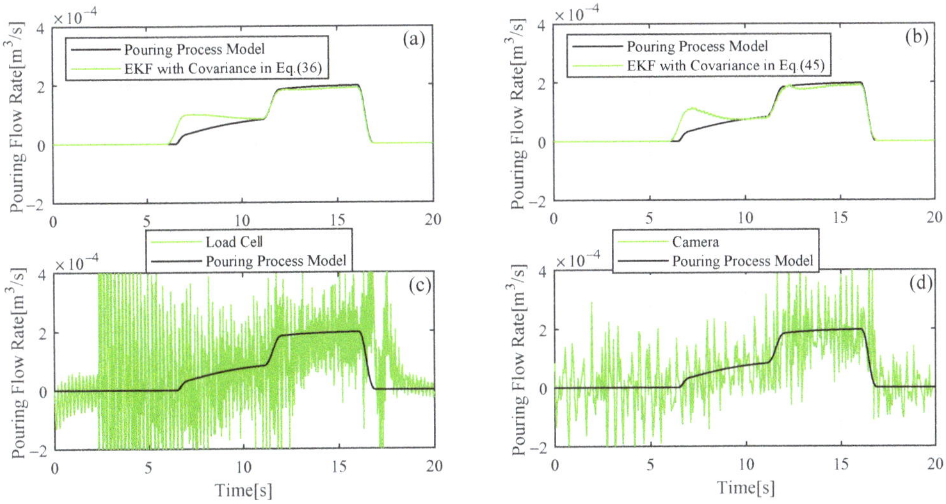

Figure 20. Comparison of flow-rate estimations in experiments with tilting angle error at the start of liquid outflow: +3 deg. (**a**) Estimated flow rate by EKF with covariance as shown in Equation (37); (**b**) Estimated flow rate by EKF with covariance as shown in Equation (46); (**c**) Estimated flow rate by differentiating weight of outflow liquid measured by load cell; (**d**) Estimated flow rate by visible camera.

We compared quantitatively the accuracies of each flow-rate estimation. Since the simulation results of the flow rate were generated by the pouring process model with the experimental conditions, we consider that the simulation results of the flow rate represented the actual flow rate in the experiments faithfully. The accuracies of the flow rate estimations were evaluated by the integral absolute error (IAE) of the estimated and the simulated flow rates as follows:

$$IAE = \sum_{i=0}^{N} |q_{esti} - q_{simi}|\Delta T, \tag{47}$$

where i is the sampling number and N is the total number of the sampling in the estimation. ΔT is the sampling interval of the flow rate estimations. q_{est} and q_{sim} are the estimated and the simulated flow rates shown in Figures 18 and 20. Table 1 presents the results of IAE. In Table 1, the IAEs in the ideal condition with no error between the ideal and the actual tilting angles at the start of the liquid outflow and the condition with +3 deg error between the ideal and the actual tilting angle at the start of the liquid outflow were shown. Furthermore, we also compared quantitatively the amounts of noises in the estimated flow rates. The amounts of the noises were evaluated by the total variation (TV) [40] of the estimated flow rates as follows:

$$TV = \sum_{i=1}^{N} |q_{esti+1} - q_{esti}|, \tag{48}$$

Table 2 presents the results of TV. This table is shown in the same manner as in Table 1. However, TV as shown in Equation (48) includes not only the amounts of the noises but also the variation of the set-point changes in the flow rate. The set points in the flow rate are changed monotonically as shown in Figure 10. Therefore, we used the modified TV (mTV) [41] for eliminating the variation of the set-point changes from TV as

$$mTV = \sum_{i=1}^{N} |q_{esti+1} - q_{esti}| - |2q_{simm} + q_{simN} - q_{sim0}|, \tag{49}$$

where q_{simm} is the extreme point of the set-point changes which has the shape of one pulse. q_{simN} and q_{sim0} are the final and initial values of the set-points in the flow rate, respectively. These parameters were obtained from the simulation result of the flow rate which does not have the noises. In this study, $q_{simm} = 2.0 \times 10^{-4}$ and $q_{sim0} = q_{simN} = 0.0$ were given. Table 3 presents the results of mTV. As seen from the comparison with TV and mTV, the evaluation values by mTVs are smaller than those by TVs because mTVs can suppress the influence of the set-points changes. Therefore, we can compare clearly the amounts of the noises by mTVs.

In the evaluations by IAEs to the condition with +3 deg error between the ideal and the actual tilting angle at the start of the liquid outflow, IAE of DKFs with the covariance of process noise in Equation (46) was slightly smaller than that in Equation (37). Thus, the estimation accuracy of the flow rate by DKFs can be improved by increasing the covariance of process noise in the liquid height on the pouring mouth. However, the amount of noise in the estimated flow rate was increased by increasing the covariance of process noise in the liquid height on the pouring mouth as shown in the evaluation by mTVs. In the application of DKFs in the flow-rate control, the covariance of process noise in the liquid height on the pouring mouth should be increased while evaluating the noise in the estimated flow rate.

Both the IAE and the TV of DKFs were the smallest in the estimation methods performed in this study because the noise in the estimated flow rate by DKFs could be suppressed. However, IAE of DKFs was increased with increasing the error between the ideal and the actual tilting angle at the start of the liquid outflow. Therefore, we consider the present flow-rate estimation by DKFs to be appropriate for estimating the flow rate in a pouring motion with small disturbances.

Table 1. Integral absolute errors (IAEs) between estimated flow rates and simulated flow rates.

Estimation Method	Tilting Angle Error at Start to Liquid Outflow +0 deg	Tilting Angle Error at Start to Liquid Outflow +3 deg
DKFs with Covariance of Process Noise in Equation (37)	4.277×10^{-5}	2.005×10^{-4}
DKFs with Covariance of Process Noise in Equation (46)	6.236×10^{-5}	1.909×10^{-4}
Differentiating Load Cell Data	2.431×10^{-3}	2.518×10^{-3}
Visible Camera	1.420×10^{-3}	1.568×10^{-3}

Table 2. Total variations (TVs) of estimated flow rates.

Estimation Method	Tilting Angle Error at Start to Liquid Outflow +0 deg	Tilting Angle Error at Start to Liquid Outflow +3 deg
DKFs with Covariance of Process Noise in Equation (37)	4.855×10^{-4}	4.746×10^{-4}
DKFs with Covariance of Process Noise in Equation (46)	8.628×10^{-4}	8.357×10^{-4}
Differentiating Load Cell Data	1.443×10^{-1}	1.486×10^{-1}
Visible Camera	3.861×10^{-2}	4.006×10^{-2}

Table 3. Modified total variations (mTVs) of estimated flow rates.

Estimation Method	Tilting Angle Error at Start to Liquid Outflow +0 deg	Tilting Angle Error at Start to Liquid Outflow +3 deg
DKFs with Covariance of Process Noise in Equation (37)	8.549×10^{-5}	7.460×10^{-5}
DKFs with Covariance of Process Noise in Equation (46)	4.628×10^{-4}	4.357×10^{-4}
Differentiating Load Cell Data	1.439×10^{-1}	1.482×10^{-1}
Visible Camera	3.821×10^{-2}	3.966×10^{-2}

8. Conclusions

In this study, we verified the appropriateness of a real-time flow-rate estimation method using the DKFs in a tilting-ladle-type automatic pouring machine. The design of this flow-rate estimation method was described in detail. This method was implemented using a laboratory-type automatic pouring machine and experimentally compared with two other flow-rate estimation methods. In the estimation using the derivative of load cell data, we were unable to recognize the flow rate due to large amounts of noise. In the estimation using a visible camera, we were barely able to recognize the flow rate. The noise in the flow rate estimated by DKFs was smallest among the approaches performed in this study. However, it was confirmed that the estimated precision of this method may be degraded by disturbances, such as uncertainty about model parameters. Although the estimated precision can be improved with increasing the covariance of process noise in the liquid height on the pouring mouth in DKFs, the amount of noise in the estimated flow rate also is increased. Accordingly, we consider flow-rate estimation by DKFs to be appropriate for estimating flow rate in a pouring motion with small disturbances. In the practical applications, it might be difficult to maintain the pouring condition with small disturbances because of varying temperature of the molten metal in the ladle and attaching the slag to the ladle. Therefore, we plan to integrate the online model parameters identification to the flow rate estimation system.

In this study, the sampling interval in the load cell data acquisition was limited to 0.02 s by the specification of the load cell amplifier with the serial communication. The noise in the load cell data might be reduced by the amplifier with a fast sampling rate and the higher order filter. In future work, we will reconstruct the sensing system in the automatic pouring machine for improving the accuracy of the flow rate estimation. Furthermore, we will try to develop the high-precision and robust state estimation system of the automatic pouring machine using the sensor fusion approach such as the integration of the load cell and visible camera.

Author Contributions: Conceptualization, Y.N.; Methodology, Y.N. and Y.S.; Software, Y.N. and Y.S.; Validation, Y.S.; Formal analysis, Y.S.; Investigation, Y.S.; Resources, Y.N.; Data curation, Y.S.; Writing—original draft preparation, Y.S.; Writing—review and editing, Y.N.; Visualization, Y.S.; Supervision, Y.N.; Project administration, Y.N.; Funding acquisition, Y.N. Both authors have read and agreed to the published version of the manuscript.

Funding: This research was funded by JSPS KAKENHI Grant No. JP25420181.

Institutional Review Board Statement: Not applicable.

Informed Consent Statement: Not applicable.

Conflicts of Interest: The authors declare no conflict of interest.

Abbreviations

The following abbreviations are used in this manuscript:

CFD	Computational Fluid Dynamics
SSKF	Steady State Kalman Filter
EKF	Extended Kalman Filter
DKFs	Decentralization of Kalman Filters
DC	Direct Current
IAE	Integral Absolute Error
TV	Total Variation
mTV	Modified Total Variation

References

1. Lindsay, W. Automatic pouring and metal distribution system. *Foundry Trade J.* **1983**, *10*, 151–176.
2. Terashima, K. Recent automatic pouring and transfer systems in foundries. *Mater. Process Technol.* **1998**, *39*, 1–8. (In Japanese)
3. Neumann, E.; Trauzeddel, D. Pouring systems for ferrous applications. *Foundry Trade J.* **2002**, *176*, 23–24.

4. Tiwari, P.; Kothawale, R.; Tadav, P. Semi automatic pouring system. *Int. Res. J. Eng. Technol.* **2017**, *4*, 2834–2837.
5. Li, L.; Wang, C.; Wu, H. Research on kinematics and pouring law of a mobile heavy load pouring robot. *Math. Probl. Eng.* **2018**, *2018*, 1–13. [CrossRef]
6. Watanabe, J.; Yoshida, K. Automatic pouring equipment for casting-Mel pore system. *Ind. Heat.* **1992**, *29*, 19–27. (In Japanese)
7. Birkhold, M.; Friedrich, C.; Lechler, A. Automation of the casting process using a model-based NC architecture. *IFAC-PapersOnLine* **2015**, *48*, 195–200. [CrossRef]
8. Yajima, T.; Noda, Y. Teaching-and-playback approach based on pouring flow rate in tilting-ladle-type pouring machine. In Proceedings of the 73rd World Foundry Congress, Cracow, Poland, 23–27 September 2018; pp. 137–138.
9. Voss, T. Optimization and control of modern ladle pouring process. In Proceedings of the 73rd World Foundry Congress, Cracow, Poland, 23–27 September 2018; pp. 497–498.
10. Yano, K.; Terashima, K. Supervisory control of automatic pouring machine. *Control Eng. Pract.* **2010**, *18*, 230–241. [CrossRef]
11. Iwamoto, K.; Tokunaga, H.; Okane, T. Training support for pouring task in casting process using stereoscopic video see-through display. In Proceedings of the 2014 1st International Conference on Information Technology, Computer and Electrical Engineering, Semarang, Indonesia, 7–8 November 2014; pp. 497–498.
12. Terashima, K.; Yano, K.; Sugimoto, Y.; Watanabe, M. Position control of ladle tip and sloshing suppression during tilting motion in automatic pouring machine. In Proceedings of the IFAC Symposium on Automation in Mining, Mineral and Metal Processing, Tokyo, Japan, 4–6 September 2001; pp. 229–234.
13. Ito, A.; Tasaki, R.; Suzuki, M.; Terashima, K. Analysis and control of pouring ladle with weir for sloshing an volume-moving vibration in pouring cut-off process. *J. Autom. Technol.* **2017**, *11*, 645–656. [CrossRef]
14. Tzamtzi, P.M.; Koumboulis, N.F. Sloshing control in tilting phases of the pouring process. *Int. Sch. Sci. Res. Innov.* **2007**, *10*, 3035–3042.
15. Tzamtzi, P.M.; Koumboulis, N.F.; Skarpetis, G.M. On the Modelling and controller design for the outpouring phase of the pouring process. *WSEAS Trans. Syst. Control* **2009**, *4*, 11–20.
16. Moriello, L.; Biagiotti, L.; Melchiorri, C.; Paoli, A. Control of liquid handling robotic systems: A feed-forward approach to suppress sloshing. In Proceedings of the 2017 IEEE International Conference on Robotics and Automation, Singapore, 29 May–3 June 2017. [CrossRef]
17. Moriello, L.; Biagiotti, L.; Melchiorri, C.; Paoli, A. Manipulating liquids with robots: A sloshing-free solution. *Control Eng. Pract.* **2018**, *78*, 129–141. [CrossRef]
18. Fukushima, R.; Noda, Y.; Terashima, K. Falling position and anti-sloshing control of liquid using automatic pouring robot. In Proceedings of the International PhD Foundry Conference, Brno, Czech Republic, 1–3 June 2009; pp. 1–10.
19. Liang, H.; Li, S.; Ma, X.; Hendrich, N.; Gerkmann, T.; Zhang, J. Making sense of audio vibration for liquid height estimation in robotic pouring. In Proceedings of the 2019 IEEE/RSJ International Conference on Intelligent Robots and Systems, Macau, China, 3–8 November 2019. [CrossRef]
20. Rozo, L.; Jimenez, P.; Torras, C. Force-based robot learning of pouring skills using parametric hidden Markov models. In Proceedings of the 9th International Workshop on Robot Motion and Control, Kuskin, Poland, 3–5 July 2013. [CrossRef]
21. Kuriyama, Y.; Yano, K.; Nishido, S. Optimization of pouring velocity for aluminium gravity casting. In *Fluid Dynamics, Computational Modeling and Applications*; IntechOpen: London, UK, 2012; pp. 575–588.
22. Noda, Y.; Terashima, K. Modeling and feedforward flow rate control of automatic pouring system with real ladle. *J. Robot. Mechatronics* **2007**, *19*, 205–211. [CrossRef]
23. Noda, Y.; Fukushima, R.; Terashima, K. Monitoring and control system to falling position of outflow liquid in automatic pouring robot. In Proceedings of the IFAC 13th Symposium on Automation in Mining, Mineral and Metal Processing, Cape Town, South Africa, 2–4 August 2010; pp. 13–18. [CrossRef]
24. Ito, A.; Noda, Y.; Tasaki, R.; Terashima, K. Outflow liquid falling position control considering lower pouring mouth position with collision avoidance for tilting-type automatic pouring machine. *Mater. Trans.* **2017**, *58*, 485–493. [CrossRef]
25. Sueki, Y.; Noda, Y.; Suzuki, M.; Ohta, K. Optimization of ladle position to minimize spilled liquid from pouring basin in automatic pouring machine. *J. Jpn. Foundry Eng. Soc. Jpn.* **2018**, *90*, 107–117. (In Japanese)
26. Sueki, Y.; Noda, Y. Optimal positioning of ladle in automatic pouring machine in consideration of pouring liquid accurately into sprue. In Proceedings of the IEEE International Conference on Advanced Intelligent Mechatronics, Munich, ND, USA, 3–7 July 2017; pp. 897–903.
27. Noda, Y.; Terashima, K. State estimation of automatic pouring system with load cell in casting process. In Proceedings of the SICE Annual Conference 2008, Tokyo, Japan, 20–22 August 2008; pp. 1548–1551.
28. Noda, Y.; Terashima, K. Estimation of flow rate in automatic pouring system with real ladle used in industry. In Proceedings of the ICROS-SICE International Conference 2009, Fukuoka, Japan, 18–21 August 2009; pp. 354–359.
29. Noda, Y.; Birkhold, M.; Terashima, K.; Sawodny, O.; Verl, A. Flow rate estimation using unscented Kalman filter in automatic pouring machine. In Proceedings of the SICE Annual Conference 2011, Tokyo, Japan, 13–18 September 2011; pp. 769–774.
30. Noda, Y.; Zeitz, M.; Sawodny, O.; Terashima, K. Flow rate control based on differential flatness in automatic pouring robot. In Proceedings of the 2011 IEEE International Conference on Control Applications, Denver, CO, USA, 28–30 September 2011; pp. 1468–1475.

31. Noda, Y.; Terashima, K. Simplified flow rate estimation by decentralization of Kalman filters in automatic pouring robot. In Proceedings of the SICE Annual Conference 2012, Akita, Japan, 20–23 August 2012; pp. 1465–1470.
32. Noda, Y.; Sueki, Y. Implementation and experimental verification of flow rate control based on differential flatness in a tilting-ladle-type automatic pouring machine. *Appl. Sci.* **2019**, *9*, 1978. [CrossRef]
33. Shinohara, K.; Morimoto, H. Development of automatic pouring equipment. *J. Soc. Automot. Eng. Jpn.* **1992**, *46*, 79–85. (In Japanese)
34. Yamaguchi, A.; Atkeson, C. Stereo vision of liquid and particle flow for robot pouring. In Proceedings of the 2016 IEEE-RAS 16th International Conference on Humanoid Robots, Cancun, Mexico, 15–17 November 2016; pp. 1173–1180.
35. Schenck, C.; Fox, D. Visual closed-loop control for pouring liquids. In Proceedings of the 2017 IEEE International Conference on Robotics and Automation, Singapore, 29 May–3 July 2017. [CrossRef]
36. Do, C.; Burgard, W. Accurate Pouring with an Autonomous robot using an RGB-D camera. *Intell. Auton. Syst.* **2019**, *15*, 210–221. [CrossRef]
37. Kennedy, M., III; Queen, K.; Thakur, D.; Kumar, V. Precise dispensing of liquids using visual feedback. In Proceedings of the 2017 IEEE/RSJ International Conference on Intelligent Robots and Systems, Vancouver, BC, Canada, 24–28 July 2017. [CrossRef]
38. Suzuki, M.; Yang, Y.; Noda, Y.; Terashima, K. Sequence control of automatic pouring system. In Proceedings of the 10th Asian Foundry Congress, Aichi, Japan, 21–24 May 2008; pp. 88–92.
39. Kabasawa, N.; Noda, Y. Model-based flow rate control with online model parameters identification in automatic pouring machine. *Robotics* **2021**, *10*, 39. [CrossRef]
40. Skogestad, S. Simple Analytic Rules for Model Reduction and PID Controller Tuning. *J. Process Control* **2003**, *13*, 291–309. [CrossRef]
41. Huba, M.; Vrancic, D. Extending the Model-Based Controller Design to Higher-Order Plant Models and Measurement Noise. *Symmetry* **2021**, *13*, 798. [CrossRef]

applied sciences

MDPI

Article

State Machine-Based Hybrid Position/Force Control Architecture for a Waste Management Mobile Robot with 5DOF Manipulator

Ionel-Alexandru Gal [1,*], Alexandra-Cătălina Ciocîrlan [1] and Mihai Mărgăritescu [2]

1 Institute of Solid Mechanics of Romanian Academy, 010141 Bucharest, Romania; alexandra.ciocirlan@imsar.ro
2 National Institute of Research and Development for Mechatronics and Measurement Technique—INCDMTM, 022404 Bucharest, Romania; mihai.margaritescu@gmail.com
* Correspondence: alexandru.gal@imsar.ro; Tel.: +40-741-575-553

Featured Application: Robots with specific joint motors, with high potential for use in Industry 4.0 automated assembly lines.

Abstract: When robots are built with state-driven motors, task-planning increases in complexity and difficulty. This type of actuator is difficult to control, because each type of control position/force requires different motor parameters. To solve this problem, we propose a state machine-driven hybrid position/force control architecture (SmHPFC). To achieve this, we take the classic hybrid position/force control method, while using only PID regulators, and add a state machine on top of it. In this way, the regulators will not help the control architecture, but the architecture will help the entire control system. The architecture acts both as a parameter update process and as a switching mechanism for the joints' decision S-matrix. The obtained control architecture was then applied to a 5DOF serial manipulator built with Festo motors. Using SmHPFC, the robot was then able to operate with position or force control depending on its designated task. Without the proposed architecture, the robot joint parameters would have to be updated using a more rigid approach; each time a new task begins with new parameters, the control type would have to be changed. Using the SmHPFC, the robot reference generation and task complexity is reduced to a much simpler one.

Keywords: hybrid control; state machine; Festo; PLC; friction force

Citation: Gal, I.-A.; Ciocîrlan, A.-C.; Mărgăritescu, M. State Machine-Based Hybrid Position/Force Control Architecture for a Waste Management Mobile Robot with 5DOF Manipulator. *Appl. Sci.* **2021**, *11*, 4222. https://doi.org/10.3390/app11094222

Academic Editor: Alessandro Gasparetto

Received: 7 April 2021
Accepted: 3 May 2021
Published: 6 May 2021

1. Introduction

Waste management is a growing concern and problem [1] around the world and especially in the European Union. To help to solve this problem, rules, policies, and guidelines were presented to the public [2], as ways of lowering the amount of recyclable material going to waste dumps [3]. Communities are trying to recover more and more recyclable materials from day-to-day garbage, even when people are not using selective waste collection methods. For this, waste recycling companies usually hire people [4] to do the tedious work of selective recycling for different types of objects and materials (plastic and glass bottles, metal cans, etc.).

In order to aid companies when the workforce is scarce or to help employees when the work conditions are bad, automated processes have been developed to try and sort the waste [5] brought by garbage trucks. To this end, Diya et al. [6] proposed an intelligent system to help communities aim towards a greener environment. This goal can also be achieved through conventional selective methods using water for paper and plastic or magnets for ferrite-containing waste, while the remaining waste was compacted and incinerated or deposited in landfills. The other methods were to use video cameras on conveyors as Kokoulin et al. [7] proposed, or to use artificial intelligence as Sousa et al. [8] showed, to detect recyclable materials. Through different methods, using an intelligent city platform created by Popa et al. [9], or by using a distributed architecture and machine learning built by Ziouzios et al. [10], recyclable waste can be moved to sorted containers, or even by combining both methods to maximize the selection efficiency. Both methods have their advantages and disadvantages [4,11], while using big and costly machinery.

An alternative to the automated conveyors and manual process of waste selective recycling is to use mobile robots that move inside a warehouse where the garbage is laid on the floor. While several robots conduct the search and recycle mission, stopping only to recharge their batteries, the process can continue practically non-stop without the need for human intervention, only requiring humans to replace the search material. This type of selective waste collection can reduce the need for human workers, therefore preventing the risks that arise from working with garbage and decomposing organic materials.

The proposed mobile robot has a mechanical arm/manipulator to grab the desired material and a video camera to detect recyclable objects. The serial mechanical manipulator has four positioning and orientation degrees of freedom and one degree of freedom for the gripper, resulting in a 5DOF mechanical arm for picking recyclable objects and placing them in a waste selective storage container.

The video camera detection method [7,8,12] is reliable and provides the position and orientation of the object with small detection errors that increase the sorting capability of recyclable material. Once the coordinates or the object is known, these are fed as a reference to the control algorithm. Having a certain robot architecture [13] and using specific Festo hardware [14], we created a hybrid position/force control method to achieve the best performance for this application. The main reason behind using the hybrid position/force control is to avoid the inverse matrix singularities of direct kinematics equations required when computing the end-effector reference values.

Starting with Raibert and Craig [15], the forefathers of the hybrid position/force control, the method has been developed [16] and used [17] in recent years to improve the interaction between robots and the environment [18], which is especially useful in our application. In 2019, Wang et al. [18] presented a hybrid position/force control used on a manipulator for close wall inspection, where the robot–environment interaction is treated as a mission consequence, and in 2020, Liu et al. [19] developed and used a hybrid position/force-controlled robot to drill and rivet metal plates with high precision ($\pm$0.1 mm).

The hybrid position/force control (HPFC) methods [20] can and have been used in multiple areas of robot environments [21], combining them with new mathematical concepts [22] or fuzzy inference systems [23]. In 2011, Vladareanu and Smarandache [22] used neutrosophic logic to improve hybrid position/force control. In 2019, Ballesteros et al. [24] designed a second-order control feedback controller using the sliding mode control (SMC) [25] for an active orthotic mechanism, by using the system's dynamics and controlling each joint through a combined force and position reference, while gathering force sensor information. This approach uses the highly complex solution of designing a dynamic regulator while maintaining stability in a Lyapunov sense, while Rani et al. [26] developed an HPFC for task-coordinated manipulators using non-modeled system dynamics. Even if the dynamic equations were not used within the control diagram, they still ensured overall stability. Using the same approach, Peng et al. [27] created a hybrid control while using joint neural network observations on uncertain parameters and external perturbations which also uses dynamic equations and the Lyapunov stability to track and follow a path reference. Others have developed parallel manipulators with redundant actuation [28], some that learn the dynamics uncertainties to ensure stability [29], and even some hybrid control methods that use a fuzzy-neural network for multiple robot control [30].

When considering the type of regulator or control method to use, we took into account the sliding mode control (SMC) [25]. One reason was the experience the authors have in developing such control methods [25], and the other was that it has very good precision in compensating robot dynamics and external disturbances. As Zhang et al. [31],

Solanes et al. [32], and Ravandi et al. [33] have proven, the sliding mode control method can be combined with other methods to create different control algorithms depending on the required application. This approach may provide high accuracy where not needed, but it adds complexity to the mathematical model used in designing and developing the final control system. Because of this, the approach used in this paper was to avoid the high complexity of the model and focus on the application results, keeping the accuracy within the required parameters.

Different research papers have developed HPFC methods for specific applications when dynamic precision was required. Gao et al. [34] developed a 6DOF hydraulic manipulator that uses force and vision for positioning, while Han et al. [35] created a manipulator with a highly compliant end-effector. Some research handles the classic approach [18,19], while others combine the usefulness of the HPFC [20] with advantages of other methods [24–30] to increase stability and precision, to prevent perturbation or to compensate for uncertainties. These approaches, as reviewed in scientific papers [16,36,37], are necessary when the robot environment interaction requires high precision in positioning tasks or in order to combine position and force to safely interact with human patients or operators. Most of the published HPFC methods use a static approach in defining the joint matrix that separates the position-controlled joints from the force-controlled joints.

One way of preventing the static use of the S-matrix is to create a state feedback decision as Pasolli et al. [38] proposed. Their approach was to switch between position and force control on certain robot joints when the system required it, thus improving upon the classic hybrid position/force control of Raibert and Craig [15] by adding a decision layer to constantly change the position and force diagonal matrix S when certain events occur. Building on this method, we have proposed the hybrid position/force control combined with a state machine which will be called state machine-based hybrid position/force control (SmHPFC). This new method will update the parameters of the S-matrix as the need arises and depending on the task state of selective waste recycling.

The proposed method architecture was combined with a stick–slip analysis of the gripper's fingers to use a low force for gripping the detected recyclable waste objects, such as plastic bottles that can deform on a higher gripping force, or glass bottles that may slip with weaker force or break with a higher one. All of these were then implemented using Festo motors, Festo motor controllers, and Festo central control unit PLC.

2. System Description

The waste sorting robot is presented in Figure 1 [39,40] and Figure 2 [40,41], and is built from several separate systems that will work as a whole. The robot is the main objective of a national research grant [40] which aims to build a working prototype of the municipal waste sorting robot. Figure 1 presents one of several proposed 3D concepts, while Figure 2 presents the 3D design of the robot that has the following main components: (1) chassis, (2) housing, (3) drive wheels, (4) free wheels, (5) XYZ handling system, (6) gripper and (7) waste container [35]. The robot structure is already defined and published in a previous paper [41] and is not the subject of this paper.

Figure 1. Initial concept of the waste sorting robot system [39,40].

Figure 2. The mobile robot 3D model with the incorporated 5DOF manipulator [40,41].

An important component of the mobile robot is the drive wheels system for navigating through the waste, which was previously laid on the ground as a uniform thin layer. Another component that is not visible in Figure 2 is the vision system, which detects recyclable objects and sends the information to the 5DOF manipulator. The manipulator is the third important component of the mobile waste sorting robot, which grabs the detected recyclable materials and stores them in the attached waste container.

The gripping system is made of a five-degree-of-freedom serial robot, mounted on the mobile chassis and it was chosen as the simple approach in building the municipal waste sorting robots' manipulator. This is intended as a simple and easy solution for replacing the human workforce in searching for scattered recyclable materials such as plastic and glass bottles or aluminum cans, and it lowers the percentage of recyclable materials with a high decomposing time that reach a landfill.

The 5DOFs of the manipulator can be separated into three categories: positioning, orientation, and interaction. The first set, required for positioning the end-effector, is composed of three linear motors. These 3DOFs move the gripper on all three axes (OX, OY and OZ), achieving a positioning system within the robot manipulators' confined workspace. In our robotic system, the horizontal XOY movement is ensured by a Festo planar surface gantry EXCM, while the OZ vertical displacement is given by a Festo electrical slide EGSK. The orientation system uses a one-degree-of-freedom Festo rotary motor ERMO to rotate the end-effector around the vertical axis OZ to align the gripper with the object/bottle. Once the position and orientation are set, the last degree of freedom is the end-effector which, in our case, is a Festo HGPLE gripper that uses a dual linear motor that moves two rigid bodies (the gripper jaws) in two opposite directions on a single axis. The first direction, which is considered to be the positive one, is when the two jaws of the gripper are closing into each other and the second is when both move away from each other in the negative direction.

To control the serial robot for selecting the recyclable materials, we chose the hybrid position/force control that allows us to position and orient the gripper within the workspace, through a position-based control law, and to grip the object using a force-based approach. Moreover, the gripper can be controlled in position for opening the jaws or in force for gripping the objects. The control type of the gripper has to switch in real-time, depending on the state of the robot and the mission objective. This is why a hybrid position/force control law with a state-machine decision system was chosen as the control algorithm that drives the 5DOF manipulator.

3. Decision Algorithm

The decision algorithm is the component that turns the static hybrid position/force control method into a dynamic control by changing the Raibert and Craig [15] static configuration of the S-matrix, and updating its values depending on a deterministic state machine. The diagram of the overall state machine is presented in Figure 3. Here, one can see multiple system states followed by one or more transitions. Table 1 presents the state description and Table 2 presents the description of the transition.

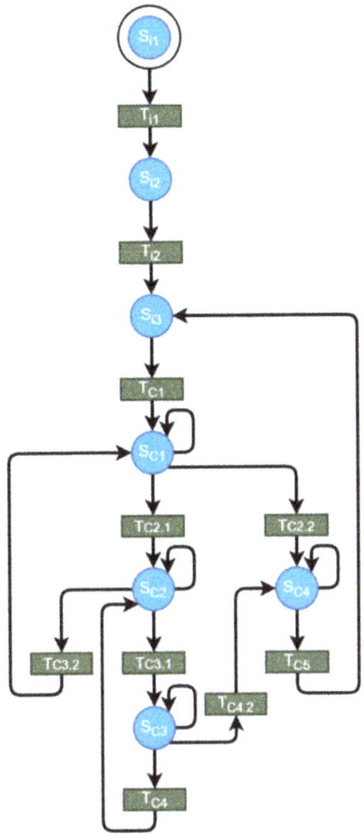

Figure 3. The main 5DOF robot state machine for waste selection.

Table 1. State machine states list and descriptions.

State	Diagram	Description
S_{i1}	Main	The initial state of the entire system.
S_{i2}	Main/Homing	Initialization complete.
$S_{i2.1}$	Homing	Homing process started.
S_{i3}	All	Homing done or motion complete.
S_{C1}	Main	Positioning in XOY plane and rotate gripper.
S_{C2}	Main/Force Control	Positioning on OZ axis (move gripper up or down) and rotate around OZ for gripper orientation.
S_{C3}	Main/Force Control	Ready to start force control (grip object).
$S_{C3.1}$	Force Control	Doing force control for selected reference.
$S_{C3.2}$	Force Control	Gripping force is stable while holding the object.
S_{C4}	Main/Position Control/Force Control	Stop force control (done). Switching to position control and open gripper.
S_{C5}	Position Control	Doing position control for selected reference.
S_{ES}	Position Control/Force Control	Emergency stop.

To keep the state machine clear and easy to understand, several components were detailed in additional diagrams presented in Figure 4a–c. These are the homing state machine, the position control state machine, and the force control state machine. Their state and transition descriptions can be also found in Table 1.

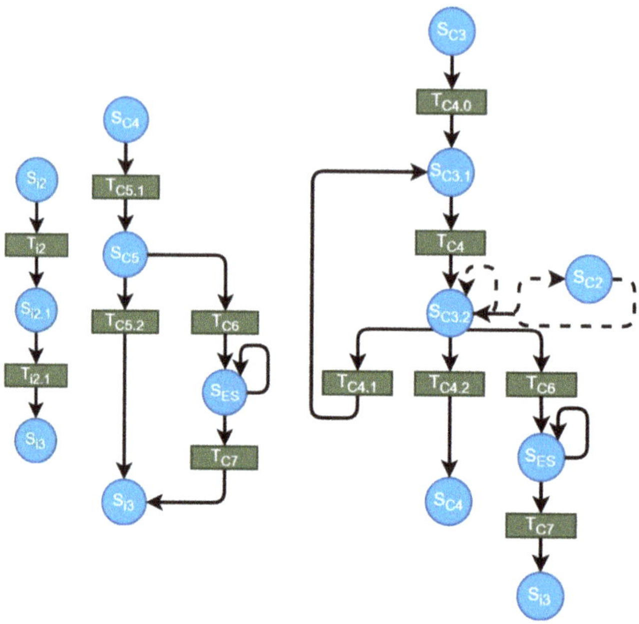

(a) (b) (c)

Figure 4. State machine sub-sections: (**a**) the homing state machine; (**b**) the position control state machine; (**c**) the force control state machine.

The homing state machine is very simple. The homing process is started sequentially on all five DOFs (state $S_{i2.1}$). When the homing process is complete, the 5DOF system transitions to the stable state S_{i3}. This state is the starting point for all following control sequences. When a new task is received (transition T_{C1}), the system begins to complete it, starting with the first two translation joints for positioning on XOY plane. Then, the vertical motion and orientation begin (S_{C2}) and on its completion, the force control takes over for the gripper joint by changing the control type for this degree of freedom within the S-matrix. Thus, by using force control, the gripper attempts to grab the target object using the reference force given as input to the control system. Figure 4c details how the force control handles different events. While the gripper force holds, the 4DOF positioning system lifts the object and position the gripper above the recycling tray. At this moment, the gripper force control ends by updating the S-matrix, and the object is dropped by opening the gripper's jaws using position control. After the object is dropped into the recycling tray, the task is complete, and the 5DOF system can start a new task.

To better explain the sequence, the decision Algorithm 1 that can be more easily understood is outlined below.

Algorithm 1 Recycling process using the hybrid position/force control

```
for (i ∈ [1; 5])
    Init (Mi)
    Homing (Mi)
while (!ES &&New Task)
    get (ObjType, Txyz, Rz)
    if (ΔP + z > εPz)
    GripperMove (Ttop)
    if (!Gopen)
    Jaws (open)
Compute (Fjaws)
GripperMove (Txy)
GripperRotate (Rz)
GripperMove (Tz)
Switch (ToForceControl)
    Update (SP, SF)
Jaws (close)
while (ΔFjaws ≤ εFjaws)
GripperMove (Ttop)
GripperRotate (Rtray)
GripperMove (Ttray)
Switch (ToPositionControl)
    Update (SP, SF)
Jaws (open)
```

where:

M_i = motor i,
ES = Emergency Stop,
Obj_{Type} = Object type (metal, glass, plastic),
T_{xyz} = target coordinates vector,
T_{top} = coordinates of top gripper position,
T_{tray} = coordinates of waste tray position,
R_Z = target rotation on the Z-axis,
R_{tray} = waste tray rotation on the Z-axis,
F_{jaws} = gripper reference force required to grab the current target,
ΔP_Z = position error on the Z-axis,
ε_{PZ} = maximum allowed positioning error on the Z-axis,
ΔF_{jaws} = gripper jaws force error,
ε_{Fjaws} = maximum allows gripper jaws force error,
S_P = position matrix joints,
S_F = force matrix joints.

4. Influence of Friction Force Analysis on the Gripper's Control Method

To pick up recyclable objects from the municipal waste collection, the gripper needs at least two fingers. For this, we created a 3D finger design which is presented in Figure 5a–c. These were designed as simple fingers that can grab different object types and shapes. From the 3D design, the fingers were 3D printed using ABS material and a Zortrax M200 3D printer.

When complete, the fingers were attached to the linear motor, resulting in a two-finger gripper mechanism presented in Figure 6. One can observe in Figure 6 a small deformation of the ABS fingers during the gripping action. This ABS finger's flexibility and deformability ensure a slower transmission of the force towards the actuation of the gripper. Certain protection of the electric actuation of the gripper is thus ensured.

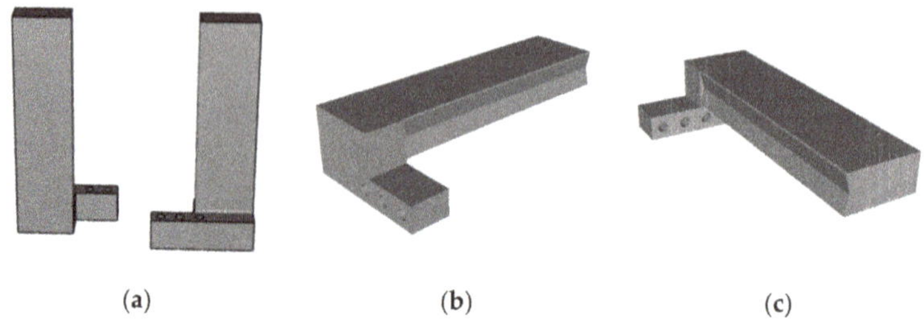

Figure 5. Gripper 3D fingers: (**a**) front view; (**b**) bottom side view; (**c**) top side view.

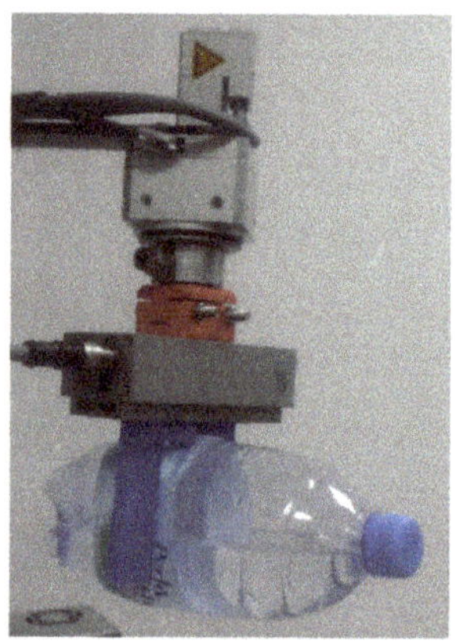

Figure 6. The gripper mechanism in action.

Once the gripper is attached, we can calculate the required force needed to grab certain recyclable materials/objects. To achieve this, we can use the stick–slip conditions [42] derived from the Coulomb friction law [43]. Figure 7 presents the forces acting between the gripper's fingers and the object being manipulated, where:

$\vec{F}_{f_L}$ = Friction force between left finger and the object
$\vec{F}_{f_R}$ = Friction force between right finger and the object
$\vec{F}_{C_L}$ = Control force on the left finger
$\vec{F}_{C_R}$ = Control force on the right finger
$\vec{G}$ = Object's weight
$\vec{N}_L$ = Normal force on left fingers' contact surface
$\vec{N}_R$ = Normal force on right fingers' contact surface

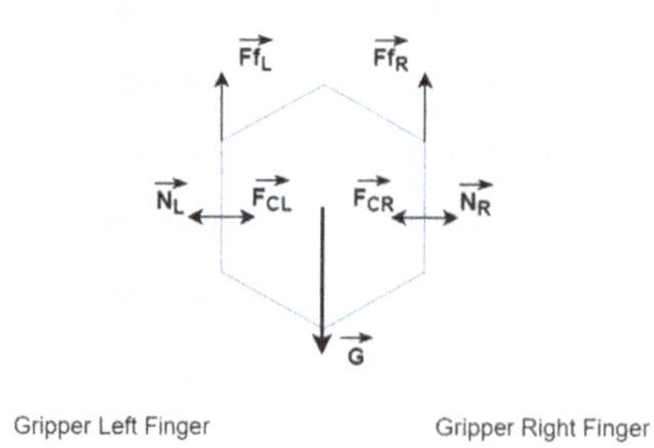

Figure 7. Forces acting on gripper fingers and the grabbed object.

Coulomb friction law states that for the object to not slide from the gripper fingers, the forces must follow Inequality (1), where the friction forces are defined in Equations (2) and (3):

$$\vec{F}_{f_L} + \vec{F}_{f_R} \geq \vec{G} \tag{1}$$

$$\vec{F}_{f_L} = \mu \vec{N}_L \tag{2}$$

$$\vec{F}_{f_R} = \mu \vec{N}_R \tag{3}$$

In our case, the control forces $\vec{F}_{C_L}$ and $\vec{F}_{C_R}$ are equal with the gripper's control force $\vec{F}_C$, resulting Equations (4) and (5):

$$\vec{F}_{C_L} = \vec{F}_{C_R} = \vec{F}_C, \tag{4}$$

$$\left.\begin{matrix} \vec{N}_L = \vec{F}_{C_L} \\ \vec{N}_R = \vec{F}_{C_R} \end{matrix}\right\} => \left\{\begin{matrix} \vec{F}_{f_L} = \mu \vec{F}_{C_L} = \mu \vec{F}_C \\ \vec{F}_{f_R} = \mu \vec{F}_{C_R} = \mu \vec{F}_C \end{matrix}\right\}. \tag{5}$$

From Equations (1)–(5) we calculate that the required condition for the gripper to hold the recyclable object is the one from the relation:

$$2\mu \vec{F}_C \geq \vec{G} \tag{6}$$

Using this inequality, we can further analyze the required force depending on the estimated weight of the object and the friction coefficient for certain types of recyclable materials. However, before analyzing the minimum required control force, we have to take into consideration the gripper motion on the same axis as the friction forces. Since the gripper will move up after the object is grabbed, the dynamic motion will add a new acceleration to the object, which will, in turn, translate into an inertial force that has the same direction as the weight of the object. This means that we need to increase the force acting on the object. Equations (7)–(9) add the new inertial force to the control force condition from Equation (6).

The maximum acceleration for the motor placed on the vertical axis, as given by the manufacturer [44] is:

$$a_i = 10\text{m/s}^2. \tag{7}$$

Then, the inertial force added to the weight is:

$$\vec{F}_i = m \times a_i. \tag{8}$$

From Relations (1) and (8) we obtain:

$$2\mu \vec{F}_C \geq \vec{G} + \vec{F}_i. \tag{9}$$

Expanding this relation, we get:

$$2\mu \vec{F}_C \geq m(g + a_i), \tag{10}$$

where:

$\vec{F}_i$ = Inertial force due to vertical motion
a_i = inertial acceleration due to vertical motion
m = Objects' mass
$g = 9.81\text{m/s}^2$ = Gravitational acceleration
μ = Friction coefficient

Knowing the maximum acceleration (Equation (11)) and the maximum force [45] (Equation (12)) that the gripper can sustain, we can compute the maximum object mass that our gripper can hold for certain materials with known friction coefficients:

$$a_{max} = g + a_i = 19.81 \text{ m/s}^2 \tag{11}$$

$$\vec{F}_{C_{max}} = 250 \text{ N}. \tag{12}$$

From Relations (10) and (12) we can compute the required force for grabbing an object and we can compare it with the maximum force the gripper can apply:

$$\vec{F}_C \geq \frac{m(g + a_i)}{2\mu}. \tag{13}$$

If we consider the recommendations presented by online references [46,47], the control force should be four times higher than the required force for gripping the object, without considering the vertical acceleration, and is given by the relation:

$$\vec{F}_C = \frac{mg}{2\mu} \times 4. \tag{14}$$

Table 2. Static friction coefficients.

Material	Friction Coefficient
ABS—PET	0.33 [48]
ABS—Glass	0.3 [49]
ABS—Aluminum	0.25 [49]

This means that we can add a safety margin and all objects regardless of the systems' accelerations will be firmly held by the gripping system. However, since Relation (13) takes into account the maximum acceleration of the vertical axis, and we do not expect any other forces to influence the system, we will use Relation (13) as the reference for computing the force needed by the system to pick up objects with different friction coefficients and weight. While the maximum acceleration is given by Relation (11), we can relate to Relation (14) and present the new force control as:

$$\vec{F}_C = \frac{mg}{2\mu} \times 2, \tag{15}$$

where the new safety margin is 2. However, in real experiments, we will use Relation (13) as it is more accurate.

In our case, the recyclable materials of interest are plastic bottles, glass bottles, and aluminum cans. Knowing this, we will continue the analysis using these materials. Thus, Table 2 presents the known static friction coefficients between the ABS fingers and PET, glass and aluminum objects, where we consider them more dry than wet.

Once we have the maximum force relation and the friction coefficients, we can compute the required force for the gripper to pick up a certain object. Figure 5 presents the computed force control value for different object mass and material types. The figure shows the maximum force limit, represented by the top dashed rectangle. Using this diagram (Figure 8), we can visualize the maximum weight of the recyclable objects that the gripper can pick up and transfer to the recycling tray. Since for all three materials the maximum weight is above 6 kg, we can presume that all recyclable objects that fit the recycling criteria can be grabbed by our 5DOF gripper.

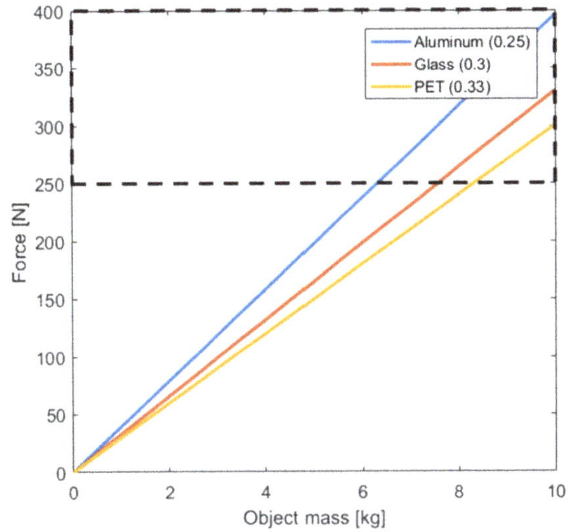

Figure 8. Gripper force by object mass and friction coefficient.

By using Equation (13) and the friction coefficients of the recyclable materials of interest (Table 2) we can now use this information in the hybrid position/force control architecture to compute the gripper reference force.

5. The 5DOF Hybrid Position/Force Control

To achieve a functional control method, the mobile robots' manipulator, used for sorting waste materials, has to be able to move at target points and then grab the recycling materials using a simple gripper. For this, the hybrid position/force control method [15] was chosen and then modified to integrate the systems' state, changing the control type of the manipulator when needed, between position and force control mode. Thus, we changed the static selection of control type for each degree of freedom to a deterministic state of the S-matrix that separates each DOF into two categories: control mode and force mode. Figure 9 presents the new hybrid position/force control architecture used to drive the 5DOF manipulator.

Figure 9 presents the data flow of information between each control block, where:

$O(p,\ type)$ = Object (position, type);

type = Glass, Aluminum, Plastic;

$O(depth)$ = Object (depth);

Depth = distance from the robot reference system to the detected object.

In the new architecture of the hybrid position/force control method of Figure 9, we have included several control blocks that group the control architecture by task, while the arrows present the data flow. The control diagram starts with the reference generation block. This block receives information from the video detection system that will provide the objects' position in XOY coordinates and the object type. The type is then used by the reference generation to compute the required force according to the friction force coefficient influence. Another input is received from the proximity sensors, which will provide the distance to the target object, which represents the third coordinate in the 3D space, the OZ axis.

The video detection system [50] uses neural networks to detect the type, position, and orientation of the recyclable objects and will send this information as input for the 5DOF manipulator, through a TCP network interface. Combining both video detection for XOY coordinates and orientation with the proximity sensors for depth distance, we get the OXYZ coordinates with θ angle orientation. The detection system was developed by our project team and is already published by Melinte et al. [50], which is why we will not focus on its results in this paper, but we are using the AI detection module as input to our control architecture. The video camera used has an RGB resolution of 1280 × 1024 pixels, which is more than enough for our AI detection system to use, with a detection confidence of 75.54% and over 95% accuracy [50], using a 9 FPS image analysis rate.

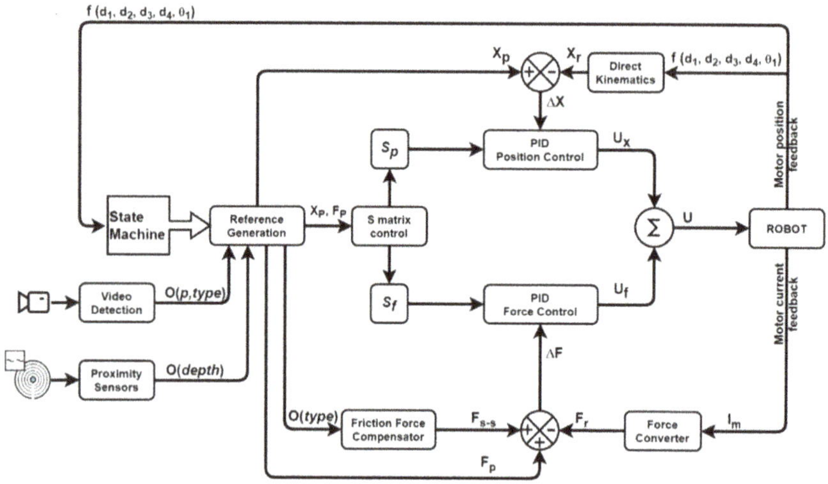

Figure 9. Hybrid position/force control diagram.

Added to the input data, the systems' state, provided by the state machine, the robot can now decide what to do next, to achieve the given recycling task. If the robot is in search mode, the manipulator is resting at the starting position. Then, when an object is detected, the systems' state switches to the start state of the pick-and-drop process. When the state of the system requires the manipulators' gripper to switch in order to force control then the "S-matrix control" block will update the S-matrix, by switching the control type of the gripper DoF. Equation (16) presents the way the S-matrix is composed out of the S_p and S_f matrix:

$$S = \begin{bmatrix} 1 & \cdots & 0 \\ \vdots & \ddots & \vdots \\ 0 & \cdots & 1 \end{bmatrix} = S_p + S_f = \begin{bmatrix} 1 & . & . & . & 0 \\ . & 1 & & & . \\ . & & 1 & & . \\ . & & & 1 & . \\ 0 & . & . & . & 0/1 \end{bmatrix} + \begin{bmatrix} 0 & . & . & . & 0 \\ . & 0 & & & . \\ . & & 0 & & . \\ . & & & 0 & . \\ 0 & . & . & . & 1/0 \end{bmatrix}. \quad (16)$$

After the reference values are calculated, the hybrid method splits into two control branches, each dealing with a specific control type: position control and force control. Each control block uses a PID to compute the joint-specific drive inputs (Equations (17)–(19)):

$$U_x = X_{PID}(S_p, \Delta X), \quad (17)$$

$$U_f = F_{PID}\left(S_f, \Delta F, F_{S-S}\right), \quad (18)$$

$$U = U_x + U_f. \quad (19)$$

One can see that while the position PID regulator depends on the positioning error between the reference and the computed value, using the direct kinematics method and Denavit-Hartenberg notation [51], the force PID regulator will have as input the force control error computed using the reference value, the real force taken as a function of motor current and the force required to compensate the object weight through the friction force compensator control block.

Equation (20) presents the function that receives the robot joint information and sends it to the state machine and direct kinematics blocks:

$$f(d_1, d_2, d_3, d_4, \theta_1). \tag{20}$$

Using the joint information, Equations (21)–(23) present the elements required to compute the positioning and force errors, which are inputs to PID regulators that, in the end, drive the 5DOF manipulators' motors.

$$U = X_{PID}(S_p, \Delta X) + F_{PID}\left(S_f, \Delta F\right), \tag{21}$$

where:

$$\Delta X = X_P(State,\ Obj(p,\ type),\ Obj(depth)) - X_r(Rob_{kin}(d_1, d_2, d_3, d_4, \theta_1)) \tag{22}$$

$$\Delta F = F_P(State) + F_{s-s}(Obj(type)) - F_r(F_{conv}(I_m)) \tag{23}$$

To obtain the PID position control values, we used the direct kinematics equations through Denavit-Hartenberg notation that are shown in Figure 10 and presented in Equation (24).

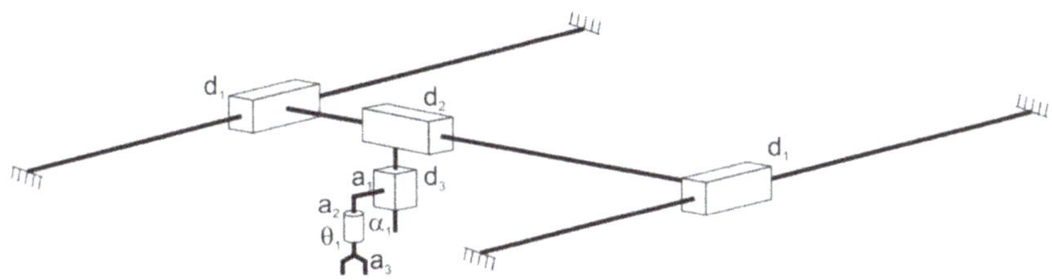

Figure 10. Kinematics diagram and D-H notations of the 5DOF, TTTRT manipulator.

Since α is $0°$, then the direct kinematic coordinates are given by Relation (24), where a_1 and a_2 are constants given by the mechanical structure dimensions. The gripper's degree of freedom is not taken into account when computing the direct kinematics matrix. This is because we had to position the gripper near the desired object, and then close the gripper's jaws (5th degree of freedom).

Using the direct kinematics matrix:

$$\begin{bmatrix} cos\theta_1 & -sin\theta_1 & 0 & -a_1 + d_1 \\ sin\theta_1 & cos\theta_1 & 0 & d_2 \\ 0 & 0 & 1 & -a_2 - d_3 \\ 0 & 0 & 0 & 1 \end{bmatrix}, \tag{24}$$

and the PID general control equation:

$$PID = K_P e(t) + K_i \int_0^t e(t)dt + K_d \frac{de(t)}{dt}, \tag{25}$$

the position and force controllers were computed following classic methods [52]. Since the PID controllers are not in the scope of this paper, we will not present their design and tuning procedure.

The hybrid method we have proposed in this paper uses hardware components that have a clear separation between position and force control. Because of this, the stability problem is avoided, since when switching the control method, the regulator method is changed for that specific motor (DoF), and the motor is in a paused state (not moving). When changing the control method, we can say that the control mission is restarted, which is why there is no stability issue at the switching moment within the control law.

6. Experiment

The hybrid position/force control is designed for the 5DOF manipulator which will be mounted on the mobile robot. The manipulator has several hardware components that build the positioning and orientation system and one linear motor for the end effector.

Figure 11 presents the interaction diagram of the hardware components. One can see that the system has only one PLC system that sends the reference information to the planar XOY movement (Festo EXCM), to the vertical OZ positioning (Festo EXCM), and the rotary motor (Festo ERMO), through CANopen, Modbus TCP or Input/Output Link (IO-Link) communication with integrated motor drivers. The gripper's motor requires a separate motor driver (Festo CCEC-X-M1) that controls the linear motor (Festo HGPLE), which forms the gripper with the attachment of two 3D printed ABS fingers.

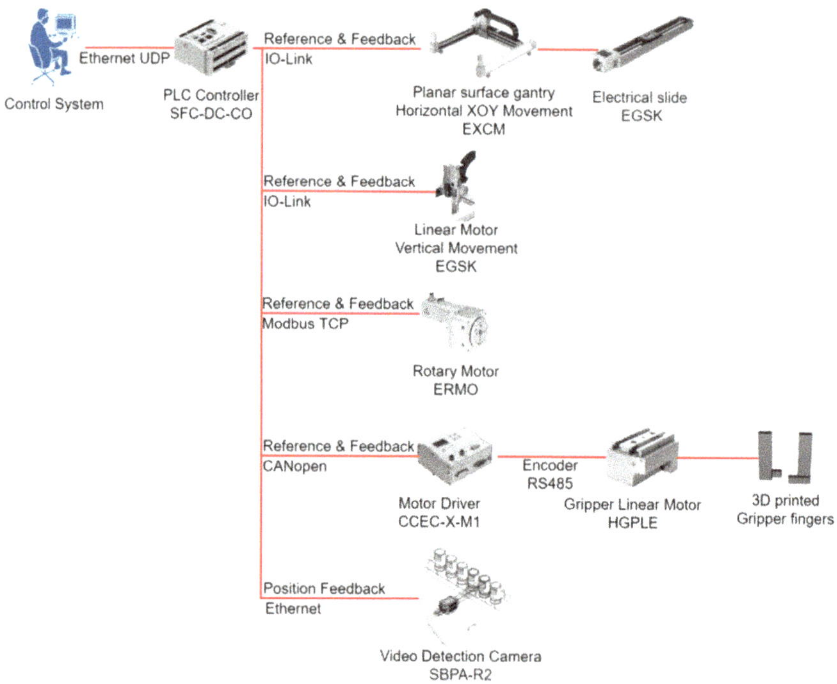

Figure 11. The robot control components interaction diagram.

The experiment uses a video detection camera, which uses a neural network [50,53] to detect different recyclable materials (glass bottles, aluminum cans, PET bottles) and sends their position and orientation to the 5DOF manipulator as input information. This information was simulated in the first stage of the experiment, since we had to prove that the hybrid position/force control has a good performance in the process of controlling the manipulator throughout the recycling operation.

Because the gripper's control was the one that will switch between position and force control, we can say that this joint (DoF) is the most important one when deciding if the hybrid control can be used for this particular task, and we will describe its process in more detail.

The 5DOF manipulator is controlled by a Festo PLC, programmed using the CoDeSys software. Using the state machine presented in Figures 3 and 4, one can see that the entire system has several control steps. The first stage is to initialize the motors and start the homing process for each motor, and then wait for the detection system to send the first signal of position, orientation, and type for the object to recycling. After the initialization is complete, the recycling system can receive the pick-and-drop commands.

To control the 5DOF mechanical system, a decision algorithm that implements the state machine was developed outside the PLC, with communication capabilities through Ethernet-UDP protocol (Figure 12). This system receives state messages from the manipulator and the video detection system, and, through a transition mechanism, sends appropriate commands to the manipulator. The messages have two parts: command name and value. The message is then decoded by the PLC through a decoding method. The decoded information is used to set motor parameters and then activate the command. While the motor is acting on the last received command, it will send status information through a transducer to the motor driver which, in turn, is received by the PLC that will send status information through Ethernet UDP to the state machine.

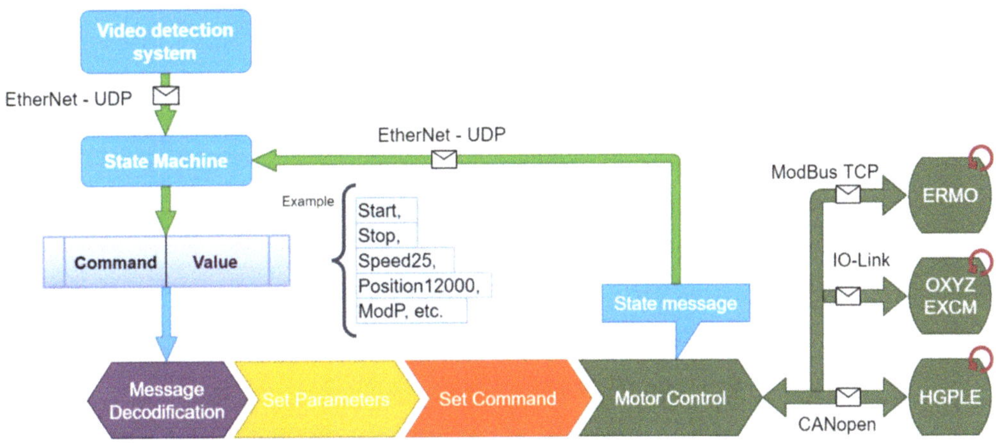

Figure 12. Motor control through PLC programming and state machine commands.

To test the gripper's commands and states, we designed a CoDeSys visualization interface that shows the state of each command or status bit.

Figure 14 presents all the main stages of the gripper's control. The left part of each figure contains the control bits, and the right part contains the status bits. As the name suggests, the control bits are the values sent to the motor driver and the status bits are the values received. Grey status means that the bit is not used by the driver, dark green means that the bit value is 0 (false) and light green means that the bit value is 1 (true). Using these control bits, we can set the motor driver states from initialization and homing to the start of positioning or force control. Thus, Figure 14a,b present the configuration bits for the initialization and homing functions of the motor driver, respectively. After the initial step and homing have been completed, the motor driver, and consequently the HGPLE motor, are in stand-by mode to receive control values. The first values that will be sent to the driver are the control mode: position control mode (Figure 14c) or force control mode (Figure 14e). After these values comes the speed reference with which the gripper will move, while trying to reach its position or force reference. The third step is to send the position and force reference values and to start the control process. Figure 14d shows position control and Figure 14f shows force control. While the positioning mode will stop after the position is reached, the force control will continuously try to hold the set force.

The force control can be stopped by setting the control force to 0, by manually stopping the control process, or by switching the control type to positioning mode.

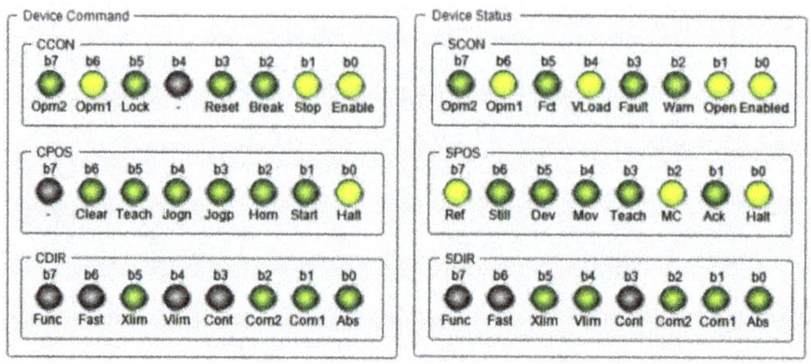

(**a**)

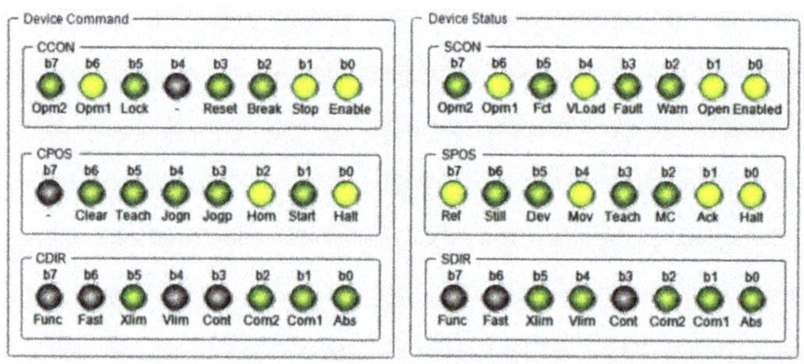

(**b**)

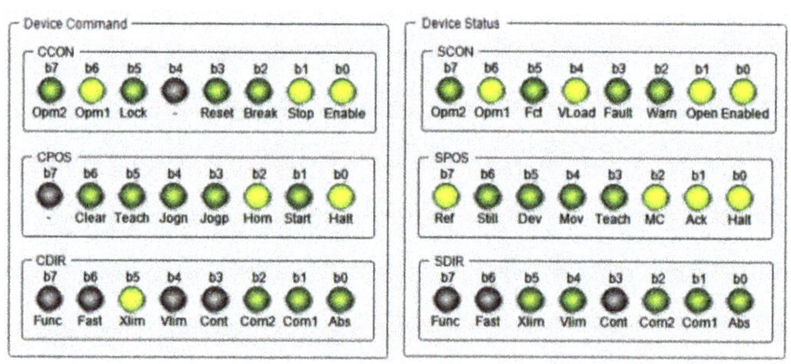

(**c**)

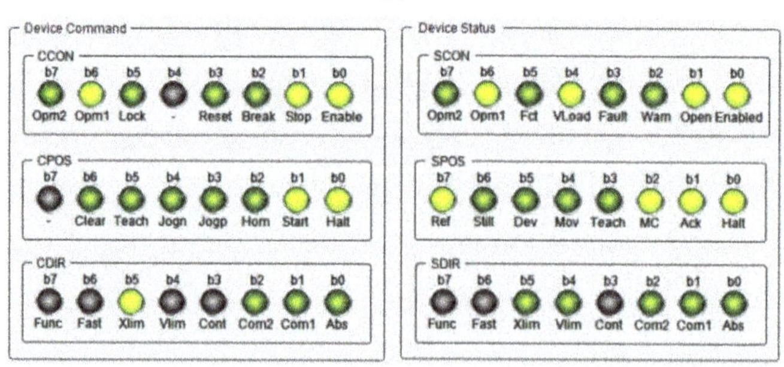

(**d**)

Figure 13. *Cont.*

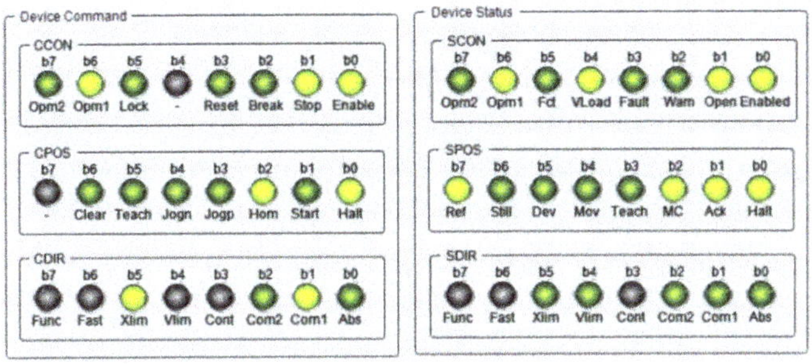

(e)

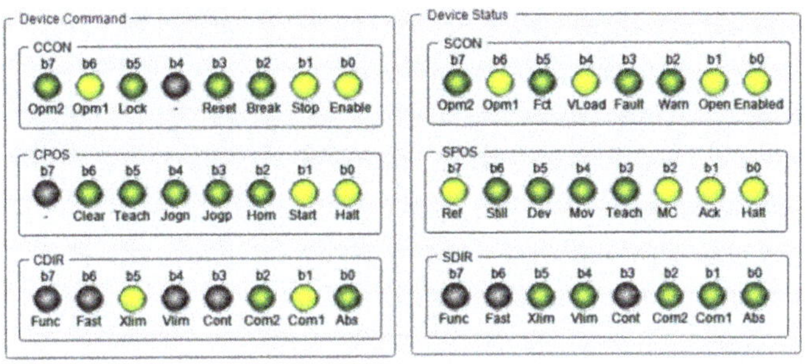

(f)

Figure 14. CoDeSys visualization interface: (**a**) initialization stage; (**b**) homing stage; (**c**) positioning mode control; (**d**) start positioning control; (**e**) force mode control; (**f**) start force control.

Using the described steps, the gripper can be used in the hybrid position/force control method for the 5DOF manipulator. While the positioning and orientation system will always be controlled in position, the gripper will switch between position and force control states depending on the system's state machine.

Figures 15 and 16 present the control phase for position control and force control, respectively, for the manipulator's gripper.

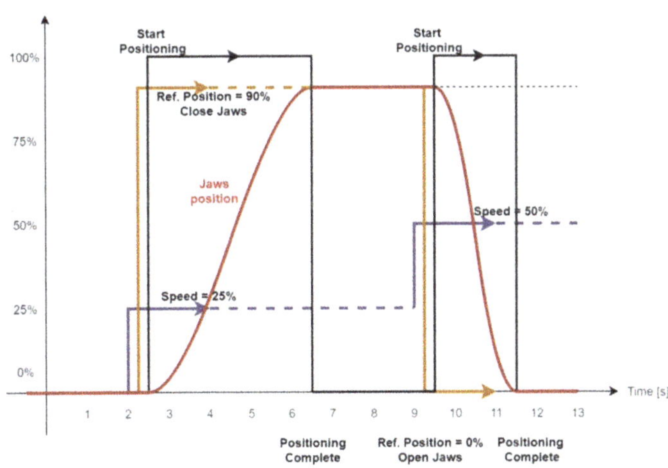

Figure 15. Gripper position control states.

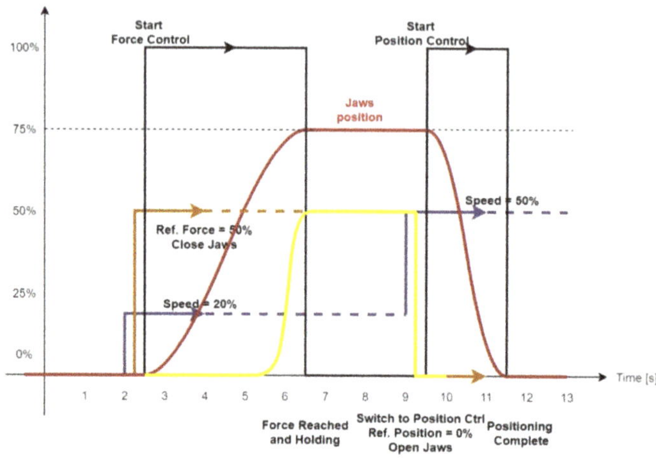

Figure 16. Gripper force control states.

The color code used in Figures 15 and 16 is:

- Purple line = speed reference;
- Brown line = force and/or position reference;
- Black line = start command and auto-stop state;
- Red line = gripper jaws position;
- Yellow line = gripper applied force.

For positioning control, Figure 15 is quite self-explanatory, where the reference velocity is the first to be set, followed by the position reference. Once these two have been set, the start command can be given, and the positioning regulator will start moving the motor to the desired position with the configured speed. The gripper jaws will stop when the target position is reached or when an alarm is detected. The alarm can be triggered by an overcurrent due to an obstacle within the jaws' path. When the position is set to 0 to open the gripper jaws, and the start command is given once again, the gripper jaws will move with the newly set speed.

For the force control, the control diagram is more complex and is presented in Figure 16. This control process starts similarly to the positioning process, with setting the speed reference. The next step is to set the target force and to start the control process. Before the target force is reached, the gripper will move until an object/obstacle is reached and the control force will increase. When the reference force is reached, the regulator will not stop and will continue to feed current to the motor. At this point, the manipulator's 4DOF positioning and orientation system can move the recyclable object. When the target OXYZ position is reached, the gripper's state is switched from force to position control, set the reference position to 0, and start the positioning control to open the gripper's jaws.

Using the described process of the hybrid position/force control to pick-and-drop recyclable waste, we tested it for several objects and materials. Figure 17a–c present three of the objects being grabbed and held until the movement state is complete and the gripper can drop the recyclable material. As one can see, Figure 17a–c present the orientation and gripper components. We tested the two degrees of freedom more thoroughly than the entire 5DOF system, since, for these two, the hybrid position/force control is more important.

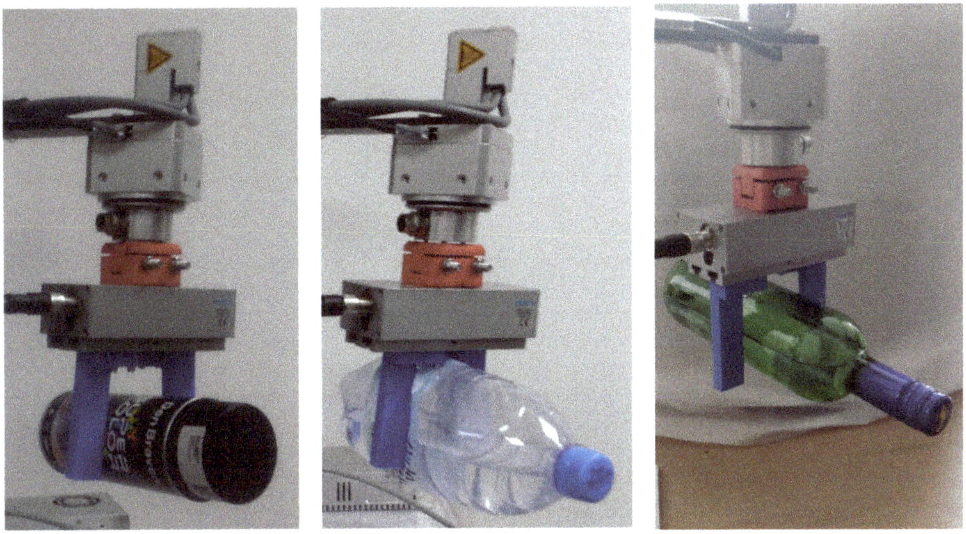

(a) (b) (c)

Figure 17. Experiments on object gripping: (**a**) experiment for aluminum pickup; (**b**) experiment for PET bottle pickup; (**c**) experiment for glass bottle pickup.

7. Discussion

In order to use a mobile robot with a 5DOF manipulator for picking up recyclable material, the first stage is to initialize the robot's components. This means that for each motor and actuator, a homing process is required when the robot starts up.

After the first initialization stage is complete, the robot is ready to start its search-and-recycle task. For that, a human operator starts the mission, and the robot will move within the mission area to search for recyclable materials, using its neural network-driven detection video camera. At this stage, the manipulator is resting in the homing-done-ready-for-task state.

When the robot detects a target object, it will compute the next stage reference for the position, orientation, and type of the object to pick up. At this moment, the state of the system's manipulator changes from resting to start-task. All reference values are, at this stage, being used to provide accurate task information for the manipulator. This data includes a position in 3D space, orientation of the target object, and estimated friction between the gripper's fingers and the target. Using these values, the robot will start its mission with a positioning task, followed by a hybrid position/force control to grab the object and place it in the recycling tray. The last task is to position the manipulator in a starting position with the gripper jaws open.

As seen from the main stages of the pick-and-drop process, the proposed hybrid position/force control architecture based on a state machine is very useful for this type of robot and mission.

For hybrid methods that require an online switch mechanism, the interest is quite high between researchers and manufacturers. However, only certain types of motors can be controlled in this way. An example is the research of Pasolli and Ruderman [38,54], which proposed a hybrid control to be used on actuators which may not require different control parameters, and should be used where the actuators' parameters rarely change, while the control values can be modeled as a single transfer function, within the control architecture diagram.

The main difference between our approach and the one used by Pasolli and Ruderman [38,54] is that our hardware has parameters that require an update every time the control mode is switched. This is because modern motors and actuators are produced with built-in regulators or fail-safe mechanisms to protect the hardware and provide the engineer with a basic motor controller. With these types of motors and actuators, you need a different type of hybrid control, one that can update the parameters of the motor while switching the control type with certain degrees of freedom, and one that can use different control methods.

Similar hybrid control with a built-in switching mechanism developed by Pasolli and Ruderman [38,54] is used by the majority of hybrid control methods [16,18,19,24,26–30,33–35,38,54,55], where the motor allows position or force control just by increasing the driving current or motor torque. These are better used with a dynamic control law integrated into the hybrid position/force control method. The built-in switching mechanism that allows the hybrid control to control a certain motor either in position or in force requires a stability analysis, but when the motor has to switch parameters to commute between position and force control, the stability analysis is not needed. Since specific motors require new parameters to switch between control types, kinematic control methods are enough to drive the robot joints.

The hybrid position/force control with an online switch is not unfamiliar to us [20], but it was considered to be unnecessary for the task at hand. We have developed other control methods that require stability analysis [20], and ones that use dynamic control methods [20,25], along with the hybrid position/force control to drive robots in their mission. Compared to our previous research, the approach presented in this paper makes it easier to implement the control method, but it has a higher complexity when using these motors. The complexity comes from the multitude of parameters required to adjust not only when commissioning the motor, but also at run time when switching the required parameters to drive the motor in either position or force/torque.

Knowing this, the proposed architecture of the state machine-driven hybrid position/force control is useful for many other robots [55,56] that are being used in different environments and for different tasks, but more importantly, for those that are being used in automated processes within Industry 4.0 type factories. These factories rely on the fact that the robots can switch tasks whenever the supervisor asks them to change the production line. Since our proposed architecture is based on a state machine, one can design a way to update the state machine for each type of mission and task the robot is given, gaining not only a hybrid position/force control method but also an adaptable robot with a set of missions to choose from.

8. Conclusions

The proposed architecture of the state machine-based hybrid position/force control (SmHPFC) can be implemented for robots and actuator-driven mechatronics, which use motors with different control parameters between position and force control. One example is the HGPLE Festo linear motor. This type of motor tends to include feedback information on the real position and force/torque that can be easily included in a feedback control loop such as the SmHPFC. Using this information, one can design a position control feedback loop, knowing that the received information is showing the actual actuator position. At the moment, when the control type is changed from position to force or torque and vice versa, the same information value must be interpreted with a different scope, which will disturb the real-time control loop since the motor control parameter has changed. In the standard approach, a motor is usually controlled in position, force/torque, or direct current, and uses different techniques to get the control value, starting from a position or force/torque reference. This approach cannot be used by a state-driven actuator such as the one used on the presented 5DOF manipulator.

One advantage of the SmHPFC architecture is that it is quite easy to integrate it into an automated assembly line or use it to change the behavior of the robot since the stability problem was mitigated due to the way the motors are built. This advantage translates into reconfiguring the automated line with ease, which is the backbone of the Industry 4.0 concept.

As an example of Industry 4.0 usage, the robots from an assembly line can be programmed to complete different tasks, while in a standard production line, the extra programmed tasks are not required. A robot can have a task to move certain components from a top shelf to the conveyor in a certain factory configuration, but can also stack products to a designated area, depending on the factory requirements. This can happen if all the robots are pre-programmed with multiple tasks and a control unit will change the robot state from a shelf-stacking robot to one that can pick up faulty products from the conveyor. Another advantage of the SmHPFC architecture is that not only can it provide easy access to multiple types of tasks for a single robot, but that these tasks can have very different control parameters and objectives: position control, contour following, force control, etc.

This type of architecture has some disadvantages. The main one is that it cannot switch between control modes while the robot is moving and it must perform a very short stop while reconfiguring the input parameters. This stage is dependent on the communication rate between the PLC and the motor, and on the time the motor can come to a stop. It takes only a few communicated messages for the motor to stop, reconfigure and start a new job. For example, to switch the Festo linear motor from position to force, we need to send just one control byte. However, since the motor parameters cannot be changed while it runs, we have to send two additional control bytes, to stop and restart the motor. On a ModBus 9600 bps communication rate, it would translate to 2.5 ms, only to send the new configuration. To this, we need to add the time it takes the motor to come to a stop if not already stopped, and a maximum of eight more bytes to send the new motor reference. This also depends on whether the motor can restart with ease. This disadvantage can turn into an advantage if the robot task requires separate motion steps during which the robot can be reconfigured to handle different control methods on a different degree of freedom.

Our proposed SmHPFC architecture does not depend on the joint regulators, and their analysis was not included in this paper. Thus, the presented applied research is meant to prove the behavior of our architecture and provide new means of control for automated Industry 4.0 lines, but also for motors that are not as easy to integrate into a fully automated robot due to restrictions in their control and parameterization functions.

Future work will focus on integrating the deterministic artificial intelligence [57] that provides clear analytic methods for reference trajectory generation on each type of control. By improving the reference aspect, we expect to improve the overall architecture, transforming the reference generation component into a more robust and reliable component that can provide the best data for the robot's current task and control type.

Author Contributions: Control architecture, I.-A.G.; conceptualization, I.-A.G.; control algorithm, I.-A.G., formal analysis, I.-A.G., A.-C.C. and M.M.; resources, I.-A.G., A.-C.C. and M.M.; writing—original draft preparation, I.-A.G. and A.-C.C.; writing—review and editing, I.-A.G.; funding acquisition, I.-A.G. and M.M. All authors have read and agreed to the published version of the manuscript.

Funding: This work was supported by a grant from the Romanian Ministry of Research and Innovation, CCCDI—UEFISCDI, project number PN-III-P1-1.2-PCCDI-2017-0086/contract no. 22 PCCDI/2018, within PNCDI III, and with the support of the Robotics and Mechatronics Department, Institute of Solid Mechanics of the Romanian Academy.

Institutional Review Board Statement: Not applicable.

Informed Consent Statement: Not applicable.

Data Availability Statement: Not applicable.

Acknowledgments: This work was supported by a grant from the Romanian Ministry of Research and Innovation, CCCDI-UEFISCDI, MultiMonD2 project number PN-III-P1-1.2-PCCDI2017-0637/33PCCDI/01.03.2018, within PNCDI III, and by the European Commission Marie Sklodowska-Curie SMOOTH project, Smart Robots for Fire-Fighting, H2020-MSCA-RISE-2016-734875 and by inter-academic project IMSAR–Yanshan University: "Joint Laboratory of Intelligent Rehabilitation

Robot", KY201501009, collaborative research agreement between Yanshan University, China and Romanian Academy by IMSAR, RO. The authors thank S.C. Festo S.R.L., Bucharest, Romania, for their technical guidance and assistance in using the Festo products concerned in this paper.

Conflicts of Interest: The authors declare no conflict of interest. The funders had no role in the design of the study; in the collection, analyses, or interpretation of data; in the writing of the manuscript, or in the decision to publish the results.

References

1. Del Borghi, A.; Gallo, M.; Del Borghi, M. A survey of life cycle approaches in waste management. *Int. J. Life Cycle Assess.* **2009**, *14*, 597–610. [CrossRef]
2. Coopers, L. *Cost-Benefit Analysis of the Different Municipal Solid Waste Management Systems: Objectives and Instruments for the Year 2000*; Office for Official Publications of the European Communities: Luxembourg, 1997.
3. Skovgaard, M.; Hedal, N.; Villanueva, A.; Andersen, F.M.; Larsen, H. *Municipal Waste Management and Greenhouse Gases*; European Environment Agency: Copenhagen, Denmark, 2008.
4. Awasthi, A.K.; Zeng, X.; Li, J. Relationship between e-waste recycling and human health risk in India: A critical review. *Environ. Sci. Pollut. Res.* **2016**, *23*, 11509–11532. [CrossRef] [PubMed]
5. Gundupalli, S.P.; Hait, S.; Thakur, A. A review on automated sorting of source-separated municipal solid waste for recycling. *Waste Manag.* **2017**, *60*, 56–74. [CrossRef] [PubMed]
6. Diya, S.Z.; Proma, R.A.; Islam, M.N.; Anannya, T.T.; Al Mamun, A.; Arefeen, R.; Rabbi, M.F. Developing an intelligent waste sorting system with robotic arm: A step towards green environment. In Proceedings of the 2018 International Conference on Innovation in Engineering and Technology, Dhaka, Bangladesh, 27–28 December 2018; pp. 1–6.
7. Kokoulin, A.N.; Uzhakov, A.A.; Tur, A.I. The Automated Sorting Methods Modernization of Municipal Solid Waste Processing System. In Proceedings of the 2020 International Russian Automation Conference, Sochi, Russia, 6–12 September 2020; pp. 1074–1078.
8. Sousa, J.; Rebelo, A.; Cardoso, J.S. Automation of Waste Sorting with Deep Learning. In Proceedings of the 2019 XV Workshop de Visão Computacional, Sao Paulo, Brazil, 9–11 September 2019; pp. 43–48.
9. Popa, C.L.; Carutasu, G.; Cotet, C.E.; Carutasu, N.L.; Dobrescu, T. Smart city platform development for an automated waste collection system. *Sustainability* **2017**, *9*, 2064. [CrossRef]
10. Ziouzios, D.; Tsiktsiris, D.; Baras, N.; Dasygenis, M. A Distributed Architecture for Smart Recycling Using Machine Learning. *Future Internet* **2020**, *12*, 141. [CrossRef]
11. Grigore, M.E. Methods of Recycling, Properties and Applications of Recycled Thermoplastic Polymers. *Recycling* **2017**, *2*, 24. [CrossRef]
12. Hua, D.; Gao, J.; Mayo, R.; Smedley, A.; Puranik, P.; Zhan, J. Segregating Hazardous Waste Using Deep Neural Networks in Real-Time Video. In Proceedings of the 10th Annual Computing and Communication Workshop and Conference, Las Vegas, NV, USA, 6–8 January 2020; pp. 1016–1022.
13. Stankovski, S.; Ostojić, G.; Šaponjić, M.; Stanojević, M.; Babić, M. Using micro/mini PLC/PAC in the Edge Computing Architecture. In Proceedings of the 19th International Symposium INFOTEH-JAHORINA, East Sarajevo, Bosnia, 18–20 March 2020; pp. 1–4.
14. Festo. Parallel Grippers HGPLE, Sturdy with Long Stroke, Electric. Available online: https://www.festo.com/cat/enus_us/data/doc_enus/PDF/US/HGPLE_ENUS.PDF (accessed on 23 October 2020).
15. Raibert, M.H.; Craig, J.J. Hybrid Position/Force Control of Manipulators. *J. Dyn. Syst. Meas. Control* **1981**, *103*. [CrossRef]
16. Ortenzi, V.; Stolkin, R.; Kuo, J.; Mistry, M. Hybrid motion/force control: A review. *Adv. Robot.* **2017**, *31*, 1102–1113. [CrossRef]
17. Rani, M.; Kumar, N.; Singh, H.P. Efficient position/force control of constrained mobile manipulators. *Int. J. Dyn. Control.* **2018**, *6*, 1629–1638. [CrossRef]
18. Wang, T.; Umemoto, K.; Endo, T.; Matsuno, F. Dynamic hybrid position/force control for the quadrotor with a multi-degree-of-freedom manipulator. *Artif. Life Robot.* **2019**, *24*, 378–389. [CrossRef]
19. Liu, Y.; Fang, Q.; Ke, Y. Hybrid Position/Force Control for Dual-Machine Drilling and Riveting System. *Math. Probl. Eng.* **2020**. [CrossRef]
20. Vladareanu, L.; Gal, I.A. A Multi-Functional Approach of the HFPC Walking Robots. In Proceedings of the 15th WSEAS International Conference on Systems, Recent Researches in System Science, Corfu Island, Greece, 14–16 July 2011.
21. Vladareanu, V.; Munteanu, R.I.; Mumtaz, A.; Smarandache, F.; Vladareanu, L. The Optimization of Intelligent Control Interfaces Using Versatile Intelligent Portable Robot Platform. *Procedia Comput. Sci.* **2015**, *65*, 225–232. [CrossRef]
22. Smarandache, F.; Vlădăreanu, L. Applications of neutrosophic logic to robotics: An introduction. In Proceedings of the 2011 IEEE International Conference on Granular Computing, Kaohsiung, Taiwan, 8–10 November 2011; pp. 607–612.
23. Vladareanu, V.; Vladareanu, L.; Munteanu, R.I.; Capitanu, L.; Dobrescu, E.M.; Radulescu, M.; Grosu, V.A. Adaptive neural network fuzzy inference system for HFC processes. *Period. Eng. Nat. Sci.* **2019**, *7*, 311–317. [CrossRef]
24. Ballesteros-Escamilla, M.; Cruz-Ortiz, D.; Salgado, I.; Chairez, I. Hybrid position/force output feedback second-order sliding mode control for a prototype of an active orthosis used in back-assisted mobilization. *Med. Biol. Eng. Comput.* **2019**, *57*, 1843–1860. [CrossRef]

25. Gal, A.I.; Vladareanu, L. Sliding mode control with bond graph modeling applied on a robot leg. In Proceedings of the 2014 International Conference on Circuits, Systems and Control, Interlaken, Switzerland, 22–24 February 2014; pp. 40–45.
26. Rani, M.; Kumar, N. A new hybrid position/force control scheme for coordinated multiple mobile manipulators. *Arab. J. Sci. Eng.* **2019**, *44*, 2399–2411. [CrossRef]
27. Peng, J.; Ding, S.; Yang, Z.; Zhang, F. Neural Network-Based Hybrid Position/Force Tracking Control for Robotic Systems without Velocity Measurement. *Neural Process. Lett.* **2020**, *51*, 1125–1144. [CrossRef]
28. Zhang, H.; Fang, H.; Zou, Q.; Song, M.; Zhu, T. Force-Position Hybrid Control of a Novel Parallel Manipulator with Redundant Actuation. In Proceedings of the WRC Symposium on Advanced Robotics and Automation, Beijing, China, 21 August 2019; pp. 128–133.
29. Rani, K.; Kumar, N. Design of Intelligent Hybrid Force and Position Control of Robot Manipulator. *Procedia Comput. Sci.* **2018**, *125*, 42–49. [CrossRef]
30. Xu, Z.; Zhou, X.; Cheng, T.; Sun, K.; Huang, D. Fuzzy-neural-network based position/force hybrid control for multiple robot manipulators. In Proceedings of the IEEE International Conference on Cybernetics and Intelligent Systems and IEEE Conference on Robotics, Automation and Mechatronics, Ningbo, China, 19–21 November 2017; pp. 94–99.
31. Zhang, H.; Li, L.; Zhao, J.; Zhao, J.; Liu, S.; Wu, J. Design and implementation of hybrid force/position control for robot automation grinding aviation blade based on fuzzy PID. *Int. J. Adv. Manuf. Technol.* **2020**, *107*, 1741–1754. [CrossRef]
32. Solanes, J.E.; Gracia, L.; Muñoz-Benavent, P.; Miro, J.V.; Vidal, C.P.; Tornero, J.; Solanes, E. Robust Hybrid Position-Force Control for Robotic Surface Polishing. *J. Manuf. Sci. Eng.* **2018**, *141*, 011013. [CrossRef]
33. Ravandi, A.K.; Khanmirza, E.; Daneshjou, K. Hybrid force/position control of robotic arms manipulating in uncertain environments based on adaptive fuzzy sliding mode control. *Appl. Soft Comput.* **2018**, *70*, 864–874. [CrossRef]
34. Gao, C.; Cong, D.; Liu, X.; Yang, Z.; Tao, H. Hybrid position/force control of 6-dof hydraulic parallel manipulator using force and vision. *Ind. Robot. Int. J.* **2016**, *43*, 274–283. [CrossRef]
35. Han, D.; Duan, X.; Li, M.; Cui, T.; Ma, A.; Ma, X. Interaction Control for Manipulator with Compliant End-Effector Based on Hybrid Position-Force Control. In Proceedings of the International Conference on Mechatronics and Automation, Takamatsu, Japan, 8–11 August 2017; pp. 863–868.
36. Shoushtari, A.L.; Dario, P.; Mazzoleni, S. A review on the evolvement trend of robotic interaction control. *Ind. Robot. Int. J.* **2016**, *43*, 535–551. [CrossRef]
37. Ruderman, M.; Iwasaki, M.; Chen, W.-H. Motion-control techniques of today and tomorrow: A review and discussion of the challenges of controlled motion. *IEEE Ind. Electron. Mag.* **2020**, *14*, 41–55. [CrossRef]
38. Pasolli, P.; Ruderman, M. Hybrid state feedback position-force control of hydraulic cylinder. In Proceedings of the 2019 IEEE International Conference on Mechatronics, Ilmenau, Germany, 18–20 March 2019.
39. Dumitriu, D.; Ionescu, M.; Melinte, D.O. New scissor-grasping tool for parallel grippers, ensuring stroke enhancement. In Proceedings of the 36th Danubia-Adria Symposium on Advances in Experimental Mechanics, Plzen, Czech Republic, 24–27 September 2019.
40. Dumitriu, D.; Melinte, O.D.; Gal, I.A. Autonomous Robot Systems for Waste Management in the Context of Smart City. Available online: https://www.incdmtm.ro/siramand/en/presentation (accessed on 14 January 2021).
41. Mărgăritescu, M.; Ancuța, P.N.; Canale, E.V.; Stanciu, D.I.; Dumitriu, D.; Brișan, C.M. Control of an Autonomous Mobile Waste Collection Robot. In Proceedings of the International Conference of Mechatronics and Cyber-Mixmechatronics, Bucharest, Romania, 5–6 September 2019; pp. 51–63.
42. Pop, N.; Vladareanu, L.; Popescu, I.N.; Ghiţă, C.; Gal, I.A.; Cang, S.; Deng, M. A numerical dynamic behaviour model for 3D contact problems with friction. *Comput. Mater. Sci.* **2014**, *94*, 285–291. [CrossRef]
43. Cho, H.; Barber, J.R. Stability of the three-dimensional Coulomb friction law. *Proc. R. Soc. Lond.* **1999**, *455*, 839–861. [CrossRef]
44. Festo. Planar Surface Gantry EXCM-30, Mechanical Installation, Dynamic Data of Stepper Motor. Available online: https://www.festo.com/net/SupportPortal/Files/327102/EXCM-30_2013--06_8025431g1.pdf (accessed on 3 November 2020).
45. Festo. Parallel Grippers HGPLE. Available online: https://www.festo.com/net/SupportPortal/Files/10495/HGPLE_ENUS.pdf (accessed on 23 November 2020).
46. Electric Grippers RoHS LES Series, Calculating Required Gripping Force. Available online: http://content2.smcetech.com/pdf/LEH.pdf (accessed on 3 November 2020).
47. SMC. Gripping Solutions, Pneumatic & Electric Grippers, Quick Overview. Available online: https://www.smc.eu/portal_ssl/WebContent/corporative/modules/Repository_e-matters/130_03--19-e-matters_pro/Grippers_leaflet_en.pdf (accessed on 3 November 2020).
48. Kurdi, A.; Chang, L. Recent Advances in High Performance Polymers—Tribological Aspects. *Lubricants* **2019**, *7*, 2. [CrossRef]
49. Tasdemir, M.; Babat, V.; Yerlesen, U. Effect of friction and wear parameters on acrylonitrile butadiene styrene/aluminum-boron carbide-glass spheres polymer composites. *Mechanics* **2014**, *20*, 407–413. [CrossRef]
50. Melinte, D.O.; Travediu, A.M.; Dumitriu, D. Deep Convolutional Neural Networks Object Detector for Real-Time Waste Identification. *Appl. Sci.* **2020**, *10*, 7301. [CrossRef]
51. Corke, P. *Denavit-Hartenberg Notation for Common Robots*; PeterCorke: Brisbane, Australia, 2014.
52. Kuyvenhoven, N. *PID Tuning Methods an Automatic PID Tuning Study with MathCad*; Calvin University Engineering Department: Grand Rapids, MI, USA, 2002; p. 315.

53. Melinte, D.O.; Vladareanu, L. Facial Expressions Recognition for Human–Robot Interaction Using Deep Convolutional Neural Networks with Rectified Adam Optimizer. *Sensors* **2020**, *20*, 2393. [CrossRef]
54. Pasolli, P.; Ruderman, M. Hybrid Position/Force Control for Hydraulic Actuators. In Proceedings of the 28th Mediterranean Conference on Control and Automation, Saint-Raphaël, France, 15–18 September 2020.
55. Vladareanu, L.; Tont, G.; Ion, I.; Vladareanu, V.; Mitroi, D. Modeling and hybrid position-force control of walking modular robots. In Proceedings of the American Conference on Applied Mathematics, Madison, WI, USA, 24–26 May 2010; pp. 510–518.
56. Wang, H.; Zhang, D.; Lu, H.; Feng, Y.; Xu, P.; Mihai, R.V.; Vladareanu, L. Active training research of a lower limb rehabilitation robot based on constrained trajectory. In Proceedings of the 2015 International Conference on Advanced Mechatronic Systems, Beijing, China, 22–24 August 2015.
57. Sands, T. *Deterministic Artificial Intelligence*; IntechOpen: London, UK, 2020.

applied sciences

MDPI

Article

Advanced Trajectory Control for Piezoelectric Actuators Based on Robust Control Combined with Artificial Neural Networks

Cristian Napole *, Oscar Barambones *, Mohamed Derbeli * and Isidro Calvo

System Engineering and Automation Deparment, Faculty of Engineering of Vitoria-Gasteiz, Basque Country University (UPV/EHU), 01006 Vitoria-Gasteiz, Spain; isidro.calvo@ehu.eus

* Correspondence: cristianmario.napole@ehu.eus (C.N.); oscar.barambones@ehu.eus (O.B.); derbelimohamed1@gmail.com (M.D.)

Abstract: In applications where high precision in micro- and nanopositioning is required, piezoelectric actuators (PEA) are an optimal micromechatronic choice. However, the accuracy of these devices is affected by a natural phenomenon called "hysteresis" that even increases the instability of the system. This anomaly can be counteracted through a material re-shape or by the design of a control strategy. Through this research, a novel control design has been developed; the structure contemplates an artificial neural network (ANN) feedforward to contract the non-linearities and a robust close-loop compensator to reduce the unmodelled dynamics, uncertainties and perturbations. The proposed scheme was embedded in a dSpace control platform with a Thorlabs PEA; the parameters were tuned online through specific metrics. The outcomes were compared with a conventional proportional-integral-derivative (PID) controller in terms of control signal and tracking performance. The experimental gathered results showed that the advanced proposed strategy had a superior accuracy and chattering reduction.

Keywords: mechatronics; hysteresis; advance trajectory control; piezoelectric; actuator; neural networks; robust control

Citation: Napole, C.; Barambones, O.; Derbeli, M.; Calvo, I. Advanced Trajectory Control for Piezoelectric Actuators Based on Robust Control Combined with Artificial Neural Networks. *Appl. Sci.* **2021**, *11*, 7390. https://doi.org/10.3390/app11167390

Academic Editors: Alessandro Gasparetto, Stefano Seriani and Lorenzo Scalera

Received: 29 June 2021
Accepted: 10 August 2021
Published: 11 August 2021

1. Introduction

A piezoelectric actuator (PEA) is a device that transduces an applied voltage into a mechanical displacement that can be in the order of nano- and micrometers. This is a huge advantage for several applications where precision is needed. For instance, optical microsurgery requires displacement capabilities from a surgeon that can be complex to achieve since the precision needed is around 10 μm [1]. An additional example is the employment of PEAs in motors where not only proper accuracy is required but also the stiffness of the actuator is a critical feature, which is another main advantage of piezoelectric actuators [2].

Regardless the benefits of PEAs, which are extensive for several uses, downsides have significant importance in the performance of these appliances. One of the main and most studied ones is hysteresis, which is a ferroelectric phenomenon related to the material poles that have arbitrary orientations that align when a voltage is applied, but the release of this action yields to a different direction [3,4]. Thus, for this reason, it is also known as a memory effect as it depends on previous history [5]. In practice, the accuracy can be reduced by up to 22% of the nominal displacement as a consequence of this anomaly [6]; another important consequence of this phenomenon is also the instability [7]. Nevertheless, since hysteresis is a natural property, available solutions comprise a material re-design or the implementation of a control algorithm [4].

Proportional-integral-derivative (PID) had been widely used in practice as a first option. Authors of [8] developed a PID for the position, which was verified in simulations and later in experiments where the results showed a tracking error reduction of 5%. A main disadvantage of PIDs is the gain scheduling, which can vary for different scenarios and or situations; thus, several ways have been suggested to tune these. For instance, the authors of [9] proposed the usage of particle swarm optimisation (PSO) in a simulation with a suitable PEA model. The obtained gains were used in an experimental platform with a commercial PEA where the results showed an improvement in the position accuracy. However, online tuning algorithms for experimental rigs can require a significant amount of computational resources, and certainly, there are other non-linear controllers that manage the compensation with better performance.

Sliding mode control (SMC) is a non-linear strategy with a discontinuity that drives the system in a sliding surface [10]. The main advantage is the robustness against uncertainties and external disturbances. The classic SMC has been implemented in PEAs for force control by the authors of [11], where they found suitable outcomes with sine signals as references. A similar approach was carried out in the research of Chouza et al. [12], where SMC based with a PID surface was implemented in a commercial PEA where they tested different reference signals, such as the ramp, constants and sine wave. In spite of the improvements of error reduction in the ramp and constant references, the sine wave showed an error of around 5%. However, in the analysed background, it was shown that the main disadvantage of SMC is the chattering that is generated by the discontinuous property. This is an unwanted effect because it increases the energy loss and also the wear in the actuator [13].

Certainly, the chattering can only be reduced and not eliminated because the discontinuity (produced by a sign function) is on of the main features of SMC. Thus, in the recent years, many proposals have been published to trim this effect. For instance, the authors of [14] changed the discontinuity by a hyperbolic function, and in comparison with conventional approaches, the enhancement was acknowledged in the results. Another advance strategy is the usage of high order sliding mode controllers (HOSMC), where high order derivatives are used in the sliding surface, and as a result, the chattering is relieved [15,16]. An example of implementation has been carried out by the authors of [17], where they used an HOSMC as an observer for error compensation in a PEA test rig. The results showed significant improvements in comparison with other conventional types of SMC strategies in simulation and experiments. Regardless of the enhancement of HOSMC over conventional SMC, the design of sliding controllers establishes that the control law is split into a switching and an equivalent term that are aimed to maintain and compensate the sliding motion [18]. The equivalent is commonly achieved through a mathematical model that describes the system, and this would imply the use of a proper hysteresis description.

Hysteresis models for PEAs are mainly classified in two main categories: physical and phenomenological [19]. The first-mentioned group is a description of the ferromagnetic effect that produces the non-linearity, although the material dependency and complex numerical solutions are the downsides of these theories [20,21]. In regards to the phenomenological, the sub-classification is related to the ones based on differential equations (Dunhem [22], Backslash [23] and Bouc-Wen [24]), operator models (Preisach [5], Prandtl-Ishlinskii [25] and Krasnoselskii-Pokrovskii [26]) and polynomial models [27]. Nevertheless, the disadvantages of these approaches are linked with complicated solutions to gather the inverse model, incapability to deal with asymmetric hysteresis, rate dependency and complex implementation [20].

Based on the research about the background that we made, we designed an HOSMC controller known as QC-Continuous (QCSMC), which provides suitable results in terms of chattering reduction in previous works [28]. Due to the drawbacks that we enumerated about hysteresis models, we decided to achieve the equivalent term of the sliding controller through means of an artificial neural network (ANN). This is possible because we could conduct experiments to acquire data for the ANN training.

The structure of this article is arranged as follows. In Section 2, we provided a description about the hardware employed for the experiments within their technical specifications, a hysteresis explanation of the PEA used, and the controllers designed within the metrics used to contrast their performance. Section 3 provides the obtained results of the trained

ANN and the control implementation outcomes. Finally, in Section 4, we recap the most important features obtained within the research.

2. Materials Furthermore, Methods

2.1. Hardware Description

The main hardware that we used is a Thorlabs PK4FYC2 PEA that was produced with several piezoelectric chips wedged with epoxy and glass beads. A maximum displacement of 38.5 μm is achieved with a driving voltage of 150 V. As the device is frequently handled for precise and micrometric displacement, the manufacturer included four strain gauges arrayed in a Wheatstone bridge so that the measurement could be obtained from resistance variation, which provides a better resolution [29].

The PEA accessories include a driver cube KPZ101, whose input signal is 0–10 V and is transformed into a 0–150 V to manage the PEA. As the PEA delivers an output voltage that is in the order of milivolts, a pre-amplifier augments this value into a 0–2 V signal. The latter is transferred into a reader cube KSG101, which transforms the voltage into 0–10 V. Further technical details of the PEA and its peripherals are summarized in Table 1.

Table 1. Thorlabs hardware used.

PEA PK4FYC2	Values	Units
Maximum displacement	38.5	μm
Blocking force	1000	N
Resonant frequency	34	kHz
Maximum error	15	%
Driver Cube KPZ101		
Output driving voltage for PEA	150	V
Input driving voltage	0–10	V
Maximum output bandwidth	1	kHz
Reader Cube KSG101		
Output range	0–10	V
Resolution	1	nm
Pre-Amplifier AMP002		
Output range	0–2	V

Since the input and output signals of the PEA are in the range of 0–10 V, we, therefore, used a proper platform for acquisition and control, such as the dSpace DS1104 board. We linked this device to a Dell Precision Workstation T3500 through a peripheral component interconnect (PCI) for monitoring and control. Moreover, the DS1104 has an external connector panel CP1104 that we connected to the PEA driver and reader.

In regards to the software, we developed the proposed strutures in Simulink by Mathworks and embedded them through the platform dSpace real-time interface (RTI). This allowed us to generate and manage a real-time control algorithm, while the RTI helped to reduce the compilation time. All the information was acquired with ControlDesk and processed with MATLAB by Mathworks. The explained relation between software and hardware is detailed in Figure 1. The chosen sampling time was 1kHz as it is a suitable match with the acquisition and the hardware physical limits.

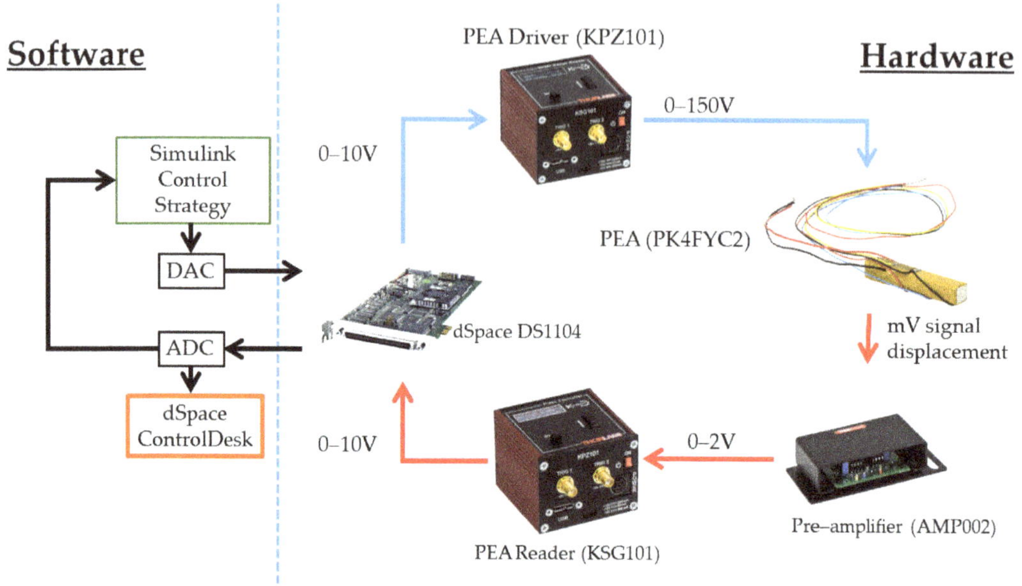

Figure 1. Hardware and software workflow.

2.2. Hysteresis Description and Reference Design

Common excitation signals that represent a suitable hysteresis graph in PEAs are triangular and sine waves [5]. The triangular waveform is a complex input for the PEA, as it not only implies that sharp slope changes are required to be followed, but also this wave is generated from high-frequency harmonics [30]. Therefore, we developed a triangular reference of a 145V amplitude, which is lower than the maximum allowed voltage so that the PEA lifetime is not affected; additionally, we chose a period of 4 s for this signal.

With the previously described triangular waveform as an output, we acquired the displacement of the PEA where the outcome is the one shown in Figure 2. The mechanics and features of the curve are developed as follows from the beginning of the applied voltage:

- **Initial point:** We calibrated the PEA so that it initially starts at a zero displacement from this.
- **Curve (1):** This is known as the initial ascending curve, which begins from the previously described point and ends at the upper target point. As the figure shows, the non-linearity is present along this path.
- **Upper target point:** At this place, the PEA reached the correspondent displacement to the specified amplitude of the triangular waveform.
- **Curve (2):** This, known as the *second ascending curve*, shows that the PEA has an asymmetric hysteresis, which is a phenomenon that creates difficulties when mathematical models need to be found to reflex.
- **Lower converging point:** Ideally, the final position could have been at the initial point when the applied voltage is null. However, in this case, the lower converging point is not the same as the initial point.
- **Curve (3):** Provided that amplitude and period are the same along the experiment, then this curve will be equal for the following ascending cycles.
- **Curve (4):** As with the curve (3), this course will be the same provided that the reference configuration is constant.

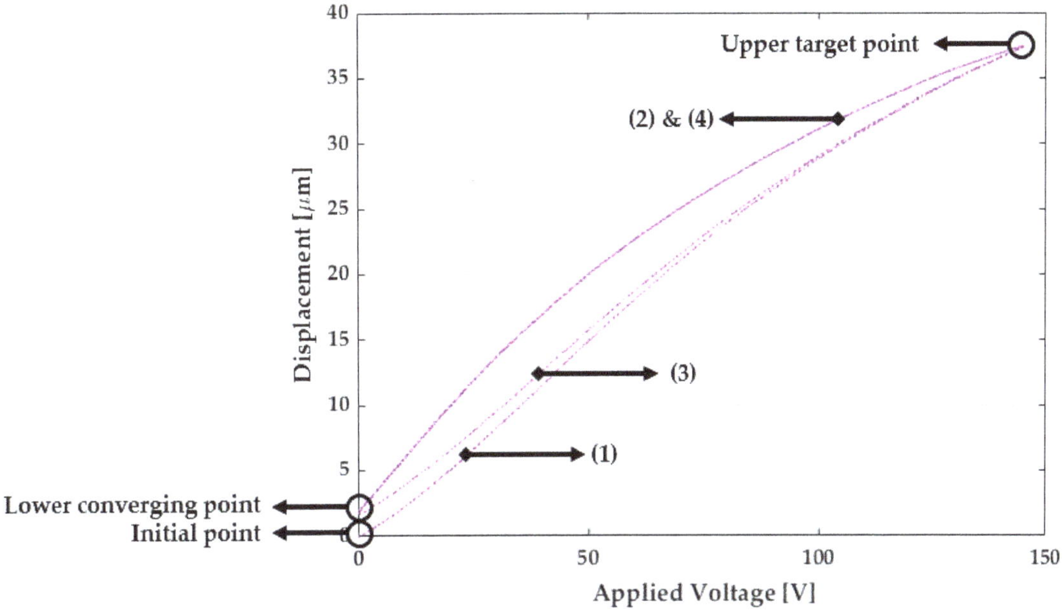

Figure 2. Hysteresis graph of the used PEA.

Taking into account the previous analysis, we designed the reference considering the applied voltage in the PEA. As a first consideration, the initial point within the first ascending curve is negligible to evaluate the performance of the PEA in a long-term experiment. Therefore, we outlined a reference between the lower converging and upper target point as linear Equation (1) reflects. In this mathematical expression, we defined X as the displacement, α the slope between the two points of interests and b the vertical offset in terms of the lower converging point.

$$X[\mu\text{m}] = \alpha \cdot V + b \tag{1}$$

2.3. Contrasted Schemes and Their Design

Based on the framework that we proposed, we decided to contrast it with a PID controller, a commonly used structure in the industry and PEA tracking performance. A simplified description of the used controllers is displayed in Figure 3. We designed the structure by taking into account the advantages of the RTI properties so that the variables of each controller could be changed in real-time. We developed the quest to find these values through the online minimization of the integral of the absolute error (IAE). Equation (2) express the IAE, where e_i is the error in at the i-th sample, Δt is the sampling time and N the number of points chosen to calculate the value.

$$IAE = \sum_{i=1}^{N} |e_i| \Delta t \tag{2}$$

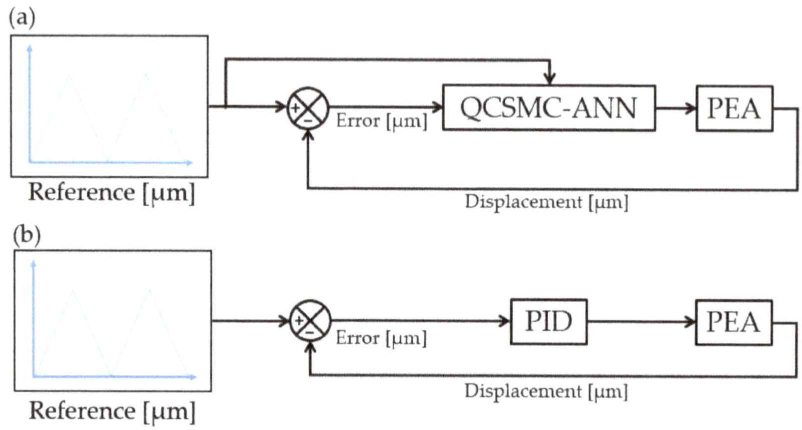

Figure 3. Control schemes used where (**a**) is the proposed QCSMC-ANN and (**b**) is the contrasted PID.

Another statistical metric used is the root-mean-square-error (RMSE), which reflects the precision of each framework in terms of the error. Equation (3) is the mathematical expression of the latter-described metric.

$$RMSE = \sqrt{\frac{1}{N}\sum_{i=1}^{N}(e_i)^2} \tag{3}$$

According to Ventura and Fridman [31], the chattering can be measured with Equation (4) in terms of the energy that is being dissipated from a nominal state $\overline{u}(t)$ of a control signal $u(t)$.

$$chatt(u) = \left[\int_{t_0}^{T}\left(\dot{u}(t) - \dot{\overline{u}}(t)\right)^2 dt\right]^{\frac{1}{2}} \tag{4}$$

2.4. *Quasi-Continuous Sliding Mode Control*

QC-SMC belongs to the HOSMC, known for the robustness and performance in tracking precision; moreover, the chattering reduction is another advantage of this controller over other structures [32]. The control law that we settled as Equation (5) comprises of the terms u_{ann} and $u_{sw}0$, which, respectively, aim to compensate the non-linearities (such as the hysteresis) and counteract uncertainties or perturbations The definition of u_{sw} is defined in Equation (6), which is dependant on a sliding surface expressed by Equation (7) and where parameters λ and γ are positively defined by the designer. The term u_{ann} is dependant on the ANN compensation, and further details are given in the following section.

$$u = u_{ann} + u_{sw} \tag{5}$$

$$u_{sw} = -\gamma\frac{\dot{s} + |s|^{1/2}sign(\dot{s})}{|\dot{s}| + |s|^{1/2}} \tag{6}$$

$$s = e + \lambda\int_0^t edt \tag{7}$$

Neural Network Compensation Design

The implementation of conventional SMC methods, where the equivalent term has to be through a mathematical model, can yield to a insufficient compensation or even increase the computational cost. Nevertheless, in recent years, ANNs have been a suitable solution for system identification. The only drawback is the time that the training algorithm requires to develop a truthful output. Hence, we used a time-delay neural network (TDNN) due to the efficiency related to training time versus accuracy obtained.

In former investigations, we tested TDNN structures to reduce hysteresis, which showed proper results in combination with conventional controllers [33]. As the name states, a TDNN is an extension of a classic multilayer perceptron (MLP) that works with time signals. The inclusion of time delays n allows the neurons to get further information about the time history of the input; this implies that the ANN will fit to a time set pattern [34]. Mathematically, this is expressed with Equation (8), where f is a non-linear function that relates the input/output of the ANN. An schematic explanation is provide in Figure 4.

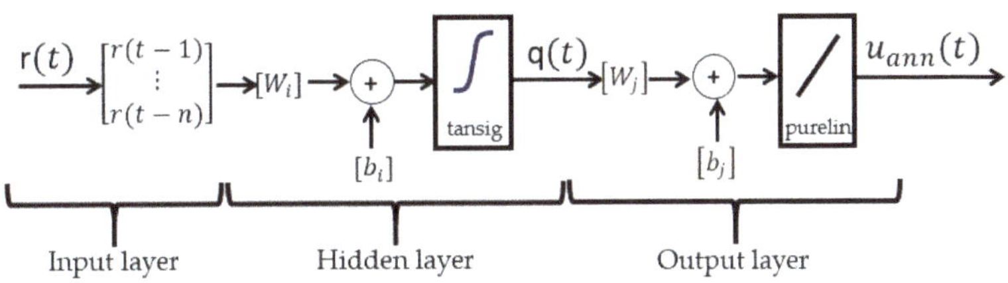

Figure 4. The ANN.

$$u_{ann} = f\Big(x(t), x(t-1), x(t-2), \ldots, x(t-n)\Big) \tag{8}$$

A further expansion of the function f is as the following Equations (9)–(12): the retarded reference inputs $r(t-n)$ are weighted with parameters W_i and bias b_i; later, this operation yields into the activation function called *tansig*. The output $q(t)$, which is the outcome of the described operation, is employed as an input into the output layer. In this case, the procedure is similar as previously, but the activation is done with a linear transfer function called *purelin*. Subsequently, the output of this layer provides the compensation voltage u_{ann}.

$$q(t) = tansig\left(\sum_{p=0}^{n} W_i \cdot r(t-n) + b_i\right) \tag{9}$$

$$tansig(x) = \frac{2}{1+e^{-2x}} - 1. \tag{10}$$

$$u_{ann} = purelin[W_j q(t) + b_j], \tag{11}$$

$$purelin(x) = x. \tag{12}$$

The calculation of the weights and bias, related to the previous explained relations, are achieved with training algorithms. In this case, we used Levenberg–Marquardt, which represents a method that guarantees the fitting of the ANN to the experimental data through an adaptive behaviour [35]. This mechanism is generated by finding the location of the minimum of a cost function is declared as the sum of square errors and the real measurements within an iterative updating.

2.5. PID Control

The ream of PID controllers in terms of design and applications is vast, which allows a robust design even with its simple structure. Even though several tuning methods are available such as Ziegler–Nichols or Cohen–Coon [36], we used, in this case, the minimization of IAE as it was explained previously. The controller expression is the one provided in Equation (13), where K_p, K_i and K_d are, respectively, the proportional, integral and derivative gains; Δt is the sampling time, and $e(k)$ is the error at a step k.

$$u(k) = K_p e(k) + K_i \sum_{i=1}^{k} e(i)\Delta t + \frac{K_d[e(k) - e(k-1)]}{\Delta t}. \tag{13}$$

3. Results

3.1. ANN Analysis Results

We recorded experimental data from the PEA where the input was a triangular signal of 145 V with a 4 s period. Furthermore, the displacement was acquired through the strain gauge reader in 40 s experiments with a sampling time of 1 kHz. From this data, we used the displacement as an input and the applied voltage as an output because the aim is to achieve an inverse model.

After several tests to achieve the best MSE, we configured the ANN in 22 neurons with 5 input delays. The data were split into 70%, 15% and 15%, respectively, for training, evaluation and testing. Finally, the performance was measured with the mean squared error (MSE) in the validation set, where the value obtained was 0.017 in 12.000 iterations made in 4 min.

Figure 5 shows the performance of the ANN to fit with the PEA hysteresis in a 4 s cycle. Although Figure 5a adapts with a decent effectiveness, Figure 5b exhibits the error where several features can be highlighted. Between 1.5 and 2.5, the error tends to increase with significant peaks; nevertheless, at 2 s, the deviation increases considerably due to the slope change as it is a complex transition to be projected by the ANN. Still, the calculated RMSE for this case provided 0.041 V in comparison to experimental data, which was acceptable.

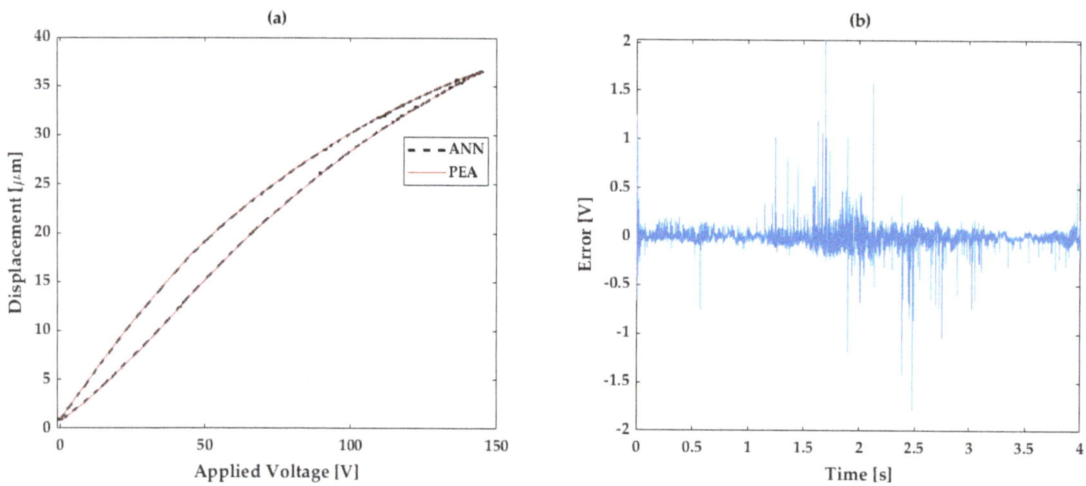

Figure 5. Ability of the ANN to fit to the PEA nonlinearity, where: (**a**) is the hysteresis graph contrast and (**b**) the error of the fitting.

3.2. Reference Tracking Results

The control structures were designed in Simulink, which was later embedded in the dSpace platform. Even though the main reference used was a triangle wave, we also used a sine signal (with same period and amplitude) and a variable amplitude triangular signal. The aim of this was to test the flexibility of the proposed structures against different references.

In regards to the parameters of each controller, these were reached through the minimization of the IAE in the experiments. The PID constants K_p, K_i and K_d gathered are, respectively, 1000, 10 and 10^{-4}. The QCSMC-ANN parameters, γ and λ, acquired are 60 and 8, respectively.

3.3. Triangular Tracking Results

The first contract was performed with a triangular wave due to its complexity to be followed. Figure 6 presents the error obtained with the QCSMC-ANN and PID in triangular cycles where several points can be highlighted. The PID controller had a variable performance since between 0 and 1 s, the error declined; however, in the following second, the error began to rise up until 2 s. Despite the QCSMC-ANN providing a high amplitude in the first second of this period, in the following, it was diminished.

Certainly, at 2 s, the first critical point appeared since it is where the slope of the triangular reference changes its sign. The PID changed suddenly from 0.1 µm to near −0.12 µm; although the controller performed an abrupt correction, the QCSMC-ANN generated a similar action but with a faster improvement in time. This is reflected after 2 s where the PID had transitory development without reaching the null value of the error. Nevertheless, the QCSMC-ANN carried with the same demeanour as previously right after the slope change.

The fourth second of this analysis exhibits another crucial point to focus as it is the following slope change at the lower converging point. The PID unveiled a similar situation as previously at 2 s but with lower amplitude and a subsequently transitory response. On the other hand, even if the QCSMC-ANN featured a peak that has a value above −0.2 µm, the later reaction shows a similar trend as previously described, where the controller aims to a mean near the null value. After 4 s, since the signal is repeated, the detailed features are mirrored.

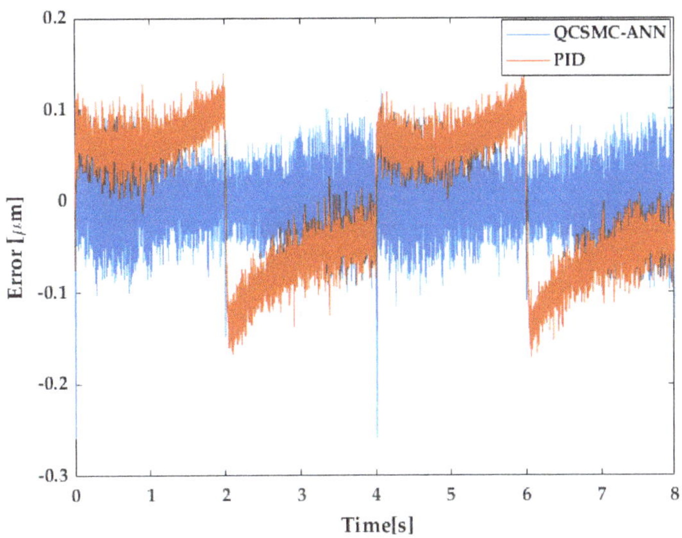

Figure 6. The error generated in 2 cycles of a triangular reference.

Aside from the error development, the control signal is an important feature to analyse because it contributes to the performance of the proposed structures. Figure 7 is a contrast of the control signal generated along the analysed error of both frameworks. As main characteristics to take into account at this point, saturations or sudden changes needed focus as these can damage the PEA driver cube. Henceforth, it can be perceived that both controllers had a suitable demeanour, and any downsides were presented in the experiments.

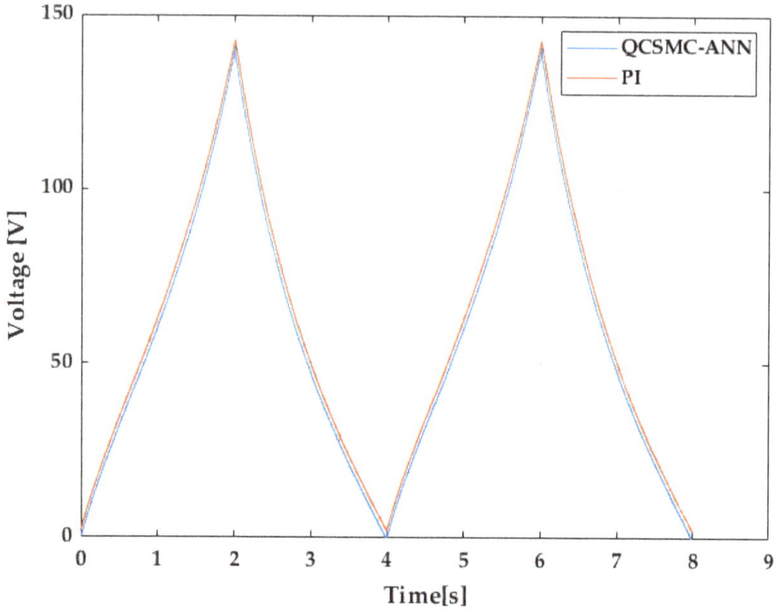

Figure 7. The control signal in 2 cycles of a triangular reference.

3.4. Sinusoidal Tracking Results

In pursuance of a suitable performance test that can show the docility of the proposed design against similar references, we inspected the development with a sine wave form with the same chosen amplitude and periods as previously. Figure 8 shows the error that the PID and the QCSMC-ANN produced in the mentioned reference signal. The PID had an inferior performance in relation to the previous test since after 1 s, the amplitude increased with peaks up to 0.15 μm. Although the slope transition is softer in 2 s, the error kept increasing afterwards, which resulted in a variation of around 0.3 μm. However, the QCSMC-ANN behaved even better than previously since the error was compensated almost equally along the test with an amplitude below 0.05 μm, which oscillates around the null value.

Finally, the control signal that is presented in Figure 9 unveils a better performance than the former analysis due to the softness of the signal. It can be seen that any harm changes were developed in the analysed time, which can lead to a damage of the involved hardware.

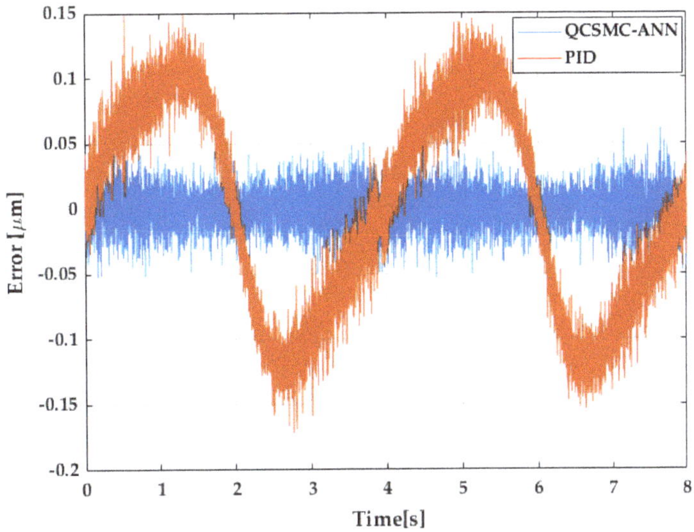

Figure 8. The error generated in 2 cycles of a sine wave reference.

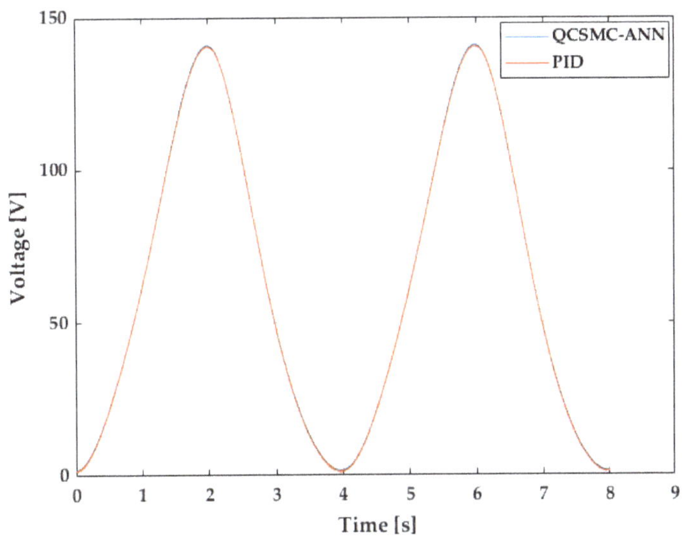

Figure 9. The control signal in 2 cycles of a sine wave reference.

3.5. Triangular Tracking Results with Variable Amplitude

Another experiment performed is the triangular reference signal with variable amplitudes that were settled randomly in 25 V and 121 V. Figure 10 shows the repercussions of the error for the considered reference. During the first 4 s, where the amplitude was 25 V, both controllers had a similar demeanour. The PID shows a perceptible shift at the upper target point in 2 s; however, the QCSMC-ANN provided a constant development along this range with the same deviation as the PID. On the other hand, the major difference can be noticeable during 121 V, where the PID behaves similarly to previously analysed triangular signals where fast corrections occur during the slope changes of the reference signal. This effect produced a brief increment of the error amplitude in the analysed controllers, but the QCSMC-ANN managed to carry this without any transients, a feature produced by the PID.

As previously, another important feature is the control signal, which is shown in Figure 11. It can be seen that any saturations or sudden corrections that can deteriorate the hardware were developed in the performed experiments. Nevertheless, effects of chattering are analysed in further details in the following section.

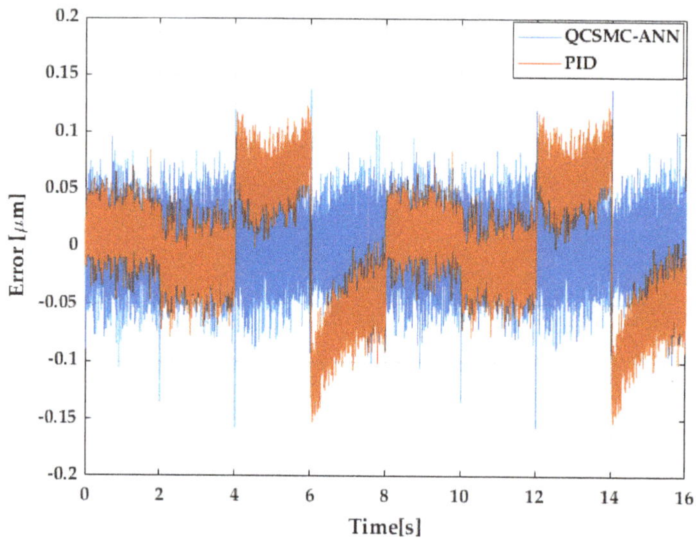

Figure 10. The error generated in 2 cycles of a triangular reference with variable amplitude.

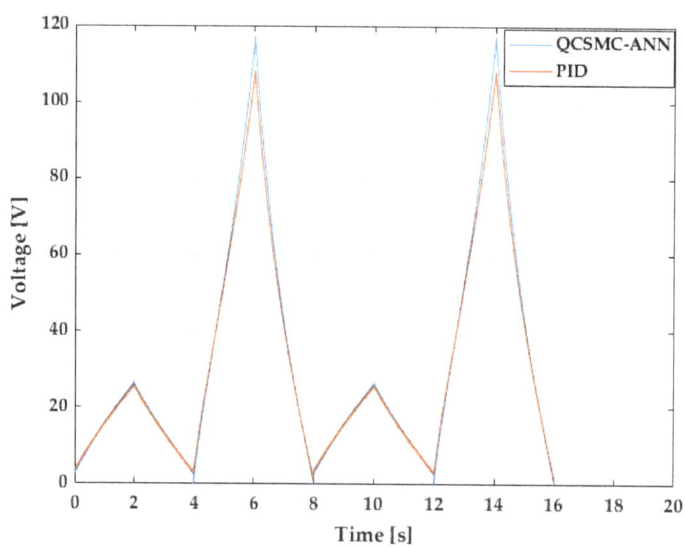

Figure 11. The control signal in 2 cycles of a triangular reference with variable amplitude.

3.6. Metrics Results

Certainly, for extra precise performance measurement and comparison, we used three tools that aided us to gather more conclusions. As previously explained, the IAE was used to tune the parameters until this value becomes minimum. The RMSE provides the accuracy in terms of the error, and because we implemented an SMC-based controller (known for the chattering generated in the control signal), we calculated this value based on the previously explained method. Numerical results of these parameters are summarized in Table 2.

Table 2. A comparison of the different metrics.

Reference	IAE			RMSE (μm)			Chatt(u) in 4 s		
	QCSMC-ANN	PID	Diff (%)	QCSMC-ANN	PID	Diff (%)	QCSMC-ANN	PID	Diff (%)
Triangle	0.1314	0.28	53.07	0.0404	0.0756	46.5	412.75	640.97	35.6
Sine wave	0.0518	0.28	81.5	0.0161	0.0795	79.7	108.8	600.8	81.8
Var. Amp.	0.2067	0.33	33.6	0.0318	0.0519	38.5	514.88	659.4	22

In regards to the IAE, the QCSMC-ANN indicated a superior performance in the triangular reference, which led to a 53.07% difference in comparison with the PID. Nevertheless, in the sine wave test, it is shown that this value was augmented, as the QCSMC-ANN reached 81.5% more. Additionally, the variable amplitude signal carried with the same trend, as the difference achieved was 33.6% favourable for the proposed algorithm. The RMSE showed a similar manner since the QCSMC-ANN kept the advantage over the PID in both signals granted 46.5%, 79.7% and 38.5% of difference with the triangular, sine and variable amplitude waves, respectively. Lastly, the measured chattering indicates the dominance of the QCSMC-ANN over the PID, and even it emblazons through values of the suitable performance of the control signal; according to the values of 35.6%, 81.8% and 22% of difference in both signals, it is clear that the QCSMC-ANN had less chattering than the PID controller.

4. Conclusions

Throughout this research, we developed a control strategy with the aim of increasing the accuracy of a commercial PEA. After an analysis of the previous investigations from other authors and based on the study that we made, we found that the hysteresis was the main non-linear phenomenon to be counteracted.

First, we made an analysis of the properties of a commercial PEA PK4FYC2 from Thorlabs. According to the manufacturer, the maximum error is 15%, which is a considerable value for applications where high precision is required. As the main reference, we used a triangular wave because it is a complex signal to be tracked due to the high-frequency harmonics and sudden slope changes.

Secondly, we proposed a robust sliding controller due to the advantages studied in the related works from the introduction. Thus, we chose to use a QCSMC, which belongs to the HOSMC, so that the chattering is reduced in comparison to classic SMCs. Commonly, sliding controllers have two terms where one of them is achieved through a mathematical model, but instead, a distinctive feature of our controller is the use an ANN. We contrasted the proposed design with a conventional PID in terms of several metrics, such as IAE (that was also used to tune the gains of each framework in experiments), MSE and chattering.

In regards to the results, at first, we analysed the performance of the ANN, which provided a suitable RMSE and fitting to the hysteresis. Later, we implemented the QCSMC-ANN and the PID in a dSpace platform for a real-time experiment. The results showed that the QCSMC-ANN generated a lower tracking error, which oscillated around the null value. The PID displayed a slow compensation and even with a disparity that had a variation with a higher amplitude that reached the 0.1 μm. Nevertheless, both schemes showed suitable control signals in the analysed graphs. Additionally, because we wanted to test the flexibility of both structures against reference changes, we used a sine wave and a triangular reference with variable amplitude. In this case, the first case exhibited a demeanour that was similar in the error and control signal because the reference was softer than the previous one. In regards to the variable amplitude signal, the QCSMC-ANN still provided a superior performance, which was verified in graphical and numerical analysis.

Finally, we calculated the mentioned metrics, which showed concrete values of the performance from the comparisons made. Thus, in terms of the IAE and RMSE, the QCSMC-ANN had a important distinction favourable for the QC-SMC in the proposed reference signals. The calculated chattering also showed a significant difference, where the QC-SMC carried the leading trend.

As future related works, we expect to study assorted options, including the usage of other ANNs configurations with several training algorithms to improve the accuracy. Certainly, this also implies taking into account the computational resources in terms of the training time. The QCSMC can also be improved through the usage of an adaptive algorithm for gain scheduling.

Author Contributions: Conceptualization, O.B. and C.N.; methodology, O.B., M.D. and C.N.; software, C.N.; validation, C.N.; formal analysis, O.B., M.D. and C.N.; investigation, O.B. and C.N.; resources, O.B.; writing—original draft preparation, C.N. and M.D.; writing—review and editing, O.B., C.N. and M.D.; supervision, O.B. and I.C.; project administration, O.B. and I.C. All authors have read and agreed to the published version of the manuscript.

Funding: This research was funded by the Basque Government through the project EKOHEGAZ (ELKARTEK KK-2021/00092), Diputación Foral de Álava (DFA) through the project CONAVANTER, and to the UPV/EHU through the project GIU20/063.

Institutional Review Board Statement: Not applicable.

Informed Consent Statement: Not applicable.

Acknowledgments: The authors wish to express their gratitude to the Basque Government, Diputación Foral de Álava (DFA) and UPV/EHU.

Conflicts of Interest: The authors declare no conflict of interest.

Abbreviations

The following abbreviations are used in this manuscript:

PEA	Piezoelectric actuator
PID	Proportional-integral-derivative
PSO	Particle swarm optimisation
SMC	Sliding mode control
HOSMC	High order sliding mode control
QCSMC	Cuasi-continous sliding mode control
ANN	Artificial neural networks
PCI	Peripheral component interconnect
RTI	Real-time interface
IAE	Integral of absolute error
RSMC	Root-mean-squared-error
TDNN	Time delay neural network
MLP	Multilayer perceptron
MSE	Mean squared error

References

1. Vander Poorten, E.; Riviere, C.N.; Abbott, J.J.; Bergeles, C.; Nasseri, M.A.; Kang, J.U.; Sznitman, R.; Faridpooya, K.; Iordachita, I. 36—Robotic Retinal Surgery. In *Handbook of Robotic and Image-Guided Surgery*; Abedin-Nasab, M.H., Ed.; Elsevier: Amsterdam, The Netherlands, 2020; pp. 627–672. [CrossRef]
2. Han, L.; Yu, L.; Pan, C.; Zhao, H.; Jiang, Y. A Novel Impact Rotary—Linear Motor Based on Decomposed Screw-Type Motion of Piezoelectric Actuator. *Appl. Sci.* **2018**, *8*, 2492. [CrossRef]
3. Tyunina, M.; Miksovsky, J.; Kocourek, T.; Dejneka, A. Hysteresis-Free Piezoresponse in Thermally Strained Ferroelectric Barium Titanate Films. *Electron. Mater.* **2021**, *2*, 17–23. [CrossRef]
4. Damjanovic, D. Hysteresis in piezoelectric and ferroelectric materials. In *The Science of Hysteresis*; Academic Press: Cambridge, MA, USA, 2006; Chapter 4; pp. 338–452. [CrossRef]

5. Xiong, R.; Liu, X.; Lai, Z. Modeling of Hysteresis in Piezoelectric Actuator Based on Segment Similarity. *Micromachines* **2015**, *6*, 1805–1824. [CrossRef]
6. Stefanski, F.; Minorowicz, B. Open loop control of piezoelectric tube transducer. *Arch. Mech. Technol. Mater.* **2018**, *38*, 23–28. [CrossRef]
7. Wang, W.; Wang, J.; Chen, Z.; Wang, R.; Lu, K.; Sang, Z.; Ju, B. Research on Asymmetric Hysteresis Modeling and Compensation of Piezoelectric Actuators with PMPI Model. *Micromachines* **2020**, *11*, 357. [CrossRef] [PubMed]
8. Dsouza, R.; Benny, B.; Sequeira, A.; Karanth, N. Hysteresis Modeling of Amplified Piezoelectric Stack Actuator for the Control of the Microgripper. *Am. Sci. Res. J. Eng. Technol. Sci.* **2016**, *15*, 265–281.
9. Ding, B.; Li, Y.; Xiao, X.; Tang, Y. Optimized PID tracking control for piezoelectric actuators based on the Bouc-Wen model. In Proceedings of the 2016 IEEE International Conference on Robotics and Biomimetics (ROBIO), Qingdao, China, 3–7 December 2016; pp. 1576–1581. [CrossRef]
10. Ma, Y.; Li, Y. Active Disturbance Compensation Based Robust Control for Speed Regulation System of Permanent Magnet Synchronous Motor. *Appl. Sci.* **2020**, *10*, 709. [CrossRef]
11. Abidi, K.; Sabanovic, A.; Yesilyurt, S. Sliding mode control based disturbance compensation and external force estimation for a piezoelectric actuator. In Proceedings of the 8th IEEE International Workshop on Advanced Motion Control, AMC '04, Kawasaki, Japan, 28 March 2004; pp. 529–534. [CrossRef]
12. Chouza, A.; Barambones, O.; Calvo, I.; Velasco, J. Sliding Mode-Based Robust Control for Piezoelectric Actuators with Inverse Dynamics Estimation. *Energies* **2019**, *12*, 943. [CrossRef]
13. Svečko, R.; Gleich, D.; Sarjaš, A. The Effective Chattering Suppression Technique with Adaptive Super-Twisted Sliding Mode Controller Based on the Quasi-Barrier Function; An Experimentation Setup. *Appl. Sci.* **2020**, *10*, 595. [CrossRef]
14. Velasco, J.; Barambones, O.; Calvo, I.; Zubia, J.; Saez de Ocariz, I.; Chouza, A. Sliding Mode Control with Dynamical Correction for Time-Delay Piezoelectric Actuator Systems. *Materials* **2020**, *13*, 132. [CrossRef]
15. Shahid, Y.; Wei, M. Comparative Analysis of Different Model-Based Controllers Using Active Vehicle Suspension System. *Algorithms* **2020**, *13*, 10. [CrossRef]
16. Saha, S.; Amrr, S.M.; Saidi, A.S.; Banerjee, A.; Nabi, M. Finite-Time Adaptive Higher-Order SMC for the Nonlinear Five DOF Active Magnetic Bearing System. *Electronics* **2021**, *10*, 1333. [CrossRef]
17. Che, X.; Tian, D.; Xu, R.; Jia, P. Finite-Time Control for Piezoelectric Actuators With a High-Order Terminal Sliding Mode Enhanced Hysteresis Observer. *IEEE Access* **2020**, *8*, 223931–223940. [CrossRef]
18. Zenteno-Torres, J.; Cieslak, J.; Dávila, J.; Henry, D. Sliding Mode Control with Application to Fault-Tolerant Control: Assessment and Open Problems. *Automation* **2021**, *2*, 1–30. [CrossRef]
19. Sabarianand, D.; Karthikeyan, P.; Muthuramalingam, T. A review on control strategies for compensation of hysteresis and creep on piezoelectric actuators based micro systems. *Mech. Syst. Signal Process.* **2020**, *140*, 106634. [CrossRef]
20. Armin, M.; Roy, P.N.; Das, S.K. A Survey on Modelling and Compensation for Hysteresis in High Speed Nanopositioning of AFMs: Observation and Future Recommendation. *Int. J. Autom. Comput.* **2020**, *17*, 1–23. [CrossRef]
21. Xue, G.; Zhang, P.; He, Z.; Li, D.; Yang, Z.; Zhao, Z. Modification and Numerical Method for the Jiles-Atherton Hysteresis Model. *Commun. Comput. Phys.* **2017**, *21*, 763–781. [CrossRef]
22. Ahmed, K.; Yan, P.; Li, S. Duhem Model-Based Hysteresis Identification in Piezo-Actuated Nano-Stage Using Modified Particle Swarm Optimization. *Micromachines* **2021**, *12*, 315. [CrossRef] [PubMed]
23. Su, C.Y.; Stepanenko, Y.; Svoboda, J.; Leung, T. Robust adaptive control of a class of nonlinear systems with unknown backlash-like hysteresis. *IEEE Trans. Autom. Control* **2000**, *45*, 2427–2432. [CrossRef]
24. Gan, J.; Zhang, X. Nonlinear Hysteresis Modeling of Piezoelectric Actuators Using a Generalized Bouc–Wen Model. *Micromachines* **2019**, *10*, 183. [CrossRef] [PubMed]
25. Li, M.; Wang, Q.; Li, Y.; Jiang, Z. Modeling and Discrete-Time Terminal Sliding Mode Control of a DEAP Actuator with Rate-Dependent Hysteresis Nonlinearity. *Appl. Sci.* **2019**, *9*, 2625. [CrossRef]
26. Xu, R.; Tian, D.; Wang, Z. Adaptive Tracking Control for the Piezoelectric Actuated Stage Using the Krasnosel'skii-Pokrovskii Operator. *Micromachines* **2020**, *11*, 537. [CrossRef]
27. Gan, J.; Zhang, X.; Wu, H. Tracking control of piezoelectric actuators using a polynomial-based hysteresis model. *AIP Adv.* **2016**, *6*, 065204. [CrossRef]
28. Silaa, M.; Derbeli, M.; Barambones, O.; Cheknane, A. Design and Implementation of High Order Sliding Mode Control for PEMFC Power System. *Energies* **2020**, *13*, 4317. [CrossRef]
29. Giordano, J.L. On the sensitivity, precision and resolution in DC Wheatstone bridges. *Eur. J. Phys.* **1997**, *18*, 22–27. [CrossRef]
30. Qin, Y.; Duan, H. Single-Neuron Adaptive Hysteresis Compensation of Piezoelectric Actuator Based on Hebb Learning Rules. *Micromachines* **2020**, *11*, 84. [CrossRef] [PubMed]
31. Ventura, U.P.; Fridman, L. Chattering measurement in SMC and HOSMC. In Proceedings of the 2016 14th International Workshop on Variable Structure Systems (VSS), Nanjing, China, 1–4 June 2016; pp. 108–113. [CrossRef]
32. Levant, A. Principles of 2-sliding mode design. *Automatica* **2007**, *43*, 576–586. [CrossRef]
33. Napole, C.; Barambones, O.; Derbeli, M.; Silaa, M.; Calvo, I.; Velasco, J. Tracking Control for Piezoelectric Actuators with Advanced Feed-Forward Compensation Combined with PI Control. *Proceedings* **2020**, *64*, 29. [CrossRef]
34. Kaiser, M. Time-delay neural networks for control. *IFAC Proc. Vol.* **1994**, *27*, 967–972. [CrossRef]

35. Du, Y.C.; Stephanus, A. Levenberg-Marquardt Neural Network Algorithm for Degree of Arteriovenous Fistula Stenosis Classification Using a Dual Optical Photoplethysmography Sensor. *Sensors* **2018**, *18*, 2322. [CrossRef] [PubMed]
36. Vilanova, R.; Alfaro, V.M. Control PID robusto: Una visión panorámica. *Rev. Iberoam. Autom. Inform. Ind.* **2011**, *8*, 141–158. [CrossRef]

applied sciences

MDPI

Article

A Theoretical Method for Designing Thin Wobble Motor Using an Electromagnetic Force and an Electropermanent Magnet for Application in Portable Electric Equipment

Sang Yong Park, Buchun Song and Yoon Su Baek *

Department of Mechanical Engineering, Yonsei University, Seoul 03722, Korea; parksy@yonsei.ac.kr (S.Y.P.); theme7749@gmail.com (B.S.)
* Correspondence: ysbaek@yonsei.ac.kr

Abstract: The thin wobble motors that are required to hold rating shafts employ an electropermanent magnet. This turns the holding force on and off by applying a momentary electrical pulse. To design the magnet devices without the need for finite element analyses, a theoretical force model is necessary for predicting the attractive force. In this paper, first, a force model is derived by estimating the permeance around the air gap. A magnetic circuit is constructed, employing a relatively simple method to build the model in clouding leakage flux. Thus, the basic structure and driving principle are also presented. Next, an analytical force model is constructed on the basis of distribution parameter analysis between the stator and the rotating shaft. The design of the electromagnet core and the control method are presented. Finally, a prototype model of the motor that is 30 mm in diameter and 7 mm in thick is fabricated. The two models are verified by comparing the results of FEM with the results of the experiments. They can properly predict the attractive force, so the thin wobble motor with holding force can be applied in portable electric equipment.

Keywords: wobble motor; permeance; magnetic circuit; leakage flux; electropermanent magnet; force model

Citation: Park, S.Y.; Song, B.; Baek, Y.S. A Theoretical Method for Designing Thin Wobble Motor Using an Electromagnetic Force and an Electropermanent Magnet for Application in Portable Electric Equipment. *Appl. Sci.* **2021**, *11*, 881. https://doi.org/10.3390/app11020881

Received: 25 December 2020
Accepted: 14 January 2021
Published: 19 January 2021

1. Introduction

Electromagnetic actuators, which have many advantages, including their small size, light weight, high precision, high accuracy, and high efficiency, have been widely used as rotary driving motors in various fields such as industrial machinery, medicine, human care device, electronic parts assembly, military equipment and mobile robots. These motors are manufactured depending on the target application in various ways. For example, new motors shapes have been developed using physical and chemical phenomena such as electromagnetic force, smart polymers, shape-memory alloy (SMA), electrostatic force, etc. Some of the materials used to drive these motors are difficult to control, and, therefore, there are limitations to their use. For this reason, we chose electromagnetically driven motors that are easy to manufacture and control and can easily calculate the generated force. However, these motors require a reduction gear device to increase their driving torque. The added reduction gear equipment increases the total volume of the driving motor, making it difficult to drive or install. In addition, it may cause mechanical backlash, added friction, or assembly errors, which may further occur to malfunctions when motor is operated. To solve these problems, there have been a number of research and development efforts aiming to increase torque through the application of various types of reduction gear device [1,2]. Several new types of stepping motors have been developed, such as a wobble motor [3–6]. In this research, we selected a wobble motor because it is simple structure, with an eccentric structure, and the torque generated per unit volume is very high due to the gear ratio while it also has the advantage of a precise control [7–9]. However, existing wobble motors have the main disadvantage that they cannot fix a rotating shaft while the shaft is halted. To fix the shaft, we introduce a new type of wobble actuator with an electropermanent magnet (EPM). It comprises semi-hard (Alnico 5) and neodymium (NdFeB) magnets, and it can be used to fix the rotating shaft.

In this paper, we introduce the two theoretical force models for application with portable electric devices and fabricate a thin electromagnetic wobble motor that is 50 mm in diameter and 7 mm thick including the EPM device. First, it is necessary to predict the magnetic force generated by the magnets of the EPM device. The most accurate method predicting force is used by commercial finite element method software. FEM is one of the ways to solve problems using numerical approximation for problems that are difficult to obtain an accurate theoretical interpretation. In addition, FEM has the advantage of being compatible with Auto CAD or other design programs, and accuracy of the model analysis and user convenience are very high, and high-quality analysis is possible even if there is a lack of professional knowledge about FEM. On the other hand, it will take several hours to obtain the precise analysis of the complex model and variables in the model. However, this comes with the problem of having to perform numerous iterations in the initial design step. On the other hand, a more systematic method is magnetic circuit analysis. An EPM device with two magnets is expressed as a model in which a magnetic flux source and an internal reluctance are connected in parallel. The magnetic flux paths, including the reluctance of the core of the magnetic material and the reluctance in the air gap, are shown as an electric circuit. The method for calculating the magnetic force in a magnetic circuit is similar to calculate the current in an electric circuit and is very simple. Since the fringing effect and the leakage flux around the air gap are not considered, the results of magnetic circuit analysis led to a considerably large degree of error. This is because it is difficult to accurately include them in magnetic circuits. Therefore, we introduce a theoretical force method into the initial design for use in the magnetic circuit. In addition, we design an EPM device including magnets with a flat plate and a constant air gap, and propose a simple 3D model for the leakage flux and fringing effect around the gap. The paths of magnetic flux are modeled using a simplified method consisting of several arcs, straight lines, and a mean flux line. Additionally, we introduce the schematic of the EPM device to validate the proposed theoretical force model, and the basic principle and structure, are presented in a straightforward manner. Its effectiveness is confirmed on the basis of a comparison with the results of FEM analysis, and experiments carried out using a prototype. Second, in order to analyze the characteristics existing between the electromagnet core and the rotating shaft, it is necessary to predict the magnetic flux density and magnetic force when changing the air gap. To this end, we introduce a distribution parameter method in order to propose a theoretical force model. This method was used to calculate the gap by mapping the air gap, which does not have a constant structure, onto the Cartesian coordinates. We also present a schematic of a thin wobble motor in order to validate the proposed force model, as well as the optimization of core shape, basic driving principle, basic structure and control method of the motor. Finally, the experimental results are presented for the EPM device and a thin wobble motor.

2. Driving Principle and Structure

2.1. Driving Principle and Structure of the Wobble Motor

Figure 1 shows the basic working principle and structure of the wobble motor, which comprises of cylindrical stator A and columnar rotating shaft rotor B. The rolling point of Figure 1a is in contact with the stator A and the rotor B. R_s, R_R, R_c and S_c denote in the radius of stator A, the radius of a rotor B, the central point of rotor B, and central point of stator A, respectively. The primary feature of the wobble motor is that rotor B rolls along stator A's inner surface as shown in Figure 1b. If the input current is supplied in a clockwise direction in a continuous manner, the force will be generated from the input current. The attractive force pulls the rotor, and the rotor rolls along the inner of the stator via the generated attractive force. The driving direction of stator A is in the opposite direction of rolling direction of rotor B [10–15].

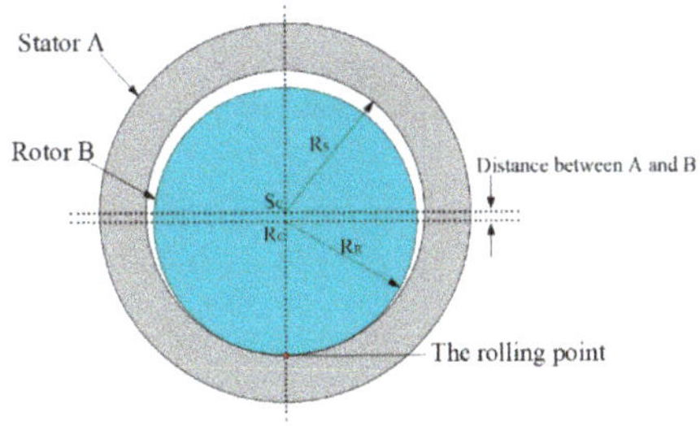

(**a**)

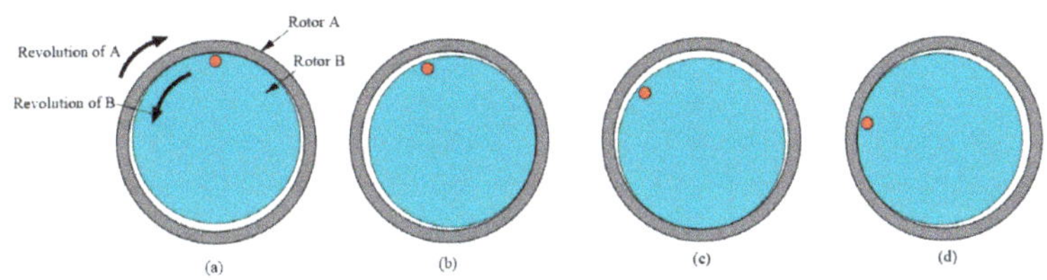

(**b**)

Figure 1. Driving principle of a wobble motor: (**a**) Schematic of a wobble motor; (**b**) Driving principle.

2.2. Driving Principle and Structure of the EPM

The EPM composes of two magnetic materials, NdFeB magnet and another Alnico5 magnet; a coil is wrapped around these, as shown in Figure 2. The basic principle driving of the EPM was showed using the MagNet 7.0 electromagnetic analysis software from Mentor. If the magnetization directions of the two magnets are oppositely combined, the magnetic flux only circulates inside connected iron as shown in Figure 2a. After a switching current pulse of the opposite polarity applied to Alinico5, the magnetization directions of the two magnets were the same, and the EPM generated an external magnetic flux as shown in Figure 2b. The magnetization direction of the NdFeB material is unchanged because it has high coercivity. In ferromagnetic materials, the phenomenon of magnetic hysteresis occurs. J.A. Ewing coined the term "hysteresis" to describe his observation that the magnetic induction in a metal behind the applied current, and he also demonstrated that the molecules of a ferromagnetic material acted as reversible permanent magnets [16]. The normal coercivity of Alnico 5 is approximately 50 kA/m, and the remnant flux density is approximately 1.2 T. The normal coercivity of NdFeB is approximately 1000 kA/m, and the remnant flux density is approximately 1.2~1.3 T [17].

When a switching current is applied to the coil wound, the magnetization direction of the Alnico 5 magnet changes, and the external magnetic force generates attractive force.

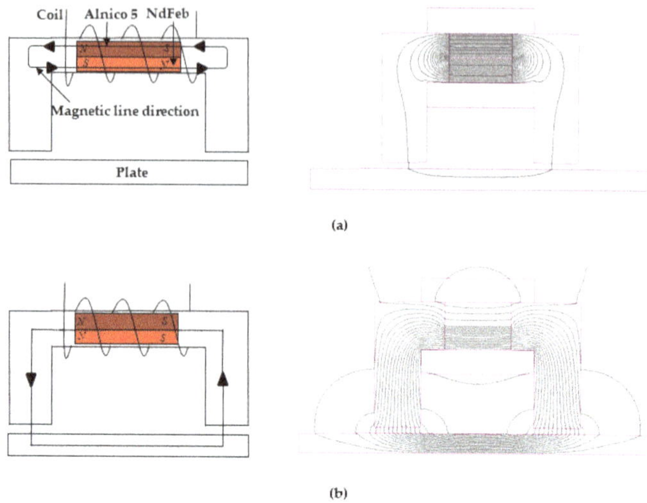

Figure 2. Basic principle driving an electropermanent magnet: (**a**) In the no current state, the two magnets are oppositely polarized; (**b**) In the switching current state, the two magnetic magnets are polarized in the same direction.

3. Derivation of Theoretical Force Model

3.1. Theoretical Force Analysis of the EPM

Figure 3 shows a full schematic of the EPM device for holding the rotating shaft of the wobble motor. Alnico 5 and NdFeB magnets are described as square magnets in the schematic, but the two magnets used in the theoretical analysis and the experiment were replaced with cylindrical magnets. To predict the attractive force, first, the structure is assumed to be a simple structure with a constant gap, and a core that contains two permanent magnets. Additionally, we assume that the rotating shaft is replaced with the plate. Second, in order to simplify the calculations for magnetic flux density, including the leakage flux, we simplified the path through which the leakage flux flows. Additionally, the fringing flux was assumed to be the leakage flux, and the path of flux was consisted as a combination of arcs and straight line, as shown in Figure 4a [18]. The fringing fluxes were considered a kind of leakage flux. It was also assumed that the paths detouring on the top outside of the surrounding air gap were connected in the quadrant with a straight line and the length of the air gap [19]. Figure 4b shows a consideration of the mean leakage flux line. The specifications of the electromagnetic elements in use and their parameters are summarized in Table 1.

For the leakage flux paths of P_{dx}, as shown in Figure 4a, the length l_x of the magnetic flux path is equal to $g + \pi x$, which is the sum of the lengths of two quadrant arcs of radius x and the length of the gap. The permeance in the path is as follows:

$$P_{dx} = \frac{1}{\Re_{dx}} = \frac{\mu_0 dA_x}{l_x} = \frac{\mu_0 b dx}{g + \pi x} \tag{1}$$

The permeance of the total leakage flux can be obtained by integrating Equation (1).

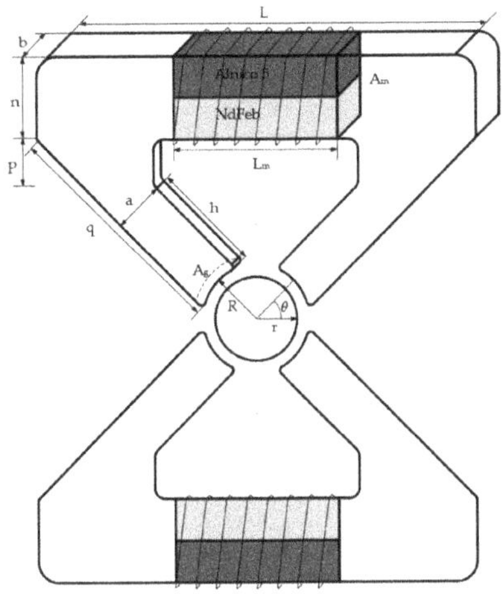

Figure 3. Schematic of the switchable eletropermanent magnet construction.

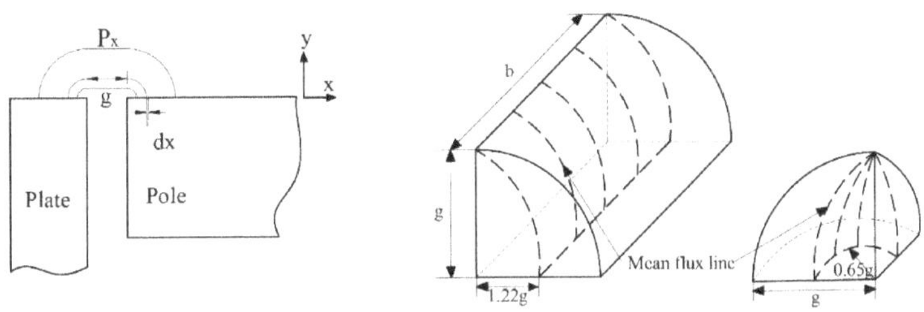

(a) (b)

Figure 4. Paths model for leakage flux: (a) Paths for the leaked flux; (b) Mean length of flux path.

Table 1. Parameters of the electropermanent magnet device.

Parameter	Symbol	Value	Unit
Thickness	b	2	mm
Length	L	22	mm
Pole length	n	4	mm
Pole length	q	11.6	mm
Pole length	h	5.5	mm
Pole height	p	2	mm
Pole width	a	2	mm
Magnet diameter	d	2	mm
Magnet length	Lm	8	mm
Air gap	g	0.1	mm
Rotor diameter	r	5	mm
Stator diameter	R	5.1	mm
Number of turns	N	120	
Angle	θ	45	deg
Applied larger than current	i	5	A
Diameter of the coil	dc	0.18	mm
Permeability of the free space	μ_0	$4\pi \times 10^{-7}$	H/m
Relative permeability of the magnet	μ_r	1.05	H/m
Cross-sectional area of the air gap	A_g	5.26	mm^2
Cross-sectional area of two magnets	A_m	6.28	mm^2

If the core and the plate have a constant air gap, as shown in Figure 5a, it is assumed that the leakage flux paths diverting to the outside of the gap are connected to two quadrants of the same magnetic flux in a straight line along the length of the air gap. Each leakage flux path indicated by a line has a reluctance, and all of them in the magnetic circuit are connected in parallel, expressing it as shown in Figure 5b. Therefore, it is possible to use a permeance that has an inverse relationship with a reluctance.

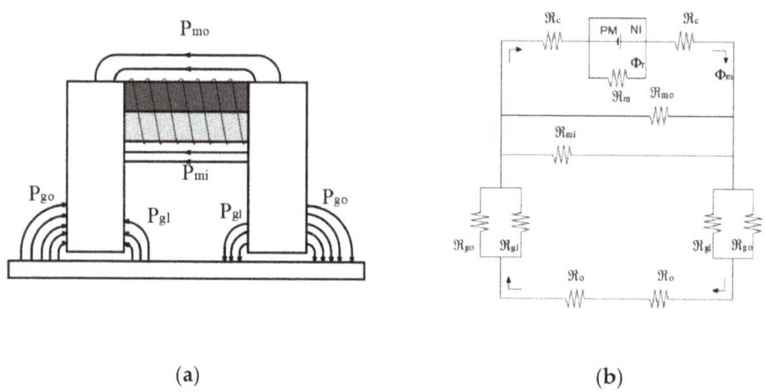

Figure 5. Analysis model consisting of an electropermanent magnet: (**a**) Flux leakage of the electropermanent magnet; (**b**) Magnetic circuit.

In Figure 5b, the reluctance $\Re_c$ and $\Re_o$ in the iron are ignored. Therefore, when the permeances of the leakage flux path are $P_{go}, P_{gl}, P_{mo}, P_{mi}$ and P_g, respectively, they can be expressed as reciprocal reluctances $\Re_{go}, \Re_{gl}, \Re_{mo}, \Re_{mi}$ and $\Re_g$, $\Re_m$, Φ_r and Φ_m are the source of the magnet, the magnetic flux of the magnetic flux passing through the air gap. To more accurately consider the leakage flux generated in the gap, it is regarded as a geometry as shown in Figure 6. The permeance of each of the flux paths through the gap can be expressed as permeance $P_{go} = P_1 + 2(P_2 + P_3) + 4P_4$ and $P_{gl} = 2(P_5 + P_6) + 4P_7$. Permeance P_1 is $\mu_0 ab/g$. The mean length of the flux path of permeance P_2 can be considered to be equal to the length of a line drawn midway between the radius and a quarter of circumference as shown in Figure 6b. Permeance P_2 and P_3 with a quadrant of the cylindrical volume are $0.528\mu_0 b$ and $0.528\mu_0 a$, respectively.

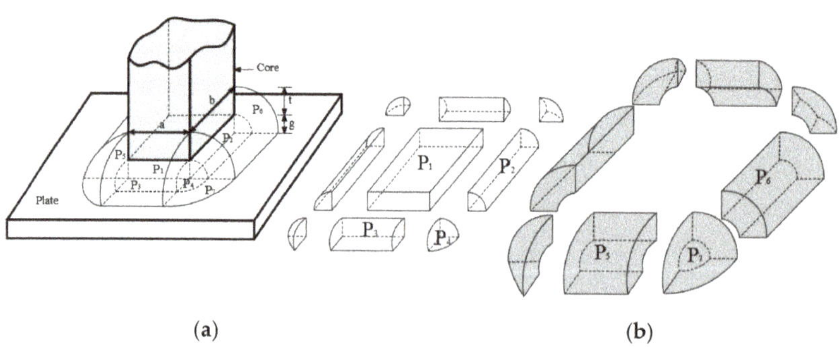

Figure 6. Permeance of flux paths through the air gap: (**a**) Permeance of parallel to end of the core; (**b**) Simple-shaped volumes of permeance.

The mean length of the flux path for permeance P_4 with one eighth of the spherical volume can be approximated by considering the mean flux line to be situated 0.65 of the way between the center of the sphere and the circumference, as shown in Figure 5b. Therefore, P_4 with one-eighth of the spherical volume is $0.308\mu_0 g$. P_5 and P_6 with one-half of the half annulus are $2\mu_0 a/\pi \ln(1+t/g)$ and $2\mu_0 b/\pi \ln(1+t/g)$, respectively. P_7 with one-half of the quadrant of the spherical shell is $0.5\mu_0 t$. Permeance P_{mo} and P_{mi} detouring into the air without going through the plate are $\mu_0 t/\pi \ln(1+\pi)$ and $\mu_0 t/2$, respectively. The reciprocal of permeance is defined as reluctance. $\Re_t$ is the total reluctance; thus, all reluctance in the magnetic circuit can be calculated. To calculate the magnetic field in consideration of the leakage flux, we used Gauss' law and a magnetic force circuit, which can be expressed as follows

$$\Re_t = \frac{1}{P_t} = \frac{1}{P_{mo} + P_{mi} + 0.5(P_{gl} + P_{go})} \tag{2}$$

$$\pi d^2 (B_a + B_r)/8 = B_g ab + \Phi_{leak}$$

$$\Phi_{leak} = (Ni - H_m L_m)/\Re_{leak} \tag{3}$$

$$H_m = \frac{(\mu_0 ab/2g + 1/\Re_{leak})Ni - \pi d^2 (B_r + B_a)/8}{\pi d^2 \mu_0/8 + (\mu_0 ab/2g + 1/\Re_{leak})L_m}$$

where H_m is the axial magnetic field intensity when combining two magnets, B_g is the magnetic field density in the air gap, and Φ_{leak} is the pole-to-pole leakage flux. The axial magnetic flux densities of the Alnico 5 and NdFeb magnets are B_a and B_r, respectively. Finally, the generated attractive force can be obtained using the Maxwell stress tensor, which can be written as

$$F = \mu_0 ab \left(\frac{Ni - H_m L_m}{2g} \right)^2 \tag{4}$$

3.2. Theoretical Force Analysis of Motor

The side view used for theoretical analysis is showed in Figure 7. An idealized model was assumed in which the end effects were neglected. The idealized model is thus similar to an idealized rotating machine. A distribution equation of the electromagnetic system can be expressed using Maxwell's equation and Ohm's law. The magnetic vector potential A is defined as B $= \nabla \times$ A. Their solutions have previously been studied in related research areas [20–23]. Therefore, we briefly describe the theoretical analysis of the flux density of the air gap. To establish analytical solutions as a function of the air gap, variations in the r-direction are ignored, and $\nabla \cdot$ A $= 0$ can also be assumed, and six terms of g, h, t, p, μ_0 and μ_r are air gap, the pole width, thickness of the stator, pole pitch, the permeability of air gap and the permeability of the material, respectively.

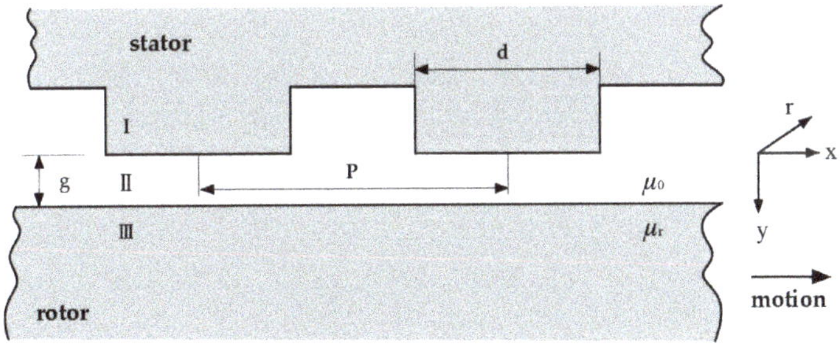

Figure 7. Geometric model for theoretical analysis.

The governing equation of the electromagnetic phenomena in the wobble motor can be calculated as

$$\nabla^2 \mathbf{A} = \mu\sigma \left\{ \frac{\partial \mathbf{A}}{\partial t} - \mathbf{v} \times (\nabla \times \mathbf{A}) \right\} \tag{5}$$

where $\mu_0 = 4\pi \times 10^{-7} N/A^2$ is the permeability of the free space, v is the speed of the moving rotor (m/s), and σ is electrical conductivity ($(\Omega \cdot m)^{-1}$). B is expressed from the Fourier series as

$$B = \sum_{n=1,3,5\cdots}^{\infty} B_m e^{j(\omega t - \beta x)} \widehat{j} \tag{6}$$

where $B_m = 4B/n\pi \sin(\beta d/2)$ and $\beta = (n\pi/p)$. β and ω are the wave number and angular velocity, respectively.

The y-direction flux density in the air gap can be calculated using the following equation:

$$\begin{aligned} \mathrm{B} = \nabla \times \mathrm{A} = \left(\tfrac{\partial A_y}{\partial y}\right)\widehat{i} - \left(\tfrac{\partial A_y}{\partial x}\right)\widehat{j} = B_x \widehat{i} + B_j \widehat{j} \\ B_{x,II} = \tfrac{\partial A_{II}}{\partial y} = \sum_{n=1,3,5,\cdots}^{\infty} (\lambda_1 C_{II} e^{\lambda_1 y} + \lambda_2 D_{II} e^{\lambda_2 y}) e^{j(\omega t - \beta x)} \\ B_{y,II} = -\tfrac{\partial A_{II}}{\partial y} = \sum_{n=1,3,5,\cdots}^{\infty} j\beta(\lambda_1 C_{II} e^{\lambda_1 y} + \lambda_2 D_{II} e^{\lambda_2 y}) e^{j(\omega t - \beta x)} \end{aligned} \tag{7}$$

where $\widehat{i}$ and $\widehat{j}$ indicate x-and y-axis unit vectors, respectively, and II represents the layer of the air gap.

where $\lambda_{1,2} = \left(\gamma \pm \sqrt{\gamma^2 + 4\alpha^2}\right)/2,\ \alpha^2 = \beta^2 + j\mu_0 \sigma s v_s \beta,\ \ \gamma = \mu_0 \sigma v_x$, and s and v_s indicate slip and a synchronous speed, respectively.

The boundary conditions are as follows:

$$B_{II}|_{y=0},\ \ B_{II,y} = B_m e^{j(\omega t - \beta x)};\ \ \ B_{II}|_{y=g},\ B_{II,g} = B_{III,g} \text{ and } \mu B_{II,x} = \mu_r B_{III,x};\ \ \ B_{III}|_{y=\infty},\ A_{III} = 0 \tag{8}$$

The coefficients in Equation (8) form boundary conditions that can be solved, and the coefficients is as follows:

$$\begin{aligned} C_{II} = \frac{B_m(\beta\mu_0 + \lambda_2\mu)e^{\lambda_2 g}}{j\beta(-e^{\lambda_1 g}\lambda_1 + e^{\lambda_2 g}\lambda_2 - \Delta)} \\ D_{II} = \frac{B_m(\beta\mu_0 + \lambda_1\mu)e^{\lambda_1 g}}{j\beta(e^{\lambda_1 g}\lambda_1 - e^{\lambda_2 g}\lambda_2 + \Delta)} \end{aligned} \tag{9}$$

where $\Delta = (e^{\lambda_1 g} - e^{\lambda_2 g})\beta\mu_0/\mu$.

Finally, in order to calculate the Equation (7) on the basis of the change in the air gap [24], the air gap of the wobble motor can be simplified by representing it as a rectangular coordinate system as shown in Figure 8.

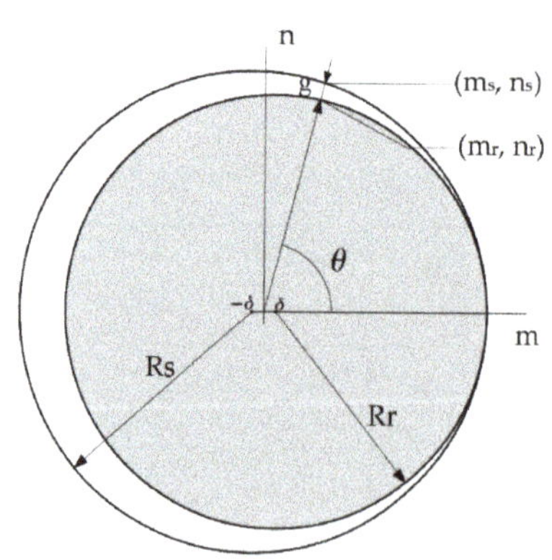

Figure 8. Simplified geometry.

The air gap, g, between the rotor and stator as a function of θ is expressed and simplified. R_s and R_r are the radius of stator and rotor, respectively. In addition, the distance of the air gap, g, between these points on the stator and the rotor was then calculated.

$$g = \sqrt{(m_s - m_r)^2 + (n_s - n_r)^2}$$
$$= \sqrt{\left(\frac{-\delta\pm\Delta}{\sec^2\theta} - \frac{\delta\pm\Psi}{\sec^2\theta}\right)^2 + \left(\frac{\tan\theta\times(-\delta\pm\Psi)}{\sec^2\theta} - \frac{\tan\theta\times(\delta\pm\Delta)}{\sec^2\theta}\right)} \quad (10)$$

where $\Delta = \sqrt{\delta^2 + \sec^2\theta(R_s^2 - \delta^2)}$, $\Psi = \sqrt{\delta^2 + \sec^2\theta(R_r^2 - \delta^2)}$.

The magnetic flux density can be calculated by substituting the valid gap obtained from Equation (10) into Equation (9). The final magnetic force can be determined using Maxwell's stress tensor, which can be written as

$$F_x = \oint_A \frac{(B_x \cdot B_y)}{\mu_0} dA, \quad F_y = \oint_A \frac{\left(|B_y|^2 - |B_x|^2\right)}{2\mu_0} dA \quad (11)$$

4. Validation of Theoretical Force Model

4.1. Modeling and Structure of Motor with EPM

To validate the proposed two force models, we developed a prototype based on a thin wobble motor with EPM 30 mm in diameter and 7 mm in thickness, the schematic of which is as shown in Figure 9. It was composed of a stator with an electromagnet core and coil, rotating shaft, and an EPM device. First, we decided to consider the shape of the core, as this was important for increasing the generated force, and then the core was simply optimized. The parameters of the core were T_1, T_2, α, β, as shown in Figure 10a. To increase the magnetic force and to decrease the leak of magnetic flux, we also carried out electromagnetic field analysis, and the original material of the core was pure iron. In this FEM analysis, the magnetic flux density was obtained by exciting two cores when the rotor was momentarily in contact with the end of the stator. The optimal shape for the core in order to increase attractive force and the magnetic flux density distribution of the core were determined through FEM, as shown in Figure 10b.

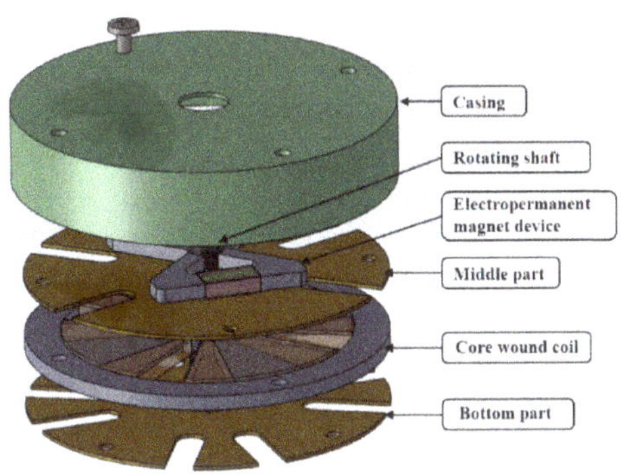

Figure 9. Schematic of the thin wobble motor with an Electropermanent magnet.

Second, we propose a wobble with EPM device so that the rotating shaft can be fixed without the continuous use of a power supply. Therefore, it is to be used to solve the difficulties presented by existing wobble motors, and the schematic is represented in Figure 11. 3D FEM simulation was conducted to estimate the flow of the magnetic flux of EPM, but the results of FEM was expressed in 2D as shown in Figure 11 to confirm the change of magnetic flux. The diameter and length of the two magnets used in FEM were 2 mm and 8 mm, respectively.

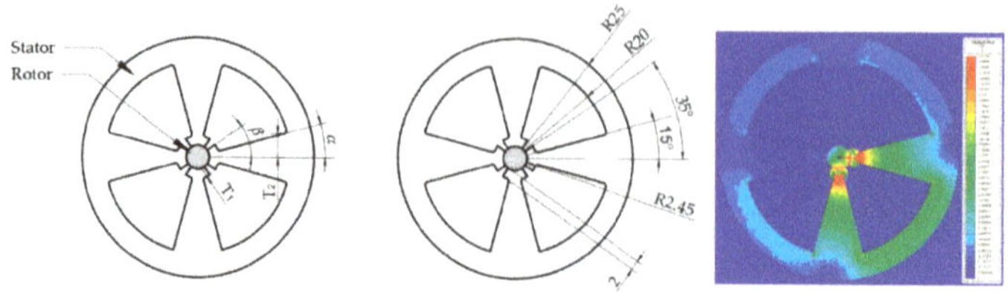

(a) (b)

Figure 10. Shape of the core: (**a**) Parameters for optimization of the core; (**b**) optimized core and the magnetic flux density distribution.

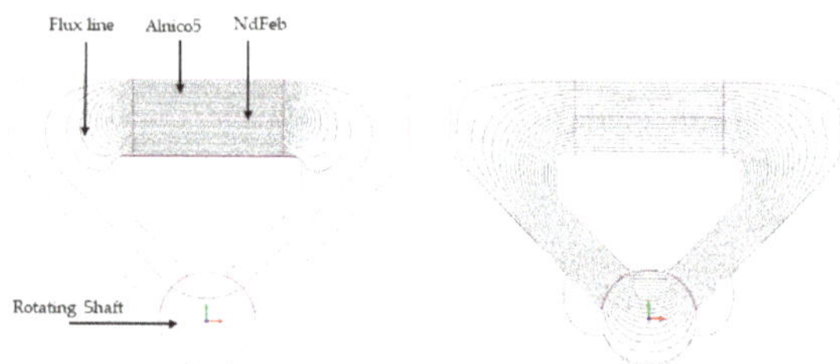

(a) (b)

Figure 11. Basic principle driving a proposed electropermanent manent magnet: (**a**) Flux path when the magnetization direction of magnets is in the opposite direction; (**b**) Flux path through air gap between iron and shaft.

When the magnetization directions of the two magnets are opposite, the magnetic flux flows only toward the inside as shown Figure 11a. It means that the rotating shaft in unaffected. When the current for changing the magnetization direction of the Alnico 5 magnet is applied, the magnetization direction is changed. However, when a current is applied to the coil wound, the magnetization direction of NdFeb does not change. Therefore, the magnetization direction of the two magnets is the same, and an external magnetic flux is generated as shown in Figure 11b.

4.2. Experimental System and Driving Principle

The driving method of the motor and the experimental system of driving motor are as shown in Figure 12. Current is applied to two electromagnets simultaneously, and the rotational speed of the wobble motor is controlled by the frequency of the controller. The U, V, W, and Z at four-phase from the signal controller matches with coil A, coil B, coil C, and coil D wrapped around the four-pole core. The coil is operated as follows: (coil A and coil B), (coil B and coil C), (coil C and coil D), and (coil D and coil A).

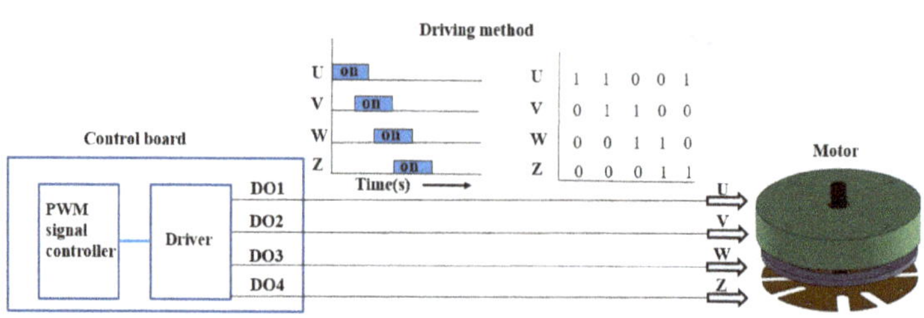

Figure 12. The driving method of the motor and total system of driving motor.

4.3. Experiment for Theoretical Model Validation

To experimentally validate the proposed theoretical model, a prototype of a thin wobble motor with an EPM device was fabricated. Pictures of the prototype are shown in Figure 13. It has the same dimensions as the one used for FEM. The magnets used for the EPM were 2 mm in diameter, 8 mm in length, and around them was wrapped a coil with a diameter of 0.18mm for 120 turns, as shown in Figure 13b. Four electromagnetic arms of core were arranged around the wobble shaft and wrapped with a coil as shown in Figure 13c. In the experiment, a core with a thickness of 1 mm, around which a coil with a diameter of 0.16 mm was wound for 150 turns, was used to create an attractive rotating shaft. To maximize the force and minimize the iron losses, the material was changed from pure iron to electrical steel. The material selected for the electrical steel was 50PN350, characterized by B_{sat} = 1.62 T and an iron loss of 3.50 Watts per kilogram from Pohang Iron and Steel Company (POSCO), and the core shape was then manufacturing by Electro Discharge Machining (EDM). Additionally, pure iron (SS400) was used as the material for the rotating shaft. This had the highest electrical conductivity, high permeability, low coercive intensity, and was also easy to purchase. We fabricated a prototype of the EPM device using a Computer Numerical Control (CNC) machine. From the FEM results, it was confirmed that a change occurs when the magnetization direction of the Alnico 5 momentarily larger than 5A. To change the magnetization direction of the Alnico 5 magnet in EPM device, we used the controller of an NI cRio-9022 from NATIONAL INSTRUMENTS and a high-performance capacitor-type driver capable of momentarily applying a current of larger than 5A. The signal controller was a TMS320F28335, and the driver was expressly made for use with four-pole cores.

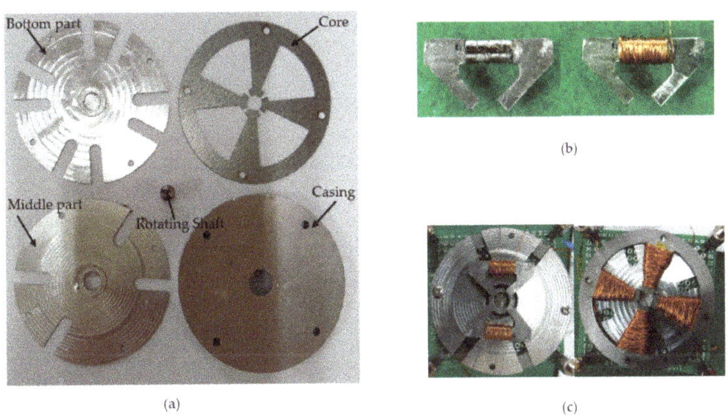

Figure 13. Pictures of a thin wobble motor with EPM. (**a**) Parts a thin wobble motor; (**b**) Fabricated EPM device; (**c**) Core with the coil wound.

4.4. Validation Results

Figure 14 shows the predictions of the magnetic circuit model in three cases, as well as the experimental results with respect to variation in the air gap of the EPM device. The black curve shows the neglecting leakage permeance; the red curve shows the 2D leakage permeance in consideration of $2 \times P_2$ and $2 \times P_6$, as shown Figure 6; the blue curve shows the 3D leakage permeance; and the green curve shows the 3D FEM results. When the air gap was set at 0, 0.1 mm and 0.2mm, the measured forces were 5.853 N, 1.762 N and 0.926 N, respectively. When the air gap was set at 0.1 mm, the results without considering magnetic flux and the FEM results showed considerable error. However, the error between the results in which the magnetic leakage flux was 3D and the FEM results was the 20 percent. Based on the plot, it can be seen that pole-to-pole leakage did not change the holding force when the air gap was, but the slope of the force vs. the distance curve changed significantly with the zero air gap. This is an important effect to consider when designing EPM devices, even those which operate over relatively small air gaps.

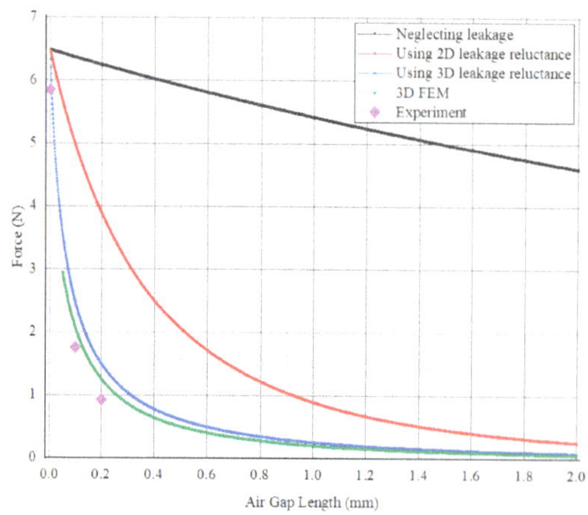

Figure 14. Experiment results, and comparisons of the force vs. air, calculated using the mantic circuit with and without leakage flux.

Table 2 summarizes the results of FEM, the experimental results, and the theoretical model results according to the applied current. An experimental test was conducted using the method shown in Figure 12. The theoretically calculated stall torque and the stall torque generated from FEM comprise the moment when the rotating shaft moves. It was confirmed that the error between the theoretical method and the FEM analysis results was 26 percent on average, and between the experiment and the FEM analysis results was 12 percent average. As the input current to the electromagnet core was increased, the stall torque of the wobble motor also increased.

Table 2. Comparison of stall torque according to current variation.

Current (A)	By Theoretical Method (mNm)	By FEM (mNm)	By Experiment (mNm)
0.1	0.166	0.134	0.118
0.2	0.672	0.523	0.456
0.3	1.488	1.173	1.034
0.4	2.64	2.088	1.843
0.5	4.128	3.264	2.857

5. Conclusions

This paper proposed the two force models in order to theoretically calculate the generated force in a thin wobble motor with the EPM device. First, the EPM can change the holding force on and off by applying a momentary electrical pulse, and the structure and driving principle are presented. It is difficult to establish an analysis model that accurately presents the magnetic leakage fluxes, but by assuming their paths to consists of straight lines and arcs, the mean length of the flux path could be obtained. Therefore, in order to increase the accuracy of the theoretical force model of the EPM device, a method considering the leakage fluxes was proposed. Comparing the FEM results and the force calculated using the proposed method, it was confirmed that there was a reduction in the error of the force generated when considering leakage fluxes were considered. However, there were differences between the theoretical and the experimental results, because the loss of friction and assembly error of the magnets and the iron influenced the experimental measurements. Despite the assembly error, it successfully generated holding force at the rotating shaft.

These results can not only be used in selecting the initial design parameters necessary for FEM analysis for the design of EPM devices, but also indicate that the proposed magnetic circuit analysis results can be used as the basis for its design. Secondly, we simply presented the optimal design of the electromagnets with respect to various parameters, as well as a partial differential equation on the basis of the distribution parameters as a function of the variation of the air gap between the stator and the rotating shaft. The theoretical force model for the generated force of the electromagnetic field using the distribution parameters was calculated by approximating the model using a linear coordinate system. We conducted experiments using a prototype motor with respect to the changing current. As shown in Table 2, it was confirmed that the theoretical method results and the FEM results were similar. Differences were observed in the experimental results owing to the loss of friction and error occurring when manufacturing the rotating shaft. However, our study shows advantages in that this method is less time consuming and systematic, because a large number of iterative analyses are not required during the initial design step. As a result, the proposed a thin motor with EPM device has great potential for commercialization as a driving part of small robots, in the medical industry, and especially in thin portable equipment in the future.

Author Contributions: FEM analysis, designed and manufactured of the motor, controller, and wrote the paper, S.Y.P.; edited and revision, analyzed the date, B.S.; investigation and validation S.Y.P. and B.S.; supervision, Y.S.B. All authors have read and agreed to the published version of the manuscript.

Funding: This research received no external funding.

Institutional Review Board Statement: Not applicable.

Informed Consent Statement: Not applicable.

Data Availability Statement: This paper is no data.

Conflicts of Interest: The authors declare no conflict of interest.

Abbreviation

EPM Electropermanent Magnet

References

1. Viviani, A. Experimental and Theoretical Study of Hypocycloidal Motors with Two-Harmonic Field Windings. *IEEE Trans. Power Appar. Syst.* **1980**, *1*, 292–300.
2. Hayashi, I.; Iwatsuki, N.; Shibata, J.; Matsunaga, H.; Wada, S. An Electromagnetic Cycloid Motor. *J. Jpn. Soc. Precis. Eng.* **1995**, *61*, 95–99. [CrossRef]
3. Nagata, T.; Suzumori, K.; Kanda, T.; Uzuka, K.; Enomoto, I. Electro Direct-Drive Stepping Motor for Robots. In Proceedings of the IEEE International Conference on Robotics and Automation, New Orleans, LA, USA, 26 April–1 May 2004; pp. 4493–4498.
4. Kimura, H.; Hirose, S.; Nakaya, K. Development of the Crown Motor. In Proceedings of the 2001 ICRA, IEEE International Conference on Robotics and Automation, Seoul, Korea, 21–26 May 2001; pp. 2442–2447.
5. Okamoto, K.; Suzumori, K.; Kanda, T.; Yamada, Y. Development of Three-Chamber Micro Pneumatic Wobble Motor. *Proc. Mach. Des. Tribol. Div.* **2006**, 295–297. [CrossRef]
6. De Cristofaro, S.; Stefanini, C.; Ng Pak, N.; Susilo, E.; Carrozza, M.C.; Dario, P. Electromagnetic wobble micromotor for microrobots actuation. *Sens. Actuators A Phys.* **2010**, *161*, 234–244. [CrossRef]
7. Jacobsen, S.C.; Price, R.H.; Wood, J.E.; Rytting, T.H.; Rafaelof, M. The Wobble Motor: Design, Fabrication and Testing of an Eccentric-Motion Electrostatic Microactuator. In Proceedings of the IEEE International Conference on Robotics and Automation, Scottsdale, AZ, USA, 14–19 May 1989; pp. 1536–1546.
8. Suzumori, K.; Hori, K. Micro Electrostatic Wobble Motor with Toothed Electrodes. In Proceedings of the IEEE Tenth Annual International Workshop, Nagoya, Japan, 26–30 January 1997; pp. 227–232.
9. Miyake, M.; Suzumori, K.; Uzuka, K. Development of a Thin Electromagnetic Wobble Motor. In Proceedings of the 2011 IEEE International Conference on Robotics and Biomimetics, Karon Beach, Phuket, Thailand, 7–11 December 2011; pp. 2523–2528.
10. Jacobsen, S.C.; Price, R.H.; Wood, J.E.; Rytting, T.H.; Rafaelof, M. The Wobble Motor: An Electrostatic, Planetary-Armature, Microactuator. In Proceedings of the IEEE Micro Electro Mechanical Systems, An Investigation of Micro Structures, Sensors, Actuators, Machines and Robots, Salt Lake City, UT, USA, 20–22 February 1989.

11. Jungreis, A.M.; Kelley, A.W. The axial air gap wobble motor—An appropriate topology for magnetic micromotors. In Proceedings of the Conference Record of the 1995 IEEE Industry Applications Conference Thirtieth IAS Annual Meeting, Orlando, FL, USA, 8–12 October 1995.
12. Mehergancy, M.; Nagarkar, P.; Senturia, S.D.; Lang, J.H. Operation of Microfabricated Harmonic and Ordinary Side-Drive Motors. In Proceedings of the IEEE Proceedings on Micro Electro Mechanical Systems, An Investigation of Micro Structures, Sensors, Actuators, Machines and Robots, Napa Valley, CA, USA, 11–14 February 1990.
13. Dhuler, V.R.; Mehregany, M.; Phillips, S.M. An experimental technique and a model for studying the operation of harmonic side-drive micromotors. *IEEE Trans. Electron Devices* **1993**, *40*, 1977–1984. [CrossRef]
14. Deng, K.; Meheregany, M. Outer-Rotor Polysilicon Wobble Micromotors. In Proceedings of the IEEE Workshop on Micro Electromechanical Systems, Oiso, Japan, 25–28 January 1994.
15. Furuhata, T.; Hirano, T.; Lane, L.H.; Fontana, R.E.; Fan, L.S.; Fujita, H. Outer Rotor Surface-Micromachined Wobble Micromotor. In Proceedings of the IEEE Micro Electro Mechanical Systems, Fort Lauderdale, FL, USA, 10 February 1993.
16. Ewing, J.A.; Klaassen, H.G. Magnetic qualities of Iron. *Philos. Trans. R. Soc. A* **1893**, *184*, 985–1039.
17. Campbell, P.; Al-Murshid, S. A model of anisotropic Alnico magnets for field Computation Magnetics. *IEEE Trans. Magn.* **1982**, *18*, 898–904. [CrossRef]
18. Hanselman, D.C. *Brushless Permanent Magnet Motor Design*; McGraw-Hill: Cranston, RI, USA, 1994.
19. Roters, H.C. *Electromagnetic Devices*, 1st ed.; John Wiley & Sons: New York, NY, USA, 1941.
20. Yamamura, S. *Theory of Linear Induction Motors*, 2nd ed.; Wiley: New York, NY, USA, 1979.
21. Wangsness, R.K. *Electromagnetic Fields*, 2nd ed.; Wiley: New York, NY, USA, 1986.
22. Griffiths, D.J. *Introduction to Electrodynamics*, 2nd ed.; Prentice Hall: Upper Saddle River, NJ, USA, 1989.
23. Nasar, S.A. *Linear Motion Electric Machines*; John Wiley & Sons: New York, NY, USA, 1976.
24. Trimmer, W.; Jebens, R. An Operational Harmonic Electrostatic Motor. In Proceedings of the IEEE Micro Electro Mechanical Systems, An Investigation of Micro Structures, Sensors, Actuators, Machines and Robots, Salt Lake City, UT, USA, 20–22 February 1989; pp. 13–16.

MDPI
St. Alban-Anlage 66
4052 Basel
Switzerland
Tel. +41 61 683 77 34
Fax +41 61 302 89 18
www.mdpi.com

Applied Sciences Editorial Office
E-mail: applsci@mdpi.com
www.mdpi.com/journal/applsci